Genetics, Evolution and Radiation

Nikolay Wladimirovich Timofeeff-Ressovsky

Victoria L. Korogodina · Carmel E. Mothersill
Sergey G. Inge-Vechtomov
Colin B. Seymour
Editors

Genetics, Evolution and Radiation

Crossing Borders, The Interdisciplinary Legacy of Nikolay W. Timofeeff-Ressovsky

United Nations
Educational, Scientific and
Cultural Organization

Springer

Editors
Victoria L. Korogodina
Joint Institute for Nuclear Research
Dubna, Moscow Region
Russia

Carmel E. Mothersill
Department of Medical Physics and Applied Radiation Sciences
McMaster University
Hamilton, ON
Canada

Sergey G. Inge-Vechtomov
Genetics and Biotechnology
St. Petersburg State University
St. Petersburg
Russia

Colin B. Seymour
Department of Medical Physics and Applied Radiation Sciences
McMaster University
Hamilton, ON
Canada

ISBN 978-3-319-48837-0 ISBN 978-3-319-48838-7 (eBook)
DOI 10.1007/978-3-319-48838-7

Library of Congress Control Number: 2016954911

Printed on acid-free paper

This Springer imprint is published by Springer Nature
The registered company is Springer International Publishing AG
The registered company address is: Gewerbestrasse 11, 6330 Cham, Switzerland

To Nikolay W. Timofeeff-Ressovsky

Foreword

Nikolay Wladimirovich Timofeeff-Ressovsky in Russia

Nikolay Wladimirovich Timofeeff-Ressovsky lived a life of hardships which reflect the crisis and tragedies of the last century. In spite of the difficulties of his private and scientific life Timofeeff-Ressovsky was an eminent biologist of the twentieth century.

N.W. Timofeeff-Ressovsky (1900–1981) was born in Moscow, where he received his education at a high school and later at Moscow State University. At the university he joined a club of young geneticists who worked together with leading Russian biologists N.K. Koltsov and S.S. Chetverikov. Among them were B.L. Astaurov, N.P. Dubinin, D.D. Romashov, A.S. Serebrovsky, and the main topic of their discussions was the origin of mutations. Also in those years N. Timofeeff-Ressovsky had his first professional contacts with N.I. Vavilov and V.I. Vernadsky.

Upon N. Koltsov's recommendation N. Timofeeff-Ressovsky was appointed for a post at the Brain Research Institute in Berlin-Buch, Germany, where he continued his studies on *Drosophila* together with his colleagues—young scientists K.G. Zimmer and M. Delbrück. Their joint work resulted in the article "On the origin of gene mutations and gene structure" published in 1935, which exerted a considerable influence on the subsequent development of molecular genetics. In the 1930s he also began his radioecological investigations. In Germany he became a member of N. Bohr's Seminar. The range of his contacts widened and included such European and American mathematicians, physicists, and geneticists as N. Bohr, A. Buzzati-Traverso, C.D. Darlington, E. Schrödinger, T. Morgan, H. Muller, D. Lee, H. Stubbe and others.

After the Second World War N. Timofeeff-Ressovsky returned to the USSR, where he was sent by the regime first to a Stalin camp near Karaganda, and then to Site No. 0215, a secret scientific laboratory in the South Urals. There he headed pioneering studies of the biological effect of radionuclides and their impact on ecosystems. Later N. Timofeeff-Ressovsky continued his research at the Institute of Biology of the Urals Branch of the USSR Academy of Sciences (Sverdlovsk) and completed it at the Institute of Medical Radiology of the USSR Academy of Medical Sciences (Obninsk). The basic theme of his studies was the determination of the coefficient of radionuclide accumulation by water and different representatives of biota. The results of his research were presented in his fundamental paper "Selected Studies in Radiation Biogeocenology" published in 1962. His conclusions were close to V. Vernadsky's ideas about the role of living matter in the migration of chemical elements in the earth's crust. Another aspect of the problems he dealt with was the use of different representatives of biosphere for purification of land and water areas contaminated with radioisotopes.

During all his creative period of life, N. Timofeeff-Ressovsky devoted himself to the research in genetics and radiobiology, radiation biogeocenology and evolution theory. His name is associated with the concept of 'hitting the target', the booster concept, mutation theory, basic aspects of microevolution, foundations of radiation biogeocenology, and other well-known branches of science developed by him together with his disciples and colleagues.

N. Timofeeff-Ressovsky's pupils and followers in Ekaterinburg, Moscow, Obninsk, St. Petersburg, Ukraine, Armenia, and Belarus have carried out and continue studies of low-dose effects on microorganisms, plants and animals (inhabiting radionuclide contaminated areas). These studies laid the basis for the measures taken against contamination in the South Urals region and after the Chernobyl accident. This research has been successfully enriched by studies on the synergism of combined effects on biota and human beings of low radiation doses and various physical and chemical factors.

Tombs of Elena Alexandrovna Timofeeff-Ressovsky and Nikolay Wladimirovich Timofeeff-Ressovsky in the cemetery of Obninsk (*left*); a memorial plaque on the house where N.W. Timofeeff-Ressovsky lived in Obninsk

Being an optimist by nature, N. Timofeeff-Ressovsky devoted much of his scientific attention to the theme "Biosphere and Humanity". He was deeply convinced that it was possible, when treated reasonably, to increase dozens of times the global productivity on the earth, which would enable dozens of times more people to be fed, as compared with the present-day calculations and preservation of the biosphere stability in general.

N. Timofeeff-Ressovsky was a full member of the German Natural Science Research Academy 'Leopoldina' in Halle, an honorary member of the American Academy of Sciences and Arts in Boston, Mendel Society in Lund (Sweden), British Genetical Society in Leeds (England), an honorary member and one of the founders of the N. Vavilov Society of Genetists and Breeders in the USSR, a member of the M. Planck Society in Germany, a member of the USSR Geographical and Botanical Societies. He was awarded with the Lazzaro Spallanzani Medal (Italy) and with the Darwin (Germany), Mendel (Czechoslovakia) and Kimber (USA) Prizes.

The centenary of the scientist was declared by UNESCO the year of N.W. Timofeeff-Ressovsky. Then the International conference "Modern problems of genetics, radiobiology, radioecology, and evolution" was organized in Dubna (2000), and later in Yerevan (2005), Crimea (2010), St. Petersburg (2015) under UNESCO aegis.

Anatoly I. Grigoriev
Russian Academy of Sciences

Nikolai V. Timofeeff-Ressovsky and the Campus Berlin-Buch

Nikolai V. Timoféeff-Ressovsky and his wife Elena Alexandrovna Timofeeff-Ressovsky were among the first scientists to work on the Campus Berlin Buch, as visiting researchers when the Kaiser-Wilhelm-Institute for Brain Research moved to its new building in Buch in 1930. Their work was pioneering at the time and laid the foundations for the research performed on campus today. Nikolai's joint paper with Max Delbrück and Karl Günter Zimmer, "On the nature of gene mutation and gene structure," was published in 1935 and has been recognized as the beginning of molecular genetics. It provided a framework for scientists like James D. Watson and Francis Crick to work on the structure of DNA. The Max-Delbrück-Center for Molecular Medicine in the Helmholtz Association (MDC) continues to use molecular genetic methods to study the basis of human diseases and is very proud of the campus' pioneering role in this field.

Memorial plaque of Nikolai V. Timofeeff-Ressovsky mounted on the structure of the gate house on Campus Berlin-Buch (1992). (Photo: Uwe Eising/Copyright MDC)

At the official opening of the MDC on October 17, 1992, a memorial plaque for Nikolai V. Timofeeff-Ressovsky was mounted on the structure that served as his residence: the gate house. A small museum was installed at his working place in the former Kaiser Wilhelm Institute for Brain Research.

On June 30, 2006, when the MDC and the Leibniz Institute for Molecular Pharmacology (FMP) opened a new laboratory building, they named it after Nikolai Timofeeff-Ressovsky. The entrance is marked by his bust, in cast stone. Sponsored by the society of friends and supporters of the MDC and the women's representative, a bi-annual lecture series named in honor of Elena Alexandrovna Timofeeff-Ressovsky. Timoféeff-Ressovsky was introduced at the MDC in 2011.

Bust of Nikolai V. Timoféeff-Ressovsky in front of the Timoféeff-Ressovsky-Haus on Campus Berlin-Buch (2006). (Photo: Uwe Eising/Copyright MDC)

Nikolai and Elena Alexandrovna Timofeeff-Ressovsky had no chance to visit Berlin-Buch again after Nikolaii's deportation to Russia in 1945, but in May 1998 their son Andrei returned for a visit to show his wife the family's residence and the institute. He was present again for a symposium on the occasion of the centennial of his father's birth, held in September 2000.

These are a few of the ways the MDC is preserving the memory of one of our most important pioneers, Nikolai V. Timoféeff-Ressovsky.

Thomas Sommer
The Max-Delbrück-Center for Molecular Medicine
in the Helmholtz Association

Nikolay V. Timoféeff-Ressovsky and American Scientists

In 2015, after 80 years, we celebrate the publication of the paper "On the nature of gene mutation and gene structure" by N.V. Timoféeff-Ressovsky, K. Zimmer, and M. Delbrück. This famous early paper was an important step in the investigation of the mutation process. T.H. Morgan and H.J. Muller stood at the foundations of these researches, and genome integrity and the mutation process remain today one of the priority efforts for geneticists.

The Russian geneticist Nikolay V. Timoféeff-Ressovsky was among the first researchers of gene structure. He was a follower of Morgan and was allied with Muller, not only in science but also as a friend. These close contacts were useful for Russian and American geneticists. In his "Recollections"[1], Timoféeff-Ressovsky described a circle of young muscovite geneticists in the 1920s, the visit of the famous American biologist Muller to Russia, and the influence of his lectures on young Russian biologists.

In 1936 Timoféeff-Ressovsky was invited to America by M. Demerec to study genetics at the Carnegie Institution of Washington, but the visit was not realized. Later Timoféeff-Ressovsky was awarded the Kimber Genetics Award of the U.S. National Academy of Sciences (1966) and was elected a Fellow Honorary Member of the American Academy of Arts and Sciences (1974).

The tradition of American geneticists visiting Russia was supported at the International Conferences on "Modern problems of genetics, radiobiology, radioecology, and evolution" which was organized in memory of Nikolay V. Timoféeff-Ressovsky. The Genetics Society of America was one of the

[1]Timofeeff-Ressovsky NV (2000) The Stories Told by Himself with Letters, Photos and Documents. Dubrovina NI (comp. & ed.). Soglasie Publ, Moscow (Russian).

co-founders of these conferences. American geneticists have presented a number of excellent reports on the fundamental concepts of the genome and the mutation process, such as "From Green Pamphlet to 2005" (J.W. Drake), "Mutation as a stress response" (S. Rosenberg), and "Evolution of the mutation rate" (M. Lynch). Some of these are presented here.

Victoria L. Korogodina
Joint Institute for Nuclear Research

John W. Drake
National Institute of Environmental Health Sciences

Preface

This book is dedicated to the great scientist and outstanding person Nikolay Wladimirovich Timofeeff-Ressovsky. Since his death 30 years have passed, but his scientific research as well as his life attract people, though there are conflicting opinions. We all appreciate his huge contribution to the science. However, Nikolay Wladimirovich himself valued human qualities more than the talents of man ("the main thing that was a good man").[2] He found the personality of man and his life on the first place. His inner freedom came herein and was manifested in all situations including the Nazi Germany, the prison and camps of the USSR. The scientific discoveries of Timofeeff-Ressovsky are the bases for later studies, and his personality will always be relevant.

To acquaint you with Nikolay Wladimirovich, we have included a few excerpts from his stories,[3] essays of persons who knew him closely, and brief papers of investigators of his life. You can find this material in Part I.

Extensive scientific research of Timofeeff-Ressovsky covered several areas. In 1935 his studies were devoted to the structure of genetic material[4]; his later works: the hit principle,[5] evolution,[6] radiation biogeocenology,[7] and finally, the paper

---

[2]Private memories of Vladimir I. Korogodin (eds).

[3]Timofeeff-Ressovsky NV (2000) The Stories Told by Himself with Letters, Photos and Documents. Dubrovina N. (ed) Soglasie, Moscow (Russian).

[4]Timofeeff-Ressovky NW, Zimmer KG, Delbrück M. (1935) Über die Natur der Genmutation und der Genstruktur, Nachrichten von der Gesellschaft der Wissenschaften zu Göttingen: Math.-Physik. Klasse, Fachgruppe VI, Biol Bd. 1(13): 189–245 (German).

[5]Timofeev-Ressovsky NW, Zimmer KG (1947) Das Trefferprinzip in der Biologie. Biophysik Bd 1 (Hirzel, 1947) (German).

[6]Timofeeff-Ressovsky NW (1939) Genetik und Evolution (Bericht eines Zoologen). J Mol Gener Genet 76(1): 158–219 (German).

[7]Timofeeff-Resovsky NW (1962) Some problems of radiation biogeocenology. Dr Sci. Thesis, Sverdlovsk (Russian).

"Biosphere and Humanity"[8] have showed his worldview. The next parts include the chapters on the main scientific directions of Timofeeff-Ressovsky: genetics, radiobiology, radiation ecology and epidemiology, and evolution.

In the "Genetic Processes" (Part II), S.G. Inge-Vechtomov analyzed the template principle in connection with central dogma of molecular biology. Evolution and expansion of genetic paradigm proceeded from Mendelism to chromosome theory and further to the concept of DNA as genetic material is presented. Replication, transcription, translation as the first-order template processes (TP I) operating with the linear templates with their universal characteristics (ambiguity and repair, or correction) are considered along with the second-order template processes (TP II), operating with spatial, or conformational protein templates. The attention to the former ones had been stimulated at the end of twentieth century by the study of prions and amyloids. Comparative characteristics of the TP I and TP II and of their interaction are discussed.

At present the protein-based inheritance and mutation rate are mainstream topics in genetic studies. Many scientists have puzzled over the mutation rate for a long time. In eukaryotic genome, it depends on some factors: nucleotides pools, conditions of replication, types of DNA damage and its repair, structure of chromatin. In this part the mechanisms of global and region-specific control of mutagenesis are presented and discussed by Y.I. Pavlov and his colleagues. But the most interesting question is "why four major groups of organisms have their own characteristic rates of spontaneous mutation?" It is clear that the evolutionary forces drive these characteristic rates. J.W. Drake discussed this mystery. The protein-based inheritance and protein assemble disorders are reviewed by A.A. Rubel with his colleagues. The authors considered the prion concept, models describing the replication and transport of prions particles, structural features and functions of the cellular PrP, the prion strain phenomenon, current developments in diagnostics of prion and potential antiprion therapies. Recent advances in wheat improvement via alien gene transfer and gene duplications are reviewed by V.K. Shumny, E.K. Khlestkina et al. with emphasis on disease resistance and new functional specialization.

In the "Radiobiology Effects and Mechanisms" (Part III), C. Mothersill and C. Seymour present evolution of radiobiological thought; they give retro- and prospective scheme of radiobiological research directions. Their chapter aims to trace the evolution of the major ideas in radiobiology from the earliest speculations about how the new rays are related to mutations, to the understanding of the relevance of indirect effects, non-targeted effects and the role of epigenetics. The authors predict shift from reductionism to a more holistic approach; it means, for example, taking into account the phenomena of adaptive response in medicine, and interactions of multiple stressors in radiation ecology.

Using a stochastic approach, V. Korogodina et al. showed that the same adaptation mechanisms of the prolonged impact and chronic exposure of plants, animals,

[8]Timofeeff-Resovsky NW (1967) Biosphere and Humanity. Proc. Geography Society USSR, Obninsk: 3–12 (Russian).

or humans lead to the essentially different effects in the populations. Radiation stress changes parameters of reproduction aimed to increase population diversity; chronic prolongation produces progressive selection of a group of the resistant objects. The chapters of O.V. Belov, A.N. Bugay and their colleagues are devoted to an analysis of the mutation processes and DNA damage repair which involve many molecular events, pathways and different genes. In their chapter, H. Abel and G. Erzgraeber noted that the "linear no-threshold hypothesis" cannot be interpreted in the sense of a dose-linear increase of all cancer-induction processes in the range of the natural ionizing radiations. B.P. Surinov considered the role of chemosignaling in the realization of the bystander effect during communications between irradiated and intact animals in mice population.

Parts IV and V are devoted to "Radiation in Ecological Systems" and "Radiation and Man". Nowadays radionuclide contamination and use of nuclear energy are widespread, and the low-dose radiation effects are extensively studied. Part IV reflects a process of data accumulation and their continued discussions.

In his chapter, President of the International Union of Radioecology (IUR) F. Brechignac stressed that assessing ecological radiation risk requires an ecosystem approach. The ecosystem approach provides a conceptual vision which integrates humans within the environment. Historically, the issue of "protection of the environment" against radiation was more bound to human health than associated with wildlife and ecosystems. Protection of the environment includes living beings and abiotic media, landscapes and ecosystems. It prevents interactions between populations of different species and other indirect cascade biological effects; that is holistic goal of protection.

Accident process, radioactivity release and ground contamination are described by T. Imanaka, who compared the radiation characteristics of the Fukushima-1 and the Chernobyl disasters in his chapter. S.A. Geras'kin et al. described radiation effects on ecosystems: changes in species dominance, reduction in productivity and changes in a community structure. The studies of T.A. Mousseau and A.P. Møller are devoted to the animals of Chernobyl and Fukushima. The authors observed the parallels between radiation effects on animals in Chernobyl and Fukushima that provide additional evidence for the significant ecological consequences of nuclear accidents, and ionizing radiation in general. There are chapters which present the radiation effects on the terrestrial (D.M. Grodzinsky, E.V. Antonova et al.) and aquatic ecosystems (D.I. Gudkov et al, S.B. Gulin and V.N. Egorov); influence of Nuclear Power Plant fallouts on soil bacteria is analyzed by G.E. Khachatryan et al. The methods of radioecology approach and models of radiation risks assessment are offered by Y.A. Kutlakhmedov et al., A.A. Cigna, S.B. Gulin and V.N. Egorov.

Part V includes research on radiation influence on humans. C. Mothersill and C. Seymour presented the fundamental concepts of radiobiology: the authors offered mechanisms of sick health and chronic fatigue syndrome suffered by atomic and gulf war veterans. Review on Chernobyl consequences for humans and evaluation of genetic radiation risks are given by I. Schmitz-Feuerhake and S. Pflugbeil; A.V. Yablokov devoted his chapter to fundamental difficulties of dose calculation. M. Rosemann described radiation-induced aging and genetic instability of mesenchymal stem cells. Despite a small number of adult stem cells in an organism, they play an important role for the long-term functionality of all organs and hence, for a healthy aging process. It has been shown that acute or chronic radiation exposure to adult stem cells can impair their genetic stability and have late health effects in various organs. I.E. Vorobtsova and A. Semenov discussed the cytogenetic effects of low-dose irradiation of people. This subsection includes the data on cancer therapy too (I.A. Zamulaeva et al.).

In Part VI ("Laws of Evolution") some systems of different levels of organization are presented, but the genetic evolution of genome was used as a basis of our brief review. Eu. V. Koonin presented analysis of the collection of genome sequences from diverse bacteria, archea, eukaryotes and viruses and his design of the "genomic universe" on the basis of features of genes. Contrary to the evolutionary stability of some genes, the genomes are dynamic on the evolutionary scale. The author demonstrates a highly dynamic picture of the evolution of the genomic universe dominated by horizontal gene transfer; but a "statistical Tree of Life" can be a strong sign of vertical evolution.

The processes of evolutionary changes are described in some chapters devoted to plant–microbe symbiotic interactions (I.A. Tikhonovich et al.), the animal domestication as an example of targeting regulatory system (N.A. Kolchanov with coworkers), and coevolution of human endogenous retroviruses with our genome (A.A. Buzdin et al.). The role of central nervous system of mammals as a mutagenic/anti-mutagenic factor is described by E.V. Daev.

We included the essay of Russian ethologist Prof. Eu. N. Panov "Roots of current concepts in the studies of social behavior in animals". He discussed the history of zoosociology development in the period from the late nineteenth century up to the 1970s. The goal was to show the principles of organization operating in the local animal populations that are considered as social organisms of complex systemic nature.

Acknowledgment Most articles were written on the reports presented at the Conferences "Modern problems of genetics, radiobiology, radioecology, and evolution" dedicated to N.W. Timofeeff-Ressovsky (2000–2015). We are grateful for support of this book issue by the Russian Academy of Sciences, the Max-Delbrück-Center (Berlin-Buch), Joint Institute for Nuclear Research (Dubna), Institute of Cytology and Genetics SB RAS (Novosibirsk). A large part had consultations of D.A. Granin, and support of A.I. Grigoriev (RAS), G. Erzgräber, D. Lafuente, M. Bader,and Rajewsky's family (MDC), E.V. Antonova and V.N. Pozolotina, E.K. Khlestkina, N.G. Gorbushin and B.P. Surinov, A.V. Tchabovsky, I.A. Kolesnik. We are grateful to the translators I. Kronshtadtova, S. Chubakova, and E. Kravchenko (JINR, Dubna).

Moscow Region, Russia — Victoria L. Korogodina
Hamilton, Canada — Carmel E. Mothersill
Saint Petersburg, Russia — Sergey G. Inge-Vechtomov
Hamilton, Canada — Colin B. Seymour

Contents

Part I
N.W. Timofeeff-Ressovsky: Science Without Borders

Some Stories Told by N.W. Timofeeff-Ressovsky

Nikolay W. Timofeeff-Ressovsky

About Muller and His Life in Moscow

Our 'coterie'—the Chetverikov[1] Circle—formed our small group into one body and in the mid-1920s it grew to about fifteen young members. The meeting place was usually not at the Institute but at Chetverikov's or in my flat. There was a very big room in my flat. So, a very easygoing circle of nice friends had formed, absolutely informal. It is obvious of course that in the 1930s such a 'coterie' would have certainly been arrested and its members would have gone to prison with a 'dime verdict' (10 years of imprisonment without rights for communicating families) each. I happened to do my time in the Lubyanka[2] jail, in 1945. My cellmates were two young students-mathematicians from Moscow University. Out either of boredom or curiosity, they had started a circle of mathematics. They had been arrested and got their 'dime verdict'. Bingo!

And in summer of 1922 the following thing happened. For the first time an outstanding foreign scientist, with radical and outspoken 'left' political beliefs, already famous at that time geneticist Hermann Muller arrived from America.

Timofeeff-Ressovsky NV (2000) The Stories Told by Himself with Letters, Photos and Documents. Dubrovina NI (comp. & ed). Soglasie Publ, Moscow (Russian).

Some stories told by N.W. Timofeeff-Ressovsky: About Muller and his life in Moscow; "Green Pamphlet"; The isolines of the female beauty (submitted by Moscow State University Lomonosov Scientific Library).

[1]Sergej Sergeevich Chetverikov (1880–1959) is an outstanding Russian and Soviet biologist, geneticist-evolutionist who pioneered synthesis in Mendel genetics and Darwin evolution theory.
[2]The State Security Committee on Lubyanka Square.

N.W. Timofeeff-Ressovsky (✉)
Moscow State University Lomonosov Scientific Library, Moscow, Russia
e-mail: nidubrovina@list.ru; APantza@nbmgu.ru

V.L. Korogodina et al. (eds.), *Genetics, Evolution and Radiation*,
DOI 10.1007/978-3-319-48838-7_1

Muller was one of the first oldest disciples of Morgan, from the so-called Ruffians' Four: Morgan, Sturtevant, Muller and Bridges. Muller arrived in Moscow by aeroplane, I mean he, of course, crossed the Atlantic ocean by steamer, in tourist class which is cheaper. From Harve he flew by aeroplane to Munich, then, as far as I know, from Munich to Warsaw and from Warsaw to Moscow. Quite a journey I would say. He brought from America a huge collection of cultures, wild cultures of various mutations and combinations of various mutations. By that time, a couple of hundreds of mutations of a glorious, actually unique, wonderful genetic object—the fruit fly *Drosophila melanogaster*—had been discovered and studied. Muller made several reports, visited our biological stations at the institute. Thus, he showed personally the technique of drosophila handling, how to work with it in laboratory. It was all entertaining, exciting and quite interesting.

I must say here that before the arrival of Muller I had to work a little with species caught in the Moscow region instead of *Drosophila melanogaster*, in the company of Dmitri Dmitrievich Romashov[3]—also a young man who had just graduated or not graduated from Moscow University in the speciality of entomologist, who later became one of outstanding geneticists of the Chetverikov Circle. On our own, according to literature, we got cooking the feed-stuff for drosophila going, as well as the reproduction technique, mercy killing flies with gas to study them under the microscope or with lens. So for us two it was not absolutely new but, nevertheless, useful. Knowing elements, we learned from Muller, in a manner, all modern for that time technology of drosophila breeding as a test object.

Muller read several reports to us. And anyway, he had quite a good time and chatted a lot. He visited both our stations: in Zvenigorod and in Anikovo. On those occasions, there were lavish sprees, even Koltsov,[4] and Serebrovsky,[5] manoeuvred and fetched a whole box of Champaign of the Abrau-Durso brand. The Champaign was delicious! And of course, pure alcohol. There were binges, and it was very exciting. Muller is a very talented and interesting man, indeed. We became big friends later.

---

[3]Dmitri Dmitrievich Romashov (1899–1963)—geneticist. Until 1942 he worked at the Institute of Experimental Biology, then at other Institutions. He was a population geneticist.

[4]Nikolai Konstantinovich Koltsov (1972–1940) is a Russian biologist, zoologist, cytologist, founder of the Russian Soviet school of Experimental Biology. Koltsov is the author of the fundamental idea of template synthesis of chromosomes. Corr. Member of the St. Petersburg Academy of Sciences from 1916 (Russian—from 1917, the Academy of Sciences of the USSR—from 1925), Acad. of the Academy of Agricultural Sciences (1935).

[5]Alexander Sergeevich Serebrovsky (1892–1948)—biologist, one of the founders of the national genetics, Corr. Member of the USSR Academy of Sciences (from 1933), Academician of the Academy of Agricultural Sciences (1935). Scientific directions of his investigations were general genetics and animal genetics. Serebrovsky formulated and experimentally confirmed the idea of the divisibility of the gene. He proposed a scheme for the linear structure of the gene and the method of determining its size, as well as a new direction in the theory of evolution—geno-geography. Serebrovsky contributed into theory of selection. He was the founder and first head of the Department of Genetics of Moscow University (1930), and one of the founders of the school of Russian geneticists.

Afterwards he came at the invitation of Vavilov—in 1934, 1935, 1936—spent three years here, first in Leningrad, then in Moscow, at the Institute of Genetics of the Academy of Sciences, in the group of Nikolai Ivanovich Vavilov. He learned to speak Russian fairly well and from Hermann Muller turned into German Germanovich, as his father was also Hermann. And then, in 1936, he showed a clean pair of heels. It was absolutely clear to him even in 1934 where everything was driving at. He put up with it until the end of 1936 and then was just in time to flee. In 1937, despite his American citizenship, it would be surely unsafe for him to stay here. At any rate, many people could have been arrested because of him. He put two and two together and took off home.

He read his reports to us in German. He had found out that only few of us knew English… very few. Even today young people do not know English, and do not have any idea about other congress languages—French, German and other. At that time back there remained a generation who had studied at illustrious schools and they spoke fluent French and German. Those who didn't have practice could not speak but could read and write and of course understood everything. English was not obligatory for us at school, and far from many pupils studied the English language. I had English at school. Muller quickly found out that nobody knew English, but he knew German and was absolutely sure that he spoke fluent German. The thing was that 'Menchen und Weibchen' he pronounced as 'Monchen und Wobschen'. Our listeners were bewildered at first what those 'Monchen und Wobschen' could the drosophila have. The words actually meant fly males and females. That's it.

Well, in 1922 a very significant thing happened: we contacted, personally, the most advanced at the time genetics group, the Morgan group, straight via Muller. Our primary task in connection with introduction of the most modern for those times experimental drosophila genetics into our circles was the necessity of thorough mastering the literature which had been absolutely unfamiliar to us before. In 1921 Koltsov received a book of Morgan "Structure basics of heredity"[6] sent to him by his friends in Germany. It had once played a great role in genetics. And it was, actually, the beginning of introduction of modern genetics into biological insight of Russian zoologists, botanists, microbiologists, etc. And only from 1922, or even 1923, scientific journals started arriving, especially on genetics, which had been in complete oblivion for us all before.

Muller brought not only live cultures of drosophila but also a great number of reprints on drosophila, corn and other papers. So, we set to an extremely important, detailed abstracting with a full critical analysis of those new genetic papers. In this regard, our coterie ceased to be a simple 'Soor' (translation from Russian: abbreviation of Russian words 'sovmestnoe' joint and 'oranie' shouting) and was entitled 'Drossoor'—joint shouting about drosophila.

Aside from everything else, for all of us it was the best school to apprehend and study scientific literature. Because our coterie was, as I have said, an informal one we could feel absolutely free and could freely annoy each lecturer with questions of

[6]The Physical Basis of Heredity by Thomas Hunt Morgan.

all sorts. So any lecturer, while delivering his report or a review of several papers, or a regular summary, had to be able to give an account of any issue we posed to him. It certainly played a big role in our further scientific progress. The leadership of Chetverikov in our completely free and democratic circle was another very important factor. Somehow he could direct all arguments, conversations that seemed to become sometimes absolutely vague, formless and chaotic into the necessary topic, but he never imposed any restrictions, either on the lecturer or the audience, in this way leaving us free for discussions and gabbing and preventing these arguments from vain talk. It seems to me now that all members of 'Drossoor' regard this 'Drossoor' school as great experience we had in 1921, 1922 and 1923.

I would like to mention once more the fact that from our juvenility to 1922 we were cut off absolutely from everything that was abroad. Genetics at that time was quite a young science and it was most interesting and viable. The period of vehement development of experimental genetics was roughly saying from 1913 to 1922–1923. That meant that we naturally could not follow its development, say nothing take part in it. We had only one or two years to catch up with it in our 'Drossoor', apprehend it and 'chew' it over. It certainly turned out right that we started parallel experimental research on this wonderful, most suitable object for experimental genetic studies, especially at that time—*Drosophila melanogaster*.

«Green Pamphlet»

Last time I tried to do my best to tell how we had joint a very exciting circle of people, mainly physicists, partially physicists-chemists, who were busy composing a new physical description of the world and new theoretical physics. In those times, just since the end of the 20-s in Copenhagen Bohr's group, first of all, they were developing the quantum theory and unification of the quantum theory with the relativity theory and the general principle of relativity. In the 30-s an absolutely new atomic and later nuclear physics began intensively to develop experimentally: neutrons were discovered and obtained a wide recognition, new and new elementary particles appeared. Thus, an extremely exciting epoch began in the history of physics.

The post-war time is often considered to be remarkable in the development of physics. I do not agree with this opinion completely. Now it is the time of the development of physics application activity. Now different discoveries in physics made at the end of 20-s and 30-s until the beginning of 40-s, are being introduced into practice. That time was indeed famous and extremely exciting. Now all this is transferred to machinery.

We have overcome the whole era of the atomic physics. What good has remained after it—is still too early to speak about because, well, several of these atomic electric power stations and the atomic station of desalting the sea water—all these are trifles, finally. But there are very many horrible things, of course: atomic bombs, nuclear bombs, hydrogen bombs which partially were realized experimentally and spoiled rather strongly the biosphere of the Earth. Now they are no explosions any more, at least, not so visibly and noticeably.

Then came the cosmic or "cosmetic" era in which we still live. Here there is even less new and unexpected going. And to what exciting things it will lead—still it is difficult to know. Maybe, the most exciting new what this "cosmetics" has brought with—is American and our long lasting but rather boring experiments on these orbital stations where some gentlemen were sitting for about 2 months and longer expecting the increase of their earth income being busy with the production of little lice and other insects. Nothing super exciting was discovered there. Unfortunately, American or our specialists almost didn't carry out gradually planned experiments and performed mainly whatever they had to do.

And again it cannot be compared with really great scientific discoveries which took place in 20s-30-s years of the 20-th century. Thus, for me, my group, the group of friends, employees, my students—it was a great luck that since the end of the 20-s and, especially, in the 30-s we managed to get into science just in the most interesting period of time, maybe, of the 20th century, in development of the natural sciences.

I very much hope that this new period of flourishing of the natural science in the 20th century will fruitfully impact on, first of all, philosophy and, second,—on many humanitarian disciplines. No doubt that some humanitarian subjects will have to change for a new manner not to turn out to be useless for all. But I believe that it will happen not so fast as the new reconstructions of the physical pictures of the world in the natural science. It will happen more gradually as a whole line of general methodological and, partially, of the philosophical principles of the modern natural sciences, not physics or biology, but just the natural sciences in general. At first, they will step by step be popularized to become "eatable" and understandable to the scientists who do not belong to the natural sciences or mathematics. Then they will penetrate the circles beyond the natural science and mathematics. Then, probably, we will enjoy the same new intensive exciting period in the development of humanitarian sciences on the Earth. It is possible. But all this belongs to the future. God knows. To predict is not good because you can "get into the sky with a finger"—that often happens.

Now I would like to speak about my own business. One of the three scientific directions which were developed in my division at Buch—as I have already told, is a quantitative study of the mutation process. And performing the study of the mutation process there was an attempt to create at least the most common picture of the origin of genes. If to get to know something substantial how something unknown to us is changing it means that we have already known something about this unknown to us. Thus, having found some regularities in the mutation process, it is possible to formulate some statements about the origin of the genes themselves whose changes are mutations. This is the main idea which was the basic of the joint discussions, discussions of genetics specialists, biologists, true biochemists and, what is the main thing,—of the physicists-theoreticians.

I have already mentioned that in my division at Buch this scientific direction was born not like a bubble on the marsh but it was the logical development of one of the directions established by my teacher Nikolai Konstantinovich Koltsov at the beginning of the century. He tried to do his best to create for himself some kind of a theoretical model of what chromosomes and genes are which are located lineally as it was known in these very chromosomes. He developed this on the basis of his experimental cytological studies where he investigated the influence of definite physical-chemical conditions on the form, structure and motion of cells as well as on the basis of the general discussions concerning the inherited elementary factors and genes. We used Koltsov's basics that chromosomes must be by definition extremely constant resistive compositions which determine all life especially of the cells and any joints of the cells. It means that it was already in those times clear that chromosomes are basics of what we call today the code of the heritage information. That is why Koltsov imagined the chromosomes as structural physical-chemical compounds, gigantic micelles, more probable, gigantic molecules of some more or less autonomic parts whose structural sub-divisions are genes lineally located in these long gigantic chromosomes. We were occupied with the studies to obtain mutations experimentally by exposure of the flies-drosophila with gamma-rays and other ionizing radiation. We—me in collaboration with physicists—theoreticians of the type of Max Delbrück as well as radiation physicists-experimentalists like my employee Zimmer and some young people who participated in this common activity which was very wide in range and huge in quantity of the data being processed. We tried to carry out the following.

Varying the conditions of exposure we tried to obtain the results from whose comparison it would be possible to conclude in the most common form what processes are the basics of the mutation origin, i.e., what do mutations mean? From physics it is exactly known that ionizing radiation can perform what other types of radiation cannot, and if to vary parameters, doses, rigidity of the ionizing radiation there must be consequences of these ionizing types of radiation. That is why during several years since we did not have other opportunities and methods we concentrated our work in this direction.

We carried out a huge volume of work. Just during those very years about, probably, ten or 15, I studied totally a couple of millions of flies and collected a rather big number of data on the direct and reverse mutations. One of the important some kind of criteria of the gene structure in the most general form, i.e., whether this structure is multi molecular or monomolecular, is an opportunity by one and the same method, say, by one and the same x-ray exposure, to cause mutation of some gene and its reverse mutation—from the mutation state back to the initial one. This is a very simple thing. Together with Muller once in some talk we formulated this way. Who did it, don't remember. Muller or me… More probable, it was Muller, in those days I was younger and sometimes felt ill at ease to chat like that but he already didn't. So, imagine this picture the following way: if the mutation were a simple quantitative damage of the gene, well, for example, a piece of the gene was bit out, then, of course, it would not have been possible to cause either direct or reverse mutations by one and the same x-ray exposure. It is like—you cannot break a window by a feast and by the same hit of the feast the window would have jumped back on its place.

From the comparison of action of different doses of the same rays and the same dose of ionizing radiation different on rigidity it is possible to clarify rather precisely whether the effect which we observe is a mono molecular or multi molecular change. We got a picture again was for the favor of a mono molecular change. That is why by the middle of the 30s we came up to some hypothesis that mutations caused by exposure are mainly relatively simple mono molecular reactions. Thus, logically it follows from the above that genes themselves must be of some what, if you like, simple physical chemical units.

Though, of course, they can be very complex. Simplicity and complexity are the notions which are rather indefinite. "Simple" I say in this case in the sense that they do not consist of combinations of different molecules forming some complex substance: some lubricant, tar, or butter, or something else. They are physical-chemical structure units, evidently, gigantic molecules or micelles or parts of a more or less autonomous, some very large micelle composing a whole chromosome which can be seen in microscopes. So, that was a simple picture we obtained—simple in the sense that it could be easily studied further. The first short overview was published by me in 1929, the second one, significantly thicker,—in 1931, and then the other one, even thicker,—in 1934 in "Cambridge philosophical bulletins". In 1935 we three—me, Zimmer, and Delbrück issued a publication in the so called "Göttingen funeral of the first class"… In Göttingen there was a famous (still exists) Göttingen Academy which is called not Academia but a Göttingen Gesellschaft der Wissenschaft, or when they feel boring they change the title for Gesellschaft der Wissenschaft zu Göttingen. They issue such green little note books where they published more or less long detailed reports made in that very society. It is still has the title of the classical one because of the respect towards our point of view concerning the mechanism of mutations.

Then after the end of the war it became clear that chromosomes and, consequently, the genes sitting in them are nucleoproteids. And the whole army of biochemists rushed to analyze and clarify the structure of those nucleoproteid compositions which form the basis of the chromosomes and then, logically, of the genes.

This was rather rapidly developing. Here the main theoretical brain work was done in England by a physicist Crick, and the main thing, I say, the chemical option was carried out in America. By the end of the 40s and in the 50s all the cream of the European science was concentrated there: everybody who managed to have snuck out there even during the war and many specialists—after the war. So, the American science began to flourish, still it is flouring till now, they say. Well, now, indeed, it is already the remains of the European big science.

In the second half of the 30s my friend and employee Max Delbrück moved to America, as I told you, a theoretician by background, but I introduced him to biology. A big number of American cytologists and European cytologists and biochemists who socialized in America fell under his theoretical influence. So, there was such an international group organized in the 50s whose core consisted of three individuals—an Englishman Crick, American Watson and a Russian physicist

Gamov, as abbreviation in Russian we called them "crick and gam" which means in Russian "shouting and noise".

Then this group was growing and growing, it meant that really famous analysis of the chemists began, true analysis, analysis of the macro molecule structures. Now by means of the Nobel prizes this splendid, indeed, organic analysis is going on, the analysis of the structure of gigantic protein molecules and nuclide acids.

There are all grounds to assume that in the near future the physical-chemical structure of the heritage information code will be determined with a rather sufficient precision. Now we are, certainly, still far from it. And only post graduates suppose that molecular genetics has already been perfectly built. In fact there is still no molecular genetics. My personal scientific and, in particular, experimental relationship with this part of genetics, with the study of the mutation process, general principles of the structure of genes and chromosomes, has been, so to say, completed. Personally since the 40s I have not been occupied with this. It is true that many people, especially, abroad consider me to be something like a granddad of this scientific direction because a new after-the-war version was issued by Delbrück: in the 30s I inserted into his brain some necessary substantial thing. So, everything started from this, from this very classical, our so called "green little notebook" of the Gottingen Society of Science. Let it be so, let it develop for good health.

The Isolines of the Female Beauty

Have I told you about the Copenhagen method—a mathematical study, precise, isokal, of isolines of the female beauty? Or haven't I? It doesn't lead to the subject beyond our topic, it is the most important option. Everything leads to micro evolution finally.

So, there [in Bologna at one of the plants of electro-technical equipment] there are absolutely amazing beauties in the assembling halls. Besides, in each hall they are dressed in a specialized uniform. In one hall all these beauties are in white uniforms, in the other—it is of the ivory color, in the third—they are in yellow, in the fourth—somewhat greenish, pink, blue. Each hall has its own color. All these clothes are sewed lovely, the waist is taken over with a little belt, they all are very nicely built, charming, good-looking faces and etc. How do these engineers live there? God knows! In this concentrate of such beautiful females! And it is not a surprise because the study, precise, of isokal has shown that one of the highest picks of the female beauty is located to the north from Florence, in Bologna itself, and in Dalmatia, in Yugoslavia. Three unreachable picks.

It was offered, I think, for the first time by a Russian physicist Gamov. He seemed to be the first one who suggested; "We all are interested … more or less, in good females, and etc. There are some strange people who insist: "Oh, there are many good-looking women in Paris." But this is assumed absolutely indefinitely, non-critically, not precisely. But the female beauty as everything else can be easily and simply studied statistically". So, a simple method was developed. Physicists-theoreticians, and in general, theoreticians like me, it means all the participants of the theoretical Copenhagen circle used such little note-books, well, like the ones used at school for foreign words. And where ever they got together and without any dependence on when they got together going or walking along the streets, being at restaurants, at cafes—it didn't matter, they put down estimations to the ladies whom they met on their way using a five-point scale with pluses and minuses. They put down the date and the place. All regions of Europe were distributed. We didn't take into account America, Africa or other continents. The Soviet Union was excluded due to political reasons: nobody was let in, nobody of the descent people gathered there, and it was unknown what was going on in the Soviet Union.

Each big region was under responsibility of one or two famous theoreticians. For example, Niels Bohr and his deputy Weiskopf were in charge of Scandinavia—Denmark, Sweden, Norway, Iceland… Then Chadwick and Blackett—two famous theoreticians and atomic specialists were in charge of England and Scotland, Ireland and, it seems to me, also of Holland. Pierre Auger and Francis Perrin, French, were in charge of France, Belgium. Then Rossetti—a wonderful Italian theoretician who knew so much about bugs and amazingly much about ommonites (archeological shellfish), he was in charge of Italy and Balkan countries.

Then Schrödinger was in charge for Austria, Czechoslovakia, Hungary and Switzerland, Heisenberg and Jordan—of Germany and Poland. Thus, all Europe was divided. It means that the leading scientists were collecting the data which were treated by absolutely first-class mathematical processing at the highest level. And the bosses—theoreticians on the basis of this processing—built isokals. For many countries it became possible already by the beginning of the Second World War, there was enough of the obtained data. Isokals—it is just the same as isobars or isotherms–isolines. Only isotherms are the lines connecting the points of the same mean temperatures, and isokals (derivative of the Greek "kallous"–"beauty") are the curves connecting the points of the same mean female beauty.

Rosetti has a study at Rome University in some ancient palazzo. It was an extremely high ceiling room and on one of its walls on its full size there was a hanging map of Italy and the attaching part of Balkans, Yugoslavia and Greece. This map was marked with the isokals. Very high picks, on the average they were a little bit lower than the "excellent" mark, but higher than the "good" mark with a plus, located also in the region of Florence and to the north of Florence in the Northern Toscana. The areas around Milano—it was also a "good" mark with a plus, on the average. The "excellent" mark with a plus was given in exceptional cases and required a special investigation "with passion". So, a very high pick—it was Bologna, then came the district Splita in Dalmatia, and—in the north from Splita in Albania.

And you who know the world mainly from the refined fiction have very often an absolutely false picture: "Oh, Italian ladies! Oh, Italian ladies!" To the south from Rome and in Rome itself, Italian women are a mixture of a frog with a monkey, speaking in general. When they are 15 years old—they are "more or less", by 25 years she weights already 100 kilos, you see, and even more "with a tail", she crawls out of all her skirts and, it is not clear what she had on her face being younger. Horrible! And among older southern Italian ladies there are, visa vice, there are absolutely dried out bones, covered with skin, just live witches. That's it! It is very sad, by the way, concerning Paris and France. Again because of our fine fiction which mixes sometimes beautiful clothes with the content in the beautiful clothes. In Paris the women's fashion is world-known concerning elegancy, and not in vain! But French women, in general, are not famous of the beauty, and sometimes, of elegancy either. Thus, do not trust the fine fiction in everything, it often just lies.

There is a very high pick in the Southern meadow Ireland, to the south from Dublin.

It was known long ago without any special proof that Irish women sometimes are amazingly beautiful. As I remember, someone put a couple of "excellent" marks with a plus in spite of the freckles on the face. This is a special phenotype—reddish hair, and even red, with green eyes, they are perfectly beautiful, just an "excellent" mark! Then there are very high picks in Norway. But in the Southern Norway there are also gaps. German women in some areas of the Southern and Western Germany are not at all good-looking, to tell the truth. But Prussian women, especially, in the north and north-east on the borderline with Poland sometimes can be given an "excellent" mark. And there the mean isokals were rather high because of this. In the Eastern Poland, also, but this is because of our Russian influence, though in Poland there are awful gaps that is why to relate the picks of the isokals directly with the country in general is very complicated. In all the more or less large countries without any exceptions there are picks and gaps, maybe, except Yugoslavia. There the highest picks are located in Dalmatia, but there is one or two picks in the old Serbia as well. There are very beautiful Serbian women with grey eyes and dark hair there like we have in the southern part of the Great Russia. So, I have presented to you the results of a large theoretical scientific investigation!

Nikolai V. Timoféeff-Ressovsky in Berlin-Buch (1925–1945)

Manfred Rajewsky, Dana Lafuente and Michael Bader

At the end of June 1925, after traveling by train from the Belorussky station in Moscow, a young Russian scientist arrived in Berlin, accompanied by his wife—a scientist like him—and their two-year-old son Dmitri. Nikolai Vladimirovich and Elena Alexandrovna Timofeeff-Ressovsky had married in 1922 and, at the invitation of the eminent brain researcher Oskar Vogt, director of the Kaiser Wilhelm Institute (KWI) for Brain Research, they were coming to his institute to do research. They had been suggested and especially recommended by Nikolai Konstantinovich Koltsov, director of the Moscow Institute of Experimental Biology,[1] with the support of the People's Commissioner (Minister) of Health, Nikolai A. Semashko. At that time an agreement for scientific exchange and cooperation existed between the German and Soviet governments, enabling the stay of Nikolai and Elena Timoféeff-Ressovsky in Berlin for a limited period.

Abridged version with minute stylistical changes (Dana Lafuente and Michael Bader) from: Manfred Rajewsky: Nikolai V. Timoféeff-Ressovsky (1900–1981), in: Geneticists in Berlin-Buch, published by the Max Delbrück Center for Molecular Medicine (MDC), May 2008.

Professor Emeritus Manfred F. Rajewsky, M.D., cell and molecular biologist with a focus on cancer research, was the founding director of the Institute of Cell Biology (Cancer Research) at the West German Cancer Center Essen, University of Essen Medical School. He died on July 26th, 2013.

Manfred Rajewsky—Deceased.

[1]Nowadays N.K. Koltsov Institute of Developmental Biology, RAS (IDB RAS).

M. Rajewsky · D. Lafuente (✉) · M. Bader
Max Delbrueck Center for Molecular Medicine in the Helmholtz Association,
Robert-Roessle-Street 10, 13125 Berlin, Germany
e-mail: d.lafuente@mdc-berlin.de

M. Bader
e-mail: mbader@mdc-berlin.de

V.L. Korogodina et al. (eds.), *Genetics, Evolution and Radiation*,
DOI 10.1007/978-3-319-48838-7_2

Nikolai V. Timoféef-Ressovsky in the early 1930s in the doorway of the genetics vivarium, an annex of the former Kaiser Wilhelm Institute (KWI) for Brain Research Berlin-Buch, now the Oskar and Cécile Vogt Building. In the course of the renovation in the nineties of the last century, the vivarium was demolished. Source: Prof. H. Bielka, Berlin

Prior to that, Vogt had acted as adviser to the Russian government concerning the establishment of an institute for brain research and the medical treatment of Lenin and later the autopsy of his brain. During this time he had become acquainted with the high scientific quality and originality of Koltsov's circle in the field of theoretical and experimental genetics, particularly with regard to the basic principles of ontogenesis, the role of mutations and population genetics. Koltsov was one of the most important figures in Russian biology during this period (Babkov 2002). Already in 1911 he had founded the world's first department of experimental biology at the Moscow City People's University named after A. L. Shanyavsky, out of which the Institute of Experimental Biology evolved in 1916. This institute was the origin of and, over a long period, home to the Russian school of genetics, population genetics and evolution biology, with its brilliant theorist Sergei Sergeevich Chetverikov. Vogt soon recognized the great significance of the

Russians' groundbreaking work for future investigations of gene characteristics and how these are influenced by other genes (gene interactions) and environmental factors, in particular with regard to identifying and influencing genetic diseases of the central nervous system. He therefore decided to integrate genetic research into his Berlin Institute for Brain Research and to recruit highly talented young scientists from Koltsov's circle to work there. At Vogt's behest, Timoféeff-Ressovsky was to establish a laboratory for genetics which was to become the nucleus of a new genetics department in the new institute building opened in Berlin-Buch in 1930. Several greenhouses ("the genetic vivarium") constituted part of the Department of Genetics. The task of building up and heading the department was assigned to Timoféeff-Ressovsky, who remained a Soviet citizen and formally was still a staff member of the Moscow Institute of Experimental Biology, but without a university diploma or PhD. In 1936 his department in the KWI for Brain Research attained autonomous status with its own budget. In 1938 Timoféeff-Ressovsky was named a scientific member of the Kaiser Wilhelm Society. The scientifically extraordinarily productive and successful chapter of his life which had begun with his arrival in Berlin ["…the most wonderful years of my life" (Rokityanskij 2005)] continued. Looking back, by far the largest part of his genetic research work forming the basis of his international acclaim was done during the years prior to World War II. In the field of history of science, many published accounts have dealt with his outstanding scientific work and the vicissitudes of his life (Babkov and Sakanyan 2002; Berg 1990; Bielka 2003a, b; Crow 1993; Deichmann 1995; Granin 1990; Glass 1990; Hossfeld 2001; Ivanov and Liapunova 1990; Korogodiin et al. 2000; Laubichler and Sarkar 2002; Medvedev 1982; Paul and Krimbas 1992; Ratner 2001; Rokityanskij 2005; Satzinger and Vogt 1999; Timoféeff-Ressovsky 1995, 2000; Tyuryukanov and Fedorov 1996; Vogt 1998; Vorontsov 1993; Winkler 2001).

[…]

Since a visit of the American geneticist Hermann Joseph Muller (Wunderlich 2008) in 1922, the Department of Genetics to which Nikolai and Elena Alexandrovna Timofeeff-Ressovsky belonged in Koltsov's Institute of Experimental Biology was headed by Chetverikov. His research group had received special lines of the fruit fly *Drosophila melanogaster*, which since 1910 were being used in Thomas Hunt Morgan's research group in the U.S. to crossanalyze chromosomal hereditary mechanisms on the basis of Mendel's Laws of Heredity. In contrast, Chetverikov and his team carried out cross-analyses using these lines to explore the emergence of species during evolution. The results were groundbreaking for the later so-called "synthetic evolution theory". Aldous Huxley characterized it as the "new synthesis", the fusion of genetics with classical evolutionary doctrine. At that time Chetverikov founded *Drozsoor* (from Russian, which means "joint shouting of drosophilists"), an extraordinarily productive discussion group which met regularly and in which Timoféeff-Ressovsky took part. He, like all of the other participants, remembered these discussions all of his life and later used them as a model for his discussion groups in Berlin. In his recollections, Timoféeff-Ressovsky also remembered as exceptional the intensive two-year internship under Koltsov, which every student had to complete. They (the students) later often said that this had been the best time of their lives.

During these years Timoféeff-Ressovsky also had his first contacts with the famous Russian biologists Nikolai I. Vavilov and Vladimir I. Vernadsky, whom he always highly admired and venerated. While still at the Moscow Institute, Timoféeff-Ressovsky published the results of his first experiments on the reversibility of spontaneous gene mutations (reverse mutations), using *Drosophila funebris* as an example. This research topic interested him, because he held the hypothesis—which later proved to be true—that mutations as the basis of evolutionary changes may not only be of a destructive nature. The research on this problem was then continued and extended in Berlin on the model of x-ray-induced mutations in *Drosophila melanogaster* immediately after Muller in 1926 had shown that x-rays could induce a high yield of mutations in *Drosophila* (Nobel Prize 1946) (Wunderlich 2008). In 1932/33 Muller was visiting professor with Vogt and Timoféeff-Ressovsky in Berlin-Buch, and from 1932 to 1936 he was member of the Board of Trustees of the KWI for Brain Research. During his Berlin period (1925–1945) Timoféeff-Ressovsky published 140 papers (among them (Möglich et al. 1944; Rajewsky and Timoféeff-Ressovsky 1939; Timoféeff-Ressovsky 1927; Timoféeff-Ressovsky and Timoféeff-Ressovsky 1927; Timoféeff-Ressovsky 1934a, b; Timoféeff-Ressovsky et al. 1935; Timoféeff-Ressovsky 1939; Timoféeff-Ressovsky and Timoféeff-Ressovsky 1940; Timoféeff-Ressovsky 1940; Timoféeff-Ressovsky and Zimmer 1947), which formed the basis of his world renown and which brought him the nomination for the Nobel Prize in Physiology and Medicine by Boris Rajewsky in 1950.

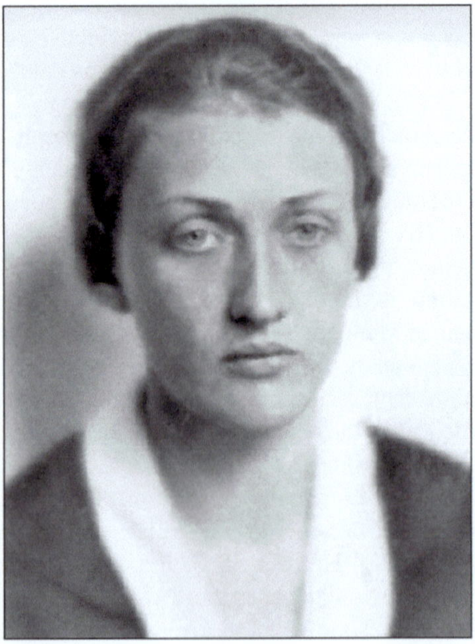

Elena Alexandrovna Timofeeff-Ressovsky. Photo: private

Not to be overlooked is the contribution of his wife, the outstanding geneticist Elena Alexandrovna Timofeeff-Ressovsky. She, too, was influenced by the Koltsov/Chetverikov research group, and with her precise, systematic approach and her balanced, unflappable way of working and her insight into human nature, she was an ideal counterpart to his quite impulsive, sometimes even chaotic nature. In the year of Elena Alexandrovna's death (1973) he was to say (Satzinger and Vogt 1999; Elena Alexandrovna Timofeeff-Ressovsky 1995): "She was a completely remarkable woman in every aspect. There are remarkable women, but they are quite rare in the world. But completely remarkable women are even rarer. My wife was such a completely remarkable woman. We worked in the same laboratory for 53 years, with four hands and two heads, and we were married for 51 years. During this time we were only separated for 21/2 years (during my imprisonment)."

Corresponding to the cosmopolitan nature of the Timoféeff-Ressovskys (and Vogt's Kaiser Wilhelm Institute for Brain Research), their Berlin apartment—first in Steglitzer Strasse (today Pohlstrasse) and, after the construction of the new building of the institute, in the gatehouse of the Berlin-Buch park premises—was always open to guests from all over the world. It was the place of social gatherings and many rounds of discussions, which often took place up into the wee hours and usually were dominated by the deep voice of the host. Besides colleagues and scientist friends, artists were frequently among the guests, especially during the twenties. Like the education of the two sons—the second son Andrei was born in Berlin in 1927—the hospitality of the Timoféeff-Ressovskys mainly rested on the shoulders of Elena Alexandrovna Timofeeff-Ressovsky. She was, besides Cécile Vogt, wife of Oskar Vogt and scientific member of the Kaiser Wilhelm Society, at that time the only female scientist at the KWI for Brain Research who combined her research activity with marriage and motherhood.

Overall considered, all of Timoféeff-Ressovsky's seminal work in experimental mutation research (mostly on *Drosophila*) was directed toward the nature of genes and mutations and their significance in evolution biology and population genetics. First, using mature spermatozoa, he confirmed the linear relationship between x-ray dosage and mutation rate and showed that the relationship between gender-related lethal mutations/visible mutations remained constant with increasing dosage, and that dose effectiveness, dose fractioning and irradiation at the same dosage level and at temperature differences in the range of 10–35 °C did not influence the mutation frequency. His classic phenogenetic studies explore the influence of the remaining genome ("genotypic milieu"), the external environmental conditions ("external milieu") and physiological variables ("internal milieu") on the expression of mutants. They were predominantly carried out on mutant *venae transversae incompletae* (*vti*) of *Drosophila funebris*. Along with Vogt, Timoféeff-Ressovsky distinguished between the penetrance, expressivity and specificity of mutated genes (Laubichler and Sarkar 2002). He found that these selected indicators of gene expression could to a certain degree vary from each other independently and that the "genotypic milieu", e.g. with respect to the influence of vti expression, can be differently active in geographically differently localized lines of a species. This all affected the way from the gene to the phene (phenogenetics), population genetics, e.g. the splitting of a species population into smaller (territorially isolated) subpopulations (microevolution processes), the generation of phenotypical characteristics in general and the genetics of ontogenesis.

Nikolai V. Timoféeff-Ressovsky (standing on the right) with his colleagues (Natascha Kromm, sitting on the left) in the dry and heated greenhouse of the genetics vivarium. Source: Prof. H. Bielka, Berlin

In the thirties it was especially important to elucidate whether recessive lethal mutations in fact represent the far most frequent kind of mutations or whether in reality innumerable other mutations are also present which do not appreciably impair the ability of an organism to survive until it reaches a reproductive age. In extraordinarily elaborate series of experiments Timoféeff-Ressovsky was able to show that xrays induce approximately twice as many mutations of the last-mentioned kind, i.e. without immediate recognizable effect, as lethal or sublethal mutations. The evolutionary significance of mutations also became clear when Timoféeff-Ressovsky analyzed the survivability of different mutations of *Drosophila funebris* at the same temperature in different combinations with other mutations. His results showed that the survivability of the mutant combination was sometimes just as good as that of the most effective mutation alone, in other cases it was just as poor as the least effective mutation alone, and in other cases it corresponded to a mean. The mutant combination could also be more effective than the most effective individual mutation or, however, less effective than the least effective. The American geneticist Bentley Glass, who in 1933 worked for half a year with Timoféeff-Ressovsky in Berlin-Buch, wrote the following about these experiments: "From the standpoint of clarifying the selective process upon the raw material of evolution, the mutations, this investigation is one of the most important ever made by anyone" (Glass 1990). In 1934, apart from "The experimental production of mutations", still today considered a classic survey and in which he—by the way—already used the term "genetic engineering", Timoféeff-Ressovsky published his most comprehensive work on phenotypical gene manifestation during the pre-war period "The link between the gene and the morphological character" (Timoféeff-Ressovsky 1934a). In his publication "Genetics and Evolution" (Timoféeff-Ressovsky 1939), which appeared in 1939 and like the aforementioned publication was also met with much acclaim, he summarized for the first time his notions on the genetic mechanisms in the processes of microevolution. In 1940, together with his wife, he reported the results of his population-genetic studies on temporal and spatial distribution in the open landscape (park area in Berlin-Buch) and on the action areas of different species of *Drosophila* (Timoféeff-Ressovsky and Timoféeff-Ressovsky 1940; Timoféeff-Ressovsky 1940). The above-named studies, including further experimental data, formed the basis for the models of microevolution which were developed as a consequence of Timoféeff-Ressovsky's research.

Due to his spectacular research work, his lectures in Germany and in other countries and, not least, his originality and his personal charisma, Timoféeff-Ressovsky soon became acquainted with leading geneticists and also with many physicists and biophysicists, who particularly appreciated him. He exchanged ideas or worked with a number of German scientists including e.g. the cytogeneticist Hans Bauer, the plant geneticist Georg Melchers, the geneticist and zoologist Alfred Kühn, the biophysicist Boris Rajewsky, the crop and radiation geneticist Hans Stubbe and the virologist Gerhard Schramm, and also the physicist and Nobel laureate Erwin Schrödinger, the physicists Karl Günter Zimmer, Max Delbrück, Robert Rompe,[2] Pascual Jordan and Friedrich Möglich (student of Max von Laue) and the chemist (Auer Society).

Outside of Germany he regularly took part in the famous seminar circle of the physicist Niels Bohr in Copenhagen along with the six-years younger Max Delbrück, as well as with Paul Dirac, Pierre Auger, Francis Perrin and William T. Astbury, and the biologists Muller, Theodosius G. Dobzhansky, Vavilov, Boris Ephrussi, Vernadsky, Cyril D. Darlington, John B. S. Haldane, Adriano Buzzati-Traverso, Torbjörn O. Caspersson and Åke Gustafsson. Supported by the Rockefeller Foundation and together with Ephrussi, he headed a small international group of eminent scientists (geneticists, physicists, chemists, cytologists, biologists and mathematicians), who before the beginning of World War II met regularly off-season in Dutch, Belgian and Danish seaside resorts to hold discussions about the most current research problems in biology. Correspondingly, in Timoféeff-Ressovsky's department in Berlin-Buch an international atmosphere prevailed, with visits, lectures and guest stays of foreign researchers, among these not seldom scientist friends from Timoféeff-Ressovsky's Moscow period, like Koltsov, Vernadsky, Vavilov, Alexander S. Serebrovsky, Yuri A. Filipchenko, Grigory A. Levitsky, Georgy D. Karpechenko, Solomon G. Levit and Chetverikov. Even the genetics department itself was multinational, although it had a certain 'Russian propensity', because e.g. Timoféeff-Ressovsky's wife Elena Alexandrovna Timofeeff-Ressovsky, the geneticist Sergei Romanovich Tsarapkin and the technical assistant Natascha Kromm worked there.

---

[2]See hear "Creating a Physical Biology: The Three-Man Paper and Early Molecular Biology", Phillip R. Sloan, Brandon Fogel (eds), Chicago and London press 2011.

The increasing mutual interest and collaboration between biologists and physicists in the twenties and thirties did not develop by happenstance. These were dictated by the general issue whether the laws of physics and biology were compatible, more specifically, whether life processes and structures obey well-known physical laws, including the quantum theory. For Timoféeff-Ressovsky, the paramount issue was the molecular nature of the gene with its exceptional stability, and connected to that, the molecular mechanisms of mutation. As Delbrück recalled in his Nobel lecture in 1969 (Delbrück 1970), the nature of the gene was at that time an issue of speculation: "From the hindsight of our present knowledge one might consider this ('that genes had a kind of stability similar to that of the molecules of chemistry') a trivial statement: what else could genes be but molecules? However, in the mid-thirties this was not a trivial statement. Genes at that time were algebraic units of the combinatorial science of genetics, and it was anything but clear that these units were molecules analyzable in terms of structural chemistry. They could have turned out to be submicroscopic steady-state systems, or they could have turned out to be something unanalyzable in terms of chemistry, as first suggested by Bohr …".

Kolyusha* hurrying from his apartment to his laboratory, drawn by Oleg Zinger.* diminutive form for Nikolai V. Timoféeff-Ressovsky. Source: Prof. H. Bielka, Berlin

In intensive discussion rounds which took place between 1932 and 1937 and in which Muller also took part in 1932 and 1933, Timoféeff-Ressovsky became Delbrück's most important teacher in genetics and quantitative mutation research. In Delbrück's words (1969) (Delbrück 1970): "Our principal teacher in the latter area (biology) was the geneticist, Timoféeff-Ressovsky, who, together with the physicist K.G. Zimmer, was at that time doing by far the best work in the area of quantitative mutation research." Timoféeff-Ressovsky began his research on the mutagenic effect of x-rays in 1928/1929, first in an x-ray laboratory in building V of the Hufeland Hospital in Buch and then continued with it from 1930 on in the "x-ray pavilion" of the new KWI for Brain Research, an annex to the connecting corridor between the institute and the hospital (Bielka 2003a, b). His first radiation-genetic publications date back to 1929–30. In 1935 the classic work "On the Nature of Gene Mutation and Gene Structure" by N. V. Timoféeff-Ressovsky, K.G. Zimmer and M. Delbrück appeared in *Nachrichten von der Gesellschaft der Wissenschaften zu Göttingen* (Timoféeff-Ressovsky et al. 1935) It also became known as the "three-man paper" or the "green pamphlet". In a masterly way, this publication summarized the results of experimental, quantitative mutation research up to that time and the model hypotheses that had been developed on the mutation process and on gene structure. The most important conclusions were, first, that a spontaneous mutation must be due to a rare and single-step (in analogy to the principles of quantum mechanics) stable molecular alteration via atomic rearrangement; and second, that mutations induced by ionizing rays are correspondingly dose-dependent, more frequent atomic rearrangements of the same kind (cumulative, direct one-hit events without threshold value, in the sense of target theory) (Timoféeff-Ressovsky and Zimmer 1947), triggered by ion pairs or small ion clusters. The possibility of an indirect triggering of a mutation through radiation-induced, short-term free radicals and/or chemical mutagens could not yet be taken into consideration at that time because these mechanisms were not discovered until years later. By contrast, first estimations—which later proved to be untenable—were already presented as to the size of a gene (derived via the volume of the mutation-triggering spatial target area). This and the notion that genes are molecules with stable atomic structure in which energy-conductive processes in turn lead to stable structural changes (mutations) of the same kind, has contributed considerably to the development of molecular genetics.

According to the activity report of the genetics department for 1937/1938 (minutes of the Board of Trustees meeting of the KWI for Brain Research from December 1938, Archives of the History of the Max Planck Society) (Satzinger and Vogt 1999), Timoféeff-Ressovsky had also at that time attempted to determine by x-ray structure analysis a crystal structure of chromosomes and to produce electron diffraction images of salivary gland giant chromosomes of *Drosophila*. The results were not published, but in principle these approaches corresponded to the later application of similar methods elucidating the structure of DNA in the 1950s. The significance of the seminal works of the Berlin-Buch research group on the nature of genes and mutation was emphasized by Schrödinger (Nobel Prize for Physics in 1933 for his contribution to quantum theory, together with Dirac) in his book "What is Life?" published in 1944 (Schrödinger 2001), and in part even served as basis for this excellently written text. Schrödinger's book strongly influenced the development of biology after the end of World War II. The international reputation of the work of Timoféeff-Ressovsky and Delbruck (who emigrated to the U.S. in 1937 and there became one of the fathers of molecular genetics) grew even more.

Nikolai V. Timoféeff-Ressovsky, 1943. Photo: private

From 1937/1938 on Timoféeff-Ressovsky's department had a powerful Philips neutron generator (linear accelerator with voltage up to 600,000 volts) at its disposal, with the aid of which radionuclides could be produced. Neutrons cause higher ionization densities in tissue than x-rays, and also in the case of neutrons the mutation rate was proportional to the radiation dose up to a saturation value, and a threshold value was not detected. Again, one ionization was postulated as a hit. Incidentally, even back then Timoféeff-Ressovsky had expressly pointed to the danger of radiation damage—including genetic damage—to humans through ionizing radiation, especially with respect to medical personnel in radiation diagnostics and therapy.

Apart from genetic and mutation research, Timoféeff-Ressovsky's department also experimented with radionuclides from about 1940 on. They were produced with the aid of the department's own neutron generator and purified by his colleagues Hans-Joachim Born (student of Otto Hahn) and Zimmer. The measurements were primarily carried out by Elena Alexandrovna Timofeeff-Ressovsky, Born, Joachim Gerlach[3] and Paul Max Wolf[4] (Auer Society) , using the radioactive

[3]See https://www.mdc-berlin.de/37810095/en/about_the_mdc/history/biography/karlzimmer.

[4]See Footnote 3.

tracer methods developed by Georg von Hevesy in the thirties. They pertained to the absorption, distribution, storage and excretion of radioactive isotopes of phosphorus, chlorine, arsenic and manganese in the mouse (in the case of radium 224 also the retention time, circulation time, in humans). After the end of the war alleged human experiments in the genetics department were an issue of controversy —quite unfairly so, because the doses used were very low and completely harmless. Experiments were also conducted on rabbits with thorium 234. It accumulated in the lymphatic system and later was used as a contrast agent (Thorotrast) in radiation diagnostics, but due to negative long-term effects (thorotrastoses, malignant thorotrastoma) it has no longer been used since the 1950s.

However, worthy of mention are also cell and molecular biological studies, which were important to establish basic methodology. An example of this is described in one of Timoféeff-Ressovsky's letters to his friend Boris Rajewsky dated March 17, 1941: "Dear Boris Nikolaevich!… Sometime ago you ran the risk of promising me schnaps, if we should succeed in radioactively labeling chromosomes or filterable viruses. I herewith have the pleasure of informing you that the latter has been successful: By biological means we have incorporated radioactive phosphorus into the tobacco mosaic virus (by breeding tobacco plants on a radioactive phosphorus-containing nutrient solution and propagating the mosaic virus on such radioactive plants). It is quite amusing! Kind regards from house to house, Yours (signature)".

As the Nazi period continued, especially with the beginning of World War II and the German attack on his fatherland on June 22, 1941, a shadow fell on Timoféeff-Ressovsky's life. The war and the responsibility for his family, staff and department increasingly became a psychological burden to him. Muller had left Berlin-Buch after the first attacks of SA troops on the KWI for Brain Research in 1933. Vogt was attacked by the Nazis because of his pacifist and cosmopolitan attitude (among other reasons because he employed women, including Jewish women, at his institute) and had received his dismissal in 1934 for political reasons. However, thanks to the support he received from Gustav Krupp von Bohlen and Halbach and Max Planck, president of the Kaiser Wilhelm Society, he remained acting director of the institute until April 1, 1937. Vogt complained at that time that the atmosphere at the KWI for Brain Research had drastically worsened due to denunciations of staff members. His successor Hugo Spatz actually intended to shut down the genetics department, which he considered to be an "alien element" in the institute. He finally had to consent to an agreement between the Kaiser Wilhelm Society and the responsible ministry, according to which the department was to remain preserved with a simultaneous budget increase. However, as far as the administration and the budget were concerned, the department was to be independent of the KWI for Brain Research. The continued existence of the genetics department was then ultimately secured in 1938 when Timoféeff-Ressovsky was named scientific member of the Kaiser Wilhelm Society at the KWI for Brain Research.

In 1932 Timoféeff-Ressovsky held a much acclaimed plenary lecture at the 6th International Congress of Genetics in Ithaca, New York, USA, in the presence of many world-renowned geneticists, among them Vavilov, Morgan and Muller. Following the conference, he was invited to work at the laboratories of the Carnegie Institution in Cold Spring Harbor for several months. Finally, in 1936 he was offered a position at the Carnegie Institution, which he rejected after much reflection. In his correspondence with Miroslav Demerec (Satzinger and Vogt 1999) he based this decision on his responsibility for his scientific and technical staff in Berlin-Buch and the fact that his sons attended the French secondary school in Berlin, and he did not want to ask them to make this adjustment. Another reason for his refusal was that in comparison to Germany, scientists had a lower status in the U.S., which Muller had pointed out to him.

In May 1937 Timoféeff-Ressovsky turned to the Soviet embassy in Berlin with the request of extending his and his family's permit to stay in Germany. This was rejected. Although this decision was very difficult for them, the Timoféeff-Ressovskys decided against returning to the USSR under the conditions prevailing there. Upon his request Timoféeff-Ressovsky had also been urgently warned beforehand by his teacher Koltsov via the Swedish embassy ("… of all the methods of suicide, you have chosen the most agonizing and difficult. And this not only for yourself, but also for your family…. If you do decide to come back, though, then book your ticket straight through to Siberia!") (Glass 1990). Vavilov, at that time president of the Academy of Agricultural Sciences of the USSR, also pointed out via Muller that only arrest and hard punishment awaited him under Stalin's rule with its waves of purges in the period of the repression of genetics and geneticists by Trofim D. Lysenko and Isaak I. Prezent. After Timoféeff-Ressovsky became a scientific member of the Kaiser Wilhelm Society, the German Minister of Science Bernhard Rust suggested to him in July 1938 that he assume German nationality. Timoféeff-Ressovsky politely refused ("I was born a Russian and see no possibility of changing this" (Paul and Krimbas 1992)). He always remained a Russian patriot (according to his assistant Natascha Kromm: "More than a patriot—a chauvinist" (Paul and Krimbas 1992)) and during the war repeatedly remarked that he was sure of Russian victory, for which he was admonished by the secretary general of the Kaiser Wilhelm Society.

For Timoféeff-Ressovsky it was especially depressing and hard to grasp that the best Russian geneticists—many of them his teachers—were arrested one after another as a consequence of the official damning of genetics in favor of the Lysenko doctrine during the waves of purges between 1929–1931 and 1936–1940. Most of them perished in prisons and labor camps. Chetverikov was denounced in 1929, arrested and banished to Sverdlovsk (today Yekaterinburg); he died in 1959. Koltsov lost his position as institute director and died in 1940. Vavilov died of starvation in prison in 1943. Karpechenko, Levitsky and Levit also died in prison. Timoféeff-Ressovsky's younger brothers Vladimir and Dmitri were likewise arrested and lost their lives, as did many of Elena Alexandrovna Timofeeff-Ressovsky's relatives. Another blow of fate followed in 1943. Timoféeff-Ressovsky's older son Dmitri had—without informing his parents in detail—at age 18 become a leading member of a young anti-Nazi resistance group which also helped prisoners of war, among them two

French pilots and East European and Western foreign workers, by providing them with hiding places and medication.

A provocateur blew their cover in 1943, and about 50 people were arrested as a consequence. Natascha Kromm had to watch from a window in the gatehouse in Berlin-Buch how Dmitri was arrested on the street. Afterwards a number of leading German scientists tried to intervene for Dmitri, but without success. The head of the Reich Security Central Office, Ernst Kaltenbrunner, wrote Timoféeff-Ressovsky in an official letter that Dmitri could not be rescued because he had worked against the Führer and the Reich. Dmitri was sent in August 1944 to the Mauthausen concentration camp, transferred later to the affiliated Melk camp and apparently perished there in 1945 a few days before the war ended. After a last sign of life in December 1944, his parents hoped for months that they would see their son again alive; Elena Alexandrovna Timofeeff-Ressovsky never gave up this hope until her death on Easter Sunday 1973. Still in July 1944 the Gestapo had offered—through Professor Julius Hallervorden, a department head of the KWI for Brain Research—to keep Dmitri in prison instead of sending him to a concentration camp, in case Timoféeff-Ressovsky declared his willingness to head the Nazi sterilization program for people of Slavic descent (Babkov and Sakanyan 2002; Bielka 2003a). Timoféeff-Ressovsky refused this categorically.

During the last years of the war the situation for Timoféeff-Ressovsky, who already was under severe psychological pressure, became increasingly perilous. Although he was a world famous geneticist, he was considered to be an "enemy alien", since he had a "consular" passport (issued outside of his native country) of the USSR. Moreover, he was neither a member of the Nazi party nor any of its organizations, and was often the target of suspicion in connection with the anti-Nazi activities of his son and his hardly reserved way of expressing his political views. Despite this, the Timoféeff-Ressovskys continued to aid many people who were in need. They hid individuals at home and in the institute who were threatened because they had Jewish relatives and helped forced laborers and prisoners of war to get jobs as temporary workers in the genetics department. The Timoféeff-Ressovskys' quick-witted, spontaneous willingness to help becomes very clear in a report of Professor Bernhard Hassenstein (Winkler 2001). He describes how Timoféeff-Ressovsky issued him a certificate (not based on fact) during his last visit in Berlin-Buch on February 10, 1945, without explaining it in more detail.

For Timoféeff-Ressovsky himself this was very risky, but for Hassenstein it was possibly life-saving in the coming chaotic period at the end of the war, which Timoféeff-Ressovsky foresaw. "Berlin-Buch, February 10, 1945 Certificate This is to certify that Mr. Bernhard Hassenstein is working as a laboratory assistant in the Department of Genetics of the Kaiser Wilhelm Institute at Berlin-Buch. Signature and department seal": "At that time I had no idea that Timoféeff-Ressovsky had saved the lives of many people in a similar fashion. Even though I never needed this certificate, it is one of the most deeply moving documents of my life."

Fighting on the Eastern Front was getting closer and closer to Berlin, and for that reason, already in 1944 the evacuation began of entire departments of the KWI for Brain Research to the western part of Germany. However, this did not apply to Timoféeff-Ressovsky's genetics department. He had decided to hold out, because he believed that he as a Russian could best negotiate with the approaching Soviet military units to preserve the department and the safety of the staff. He stuck to this decision, although many friends and colleagues urged him to evacuate to the West. Among these was also Boris Rajewsky from Frankfurt am Main, who went to see him in Berlin-Buch with the same intention right before the end of the war, but without success. It would be wrong to believe that Timoféeff-Ressovsky was sure of his course. On the contrary, he had been in a desperate state of mind for months—abstracted and hardly approachable (N. Kromm, personal communication).

In Berlin during the last weeks before the end of the war, continual bombardments, destruction and chaotic conditions prevailed. Berlin-Buch, which had been comparatively spared, was overrun on April 21, 1945 by a first wave of attack of the Soviet Army and then occupied. Following a conversation with General A. P. Savenyagin, deputy to the People's Commissioner for Internal Affairs of the USSR and responsible for certain areas of Soviet research, Timoféeff-Ressovsky was named acting director of an institute of genetics and biophysics, his former Department of Genetics, by a department of the Soviet Secret Service NKVD responsible for atomic research. With that, the institute came under the authority of the Soviet Military Administration. Besides that, the Soviets put him in the post of mayor of Berlin-Buch. A Soviet guard was stationed on the institute's premises to preserve law and order.

Five months later, however, in the night from the 14th to the 15th of September 1945, Timoféeff-Ressovsky was "requested" by another department of the NKVD to come to an approximately one-hour meeting and was transported away in a black limousine. A follower of Lysenko, Nikolai I. Nuzhdin, had denounced Timoféeff-Ressovsky. His Russian colleague, the geneticist Tsarapkin, was also arrested. The institute equipment was dismantled and taken to the USSR. Elena Alexandrovna Timofeeff-Ressovsky remained in Berlin-Buch with her son Andrei and like everyone in Berlin-Buch, did not receive news that her husband was still alive until two years later when word came from a secret research station in Sungul in the southern Ural Mountains. Soon afterwards, Elena Alexandrovna Timofeeff-Ressovsky and Andrei were allowed to join him there. In the meantime she had first been out-of-work in Berlin-Buch, supported by Care packages sent by American colleagues. However, from May 1946 on she was employed as an assistant to Hans Nachtsheim in the Institute of Zoology of the University of Berlin. Andrei Timoféeff was able to begin studying physics at the same university.

[…]

Acknowledgment This work was done with financial support from Program of the Russian Foundation on Basic Research (projects 15-04-01023 and 15-34-20639), partly from Program of UB RAS (project 15-2-4-21) and the Federal Special Scientific and Technical Program in support of leading scientific schools (SS-2840.2014.4). I would like to extend special thanks to Professor

Volker Wunderlich and Barbara Bachtler for their manifold support and constructive criticism in writing this manuscript as well as to Professor Ute Deichmann, Dr. Gudrun Erzgräber, Professor Heinz Bielka and Professor Erhard Geißler for their suggestions of important references. I would like to thank all of the above named for valuable conversations and the Berlin colleagues for visits to buildings in Berlin-Buch and help in the selection of illustrations. The personal memories of my aunt Natascha Kromm (1911–2006) regarding the period between 1930–1945 in the Timoféeff-Ressovsky research group have also found consideration. I dedicate this text to her, my parents Olga (Natascha's sister) and Boris Rajewsky, who were good friends of the Timoféeff-Ressovsky family, and to my nephew Nikolaus Rajewsky at the new Timoféeff-Ressovsky Laboratory for Medical Genome Research on the Berlin-Buch campus.

References

Babkov VV (2002) The theoretical-biological concept of Nikolai K. Kol'tsov. Russ J Dev Biol 33:255–262

Babkov VV, Sakanyan ES (2002) Nikolay Vladimirovich Timoféeff-Ressovsky 1900–1981. In: Sokolov BS (ed) Ser. Pamyatniki istoricheskoi mysli (A Collection of Historic Ideas). Moskau

Berg RL (1990) The grim heritage of Lysenkoism: four personal accounts. V. In Defense of Timoféeff-Ressovsky. Q Rev Biol 65:457–479

Bielka H (2003a) Das Leben der Timoféeff-Familie in Berlin-Buch 1930 bis 1945. In: Bielka H (ed) Diskurse über Medizin und Biologie in Berlin-Buch. Druckerei & Verlag Götze, Berlin, pp 20–32

Bielka H (2003b) Die wissenschaftlichen Arbeiten von N.W. Timoféeff-Ressovsky in Berlin-Buch, mit Anmerkungen über sein Verhältnis zur nationalsozialistischen Macht. In: Bielka H (ed) Diskurse über Medizin und Biologie in Berlin-Buch. Druckerei & Verlag Götze, Berlin, pp 33–56

Crow JF (1993) N.I. Vavilov, martyr to genetic truth. Genetics 134:1–4

Deichmann U (1995) Biologen unter Hitler. Porträt einer Wissenschaft im NS-Staat. Fischer Taschenbuch Verlag, Frankfurt am Main

Delbrück M (1970) A physicist's renewed look at biology: twenty years later. Science (Wash) 168:1312–1315

Granin D (1990) The Bison: A Novel about the Scientist Who Defied Stalin. Doubleday, New York. In German: (1988) Der Genetiker. Das Leben des Nikolai Timofejew-Ressowski, genannt Ur. Pahl-Rugenstein, Köln

Glass B (1990) Timoféeff-Ressovsky, Nikolai Vladimirovich. In: Gillispie ChC (ed) Dictionary of scientific biography 18 (II). Charles Scribner's Sons, New York, pp 919–926

Hossfeld U (2001) Im unsichtbaren Visier: Die Geheimdienstakten des Genetikers Nikolaj V. Timoféeff-Ressovsky. Med hist J 36:335–367

Ivanov VI, Liapunova NA (1990) Nikolay W. Timoféeff-Ressovsky (1900–1981). An essay on his life and scientific achievements. JINR, Dubna

Korogodin VI, Korogodina VL, Reformatskaya MA (eds) (2000) N.V. Timoféeff-Ressovsky. JINR, Dubna

Laubichler MD, Sarkar S (2002) Flies, genes, and brains: Oskar Vogt, Nikolai Timoféeff-Ressovsky, and the origin of the concepts of penetrance and expressivity. In: Parker LS, Ankeny RA (eds) Mutating concepts, evolving disciplines: genetics, medicine and society. Kluwer Academic Publishers, Dordrecht, pp 63–85

Medvedev ZA (1982) Nicolai Wladimirovich Timofeeff-Ressovsky. Genetics 100:1–5

Möglich F, Rompe R, Timoféeff-Ressovsky NW (1944) Über die Indeterminiertheit und die Verstärker-Erscheinungen in der Biologie. Naturwiss 32:6–10

Paul DB, Krimbas CB (1992) Nikolay V. Timoféeff-Ressovsky. Sci Am 266:86–92

Rajewsky BN, Timoféeff-Ressovsky NW (1939) Höhenstrahlung und die Mutationsrate von *Drosophila melanogaster*. Z. Indukt. Abstammungs- und Vererbungslehre 77:488–500
Ratner VA (2001) Nikolay Vladimirovich Timoféeff-Ressovsky (1900–1981): Twin of the Century of Genetics. Genetics 158:933–939
Rokityanskij YG (2005) N.V. Timoféeff-Ressovsky in Germany (July, 1925 September, 1945). J Biosci 30:573–580
Satzinger H, Vogt A (1999) Elena Alexandrovna und Nikolaj Vladimirovic Timoféeff-Ressovsky (1898–1973); (1900–1981). MPI für Wissenschaftsgeschichte Berlin, Preprint 112
Schrödinger E (2001) Was ist Leben?, 5th edn. Piper, München, Zürich
Timoféeff-Ressovsky NW (1927) Studies on the phenotype manifestation of hereditary factors. I. On the phenotypic manifestation of the genovariation radius incompletus in Drosophila funebris. Genetics 12:128–198
Timoféeff-Ressovsky NW (1934a) Verknüpfung von Gen und Außenmerkmal (Phaenomenologie der Genmanifestierung). Wiss Woche Frankfurt a.M. 1:92–115
Timoféeff-Ressovsky NW (1934b) The experimental production of mutations. Biol Rev (Cambridge) 9:411–457
Timoféeff-Ressovsky NW (1939) Genetik und Evolution (Bericht eines Zoologen). Z Indukt Abstammungs-Vererbungsl 76:158–218
Timoféeff-Ressovsky NW (1940) Mutations and geographical variation. In: Huxley J (ed) The New Systematics. Oxford University Press, Oxford, pp 73–136
Timoféeff-Ressovsky NV (1995) Vospominanija (Memories). Progress, Moskva
Timoféeff-Ressovsky NV (2000) The stories told by himself, with letters, photos and documents. Soglasie, Moscow
Timoféeff-Ressovsky HA, Timoféeff-Ressovsky NW (1927) Genetische Analyse einer freilebenden Drosophila melanogaster-Population. W Roux' Arch Entw Mech 109:70–109
Timoféeff-Ressovsky NW, Timoféeff-Ressovsky HA (1940) Populationsgenetische Versuche an Drosophila. I – III. Z Indukt Abstammungs und Vererbungslehre 79:28–49
Timoféeff-Ressovsky NW, Zimmer KG (1947) Biophysik I. Das Trefferprinzip in der Biologie. Hirzel Verlag, Leipzig
Timoféeff-Ressovsky NW, Zimmer KG, Delbrück M (1935) Über die Natur der Genmutation und der Genstruktur. Nachr v d Ges f Wiss zu Göttingen, Mathem-physikal Klasse Fachgruppe VI, Biologie, Neue Folge 1:189–245
Tyuryukanov AN, Fedorov VM (1996) N.V. Timoféeff-Ressovsky: Biospheric Thoughts. Moscow
Vogt A (1998) Ein Russisches Forscherehepaar in Berlin-Buch. Edition Luisenstadt, Berlin
Vorontsov NN (ed) (1993) Nikolay Vladimirovich Timoféeff-Ressovsky. Essays. Memoires. Materials. Nauka, Moscow
Winkler R-L (2001) Das Forscherpaar Timoféeff-Ressovsky – wissenschaftssoziologische Impressionen. Lomonossow (deutsche Assoziation der Absolventen u. Freunde der Moskauer Lomonossow-Universität (DAMU). Lomonossow 1:20–25 (with attachments)
Wunderlich V (2008) Hermann Joseph Muller, One of the Foremost Pioneers of Genetics, Seventy-Five Years Ago in Berlin-Buch; in Geneticists in Berlin-Buch, published by the Max Delbrück Center for Molecular Medicine (MDC), pp 40–60

Contribution of N.W. Timoféeff-Ressovsky to Biology and Methodology of Science

Alexey Yablokov

Science Contribution

***Concepts of expressivity* and *penetrance*.** In the 1920s, in the work carried out under the supervision of N.K. Koltsov, N.W. Timofeeff-Ressovsky formulated two fundamental concepts in classical genetics—the concepts of *expressivity* and *penetrance*. Expressivity is the degree of phenotypic expression of a gene, the degree of development of the phenotypic trait. It depends on the influence of other genes in the development of individuals and on environmental factors. Penetrance is the frequency of gene expression; in other words, it is the proportion of individuals in the population who have this gene phenotypically manifested. With complete penetrance, dominant or homozygous recessive allele appears in each individual in a population, and with incomplete—only in some ones. The concepts of penetrance and expressivity have become the fundamental ones in phenogenetics, medical genetics, and practical breeding of animals, plants and microorganisms. Working in Germany, Nikolay Wladimirovich found that the combination of recessive mutations may increase viability; it deepened the understanding of the evolutionary significance of the phenomena of recessivity and dominance.

***Biophysical model of the structure of the gene*.** In the 1930s, developing N.K. Koltsov's ideas of "hereditary molecules", Nikolay Wladimirovich, together with M. Delbrück and K. Zimmer, created the first *biophysical model of the structure of the gene* as a macromolecule encoding on the template principle. As a Nobel laureate for the development of this model, M. Delbrück always stressed the fun-

Advanced text of the speech at the Round Table "N.W. Timofeeff−Ressovsky: Science without border" Fourth Int. Conf after N.W. Timofeeff-Ressovsky SPb June 5 2015.

A. Yablokov (✉)
N.K. Koltzov Institute of Developmental Biology RAS, Moscow, Russia
e-mail: alexey.ablokov@gmail.com

V.L. Korogodina et al. (eds.), *Genetics, Evolution and Radiation*,
DOI 10.1007/978-3-319-48838-7_3

damental importance of Timofeeff-Ressovsky's ideas underlying molecular biology.

Hit principle and target model. Nikolay Wladimirovich is one of the founders of modern radiobiology. In 1930 he, together with K. Zimmer, formulated the "hit principle" and "target" theory; however, even rare hits in "target" can cause cell death or mutation, the frequency of which will increase with the dose. These principles form the basis of the quantitative biophysics of ionizing radiation and have not lost their significance today, after the discovery of other biological effects which can be induced by ionizing radiation (e.g., "bystander effects").

Theory of microevolution. It is difficult to overestimate the contribution of Timofeeff-Ressovsky to evolutionary and population biology. He created the *theory of microevolution*—evolution processes taking place at the level of populations and ending speciation. Nikolay Wladimirovich defined elementary factors in the process of microevolution (stochastic: mutation process, the insulation, the waves of life, and natural selection as the only directional factor), elementary unit (population), elementary evolutionary event (genotypic changes in the composition of the population), elementary evolutionary material (mutations of various types). Long known apart, all these processes and phenomena, embedded in a clear pattern of interaction, made the section of microevolution most "advanced" in evolutionary theory. His work on *adaptation polymorphism* in natural populations became classic. In population biology, Timofeeff-Ressovsky initiated the creation of yet another frontier of biological science—phenetics of natural populations, which is based on the doctrine of phens (elementary signs of phenotype).

Concept of levels of organization of living matter. Up to now, the contribution of the scientist to general biology, his concept of *levels of organization of living matter, is not fully appreciated*. Each of the four levels selected by Timofeeff-Ressovsky (molecular genetics, developmental, population-specific and biogeocenotic) is characterized by the presence of specific elementary structures and phenomena.

Radiation biogeocenology. Together with his wife Elena Alexandrovna Timofeeff-Ressovsky, Nikolay Wladimirovich laid the foundations for another frontier science—*radiation biogeocenology*. This scientific direction has become urgent in our time because of the widespread pollution of the biosphere by a variety of anthropogenic radionuclides as a result of careless and irresponsible handling of the atom by man.

Over the past decade it has become clear that the scientific directions which were established or advanced by Timofeeff-Ressovsky have a different fate. Something has become a cornerstone in the building of modern biology (for example, the theory of microevolution), something pushed colleagues to the outstanding achievements (gene structure), something required additions and development (as it turned out, the target theory describes only part of the complex picture of the radiation effects on living matter). The concept of the levels of the living organization, as well as phenetics of populations, is still waiting for its development; they may be in greater demand in the near future than today.

Methodology of Science

Methodological principles. The place of Timofeeff-Ressovsky in science is determined not only by the areas of specific scientific disciplines briefly listed above, but also by the development of the methodology of scientific knowledge. As Timofeeff-Ressovsky had once said, he thought the most important thing he made in science was a formulation of the enhancer principle (it was described by him in the 1930s together with R. Rompe). According to this principle, a single change at the level of a gene can change the properties of the whole organism and activate the power several orders of magnitude greater for the energy used, or, in more general terms, energetically insignificant events can lead in nature to the consequences of many orders of magnitude larger. The enhancer principle has proven to be one of the five axioms of theoretical biology ("random change in genetic programs during the formation of the phenotype is magnified").

Below are listed four methodological principles that were widely used by Nikolay Wladimirovich, in his characteristic aphoristic form of words.

1. ***It is always necessary to "distinguish the essential from the inessential"***. It was the main principle of Timofeeff-Ressovsky. He formulated this principle in different variations: "*it is not necessary to study the 40th leg of the centipede*", and, saying that someone "*does not spoil the starry sky*", he meant that science should try to "spoil" the starry sky: try to solve the major challenges. Nikolay Wladimirovich used very carefully the word and the concept of "scientific theory". The "theory" is the highest achievement of scientific thought. He applied this term only to Darwinism and classical genetics (Mendelism). According to Timofeeff-Ressovsky, the structure of scientific research looked like this: interesting facts—empirical generalizations—a hypothesis—a concept —a theory. If the hypothesis found confirmation of new facts, it could turn into a concept. Behind such a formal approach lies clarity (even stiffness) of scientific thinking peculiar to Nikolay Wladimirovich, which allowed him always to distinguish the essential from the inessential.
2. ***All must be "sorted out"***. Hence follows the need for finding basic structures and factors (structuring), establishing a hierarchy of events and phenomena. This principle is adjacent to the principle of "*non-jellylike world*"—discreteness of all things, and the saying "let me pause to think of it", which he used as a humorous analogue of the so-called "system approach", having then the general vogue.
3. ***"Do not do what the Germans are doing better than you"***. The principle is multi-faceted. It is not necessary to conduct research in a certain field if someone is well under way in this way. Try to be at the forefront of science.
4. ***"You can pick up an example for any rubbish"***. With his piercing theoretical thinking, Nikolay Wladimirovich was skeptical about various "inconvenient" facts and examples. With constant imperfection of our knowledge, one can discover some apparent confirmation of any extravagant hypothesis. Therefore, Timofeeff-Ressovsky was in no hurry to abandon the considered constructs, just

> because some facts fall beyond the general concept. In all such cases known to me, he was, after all, right, and the facts which seemed suicidal for a concept were then adequately explained.

He was a born Teacher and methodologist. He was open to all, and especially to young people. It gave him obvious pleasure to discuss directions for future research. Thus, on the one hand, he was always markedly polite in such discussions, but sometimes finished literally with utter rout of the proposed scheme, or areas of research, showing weaknesses and irrationality, and offering other ways of solving the issue. Perhaps it is this openness to everyone who addressed him and the ability to help everyone understand what and how to do something interesting in science that attracted a lot of young—and elderly—researchers.

Timofeeff-Ressovsky fumed when someone began to justify the importance of the proposed research areas referring to the importance of one thing and another. He often said that a sufficient basis for the development of a topic in science is the principle "*because it's fun*". Based on many conversations with Timofeeff-Ressovsky, I am sure that this seemingly unpretentious expression is associated with the deep philosophical principles of free will.

When he saw the man perceiving his criticism, he started "smashing" seriously, not mincing words. Oddly enough, such a dressing down did not depress people, rather it inspired them. When talking about classes in science, he often repeated the phrase, which later also became winged: "*Science needs to be addressed without wild beast's seriousness*".

N.W. Timofeeff-Ressovsky is not only one of the greatest biologists (who thoroughly "spoiled" the starry sky of the biology of the twentieth century), but also an outstanding science methodologist.

Personal Recollections About N.W. Timoféeff-Ressovsky and His Action for Radiation Biophysics in Berlin-Buch and Dubna

Helmut Abel and Gudrun Erzgräber

Helmut Abels Memories

In 1957 I had my first business trip to Leningrad. Professors Walter Friedrich and Robert Rompe implicitly encouraged me to inquire about Timoféeff-Ressovsky. But all my efforts were unsuccessful. This went on year after year, at some points I was even warned that any question about Timoféeff-Ressovsky—were bad questions.

Our Department of Biophysics in Rossendorf, which was established in 1966, has aspired cooperation with Soviet institutions, so we decided to contact Dubinin. Of course, for our interests and ideas in radiobiology the contact with Timoféeff-Ressovsky were most important. The trip to Dubinin to Moscow already took place in 1966. Dubinin, answering the question about Timoféeff-Ressovsky, has advised me to contact the Obninsk Institute of Medical Radiology.[1] We have applied for this trip, on the ground that scientific cooperation should be initiated. This was approved, which allowed me to go to Obninsk in 1966, where I first met Timoféeff-Ressovsky.

We talked for hours about one of our newly published papers with research from Rossendorf. Frankly speaking, it was more of an interrogation than a discussion. He asked questions about my graduate and doctoral theses, and where they were performed, and who my (doctoral) supervisors were. Having mentioned the names of Rompe and Friedrich and my temporary work in 1952 in Berlin-Buch, I handed

[1]Nowadays Medicine Radiology Scientific Center after A.F. Tcyb, Russian Ministry of Health.

H. Abel (✉) · G. Erzgräber
Former Zentralinstitut für Krebsforschung and Max-Delbrück-Centrum, Berlin-Buch, Germany
e-mail: helmut.abel@gmx.de

G. Erzgräber
e-mail: gerzgra@gmx.de

V.L. Korogodina et al. (eds.), *Genetics, Evolution and Radiation*,
DOI 10.1007/978-3-319-48838-7_4

him a letter and photos from Rompe. Timoféeff-Ressovsky suddenly became very excited, rushed out of the room and came back with his wife Elena Alexandrovna Timofeeff-Ressovsky. He introduced me to her with the words: “We have a visitor from Buch”. We then immediately drove to their apartment. Not another word was said on radiation biophysics, we only spoke about Berlin-Buch, Robert Rompe and his family, on literature, painting, music and the political developments in the past and present. These conversations went on late into the night.

The planning of our cooperation was discussed then in the following days in the institute and every evening up to deep in the night in Timoféeff's apartment with himself, his wife Elena Alexandrovna Timofeeff-Ressovsky and his young colleagues. But we talked not only about possible topics for our cooperation. He was literally by leaps and bounds back and forth. Just were talking about experiments on Drosophila, then about Leonardo da Vinci, then about Heisenberg's uncertainty principle. These jumps and broad range of scientific as well as cultural interests impressed me very much.

During the week that I spent with him in Obninsk we were able to design a fairly extensive program of our collaboration in writing. Our plan included 2 stages: A 1st stage was aimed at the investigations of radiation-induced mutations in *Drosophila* and *Arabidopsis*; irradiation in Rossendorf and evaluation in Obninsk. The first goal was the continuation of their previous work in Berlin-Buch, and the second was to clarify some contradictory results in the literature. A 2nd stage was dedicated to the quantitative analysis of radiation-induced damage of isolated DNA and its possible intracellular repair mechanisms and biological consequences thereof.

In 1967 our biophysical department in Rossendorf was re-organized as an external part of the Institute of Biophysics in Berlin-Buch, in which I was invited to build up a department of radiation biophysics. Our cooperation has thus led Timoféeff-Ressovsky indirectly to Berlin-Buch.

The first stage of our cooperation covered the years 1967–1972 with a total of 7 publications.

In 1972, Timoféeff-Ressovsky was fired from the Institute in Obninsk. However, he gave me a phone number of a friend of him in Moscow, and through this person we always could reach him. After 1972, we met several times for in-depth discussions of the second stage in Moscow. It was planned to perform comparative experiments with beams of different ionization density, gamma-rays, protons, neutrons and heavy ions. Experiments with heavy ions, however, could only be conducted in the nuclear research center in Dubna. Timoféeff suggested that I should give a lecture in Dubna and handed over its written text as an application for approval of experiments in Dubna. The idea was that he would certainly be asked by Dubna for consulting and expertise. And it happened exactly like he said.

After initially only sporadic experiments in 1973–1975 we received our own laboratory in Dubna,[2] which existed from 1976 to 1983 and was built by Gudrun Erzgräber as a branch of our Berlin-Buch Radiation Biophysics Department.

[2]Joint Institute for Nuclear Research, Laboratory of Nuclear Problems.

Thanks to the support and the many discussions with Timoféeff-Ressovsky we have performed and published a large number of studies on radiation-induced cell damage and intracellular DNA repair.

In 1987 the Soviet military jurisdiction asked the GDR for mutual legal assistance regarding the applications for the rehabilitation of Timoféeff-Ressovsky. The Academy of Sciences of the GDR fulfilled this request. After a detailed analysis of available documents, lectures and publications of Timoféeff-Ressovsky we came to the conclusion that his research at no time served fascism in Germany or provided materials for the warfare. In June 1992, posthumously, Timoféeff-Ressovsky was rehabilitated officially.

Robert Rompe expressed the opinion of many colleagues when he said: "His rehabilitation came too late for him. Yet it gave to all who knew, admired and worshiped him a sense of joy about a fair decision".

Gudrun Erzgräbers Memories and Activities in Buch to Commemorate Timoféeff-Ressovsky

I met Prof. Nikolai Vladimirovitch Timoféeff-Ressovsky in March 1967 in Obninsk. This was preceded by a long story, as described by Helmut Abel, my teacher and leader of radiation biophysics in GDR.

I came to Obninsk to Nikolai Vladimirovitch for a 10-day working visit with the aim to manage the planned collaboration between Helmut Abel and Timoféeff-Resovsky. The purpose of the planned experiments was to study the effect of different kinds of radiation on mutation frequency in different biological objects, especially in *Drosophila melanogaster* and in *Arabidopsis thaliana.*

Nikolai Vladimirovitch asked a lot of questions at the same time, was impatient regarding the start of joint experiments. There were two other colleagues of Helmut Abel together with me in Obninsk. On the next days we were preparing common reports which were supposed to be presented in front of the Timoféeff-Ressovsky team.

In 1968 I visited Nikolai Vladimirovitch for the second time. I had a special mission. Due to the precarious situation of our Department of Biophysics at the former Institute for Nuclear Research in Rossendorf our cooperation which has just begun was at risk to be terminated, so I came to ask Timoféeff-Ressovsky for help.

The hospitality of the Timoféeff-Ressovsky family was very warm; we have already carried out the first joint experiments and also the first close colleagues of Timoféeff-Ressovsky visited us in Rossendorf.

Most impressive events of this week for me were the evenings at his home, in which I was integrated not only as a guest like during my first stay, but more like a good friend. Just 4 years ago I graduated from the University and for me this kind of professor-student relationship which I could observe there was extremely fascinating: in the hallway of the small apartment there were at least 10 pairs of slippers of those who visited Timoféeff-Ressovsky every day, no, they were actually at home there. The discussions over tea and biscuits were very loud, everything what was interesting was discussed very openly and frankly: art, science, philosophy, not so often Institute's problems. However, if these came up Timoféeff-Ressovsky was never shy to call a spade a spade. For me as a young graduate, this was the first serious and loud criticism of conditions during Socialism, which I heard. Only later, when I learned more about his life and have been working and living in Russia for many years, I wondered how he could talk like that without being punished.

The result of my diplomatic mission was a letter from Timoféeff-Ressovsky to his old friend Prof. Rompe, the head of the Department of Physics North of the Academy of Sciences of GDR. This was of great help, and the cooperation with Timoféeff-Ressovsky could be continued. During period of the joint work with Timoféeff-Ressovsky in 1967–1971 we met three of his students in Rossendorf: Vladimir I. Ivanov,[3] Eugene K. Ginter and Nikolay V. Glotoff who worked each several times for several weeks with us. Seven joint publications in the "studia biophysica" resulted from our cooperation.

At the beginning of the seventies the Department Timoféeff-Ressovsky was dissolved and his students were stuffed into different places in Russia. Thus, our contact with him was broken for the time being. In 1973 and 1974 Helmut Abel succeeded to get in touch with Timoféeff-Ressovsky again, through colleagues from the Intercosmos—Institute in Moscow. However, the meetings had conspiratorial character and took place in Moscow apartments.

Nevertheless, this illegal contact was very helpful for us. In 1976, thanks to the support of Timoféeff-Ressovsky, we were able to organize radiobiological working group at the nuclear research center in Dubna.[4] So I also owe Timoféeff-Ressovsky my almost eight-year stay in Dubna.

On the day of his death Helmut Abel and I were in Dubna. We worked there since 1978 together with V.I. Korogodin, a student of Timoféeff-Ressovsky. Unfortunately, our urgent appeals to allow us to travel to Obninsk to attend the funeral were rejected.

After the reunification of East and West Germany Professor Ganten and Professor Bielka, the new and former directors of our institute, supported the memory and the respect of scientific traditions on the Campus Berlin-Buch.

So we set up a small Timoféeff-Ressovsky-museum in one of the laboratory buildings with many exhibits from the old laboratory of Timoféeff-Ressovsky time. On open house days on the Campus Berlin-Buch there are regular guided tours of the Museum. In 2000, on the occasion of the 100th birthday of Timoféeff-Ressovsky, we organized in Berlin-Buch a symposium at which three

---

[3]See "Vladimir Il'ich Ivanov (1932–2011)" In: Korogodina VL, Mothersill C, Seymour C (eds). Proceedings of the Third Int. Conf after N.W. Timofeeff–Ressovsky (2012). JINR Publ, Dubna: 230–238 (Russian).

[4]Joint Institute for Nuclear Research, Laboratory of Nuclear Problems.

former students of Timoféeff-Ressovsky took part: Glotoff, Ginter and Ivanov. At this symposium, we also showed our own summary of the movies directed by Sakanyan.

In 2006 we named the new laboratory building for genomics after Timoféeff-Ressovsky. A close friend of him, Prof. Avakyan,[5] who unfortunately could not attend, sent a message of greeting, which I read and added with my own personal memories on Timoféeff-Ressovsky. Furthermore, we organized some commemorative readings in particular for Elena Alexandrovna Timofeeff-Ressovsky with the participation of high-level scientists.

With all our activities in Berlin-Buch we hope to contribute to honor the memory of Timoféeff-Ressovsky and his scientific achievements and to give the possibility for young scientists and other interested people to learn more about Timoféeff-Ressovsky personality.

[5]See "Vladimir Il'ich Ivanov (1932–2011)" In: Korogodina VL, Mothersill C, Seymour C (eds). Proceedings of the Third Int. Conf after N.W. Timofeeff-Ressovsky (2012). JINR Publ, Dubna: 204–211 (Russian).

Part II
Genetic Processes

Template Principle in Biology

Sergey G. Inge-Vechtomov

Origin of the Template Principle

Evolution and expansion of genetic paradigm proceeded from Mendelism to chromosome theory and further to the concept of DNA as genetic material is presented at Fig. 1 (Inge-Vechtomov 2013) . The idea of protein inheritance, initiated by the study of infectious prion proteins (Prusiner 1994), especially in lower eukaryotes (Wickner 1994; Wickner et al. 1995) , had been shaped just to the end of the XX century. It raised the new problems in description of inheritance and variability phenomena from the unitary position. As we suppose the new paradigm based on the template principle overcomes those difficulties. The essence of the template principle in this context offered N.V. Timofeev-Ressovsky at the beginning of the 30-ies of the XX century in his idea of "convariant reduplication", which he posed to describe reproduction and variability of the genetic material as the single process (see: Timofeev-Ressovsky 1980).

The roots of the template principle may be followed from R. Virhow ("Omnis cellula e cellula") and L. Pasteur who postulated at the middle of XIX century the uninterruptable continuity of reproduction of living matter. The principle had been finally shaped by N.K. Koltsov (Fig. 2) in the period of chromosome theory establishment, even before recognition of DNA as universal genetic material: "Omnis molecula e molecula" (Koltsov 1936). We can go on in discussion about whether Koltsov's view was adequate to our contemporary knowledge of template processes when a molecule builds a copy of itself. Koltsov, as the majority of his colleges, considered proteins as genetic material. Nevertheless we should stress the general historical trend which had been expressed by Koltsov.

S.G. Inge-Vechtomov (✉)
Genetics and Biotechnology,
St. Petersburg State University, St. Petersburg, Russia
e-mail: ingevechtomov@gmail.com

V.L. Korogodina et al. (eds.), *Genetics, Evolution and Radiation*,
DOI 10.1007/978-3-319-48838-7_5

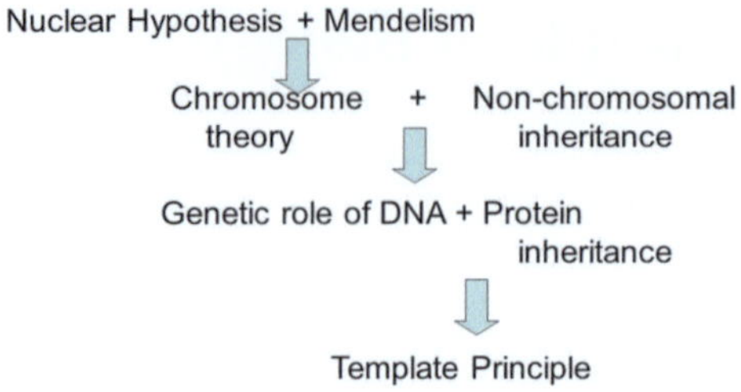

Fig. 1 Paradigm evolution in genetics

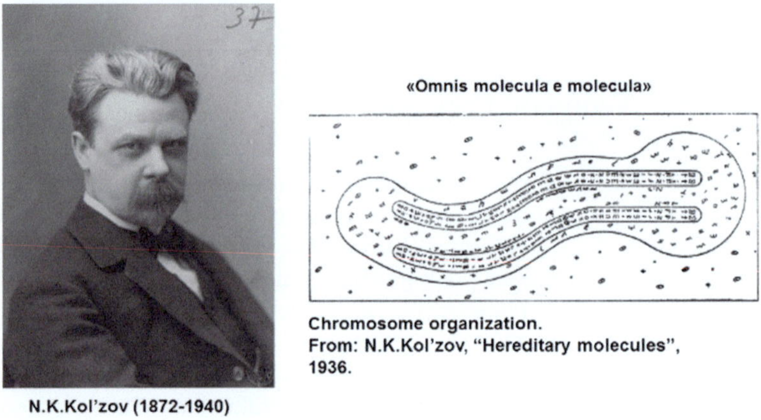

Fig. 2 N.K. Koltsov—founder of Moscow school of experimental biology

Fig. 3 N.V. Timofeeff-Ressovsky 1960s

Fig. 4 (From left to right) N. Timofeeff-Ressovski, M. Delbrück, K. Zimmer (Goldman 2011)

N.V. Timofeeff-Ressovsky (Fig. 3) was one of the brightest representatives of Koltsov's team. He could not avoid the influence of his teacher and finally it was expressed in his formula of "con-variant reduplication".

Timofeeff studied the problem of the gene with K. Zimmer and M. Delbrück (Fig. 4), utilizing radiation induced mutagenesis in *Drosophila*. Their famous paper —so-called "Green pamphlet" (Timofeeff-Ressovsky et al. 1935), in which target theory had been offered for explanation of the primary events in gene mutations, became the basis of the future molecular biology. In spite of the fact that the principle was not fully adequate for the explanation of the mechanism of mutagenesis as we shall see further, the biological phenomenology had been offered in a

Fig. 5 Nobel Prize winners (1962), from left to right: Francis Harry Compton Crick, James Dewey Watson, Maurice Hugh Frederick Wilkins

Fig. 6 J. Watson and F. Crick. Cavendish Laboratory, Cambridge, 1953

form amenable to physicists. It was the first attempt to reduce biological phenomenology to the level of molecular mechanisms.

We know that the "Green pamphlet" highly impressed the physicist E. Schrödinger, who used it as a basis of his lectures and later on of his book "What is Life? The Physical Aspect of the Living Cell" (Schrödinger 1944). After reading of this book a number of physicists turned to biology. We can remain M. Wilkins (Nobel Prize 1962 with J. Watson and F. Crick) (Figs. 5, 6).

Central Dogma and Protein Inheritance

The further progress in description of replication of genetic material (DNA) and expression of genetic information via transcription and translation had been summarized in the Central Dogma (CD) of molecular biology by F.H.C. Crick at 1958 and acquired the widely excepted form (Fig. 7) to 1970 (Crick 1958, 1970). Numerous efforts to revise this idea failed. Opponents of CD inadequately exploded the term "information" (genetic information). F. Crick partially also was responsible. Nevertheless, if to read attentively the essence of his works, but not simply the words, one can be convinced that the template processes are really described: replication, transcription, translation, when DNA serves as the template for DNA and RNA and mRNA serves as the template for the polypeptides.

The only problem met by CD was connected with the discovery of the so-called protein inheritance, or "replication" of infectious proteins—prions (Prusiner 1994) ,

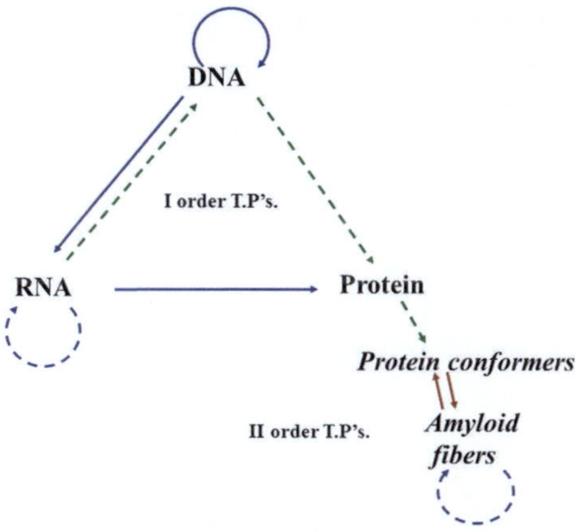

Fig. 7 Central dogma of molecular biology as the template principle reflection (Crick 1958, 1970) with addition

which had been found not only in mammals, but in lower eukaryotes also, in which they are especially abundant (about 10) (see: Inge-Vechtomov 2003; Inge-Vechtomov et al. 2012). In the latter case (but not in mammals) the prions serve as cytoplasmic hereditary factors. Short-lived sensation about proteins self replication through direct copying of their primary structure had been closed pretty soon. The synthesis of prion producing proteins proceeds on ribosomes as usually. Meanwhile, modified conformation (spatial folding) of some proteins, followed by their aggregation, may be transferred to homologous and even to some heterologous polypeptides. The idea of spatial or conformational templates was born this way.

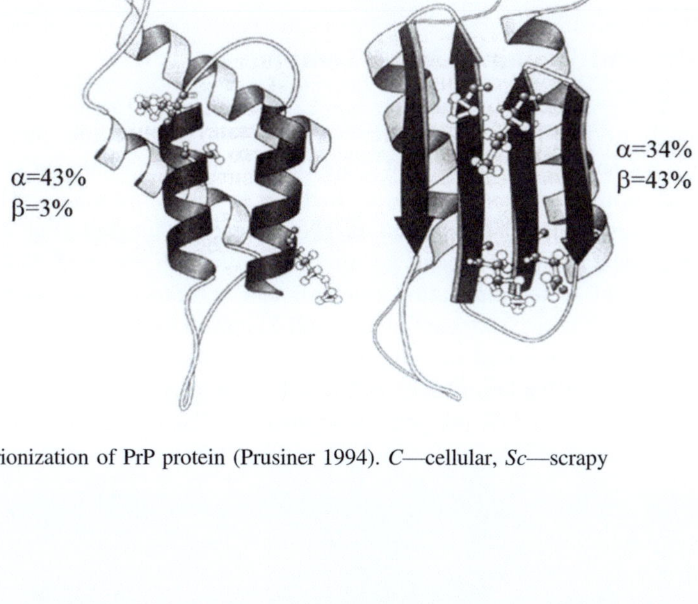

Fig. 8 Prionization of PrP protein (Prusiner 1994). *C*—cellular, *Sc*—scrapy

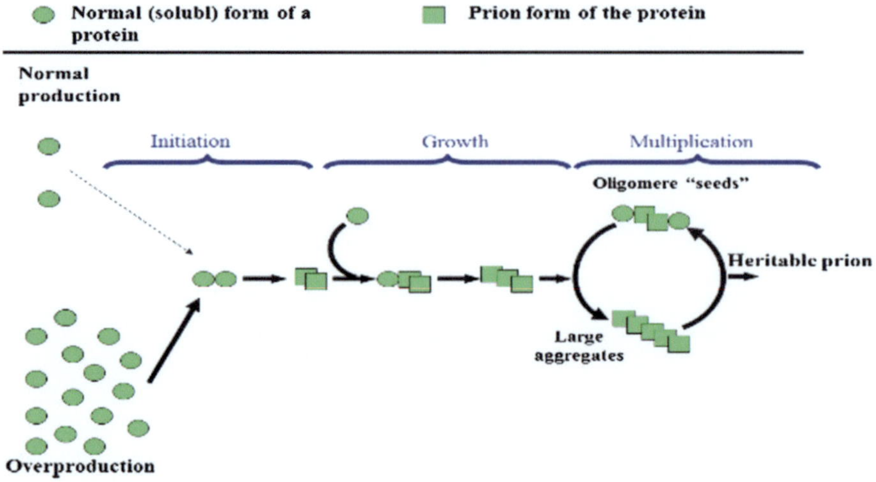

Fig. 9 Stages of prion formation (Borchsenius et al. 2003)

Moreover, there is an idea of probiotic origin of primary replicating structures as of amyloidogenic peptides with β-sheet structures (Maury 2015). So if we would present CD as the essence of the template principle we should accomplish it with the second order templates, protein conformational templates (Fig. 7). This addition to the CD must be more detailed comparable with the original Crick's description, though there is no satisfactory knowledge of the mechanism so far.

Conformational modifications of protein from prion precursor to prion presented by S. Prusiner are shown on Fig. 8 (Prusiner 1994). The whole process of prionization followed by appearance of protein aggregates and of their "multiplication" by fragmentation and further olygomerization—sort of biological crystallization is illustrated by Fig. 9 (Borchsenius et al. 2003).

Ambiguity and Repair (Linear Templates)

Coming back to the first order template processes (replication, transcription, translation), we should say that all of them have a common characteristics—ambiguity, or as it used to be said, they are inclined to make "mistakes". The ambiguity level of every template process is evolutionary optimized (Fig. 10). It is evident from the fact that you can get mutants with both elevated and lowered ambiguity level of either replication (mutator and antimutator mutants) or transcription and translation. Sturtevant (1937) and Shapiro (1938) offered to consider mutability as an adaptive trait of a species. J. Drake spotted that different haploid microorganisms from bacteriophages to fungi have a general frequency of spontaneous mutability about 1 % per genome replication. Taking in consideration that the genomes of these objects varies in a limit about 1000, mutation frequency per

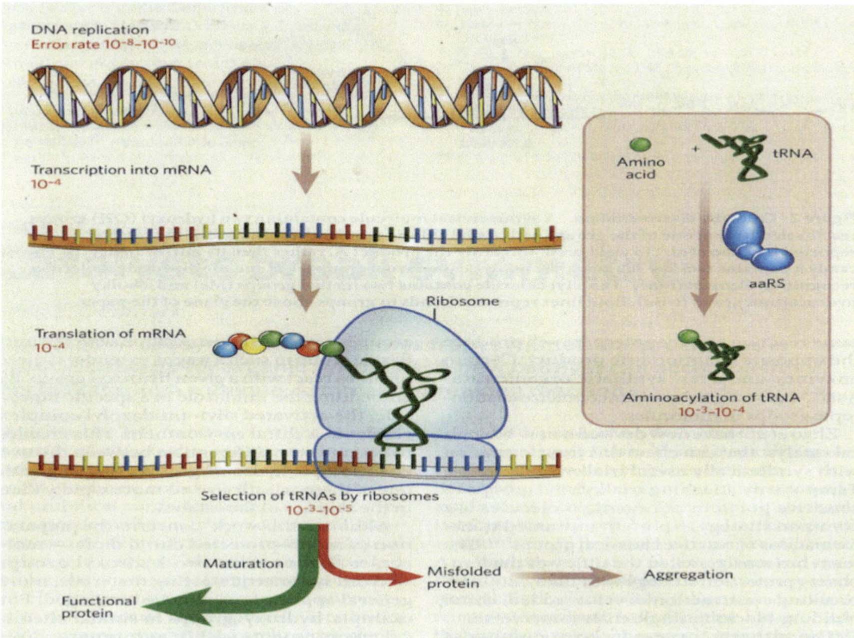

Fig. 10 Accuracy in gene expression (After: Roy and Ibba 2006)

nucleotide pair should vary inversely proportional to this value (Drake 1991). It means that there should operate a mechanism which controls such an optimization. This one is the mechanism of repair or correction.

The two terms "mutation" and "repair" were put together by M.E. Lobashev (Fig. 11) in his doctoral theses about physiological hypothesis of mutation process at 1946 (Lobashev 1946, 1947). Lobashev described the very process of mutation origin which was preconditioned by the primary or pre-mutational lesions in genetic material. Whether they would be either converted into mutations (we shall add: or into the other genetic changes, e.g. recombination) or they would be erased without any trace, is defined by numerous repair systems, described now. This connection between mutations and repair became evident to the end of the 1960s (Von Borstel 1969).

So, the whole story started with "convariant reduplication", from the study of ambiguity of replication and of repair (correction), which balanced the ambiguity. These two characteristics appeared to be universal for all three first order template processes. Their common features are presented at Fig. 12. All of the processes proceed through three stages: initiation, elongation (copying of the template per se), termination. The universal characteristics: ambiguity and correction ability are expressed at all these stages. We mentioned replication already and we have no possibility to describe here different DNA repair mechanisms and would refer to Figs. 13, 14 as illustrations.

Fig. 11 N.V. Timofeeff-Ressovsky and M.E. Lobashev (*Photo* by V.D. Simonenko, 1960-s)

***Stages*:**	***Characteristics*:**
Initiation	**Ambiguity**
Elongation (copying)	**Correction (repair)**
Termination	

Fig. 12 Common features of all first order template processes

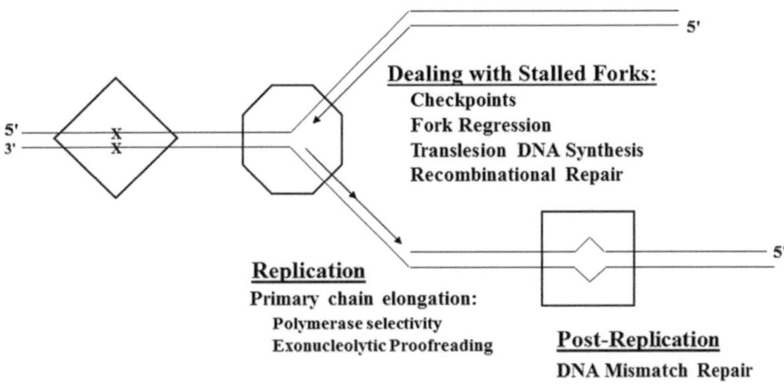

Fig. 13 Determinants of DNA Replication Fidelity (Cortesy of Prof. T. Kunkel)

Ambiguity and correction are immanent features of transcription as well (Fig. 15). RNA polymerase (all three polymerases in eukaryotes) may include a "wrong" nucleotide. It is followed by pausing of polymerase processing, removal the "wrong" nucleotide by reverse movement and further inclusion of the correct nucleotide (Sydow and Cramer 2009). Ambiguity and correction (do not mix it with editing) studied only recently, possibly because of difficulty to register their phenotypic expression besides the so-called transcriptional mutagenesis (Bregeon

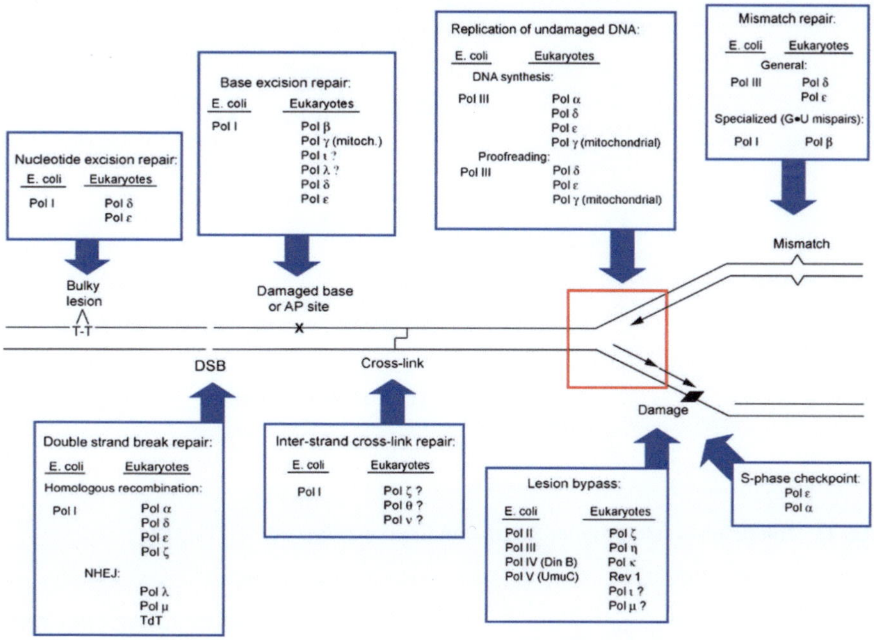

Fig. 14 Putative functions of DNA polymerases (Bebenek and Kunkel 2004)

et al. 2009). That is why the transcriptional characteristics mentioned above had been described poorly, comparing to the other first order template processes.

Study of translation ambiguity had been started in 1960s–1970s by L. Gorini basing on such a reliable criterion as phenotypic suppression. Already then Gorini pointed that the ribosome is inclined to "goof", when reading mRNA. Various external effectors only elevate or suppress this tendency. He identified the ribosomal ambiguity center (Gorini 1974). Later on there was discovered the factor of ribosomal correction—elongation factor EF-Tu (in bacteria) and its ortolog EF-1A (in eukaryotes). The schematic presentation of its activity is on Fig. 16. This factor controls interaction between anticodone of aminoacyl-tRNA and codone of mRNA. If there is a non-cognate codone-anticodone pairing, the wrong aminoacyl-tRNA would be removed by EF-TU (Thompson 1981; Thompson and Stone 1977; Zaher and Green 2009).

Here we need a comment on the term repair, or correction. There is no room for the anthropomorphic question "What is the truth?" Correction in any template process proceeds via physic-chemical mechanisms providing synthesis of a stable (in given conditions) reaction product. By this reason correction in replication may either remove a primary lesion from DNA or may convert it into mutation or process it via recombination. The same way in translation, correction may e.g. preserve the triplet reading frame or may provide a regular frame shift at some specific situation. There are several examples of regular reading frame shifting in

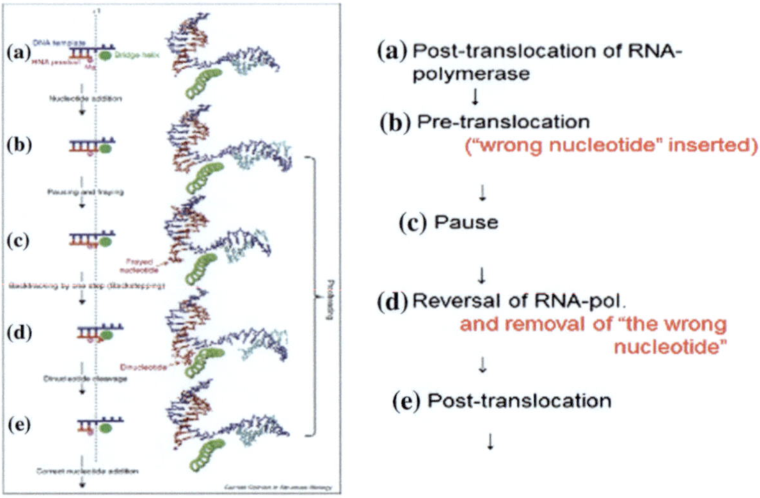

Fig. 15 Transcription fidelity (after Sydow and Cramer 2009)

Fig. 16 EF-Tu as a factor of ribosomal correction: (I) incorporation of aminoacid residue into growing polypeptide chain (1, 2, 4); (II) ribosomal correction (1, 2, 3)

(I) aa-tRNA.EF-Tu.GTP + Rs.mRNA
1 ↓↑
aa-tRNA.EF-Tu.GTP.Rs.mRNA
2 ↓↓
↓∟→ EF-Tu.GDP
↓
3 ↓
(II) aa-tRNA + Rs.mRNA ← aa-tRNA.Rs.mRNA
4 ↓
incorporation

retrovirus genome translation, like in Ty1 retrotransposon of *Saccharomyces* yeast (Farabaugh 1996).

Discussing ambiguity and correction in the template processes we used to think about their elongation stage. In reality ambiguity and correction may be demonstrated at the stages of initiation and termination as well. Specific nucleotide sequences serve as signals of beginning and finishing for replication, transcription and translation. At replication and transcription in eukaryotes the chromatin structure is also essential besides nucleotide sequences. It is essential that these signals vary near some consensus sequences. As an example you can see variations in the element of bacterial and phage promoters at position -10, shown in Fig. 17. The variability of initiation signals of template processes definitely shows their ambiguity. Besides that ambiguity of initiation process is adequately illustrated by different efficiency of either promoters of transcription or translation initiation regions, for example, variation of bacterial Shine-Dalgarno sequence necessary for proper translation initiation (Fig. 18) (Steitz 1979).

5'- $T_{82}A_{89}T_{52}A_{59}A_{49}T_{89}$ – 3'

Fig. 17 "-10 element" consensus of prokaryotic promoter. *Small figures*—frequency (%) of the nucleotide in this position from several thousand sequences (courtesy of Prof. Mironova L.N.)

Fig. 18 Examples of Shine-Dalgarno sequence in *E.coli* (from: Steitz 1979)

trp E 5'-...UA**G AG**A AUAACA **AUG**... - 3'
trp A **AGG GG**A AAU CUG **AUG**
lac Z C**AG GA**A ACA GCU **AUG**
gal E AU**G GAG** CGAAUU **AUG**

Shine-Dalgarno consensus: **GGAGG**

Start codon **AUG** is also shown.

We can think about degenerate trait of initiation signals in transcription and in translation which is a reminiscence of degenerate genetic code. Different efficiency of initiation signals for the first order template processes is controlled by many factors, such as: nucleotide context, secondary structure of polynucleotides, their location within interphase nucleus (in eukaryotes), nucleic acids and chromatin modifications and surely—proteins, including repressors, inducers and the other multiple transcription factors. Function of all these factors is to correct or to balance ambiguity of initiation and to provide the function of these signals with defined efficiency. By analogy we can consider the ambiguous reading of termination signals in the template processes. In other words it is canalization of ambiguity or regulation, which is presented here in somewhat unusual aspect.

Ambiguity and Repair (Conformational Templates)

Until now we have been considering predominantly the linear templates or the first order templates. What about the second order templates (Fig. 7)? Can we attribute the characteristics of the first order template processes to them? Honestly speaking there are more questions than answers. Nevertheless there are possibly also three stages: initiation of prions (and of amyloids in general), their growth which corresponds to elongation. The questionable is the stage of termination. Does it exist or not? Or is there a simple fragmentation of the long aggregate? The expression of ambiguity at least at the stage of initiation serves the existence of prion strains different by their characters, but originated from the same prionogenic protein, as it is shown for the yeast [PSI+] prion, which is a derivative of translation termination factor—eRF-3 (Bradley et al. 2002). Is there ambiguity at the elongation stage is under the question now. The same problem exists with the process of correction now. Anyway it is known that the proteins may exist as several conformers but it is not known which of them among potentially prionogenic ones (or any) are capable of prionization. The most plausible candidates for fulfilling repair or correction functions in the second order template processes are some chaperones.

There is an additional question about interaction between the first and the second order template processes. No doubt that there exists this type of interactions. For example, among 9 prionogenic proteins known for this moment in yeast *S. cerevisiae*, 5 are transcription factors and 1 is factor of translation. At least for some of them it was shown that their prionization is followed with the wide pleotropic effect, e.g. prionization of transcription factor Sfp (prion [ISP]) is followed by modification of more than 300 genes transcription (Drozdova et al. 2014).

Conclusion

Coming to the end of this report I would like to get back to the beginning. Fast progress of genetics, as it was shown at Fig. 1, ruins the accustomed system of ideas. Our intentions to insert new facts into the system of classical genetics are useless. It is necessary to remember biological principle of uncertainty which reflects the very multilevel organization of biological systems. All the mechanisms of biological processes are of molecular nature, though biological phenomenology is expressed at cellular, organism and population level. It means that the same primary mechanism may be responsible for different phenomena and vice versa. E. Sverdlov was the first, who offered the idea of biological principle of uncertainty, though we can discuss whether it was adequately used in the context (Sverdlov 2009). So we need a new paradigm. I suppose that the best candidate for this paradigm is the template principle, in development of which considerable contribution made N.V. Timofeeff-Ressovsky to whose 115 anniversary we dedicate this symposium.

Acknowledgments I am thanking my colleagues at Dept of Genetics and Biotechnology of St-Petersburg State University and Spb Branch Vavilov Institute of General Genetics RAS: Prof. Mironova L.N., Prof. Sambuk E.V., Dr Stepchenckova E.I., Dr Sopova Ju.V., Dr Nizhnikov A.A., Antonets K.S. and the others, who took part in preparation and in discussion of the paper. The work was supported by: Grant from the President of the Russian Federation for governmental support of leading scientific schools (Grant No. NSH- 5115.2014.4); Federal Target Program "Scientific and scientific-pedagogic cadres of innovation Russia"(2009–2013), Contract 8045; Sankt-Petersburg University Grant No 1.50.2218.2013

References

Bebenek K, Kunkel TA (2004) Functions of DNA polymerases. In: Yang W (ed) DNA repair and replication, advances in protein chemistry, vol 69. Elsevier Press, Amsterdam, pp 135–167

Borchsenius A, Zharkevich S, Ryabinkova N, Alenin V, Sasnauskas K, Newnam G, Chernoff Y, Inge-Vechtomov S (2003) Induction of the yeast Saccharomyces cerevisiae prion [PSI+] by chimeric protein Sup35 N-Ade2.Yeast. In: XXI International conference on yeast genetics and molecular biology. Göteborg, vol 20, Suppl 1, Sweden, p 203, July 7–12, 2003

Bradley ME, Edskes HK, Hong JY, Wickner RB, Liebman SW (2002) Interactions among prions and prion "strains" in yeast. Proc Natl Acad Sci USA 99:6392–16399
Bregeon D, Peignon AS, Sarasin A (2009) Transcriptional mutagenesis by 8-oxoguanine in mammalian cells. PLoS Genet 5(7):e1000577
Crick HFC (1958) On protein synthesis. Symp Soc Exptl Biol 12:138–163
Crick F (1970) Central dogma of molecular biology. Nature 227:561–563
Drake JW (1991) A constant rate of spontaneous mutation in DNA-based microbes. Proc Natl Acad Sci 88:7160–7164
Drozdova P, Rogoza T, Radchenko E, Lipaeva P, Mironova L (2014) Transcriptional response to the [ISP+] prion of Saccharomyces cerevisiae differs from that induced by deletion of its structural gene, SFP1. FEMS Yeast Res 14:1160–1170
Farabaugh PJ (1996) Programmed translational frameshifting. Microbiol Rev 60:103–134
Goldman MA (2011) Molecular biology: seed of revolution. Nature 480:317
Gorini L (1974) Streptomycin and misreading of the genetic code. In: Nomura M, Missieres A, Lengyel P (eds) Ribosomes. Academic Press, NY, pp 791–803
Inge-Vechtomov SG (2003) Template principle in biology (past, contemporary, future?). Ecol Genet 1:6–15 (Russin)
Inge-Vechtomov SG (2013) Template principle: paradigm of modern genetics. Russ J Genet 49: 4–9
Inge-Vechtomov SG, Galkin AP, Sopova JV, Rubel AA (2012) Prions, "protein inheritance" and epigenetics. In: Zakiyan SM, Vlasov VV, Dement'eva EV (eds) Epigenetics. Publ SD RAS, Novosibirsk, pp 481–496 (Russian)
Koltsov NK (1936) Hereditary molecules. In: Organization of the cell. GosIzdat, Moscow-Leningrad, pp 585–622 (Russian)
Lobashev ME (1946) On the nature of external factors action on the dynamics of the mutational process. Doctoral Thesis, Leningrad University (Russian)
Lobashev ME (1947) Physiological (paranecrotic) hypothesis of mutational process. Vestnik Leningrad Univ 8:10–29 (Russian)
Maury CPJ (2015) Origin of life. Primordial genetics: information transfer in a pre-RNA world based on self-replicating beta-sheet amyloid conformers. J Theor Biol 382:292–297
Prusiner SB (1994) Inherited prion diseases. Proc Natl Acad Sci USA 91:4611–4614
Roy H, Ibba M (2006) Phenylalanyl-tRNA synthetase contains a dispensable RNA-binding domain that contributes to the editing of noncognate aminoacyl-tRNA. Biochemistry 45:9156–9162
Schrödinger E (1944) What is life? The physical aspect of the living cell. The Macmillan Co, Cambridge
Shapiro NI (1938) Mutational process as an adaptive trait of the species. Russ Zoolog J 17:592–601 (Russian)
Steitz JA (1979) Genetic signals and nucleotide sequences in messenger RNA. In: Goldberger RF (ed) Biological regulation and development, vol 1. Steitz Publ, NY, pp 349–399
Sturtevant AH (1937) Essays on evolution. I. On the effects of selection on mutation rate. Quart Rev Biol 12:464–467
Sverdlov ED (2009) A glance at life trough the genome window (1) Nauka, Moscow, p 525 (Russian)
Sydow JF, Cramer P (2009) RNA polymerase fidelity and transcriptional proofreading. Curr Opin Struct Biol 19:732–739
Thompson RC (1981) A GTPase reaction accompanying the rejection of leucyl-tRNAs by UUU-programmed ribosomes. J Biol Chem 256:81–86
Thompson RC, Stone PJ (1977) Proofreading of the codon-anticodon interaction on ribosomes. Proc Natl Acad Sci USA 74:198–202
Timofeev-Ressovsky NV (1980) Genetics, evolution and theoretical biology. Priroda 9:62–65 (Russian)
Timofeeff-Ressovsky NW, Zimmer KG, Delbrück M (1935) Über die Natur der Genmutation und der Genstruktur. Nachr Ges Wiss Göttingen. Fachr. VI. N. F. Bd 1:189–245

Von Borstel R (1969) On the origin of spontaneous mutations. Jpn J Genet 44(Suppl):102–105
Wickner RB (1994) [URE3] as an altered URE2 protein: evidence for a prion analog in Saccharomyces cerevisiae. Science 264:566–569
Wickner RB, Masison DC, Edskes HK (1995) [PSI] and [URE3] as yeast prions. Yeast 11:1671–1685
Zaher HZ, Green R (2009) Quolity control by the ribosome following peptide bond formation. Nature 457:161–166

Mechanisms of Global and Region-Specific Control of Mutagenesis

Youri I. Pavlov, Artem G. Lada, Corinn Grabow and Elena I. Stepchenkova

DNA Damage, Repair, Replication and Mutagenesis

Evolutionary determined low mutations rates are inherent property of living organisms (Drake 1999; Lynch 2010). Complex systems ensure repair of damaged DNA and its accurate replication (Kunkel 2004; Sancar et al. 2004; Kunkel and Erie 2005; Hanawalt 2007), Fig. 1.

Y.I. Pavlov (✉) · A.G. Lada · C. Grabow · E.I. Stepchenkova
Eppley Institute for Research in Cancer and Allied Diseases,
University of Nebraska Medical Center, Omaha, USA
e-mail: ypavlov@unmc.edu

A.G. Lada
e-mail: artem.g.lada@gmail.com

C. Grabow
e-mail: cegrabow@unmc.edu

E.I. Stepchenkova
e-mail: stepchenkova@gmail.com

Y.I. Pavlov
Department of Biochemistry and Molecular Biology,
University of Nebraska Medical Center, Omaha, USA

Y.I. Pavlov
Department of Pathology and Microbiology,
University of Nebraska Medical Center, Omaha, USA

A.G. Lada
Department of Microbiology and Molecular Genetics,
University of California, Davis, USA

C. Grabow
Human Genetics Laboratory, Munroe-Meyer Institute for Genetics
and Rehabilitation, UNMC, Omaha, USA

V.L. Korogodina et al. (eds.), *Genetics, Evolution and Radiation*,
DOI 10.1007/978-3-319-48838-7_6

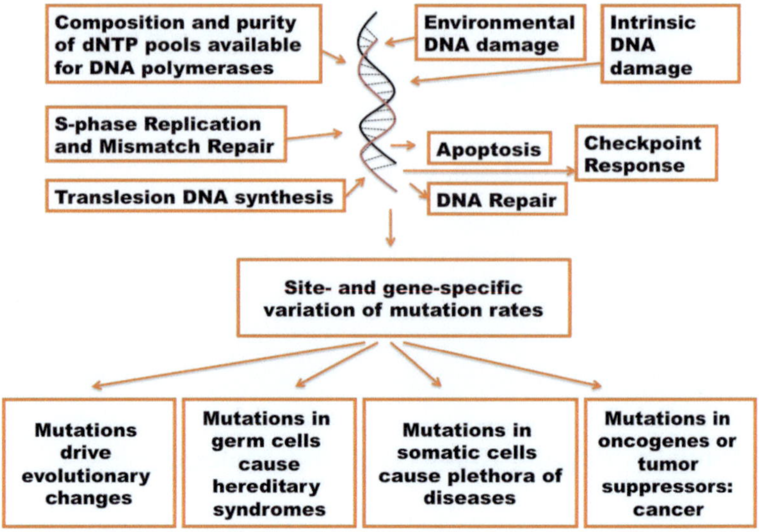

Fig. 1 Sources of DNA damage, cellular responses, origin of mutations and their consequences. DNA is copied with a certain level or errors on undamaged templates by replicative DNA pols and with a high level of errors on templates damaged by environmental or intrinsic mutagens by TLS pols. Quality of nucleotide pool significantly contributes to mutagenesis. Altered DNA elicits a wide range of cellular responses, from DNA repair to apoptosis. After all these transactions the daughter DNA molecules contain non-randomly distributed sequence alterations that have various genetic consequences

When elements of these systems are impaired by mutations or by environmental factors, genomic mutations rates sharply increase, causing genetic catastrophe leading to various diseases (Miller 1996; Salk et al. 2010; Waisertreiger et al. 2012). The global mutators introduce numerous mutations inevitably damaging some housekeeping genes; thus they can rarely account for evolutionary and developmental processes in eukaryotes (Herr et al. 2011; Lada et al. 2013; Shendure and Akey 2015). Current evidence suggests that mutation rates differ significantly along the genome (Lang and Murray 2011; Lawrence et al. 2013; Lujan et al. 2014; Supek and Lehner 2015). The most striking example of such variation comes from immunology. It has been known for a long time that in specialized cells responsible for antibody production, the variable regions of immunoglobulin genes have six orders of magnitude higher mutation rates than other genes (Wagner and Neuberger 1996; Kato et al. 2012), Fig. 2.

The mechanism of this localized mutability, called somatic hypermutation (SHM), has been found only at the dawn of this century (Muramatsu et al. 2000). It was discovered that, surprisingly, in addition to faithful repair, human cells are equipped

E.I. Stepchenkova
Saint Petersburg Branch of Vavilov Institute of General Genetics,
St Petersburg University, St Petersburg, Russia

E.I. Stepchenkova
Department of Genetics, St Petersburg University, St Petersburg, Russia

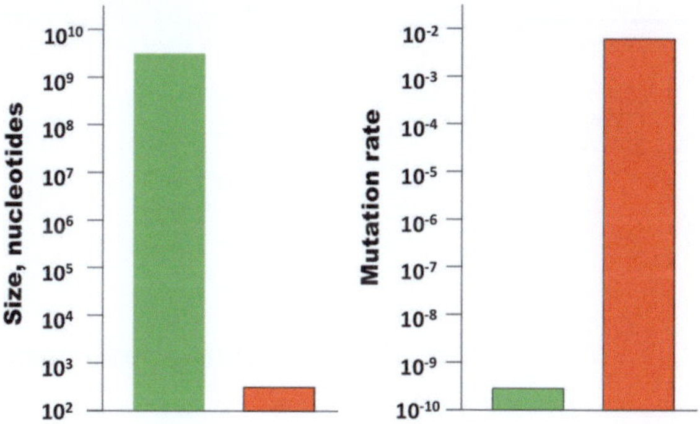

Fig. 2 Inverse correlation between sizes and mutation rates in genome and immunoglobulin somatic hypermutation target region. *Bars* relevant to human genome are in *green* and relevant to SHM target, variable region of IG gene, are in *red*. *Left* panel shows physical sizes, *right* panel—mutation rates

with powerful mutator machines—proteins that act in highly mutagenic way. Most prominent are DNA/RNA editing cytosine deaminases of AID/APOBEC family (Neuberger et al. 2003) and inaccurate translesion synthesis DNA polymerases (pols) (Rogozin et al. 2001; Zeng et al. 2001). The availability of intrinsic mutators provides an opportunity to create variability "on demand" for developmental programs and adaptive responses (Honjo et al. 2005), but poses a threat to genome integrity in case of their faulty regulation. The unleashed deaminases are suspect factors in etiology of cancer and many other diseases (Roberts and Gordenin 2014).

With or without DNA damage, hereditary changes happen only when DNA adducts or replication errors are converted into new information in both strands of duplex DNA. Therefore, all mutagenesis is mediated by DNA pols and even highly accurate enzymes take responsibility for the origin of mutations, participating in their fixation. The error spectrum of the individual pols determines the accuracy of synthesis of specific DNA patches in the genome. The mechanisms of mutagenesis, however, are not a mere sum of transactions by pols. Various additional factors affect the mutation rates and overall shape of mutation landscape of the eukaryotic genome, from nucleotide pools (Mathews 2006), unusual DNA structures (Shah et al. 2012) to chromatin structure (Kadyrova et al. 2011; Schopf et al. 2012; Liu et al. 2013). In our review we analyze factors affecting specificity and distribution of mutations along genomes and compare classical data with results of the whole genome sequencing approaches for mutation analysis, a break-through in analysis of mutagenesis in humans, Fig. 3.

Many DNA Pols Operate at the Replication Fork

Replication and repair of DNA and successful cell proliferation relies on the four multi-subunit pols, pol α, pol ε, pol δ and pol ζ, which belong to the B-family (Johansson and Macneill 2010; Pellegrini 2012; Tahirov 2012; Makarova and

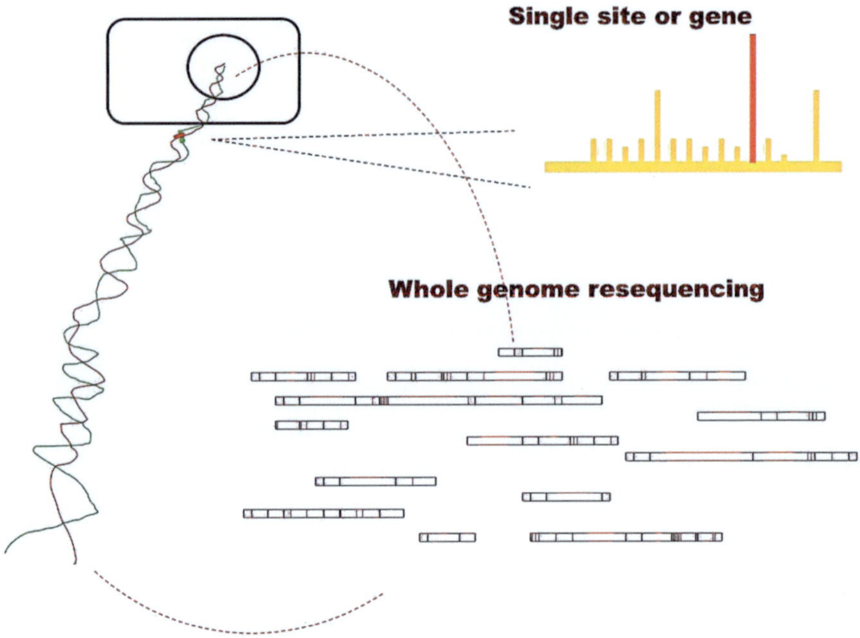

Fig. 3 "Gene reporter" *versus* "genome-wide" approaches to study mutagenesis. A cell with a nucleus is in the *upper left* corner with genomic DNA schematically protruding below. Classical approach for study mutations was to determine mutation frequencies and distributions in a reporter gene or even one site of the reporter (reversion assays). The results are frequently presented as a spectrum of mutations (*upper right*), where the height of bars equals numbers of mutations in the particular site. It our example the most prominent hot spot is marked by *red* color. In this case we gain information on mutation processes occurring only in a small portion of genome (*small rectangle* on genomic DNA and *dashed lines* leading to spectrum). Whole genome sequencing allows for the study of the full distribution (*red dashed lines*) of mutations (*vertical lines* on *long rectangles* representing individual chromosomes) in the genome

Burgers 2015). The asymmetric nature of the DNA duplex poses topological problems for the replication of the anti-parallel strands by a unidirectional fork, therefore the replication of the two strands is inherently different. Under normal circumstances, chromosomal replication in eukaryotes requires three DNA polymerases, pol α, pol ε and pol δ (Garg and Burgers 2005; Pavlov et al. 2006b; Kunkel and Burgers 2008). Yeast can use only pol α and pol δ polymerase activities to replicate their whole genome, but at the expense of a poor growth rate (Kesti et al. 1999) and genome instability (Ohya et al. 2002). Pol ζ is backing up the major pols when nucleotide pools are depleted or when pols experience other difficulties copying the template (Northam et al. 2006, 2010). In addition, many specialized polymerases assist replicative polymerases after DNA damage (Prakash et al. 2005; Zhong et al. 2006; Yang and Woodgate 2007).

The process of recently reconstituted replication of the eukaryotic genome (Yeeles et al. 2015) is started at multiple sites by origin recognition and the assembly of an active helicase (Araki 2011). Pol ε and its accessory factors are obligatory components of the process and thus it is the first DNA pol arriving at the nascent replication fork. Its pol activity is not needed at this time, because there are no primers available for the synthesis. Pol α-primase plays the main role in the actual start. Pol α synthesizes a small amount of DNA in comparison to other replicative DNA polymerases—less that 1/10 of Okazaki fragment (Waga and Stillman 1998; Smith and Whitehouse 2012; Waisertreiger et al. 2012). Thus, it is not expected to contribute much to the accuracy of genomic DNA replication. Moreover, patches synthesized by pol α are removed with RNA primer during Okazaki fragment maturation and errors by pol α are the subject of correction by proofreading activity of pol δ (Pavlov et al. 2006a), by mismatch repair (MMR) (Niimi et al. 2004; Pavlov et al. 2006a) or Okazaki fragment processing machinery (Rumbaugh et al. 1999; Liu et al. 2015). Despite all of this, mutations reducing base selectivity of pol α lead to a very strong mutator phenotypes in certain genetic backgrounds (Pavlov et al. 2001, 2004; Gutierrez and Wang 2003; Niimi et al. 2004; Tanaka et al. 2010). Mutations generated by pol α localize at the junctions of Okazaki fragments and such sites could be hotspots of mutations in genomes at sites with synchronous initiations (Waisertreiger et al. 2012). Thus, the nucleotide selectivity of pol α plays an important role in maintaining genome stability, and most mutations are left in the lagging strand, because of initiation events at every Okazaki fragment (Niimi et al. 2004; Tanaka et al. 2010; Reijns et al. 2015).

RNA-DNA primers are extended by accurate proofreading-proficient pols. Pol δ continues the synthesis of the lagging strand until it meets the start of another Okazaki fragment (Garg et al. 2004). Solid genetic evidence and direct biochemical experiments suggest that pol δ is operating on the lagging strand (Garg et al. 2004; Jin et al. 2005; Pavlov et al. 2006a; Nick McElhinny et al. 2008). This pol is well suited for this, because, contrary to pol ε, it can idle at the nick until it is ligated. The size of Okazaki fragments in eukaryotes is ten times smaller than in prokaryotes. The current estimate of the Okazaki fragment in yeast directly links the size of the fragments to the size of the nucleosomes, 165 base pairs (Smith and Whitehouse 2012; Waisertreiger et al. 2012).

We favor the model where the leading DNA strand in eukaryotes is replicated both by pol ε (near the origins) and pol δ (away from origins, most of the strand), while the lagging DNA strand is replicated mostly by pol δ (Pavlov and Shcherbakova 2010). One of the first models of the replication fork, which was proposed by Dr. Akio Sugino after the discovery of pol ε, postulated that pol ε is responsible for copying exclusively the leading strand DNA template, and pol δ is responsible for the lagging strand replication (Morrison et al. 1990). The model received ample support, because when transactions of pol ε are tracked genetically, errors attributable to this pol are found predominantly on the leading DNA strand (Morrison and Sugino 1994; Shcherbakova and Pavlov 1996; Karthikeyan et al. 2000; Kunkel and Burgers 2008; Miyabe et al. 2011). We discussed the concept of the "division of labor" (Nick McElhinny et al. 2008) and the evidence for and against this broadly accepted model in depth in (Pavlov and Shcherbakova 2010). One of the main arguments against Sugino's model is the viability of mutants with deletions of the polymerization-proficient half of the catalytic subunit of pol ε (Kesti et al. 1999). Newer whole genome sequencing data in general support the model but traces of mutations attributable to mutator pol ε fade with an increase in distance from replication origins (Larrea et al. 2010; Lujan et al. 2014). Unexpected to many, recent work utilizing classical reversion approach presented evidence that pol δ solely works on both strands (Johnson et al. 2015) reviving an intrigue in the problem of pols in leading/lagging strand replication (Stillman 2015). Obviously, much work has to be done, as currently there are no direct insights into the mechanism of coordinated work of the two pols at the moving fork.

The fourth pol at the replication fork is pol ζ (Pavlov and Shcherbakova 2010). It is critical for replication at difficult template sites or when replicative DNA Pols are compromised or deprived from appropriate levels of dNTPs (Northam et al. 2010) and during DNA damage bypass (Pavlov et al. 2006b; Gan et al. 2008; Makarova and Burgers 2015). Induced mutagenesis associated with bulky lesions completely depends on pol ζ, therefore, this pol is finalizing the bypass of damaged sites on both leading and lagging DNA strands. Because pol ζ is less accurate than other members of B-family and does not have proofreading exonuclease activity, it leaves patches with additional mutations beyond the damage site and the length of these patches can exceed the size of Okazaki fragments (Kochenova et al. 2015).

Nucleotide Pools and Mutagenesis

It is well known that concentrations and ratios of nucleotides affect mutagenesis outcomes (Mathews and Ji 1992; Kumar et al. 2011). The altered pools increase mutagenesis evenly on leading and lagging strands but with preference to late replicating regions at the scale of the whole genome (Watt et al. 2015). One serious problem is the overlap of NTP and dNTP pools. Ribonucleotides are a regular part of nascent DNA chains. If the machinery of their removal fails, the remnants of these primers can be left in DNA (Reijns et al. 2012). In addition, under certain conditions, DNA pols can directly incorporate ribonucleotides into DNA, causing genetic instability (Nick McElhinny et al. 2010; Miyabe et al. 2011). Hopefully, specialized repair system deals with these lesions (Vaisman and Woodgate 2015).

Elevation of mutagenesis is caused by tainting of pools with deoxynucleoside triphosphosphate forms of base analogs generated by environmental factors, biochemical reactions during oxidative stress and inflammation (Simandan et al. 1998; Colussi et al. 2002; Dedon and Tannenbaum 2004). The analogs deceive DNA polymerases and become drivers of replication errors, lead to elevated mutation rates and chromosome instability (Abolhassani et al. 2010), elevate risk of cancer (Sekiguchi and Tsuzuki 2002), and cause developmental problems (Behmanesh et al. 2009). One of the best studied base analog in the nucleotide pool is 8-oxo deoxyguanosine triphosphate (8-oxoG) (Maki and Sekiguchi 1992). An accumulation of oxyguanine in DNA leads to several diseases, including cancer (Kovtun et al. 2007; D'Errico et al. 2008; De Luca et al. 2008).

Cleansing of the precursor dNTP pool from potentially mutagenic nucleotide analogs is an important prerequisite for high fidelity DNA replication (Hochhauser and Weiss 1978; Ames and Gold 1991; Michaels and Miller 1992). Interesting, some tumors generate excess of base analogs, and inactivation of protective systems can specifically kill tumors by provoking the destruction of their genomes (Gad et al. 2014).

IMP and XMP are central intermediates in purine metabolism. Inadvertent activation of these compounds into triphosphates results in their incorporation into DNA and leads to genotoxicity (Budke and Kuzminov 2009; Pang et al. 2012). Under certain conditions, the amount of ITP in the pool reaches 10 % of the amount of adenine (Sakumi et al. 2010). A related compound, base analog 6-N-hydroxyl aminopurine, (HAP), can arise from hydroxylation of adenine and can be converted to adenine or hypoxanthine (Kozmin et al. 2008; Sakumi et al. 2010). Deoxy- and ribonucleoside triphosphates (XTP, ITP and HAPTP) are destroyed by the same enzyme inosine triphosphate pyrophosphatase (ITPA). HAP incorporations are easy to track because they are highly mutagenic. That makes HAP a biologically relevant tool for the study of DNA replication and as a model compound to interrogate the possible effects of the natural base analogs, hypoxanthine and xanthine. HAP is a potent universal base analog mutagen with ambiguous base pairing capacity (Pavlov et al. 1991; Kozmin et al. 1998). In an amine state, it forms two hydrogen bonds with "T", and in an imine state it makes two hydrogen bonds with "C" and induces G-C to A-T and A-T to G-C transitions (Shcherbakova and Pavlov 1993; Kulikov et al. 2001). The data with classical reporters established that mutations at G-C pairs are more frequent.

We would like to emphasize the interdependence of mutagenesis on DNA pol fidelity and on the quality of nucleotide pools. In vivo, DNA pols utilize natural pools with certain cell-cycle regulated concentrations and ratios of individual precursors that also may have various contaminants. Therefore, the cases of correlation between the fidelity of DNA pols and their inaccurate variants in vitro, when the pols are highly controlled by experimenters, and mutational signatures of the same pols in vivo (Pavlov et al. 2002b; Shcherbakova et al. 2003; Pursell et al. 2007), could be regarded as exceptions rather than the rule. Moreover, some inaccurate pol mutants induce drastic elevation of concentration of nucleotides in cells that further enhance their infidelity, creating a "vicious circle" leading to hypermutagenesis with a new specificity (Mertz et al. 2015).

Mutagenesis During Synthesis of Leading and Lagging DNA Strands in the Presence of Base Analogs

In the case of spontaneous base-pair substitutions, it is difficult to determine the DNA strand where the initial error occurred. The change of a G-C base pair to an A-T base pair could be initiated by a G-dTTP mispair on one strand or a dATP-C mispair on the other strand. In this case, strand assignment is possible in when the ratio of these reciprocal mispairs is different (Iwaki et al. 1996; Fijalkowska et al. 1998; Pursell et al. 2007; Nick McElhinny et al. 2008). Assignment of a DNA strand where a mutation occurs is also possible with strand-specific DNA damage (Veaute and Fuchs 1993), the use of base analogs (Shcherbakova and Pavlov 1996) and nucleotide pools imbalances (Roberts et al. 1994). We are advocates of the base analog approach, because analogs, for example HAP, are not as toxic as many mutagens or pool imbalances and are mutagenic even in wild-type strains, so no specific mutational alteration of replication apparatus is required (Kozmin et al. 1998).

In earlier studies of site-specific reversions HAP-induced errors occurred preferentially in one DNA strand and HAP-induced errors were proofread by pol δ and by pol ε on opposite DNA strands (Shcherbakova and Pavlov 1993, 1996). To demonstrate that this phenomenon is a general property of replication and not a specifics of the basic strain CG379, we changed the genetic background of the strain by series of crosses to a mutator strain as well as to standard strains of different genealogy and found that the strand bias remained exactly the same (Pavlov et al. 2001, 2002a). In order to study this phenomenon in more detail, we studied reversions of reporter in two orientations (Fig. 4a) after HAP treatment or in strains unable to excise 8-oxoG (Fig. 4b) at multiple locations relative to early replication origins within chromosome III (Pavlov et al. 2002a) (Fig. 4c, d).

The rate of G-C to A-T transitions, resulting from HAP misincorporation, varied by three- to ten-fold, depending on the reporter orientation and its distance from the flanking replication origins, reflecting higher mutagenesis associated with replication of the leading strand DNA template. The mechanism of the bias is unknown. It could be connected to the intrinsic property of asymmetric replication, to the identity of pols replicating leading and lagging DNA strands or the efficiency of MMR on different strands. The possible role of polymerases is supported by observation that pol ε has a more profound propensity to misincorporate wrong nucleotides and base analogs dHAPTP, dITP and dXTP in vitro (Fig. 5). On the other hand, we found that MMR is the most important contributor to strand bias for 8-oxoguanine (Pavlov et al. 2003). Further study of the leading/lagging bias with HAP was done using a forward mutation system and contrary to our expectations based on the reversion system, no orientation bias was detected (Waisertreiger et al. 2012). It appeared that initial HAP incorporations were predominant when replicating the transcribed DNA strand, irrespective of the direction of replication. To consolidate the data with the previous results, we proposed that there are two components of HAP-induced mutagenesis, contributing to the final mutation spectra. The minor component, which we were lucky to reveal in reversion system (Shcherbakova and Pavlov 1996; Pavlov et al. 2002a), depends on replication direction (Waisertreiger et al. 2012). The major component depends on transcription or the peculiarity of the transcribed DNA strand of our reporter gene. The results prompted further analysis of HAP-induced mutations by next generation sequencing, described in the next section.

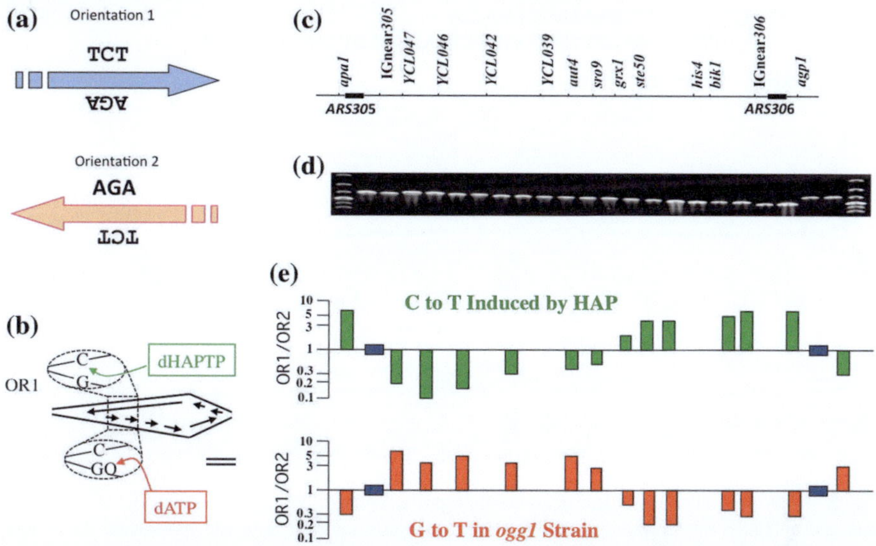

Fig. 4 Leading DNA strand is more susceptible to site-specific reversions induced by base analogs. **a** Experimental system: reversions of the *ura3-29* mutation were examined in two orientations relative to ARS. **b** Replication mistakes with base analogs HAP and 8-oxoG leading to the *ura3-29* reversion occur on opposite DNA strands. Bidirectional replication fork initiates at ARS in the center, *inset* shows initiation of reversion by base analogs on the fork moving to the left from ARS (corresponding to locations to the left from the *ARS306* in next panel). **c** Map of insertions of the *ura3-29* reporter construct in chromosome III region with two strong ARS elements (black rectangles). **d** Verification of proper insertion of the reporter cassette by PCR analysis using primers complementary to chromosome regions flanking the insertion. The primers were selected in such a way that PCR for every individual insertion will give slightly different fragment size. **e** Ratio of reversion frequencies in the two orientations at different locations by HAP (*upper*) and in *ogg1* strains unable to repair 8-oxoG

Novel Features of Genome-Wide Base-Analog Induced Mutagenesis in Diploids

Mutations in haploids and diploids have different consequences. They immediately result in a phenotypic change in haploids but will be detectable in diploids only if they are dominant. It was established in classical studies that mutations in diploids arising spontaneously, or induced by UV light or γ-irradiation, appear by a two-step mechanism: mutation in one homologue and its mitotic segregation (Gordenin and Inge-Vechtomov 1981). Most mutagens induce point mutations and lead to an acute increase of mitotic recombination or chromosome loss events. Increases in the error rates during inaccurate replication often are highly mutagenic, but do not lead to the induction of recombination. In such a case, mutations in diploids can arise by rare events of independent mutations in both homologous genes. The high levels of

(a)

```
5'-TYE665-CTTGAAAACATAGCGA
  3'-CTTAGGAACTTTTGTATCGCTXACGAATCTAATTCTGCGA

X=C, T or I
```

(b)

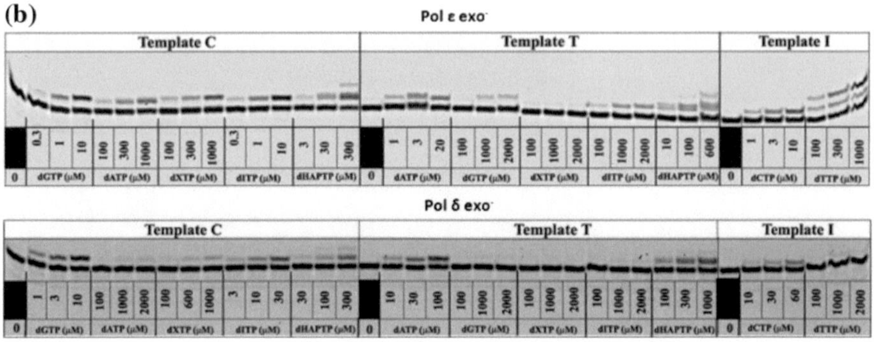

Fig. 5 DNA polymerase ε is less accurate than pol δ with wrong natural nucleotides or with base analogs. **a** Nucleotide sequences of fluorescent primer/unlabeled template pairs used for kinetic analysis. **b** Incorporation of correct and wrong nucleotides by exonuclease-defective variants of pol ε (catalytic subunit is encoded by allele *pol2-4*, provided by Dr. E. Johansson, Umea University, Sweden) and pol δ (catalytic subunit is encoded by allele *pol3-5DV*, provided by Dr. P. Burgers, Washington University, Saint-Louis, USA) under reaction conditions described elsewhere (Shcherbakova et al. 2003; Fortune et al. 2005)

replication errors cause, however, an amazingly high level of mutagenesis in diploids (Pavlov et al. 1988, 1991; Tran et al. 1999). As a result, genomes of diploids accumulate many more mutations induced by the replicative mutagen HAP, by the editing deaminase or during inaccurate replication (Lada et al. 2013; Lujan et al. 2014), Fig. 6a, b.

To explain this we hypothesized that the cells in culture have different levels of mutability and most mutable cells die in haploids but survive in diploids, thus contributing to a much higher estimated mutation rate. According to our results, one out of eight genomes of unselected clones grown from a cell treated with HAP was hypermutated, suggesting that the fraction of hypermutable cells is quite large (Fig. 6c). We proposed that the reason for hypermutable state is transient malfunction of systems protecting from mutagens in the cell that gave rise to the sequenced clone (Lada et al. 2013). Interestingly, our analysis of mutation biases near the origins did not reveal substantial preference for HAP-induced mutations in the leading strand, confirming our results with forward mutation assay discussed previously. There is a possibility that the laws of mutagenesis are unique in hypermutable cells whose progeny we analyzed.

During the genome-wide study of HAP-induced mutations we found previously unseen effect, indicating that not only the levels of mutagenesis but its specificity varies between different cells. As we discussed, mutations at C-G pairs result from misincorporation of HAP, and mutations at A-T pair results from its misrepication

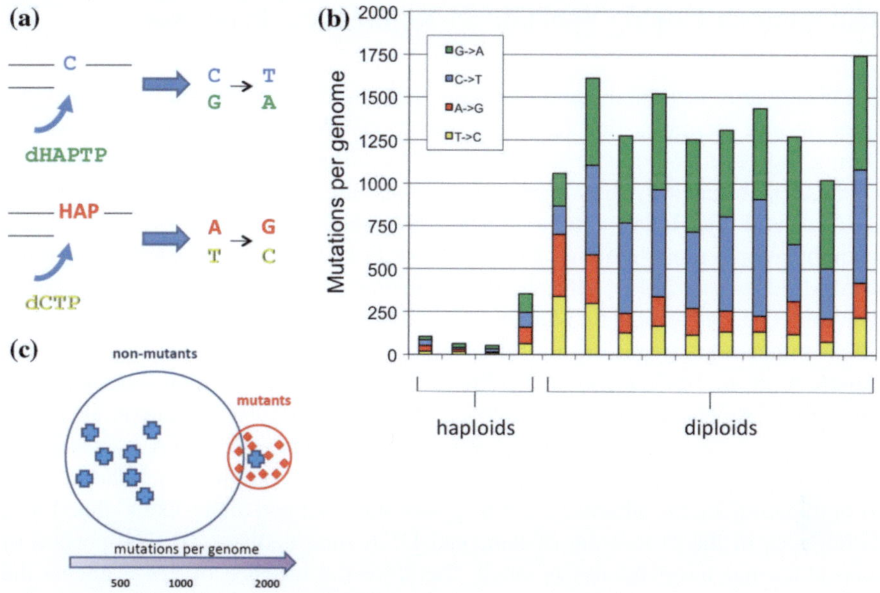

Fig. 6 Whole-genome sequencing studies provide a snapshot of the mutagenic process in individual cells. **a** Mechanism of induction of mutations by HAP when it is misincorporated opposite C (*upper row*, GC → AT transition) or when it is correctly incorporated opposite T but them misreplicated with insertion of dC (*lower row*, AT → GC transition). **b** Genomic mutation loads in individual Can^r mutants induced by HAP in haploids and diploids (Lada et al. 2013). Each bar represents mutations detected in one genome. **c** Randomly sequenced HAP treated cells (non-mutants) in diploids accumulate less mutations in the genome than mutants selected for the Can^r phenotype (Lada et al. 2013). *Blue crosses*—unselected clones after HAP treatment, *red diamonds*—Can^r mutants after HAP. The position in the sets roughly reflects the number of mutations found in the corresponding genomes

(Fig. 6a). The ratio of mutations in C-G pairs to mutations in A-T pairs in genomes of yeast diploids varied widely between different HAP-treated clones (Lada et al. 2013) (Fig. 6b). Moreover, one HAP-induced diploid mutant possessed a non-typical bias toward A-T to G-C transitions, whereas in other sequenced clones and in published data G-C to A-T transitions are more frequent. Thus, in most cells, mutations occurred when dHAPTP was misincorporated opposite “C”, while in a fraction of cells mutations were caused by misreplication of HAP previously incorporated into DNA opposite “T”. This apparently could have happened if in the unusual cell during first replication virtually all HAP was in amine form and behaved as “A”, while in next replication most of HAP was in imine form and mimicked “G”. The use of genome sequencing for the first-time enabled detection of the two unusual fractions of diploid cells: hypermutable cells and cells with altered specificity of mutagenesis.

Mutagenesis During Specialized Replication Processes

The observation that the mutation rates are higher in meiosis than in mitosis led to a proposal that recombination can cause mutations (Magni and von Borstel 1962). During the repair of double strand breaks by homologous recombination, DNA synthesis is initiated by the invasion of nucleofilament of the 3' end of the broken DNA coated by Rad51 into a donor DNA sequence. DNA synthesis during recombinational repair required either Pol δ or Pol ε (Wang et al. 2004) or by pol δ solo (Miyabe et al. 2015). It has been demonstrated that DNA synthesis associated with this repair in yeast is two orders of magnitude less accurate than during normal DNA replication, suggesting the involvement of an inaccurate DNA polymerase (Strathern et al. 1995; Hicks et al. 2010). The mutagenesis during recombination was largely dependent on Pol ζ (Holbeck and Strathern 1997; Rattray et al. 2002). This suggests that Pol ζ can be recruited to perform some DNA synthesis on substrates that are generated during homologous recombination. The rate of recombination in the absence of Pol ζ was not changed. It is likely that Pol ζ participates in the processing of damaged DNA intermediates. The proportion of such abnormal intermediates is small, therefore the absence of Pol ζ affects the frequency of mutations, but not the overall rate of recombination.

Synthesis of DNA in specialized break-induced replication driven by double strand breaks is very inaccurate over long distances (hundreds of kilobases) (Deem et al. 2011). Replication established under these conditions stunningly deviates from classics paradigm: it is conservative and cells suffer from moderate but numerous limitations: elevated dNTP pools, somewhat reduced proofreading and difficulties with mismatch repair because both strand are newly synthesized (Malkova and Ira 2013; Saini et al. 2013a). It was also proposed that this unusual recombination creates clustered mutations (Sakofsky et al. 2014).

Unusual DNA Structures Predispose to Mutation and Are Contagious

Genome stability is compromised not only by DNA damage. Some DNA sequence contexts can impede DNA replication or repair (Gordenin and Resnick 1998). A classic example of genomic instability caused by problems in replicating an unusual DNA template is repeat expansions. These so-called "dynamic mutations" are the cause of more than 40 human disorders with a wide range of manifestations: mental retardation, muscular atrophy, cranial dysplasia and others (Pearson et al. 2005). The current models of triplet instability predict that the maximal size of repeat expansion depends on the size of the Okazaki fragments on the lagging strand of DNA (Pearson et al. 2005; Shishkin et al. 2009; Shah et al. 2012). The adverse effects of the repeated DNA on replication and repair are linked to the ability of these sequences to form aberrant DNA structures, such as intra-strand hairpins, and triple- and quadruple-stranded DNA (Kovtun and McMurray 2001; Lahue and Slater 2003; Cleary and Pearson 2005; Mirkin 2006), which are difficult to replicate by Pols δ and ε (Abdulovic et al. 2011; Korona et al. 2011). Recent findings suggest that the effects of unusual structures can be spread to adjacent nearby normal regions of DNA (Shah et al. 2012; Saini et al. 2013b; Tang et al. 2013). These passenger mutations are linked to the formation of gaps in DNA that are repaired with the use of inaccurate pol ζ.

APOBEC Editing Deaminases Induce Biased Mutation Distributions and *Kataegis*

The enzymes of the AID/APOBEC superfamily of RNA/DNA editing deaminases are among the most astonishing discoveries of this century. They play an important role in developmental and adaptive reactions in eukaryotes by modifying genetic material directly or altering gene expression (Conticello 2012; Schmitz and Petersen-Mahrt 2012; Franchini and Petersen-Mahrt 2014). Cytosine deaminase APOBEC1 edits mRNA. Activation-induced cytosine deaminase, AID, is involved in SHM and class switch recombination in the affinity maturation of antibodies in jawed vertebrates (Petersen-Mahrt and Neuberger 2003; Neuberger and Rada 2007). Deaminase PmCDA1 from sea lamprey is indispensable for unique rearrangements of variable lymphocyte receptors in immune response characteristic for jawless fish (Rogozin et al. 2007; Hirano 2015). The enzymes of the APOBEC3 subfamily play a role in innate immunity by protecting against retroviruses (Refsland and Harris 2013). Deaminases are also implicated in gene regulation by participating in the active demethylation (Franchini and Petersen-Mahrt 2014). Members of the AID/APOBEC superfamily have been shown to be mutagenic in yeast and bacteria (Petersen-Mahrt et al. 2002; Poltoratsky et al. 2004) , reviewed in (Lada et al. 2011a). Deaminases catalyze the conversion of cytosine to the uracil in single-stranded DNA (ssDNA) or RNA. Deaminases in vitro work processively on ssDNA and produce clustered mutations (Pham et al. 2003; Lada et al. 2011b). Their in vivo targets could be as small as one nucleotide in one particular mRNA (APOBEC1), or confined to specific loci (variable regions of immunoglobulin genes, AID) or foreign genomes (e.g. retroviruses, APOBEC3s).

Despite that all deaminases perform the same basic reaction, cytosine deamination, their molecular specificity in inducing C-G to T-A transitions or transversions of C-G pairs in natural and heterologous hosts is very different, as seen at first by reporter systems (Lada et al. 2011a) and then elucidated with more precision by whole genome sequencing (Lada et al. 2012, 2013; Taylor et al. 2013, 2014; Chan et al. 2015) (Fig. 7).

The choice of transition versus transversion mutations is regulated by the base-excision repair enzyme uracil-DNA-glycosylase, which is responsible for the excision of uracil from DNA and creation of abasic site, and, as was shown in yeast, depend on translesion DNA polymerases Rev1 and the mutual subunit of pols ζ and

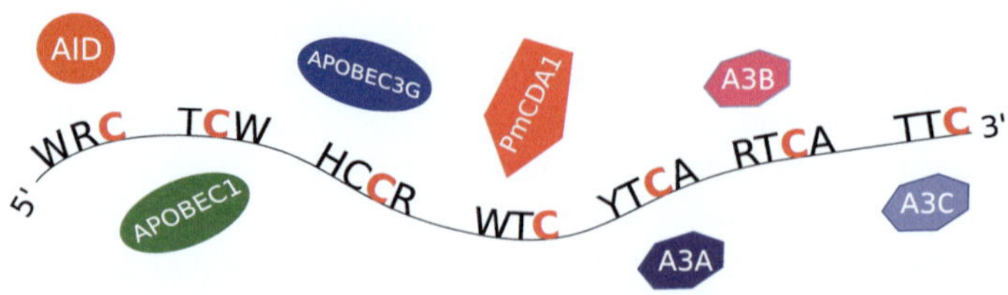

Fig. 7 DNA sequence-specifity of cytosine deaminases of APOBEC family members. Deaminated cytosine is shown in *bold red font*. DNA sequence motifs where changes of these cytosines preferably happen are shown nearby to the corresponding protein. Polymorphic nucleotides designations: W = A or T; H = C, T, or A; Y = T or A; R = G or A. Sequence context preferences summarized from (Rogozin et al. 2007; Lada et al. 2011a, 2011b, 2013, 2015; Taylor et al. 2013; 2014; Chan et al. 2015)

δ, Pol32 (Poltoratsky et al. 2010; Taylor et al. 2013). Replication past uracil generates transitions; replication past the abasic sites often results in transversions during translesion synthesis (Sale 2013; Hirota et al. 2015). The absence of uracil DNA glycosylase increases frequency of transitions induced by deaminases but, at the same time, abrogates the induction of mitotic recombination (Di Noia and Neuberger 2004; Poltoratsky et al. 2004) .

PmCDA1 from sea lamprey is strongly mutagenic in yeast and bacteria and induces changes in G-C pairs predominantly in 5'TC motifs, similar to APOBEC3A and 3B (Rogozin et al. 2007; Lada et al. 2011a, 2013), Fig. 7. PmCDA1 exerts the strongest mutagenic effect in yeast among all studied AID/APOBEC family members (Lada et al. 2011a), induces high frequency of mutation in diploids and *kataegis* (Lada et al. 2012, 2013), similar to other editing deaminases (Taylor et al. 2013, 2014; Chan et al. 2015).

When the mutagenic effect of editing deaminases was first found, it was proposed that AID/APOBECs can aberrantly edit the genome, thereby contributing to the carcinogenesis (Neuberger et al. 2003). In the last several years, the elevated expression of APOBECs and their mutational signatures have been found in multiple tumors advocating for their connection with the tumor development (Nik-Zainal et al. 2012; Alexandrov et al. 2013; Burns et al. 2013; Roberts et al. 2013; Roberts and Gordenin 2014; Walker et al. 2015). The most notable peculiarity of APOBEC signatures, *kataegistic* clustered strand-specific mutations in TC motifs (Fig. 7) in breast cancers, could be explained by processive action of deaminases in regions with exposed ssDNA (Roberts et al. 2012). The direct link between deaminases and cancer is lacking and the overall understanding of AID/APOBEC roles in carcinogenesis is at an early stage. Most critical, it is under debate whether the mutations with specific APOBEC signatures are a cause, a consequence of the tumorigenesis or even a form of protection from it (Burns et al. 2013; Cescon et al. 2015; Wu et al. 2015).

Preference of editing deaminases for ssDNA predicts that their action will depend on transcription, repair, recombination and replication, the processes with the ss DNA is an intermediate product. The relation is not straightforward, because ssDNA binding protein RPA protects this form of DNA from deaminases

(Pham et al. 2008; Lada et al. 2011b). As a result of this (and probably other) complexities, different studies found that effect of deaminases is mediated by one type of DNA metabolic processes. It is well documented that mutagenic effects of deaminases in native environment in many cases correlates with active transcription (Storb 2014). The dependence on the transcriptional profiles in model systems is quite strong and peculiar (Taylor et al. 2014; Lada et al. 2015): most mutations originate from deaminations in the non-transcribed strand and are concentrated at 5'UTRs and this bias drastically changes in mutants with a deletion of transcription regulator *sub1* gene (Lada et al. 2013, 2015). DNA cleavage during uracil repair and the accompanying recombination undoubtedly stimulates mutations induced by deaminases in heterologous yeast host (Poltoratsky et al. 2010; Taylor et al. 2013). These recombination events are thought to be responsible for megabase *kataegistic* events in breast cancer (Nik-Zainal et al. 2012; Taylor et al. 2013). Mutagenesis by editing deaminases in humans leads to unusual dependence of mutations on replication timing: most characteristic mutations in human cancer are found in early repicating regions while other typically mutations are more frequent in late replicating regions (Kazanov et al. 2015).

Determination of the landscape, gene specificity and mechanisms of genomic changes induced by intrinsic mutagenic factors is on the way and will advance our understanding of mutation processes leading to evolutionary change, development and diseases.

Conclusions

Current developments suggest that mutagenesis is a highly non-random process regulated by a plethora of cellular factors.

Acknowledgments We are thankful Dr. V.L. Korogodina and Dr. S.G. Inge-Vechtomov for the invitation to present the lecture that was a starting point of work on the current review at the IV International meeting dedicated to the 115th anniversary of the birth of N.W.Timofeeff-Ressovsky and his international scientific school in Saint-Petersburg. We are grateful to Dr. Erik Johansson and Dr. Peter Burgers for sharing preparations of yeast DNA pols. The work was supported by UNMC Eppley Institute Pilot grant in 2014–2015, grant RFBR # 15-04-08625 and Research Grant of SPbU #1.38.426.2015

References

Abdulovic AL, Hile SE, Kunkel TA, Eckert KA (2011) The in vitro fidelity of yeast DNA polymerase delta and polymerase epsilon holoenzymes during dinucleotide microsatellite DNA synthesis. DNA Repair 10(5):497–505

Abolhassani N, Iyama T, Tsuchimoto D, Sakumi K et al (2010) NUDT16 and ITPA play a dual protective role in maintaining chromosome stability and cell growth by eliminating dIDP/IDP and dITP/ITP from nucleotide pools in mammals. Nucleic Acids Res 38(9):2891–2903

Alexandrov LB, Nik-Zainal S, Wedge DC, Aparicio SA et al (2013) Signatures of mutational processes in human cancer. Nature 500(7463):415–421

Ames BN, Gold LS (1991) Endogenous mutagens and the causes of aging and cancer. Mutat Res 250(1–2):3–16

Araki H (2011) Initiation of chromosomal DNA replication in eukaryotic cells; contribution of yeast genetics to the elucidation. Genes Genet Syst 86(3):141–149

Behmanesh M, Sakumi K, Abolhassani N, Toyokuni S et al (2009) ITPase-deficient mice show growth retardation and die before weaning. Cell Death Differ 16(10):1315–1322

Budke B, Kuzminov A (2009) Production of clastogenic DNA precursors by the nucleotide metabolism in *Escherichia coli*. Mol Microbiol 75(1):230–245

Burns MB, Lackey L, Carpenter MA, Rathore A et al (2013) APOBEC3B is an enzymatic source of mutation in breast cancer. Nature 494(7437):366–370

Cescon DW, Haibe-Kains B, Mak TW (2015) APOBEC3B expression in breast cancer reflects cellular proliferation, while a deletion polymorphism is associated with immune activation. Proc Natl Acad Sci USA 112(9):2841–2846

Chan K, Roberts SA, Klimczak LJ, Sterling JF et al (2015) An APOBEC3A hypermutation signature is distinguishable from the signature of background mutagenesis by APOBEC3B in human cancers. Nat Genet 47(9):1067–1072

Cleary JD, Pearson CE (2005) Replication fork dynamics and dynamic mutations: the fork-shift model of repeat instability. Trends Genet 21(5):272–280

Colussi C, Parlanti E, Degan P, Aquilina G et al (2002) The mammalian mismatch repair pathway removes DNA 8-oxodGMP incorporated from the oxidized dNTP pool. Curr Biol 12(11):912–918

Conticello SG (2012) Creative deaminases, self-inflicted damage, and genome evolution. Ann NY Acad Sci 1267:79–85

D'Errico M, Parlanti E, Dogliotti E (2008) Mechanism of oxidative DNA damage repair and relevance to human pathology. Mutat Res 659(1–2):4–14

De Luca G, Russo MT, Degan P, Tiveron C et al (2008) A role for oxidized DNA precursors in Huntington's disease-like striatal neurodegeneration. PLoS Genet 4(11):e1000266

Dedon PC, Tannenbaum SR (2004) Reactive nitrogen species in the chemical biology of inflammation. Arch Biochem Biophys 423(1):12–22

Deem A, Keszthelyi A, Blackgrove T, Vayl A et al (2011) Break-induced replication is highly inaccurate. PLoS Biol 9(2):e1000594

Di Noia JM, Neuberger MS (2004) Immunoglobulin gene conversion in chicken DT40 cells largely proceeds through an abasic site intermediate generated by excision of the uracil produced by AID-mediated deoxycytidine deamination. Eur J Immunol 34(2):504–508

Drake JW (1999) The distribution of rates of spontaneous mutation over viruses, prokaryotes, and eukaryotes. Ann NY Acad Sci 870:100–107

Fijalkowska IJ, Jonczyk P, Tkaczyk MM, Bialoskorska M et al (1998) Unequal fidelity of leading strand and lagging strand DNA replication on the *Escherichia coli* chromosome. Proc Natl Acad Sci USA 95(17):10020–10025

Fortune JM, Pavlov YI, Welch CM, Johansson E et al (2005) *Saccharomyces cerevisiae* DNA polymerase delta: high fidelity for base substitutions but lower fidelity for single- and multi-base deletions. J Biol Chem 280(33):29980–29987

Franchini DM, Petersen-Mahrt SK (2014) AID and APOBEC deaminases: balancing DNA damage in epigenetics and immunity. Epigenomics 6(4):427–443

Gad H, Koolmeister T, Jemth AS, Eshtad S et al (2014) MTH1 inhibition eradicates cancer by preventing sanitation of the dNTP pool. Nature 508(7495):215–221

Gan GN, Wittschieben JP, Wittschieben BO, Wood RD (2008) DNA polymerase zeta (pol zeta) in higher eukaryotes. Cell Res 18(1):174–183

Garg P, Burgers PM (2005) DNA polymerases that propagate the eukaryotic DNA replication fork. Crit Rev Biochem Mol Biol 40(2):115–128

Garg P, Stith CM, Sabouri N, Johansson E et al (2004) Idling by DNA polymerase d maintains a ligatable nick during lagging-strand DNA replication. Genes Dev 18(22):2764–2773

Gordenin DA, Inge-Vechtomov SG (1981) Mechanism of mutant induction in the ade2 gene of diploid *Saccharomyces cerevisiae* yeasts by ultraviolet rays. Genetika 17(5):822–831
Gordenin DA, Resnick MA (1998) Yeast ARMs (DNA at-risk motifs) can reveal sources of genome instability. Mutat Res 400(1–2):45–58
Gutierrez PJ, Wang TS (2003) Genomic instability induced by mutations in *Saccharomyces cerevisiae POL1*. Genetics 165(1):65–81
Hanawalt PC (2007) Paradigms for the three Rs: DNA replication, recombination, and repair. Mol Cell 28(5):702–707
Herr AJ, Ogawa M, Lawrence NA, Williams LN et al (2011) Mutator suppression and escape from replication error-induced extinction in yeast. PLoS Genet 7(10):e1002282
Hicks WM, Kim M, Haber JE (2010) Increased mutagenesis and unique mutation signature associated with mitotic gene conversion. Science 329(5987):82–85
Hirano M (2015) Evolution of vertebrate adaptive immunity: immune cells and tissues, and AID/APOBEC cytidine deaminases. BioEssays 37(8):877–887
Hirota K, Yoshikiyo K, Guilbaud G, Tsurimoto T et al (2015) The POLD3 subunit of DNA polymerase delta can promote translesion synthesis independently of DNA polymerase zeta. Nucleic Acids Res 43(3):1671–1683
Hochhauser SJ, Weiss B (1978) *Escherichia coli* mutants deficient in deoxyuridine triphosphatase. J Bacteriol 134(1):157–166
Holbeck SL, Strathern JN (1997) A role for *REV3* in mutagenesis during double-strand break repair in *Saccharomyces cerevisiae*. Genetics 147(3):1017–1024
Honjo T, Nagaoka H, Shinkura R, Muramatsu M (2005) AID to overcome the limitations of genomic information. Nat Immunol 6(7):655–661
Iwaki T, Kawamura A, Ishino Y, Kohno K et al (1996) Preferential replication-dependent mutagenesis in the lagging DNA strand in *Escherichia coli*. Mol Gen Genet 251(6):657–664
Jin YH, Garg P, Stith CM, Al-Refai H et al (2005) The multiple biological roles of the 3'→5' exonuclease of *Saccharomyces cerevisiae* DNA polymerase δ require switching between the polymerase and exonuclease domains. Mol Cell Biol 25(1):461–471
Johansson E, Macneill SA (2010) The eukaryotic replicative DNA polymerases take shape. Trends Biochem Sci 35(6):339–347
Johnson RE, Klassen R, Prakash L, Prakash S (2015) A major role of DNA polymerase delta in replication of both the leading and lagging DNA strands. Mol Cell 59(2):163–175
Kadyrova LY, Blanko ER, Kadyrov FA (2011) CAF-I-dependent control of degradation of the discontinuous strands during mismatch repair. Proc Natl Acad Sci USA 108(7):2753–2758
Karthikeyan R, Vonarx EJ, Straffon AF, Simon M et al (2000) Evidence from mutational specificity studies that yeast DNA polymerases delta and epsilon replicate different DNA strands at an intracellular replication fork. J Mol Biol 299(2):405–419
Kato L, Stanlie A, Begum NA, Kobayashi M et al (2012) An evolutionary view of the mechanism for immune and genome diversity. Journal of Immunology 188(8):3559–3566 (Baltimore)
Kazanov MD, Roberts SA, Polak P, Stamatoyannopoulos J et al (2015) APOBEC-induced cancer mutations are uniquely enriched in early-replicating, gene-dense, and active chromatin regions. Cell Rep 13(6):1103–1109
Kesti T, Flick K, Keranen S, Syvaoja JE et al (1999) DNA polymerase ε catalytic domains are dispensable for DNA replication, DNA repair, and cell viability. Mol Cell 3(5):679–685
Kochenova OV, Daee DL, Mertz TM, Shcherbakova PV (2015) DNA polymerase zeta-dependent lesion bypass in *Saccharomyces cerevisiae* is accompanied by error-prone copying of long stretches of adjacent DNA. PLoS Genet 11(3):e1005110
Korona DA, Lecompte KG, Pursell ZF (2011) The high fidelity and unique error signature of human DNA polymerase epsilon. Nucleic Acids Res 39(5):1763–1773
Kovtun IV, McMurray CT (2001) Trinucleotide expansion in haploid germ cells by gap repair. Nat Genet 27(4):407–411
Kovtun IV, Liu Y, Bjoras M, Klungland A et al (2007) OGG1 initiates age-dependent CAG trinucleotide expansion in somatic cells. Nature 447(7143):447–452

Kozmin SG, Schaaper RM, Shcherbakova PV, Kulikov VN et al (1998) Multiple antimutagenesis mechanisms affect mutagenic activity and specificity of the base analog 6-N-hydroxylaminopurine in bacteria and yeast. Mutat Res 402(1–2):41–50

Kozmin SG, Leroy P, Pavlov YI, Schaaper RM (2008) YcbX and yiiM, two novel determinants for resistance of *E. coli* to N-hydroxylated base analogues. Mol Microbiol 68(1):51–65

Kulikov VV, Derkatch IL, Noskov VN, Tarunina OV et al (2001) Mutagenic specificity of the base analog 6-N-hydroxylaminopurine in the *LYS2* gene of yeast *Saccharomyces cerevisiae*. Mutat Res 473(2):151–161

Kumar D, Abdulovic AL, Viberg J, Nilsson AK et al (2011) Mechanisms of mutagenesis in vivo due to imbalanced dNTP pools. Nucleic Acids Res 39(4):1360–1371

Kunkel TA (2004) DNA replication fidelity. J Biol Chem 279(17):16895–16898

Kunkel TA, Erie DA (2005) DNA mismatch repair. Annu Rev Biochem 74:681–710

Kunkel TA, Burgers PM (2008) Dividing the workload at a eukaryotic replication fork. Trends Cell Biol 18(11):521–527

Lada AG, Frahm Krick C, Kozmin SG, Mayorov VI et al (2011a) Mutator effects and mutation signatures of editing deaminases produced in bacteria and yeast. Biochemistry 76:131–146 (Moscow)

Lada AG, Waisertreiger IS, Grabow CE, Prakash A et al (2011b) Replication protein A (RPA) hampers the processive action of APOBEC3G cytosine deaminase on single-stranded DNA. PLoS ONE 6(9):e24848

Lada AG, Dhar A, Boissy RJ, Hirano M et al (2012) AID/APOBEC cytosine deaminase induces genome-wide kataegis. Biol Direct 7:47

Lada AG, Stepchenkova EI, Waisertreiger IS, Noskov VN et al (2013) Genome-wide mutation avalanches induced in diploid yeast cells by a base analog or an APOBEC deaminase. PLoS Genet 9(9):e1003736

Lada AG, Kliver SF, Dhar A, Polev DE et al (2015) Disruption of transcriptional coactivator Sub1 leads to genome-wide re-distribution of clustered mutations induced by APOBEC in active yeast genes. PLoS Genet 11(5):e1005217

Lahue RS, Slater DL (2003) DNA repair and trinucleotide repeat instability. Front Biosci 8:s653–s665

Lang GI, Murray AW (2011) Mutation rates across budding yeast chromosome VI are correlated with replication timing. Genome Biol Evol 3:799–811

Larrea AA, Lujan SA, Nick McElhinny SA, Mieczkowski PA et al (2010) Genome-wide model for the normal eukaryotic DNA replication fork. Proc Natl Acad Sci USA 107 (41):17674–17679

Lawrence MS, Stojanov P, Polak P, Kryukov GV et al (2013) Mutational heterogeneity in cancer and the search for new cancer-associated genes. Nature 499(7457):214–218

Liu L, De S, Michor F (2013) DNA replication timing and higher-order nuclear organization determine single-nucleotide substitution patterns in cancer genomes. Nat Commun 4:1502

Liu S, Lu G, Ali S, Liu W et al (2015) Okazaki fragment maturation involves alpha-segment error editing by the mammalian FEN1/MutSalpha functional complex. EMBO J 34(13):1829–1843

Lujan SA, Clausen AR, Clark AB, MacAlpine HK et al (2014) Heterogeneous polymerase fidelity and mismatch repair bias genome variation and composition. Genome Res 24(11):1751–1764

Lynch M (2010) Evolution of the mutation rate. Trends Genet 26(8):345–352

Magni GE, von Borstel RC (1962) Different rates of spontaneous mutation during mitosis and meiosis in yeast. Genetics 47:1097–1108

Makarova AV, Burgers PM (2015) Eukaryotic DNA polymerase zeta. DNA Repair 29:47–55

Maki H, Sekiguchi M (1992) MutT protein specifically hydrolyses a potent mutagenic substrate for DNA synthesis. Nature 355(6357):273–275

Malkova A, Ira G (2013) Break-induced replication: functions and molecular mechanism. Curr Opin Genet Dev 23(3):271–279

Mathews CK (2006) DNA precursor metabolism and genomic stability. FASEB J 20 (9):1300–1314

Mathews CK, Ji J (1992) DNA precursor asymmetries, replication fidelity, and variable genome evolution. BioEssays 14(5):295–301

Mertz TM, Sharma S, Chabes A, Shcherbakova PV (2015) Colon cancer-associated mutator DNA polymerase delta variant causes expansion of dNTP pools increasing its own infidelity. Proc Natl Acad Sci USA 112(19):E2467–E2476

Michaels ML, Miller JH (1992) The GO system protects organisms from the mutagenic effect of the spontaneous lesion 8-hydroxyguanine (7,8-dihydro-8-oxoguanine). J Bacteriol 174 (20):6321–6325

Miller JH (1996) The relevance of bacterial mutators to understanding human cancer. Cancer Surv 28:141–153

Mirkin SM (2006) DNA structures, repeat expansions and human hereditary disorders. Curr Opin Struct Biol 16(3):351–358

Miyabe I, Kunkel TA, Carr AM (2011) The major roles of DNA polymerases epsilon and delta at the eukaryotic replication fork are evolutionarily conserved. PLoS Genet 7(12):e1002407

Miyabe I, Mizuno K, Keszthelyi A, Daigaku Y et al (2015) Polymerase delta replicates both strands after homologous recombination-dependent fork restart. Nat Struct Mol Biol 22 (11):932–938. doi:10.1038/nsmb.3100

Morrison A, Sugino A (1994) The 3'→ 5' exonucleases of both DNA polymerases delta and epsilon participate in correcting errors of DNA replication in *Saccharomyces cerevisiae*. Mol Gen Genet 242(3):289–296

Morrison A, Araki H, Clark AB, Hamatake RK et al (1990) A third essential DNA polymerase in *S. cerevisiae*. Cell 62(6):1143–1151

Muramatsu M, Kinoshita K, Fagarasan S, Yamada S et al (2000) Class switch recombination and hypermutation require activation-induced cytidine deaminase (AID), a potential RNA editing enzyme. Cell 102(5):553–563

Neuberger MS, Rada C (2007) Somatic hypermutation: activation-induced deaminase for C/G followed by polymerase eta for A/T. J Exp Med 204(1):7–10

Neuberger MS, Harris RS, Di Noia J, Petersen-Mahrt SK (2003) Immunity through DNA deamination. Trends Biochem Sci 28(6):305–312

Nick McElhinny SA, Gordenin DA, Stith CM, Burgers PM et al (2008) Division of labor at the eukaryotic replication fork. Mol Cell 30(2):137–144

Nick McElhinny SA, Kumar D, Clark AB, Watt DL et al (2010) Genome instability due to ribonucleotide incorporation into DNA. Nat Chem Biol 6(10):774–781

Niimi A, Limsirichaikul S, Yoshida S, Iwai S et al (2004) Palm mutants in DNA polymerases α and η alter DNA replication fidelity and translesion activity. Mol Cell Biol 24(7):2734–2746

Nik-Zainal S, Alexandrov LB, Wedge DC, Van Loo P et al (2012) Mutational processes molding the genomes of 21 breast cancers. Cell 149(5):979–993

Northam MR, Garg P, Baitin DM, Burgers PM et al (2006) A novel function of DNA polymerase zeta regulated by PCNA. EMBO J 25(18):4316–4325

Northam MR, Robinson HA, Kochenova OV, Shcherbakova PV (2010) Participation of DNA polymerase zeta in replication of undamaged DNA in *S. cerevisiae*. Genetics 184(1):27–42

Ohya T, Kawasaki Y, Hiraga S, Kanbara S et al (2002) The DNA polymerase domain of pol(e) is required for rapid, efficient, and highly accurate chromosomal DNA replication, telomere length maintenance, and normal cell senescence in *Saccharomyces cerevisiae*. J Biol Chem 277 (31):28099–28108

Pang B, McFaline JL, Burgis NE, Dong M et al (2012) Defects in purine nucleotide metabolism lead to substantial incorporation of xanthine and hypoxanthine into DNA and RNA. Proc Natl Acad Sci USA 109(7):2319–2324

Pavlov II, Noskov VN, Chernov Iu O, Gordenin DA (1988) Mutability of *LYS2* gene in diploid *Saccharomyces* yeasts. II. Frequency of mutants induced by 6-N-hydroxylaminopurine and propiolactone. Genetika 24(10):1752–1760

Pavlov YI, Shcherbakova PV (2010) DNA polymerases at the eukaryotic fork-20 years later. Mutat Res 685(1–2):45–53

Pavlov YI, Noskov VN, Lange EK, Moiseeva EV et al (1991) The genetic activity of N6-hydroxyadenine and 2-amino-N6-hydroxyadenine in *Escherichia coli*, *Salmonella typhimurium* and *Saccharomyces cerevisiae*. Mutat Res 253(1):33–46

Pavlov YI, Shcherbakova PV, Kunkel TA (2001) In vivo consequences of putative active site missense mutations in yeast replicative DNA polymerases α, ε, δ and ζ. Genetics 159(1):47–64

Pavlov YI, Newlon CS, Kunkel TA (2002a) Yeast origins establish a strand bias for replicational mutagenesis. Mol Cell 10(1):207–213

Pavlov YI, Rogozin IB, Galkin AP, Aksenova AY et al (2002b) Correlation of somatic hypermutation specificity and A-T base pair substitution errors by DNA polymerase eta during copying of a mouse immunoglobulin k light chain transgene. Proc Natl Acad Sci USA 99 (15):9954–9959

Pavlov YI, Mian IM, Kunkel TA (2003) Evidence for preferential mismatch repair of lagging strand DNA replication errors in yeast. Curr Biol 13(9):744–748

Pavlov YI, Maki S, Maki H, Kunkel TA (2004) Evidence for interplay among yeast replicative DNA polymerases alpha, delta and epsilon from studies of exonuclease and polymerase active site mutations. BMC Biol 2(1):11

Pavlov YI, Frahm C, McElhinny SA, Niimi A et al (2006a) Evidence that errors made by DNA polymerase alpha are corrected by DNA polymerase delta. Curr Biol 16(2):202–207

Pavlov YI, Shcherbakova PV, Rogozin IB (2006b) Roles of DNA polymerases in replication, repair, and recombination in eukaryotes. Int Rev Cytol 255:41–132

Pearson CE, Edamura KN, Cleary JD (2005) Repeat instability: mechanisms of dynamic mutations. Nat Rev Genet 6(10):729–742

Pellegrini L (2012) The Pol alpha-primase complex. Sub-cell Biochem 62:157–169

Petersen-Mahrt SK, Neuberger MS (2003) In vitro deamination of cytosine to uracil in single-stranded DNA by apolipoprotein B editing complex catalytic subunit 1 (*APOBEC1*). J Biol Chem 278(22):19583–19586

Petersen-Mahrt SK, Harris RS, Neuberger MS (2002) AID mutates *E. coli* suggesting a DNA deamination mechanism for antibody diversification. Nature 418(6893):99–103

Pham P, Bransteitter R, Petruska J, Goodman MF (2003) Processive AID-catalysed cytosine deamination on single-stranded DNA simulates somatic hypermutation. Nature 424 (6944):103–107

Pham P, Smolka MB, Calabrese P, Landolph A et al (2008) Impact of phosphorylation and phosphorylation-null mutants on the activity and deamination specificity of activation-induced cytidine deaminase. J Biol Chem 283(25):17428–17439

Poltoratsky VP, Wilson SH, Kunkel TA, Pavlov YI (2004) Recombinogenic phenotype of human activation-induced cytosine deaminase. J Immunol 172(7):4308–4313

Poltoratsky V, Heacock M, Kissling GE, Prasad R et al (2010) Mutagenesis dependent upon the combination of activation-induced deaminase expression and a double-strand break. Mol Immunol 48(1–3):164–170

Prakash S, Johnson RE, Prakash L (2005) Eukaryotic translesion synthesis DNA polymerases: specificity of structure and function. Annu Rev Biochem 74:317–353

Pursell ZF, Isoz I, Lundstrom EB, Johansson E et al (2007) Yeast DNA polymerase epsilon participates in leading-strand DNA replication. Science 317(5834):127–130

Rattray AJ, Shafer BK, McGill CB, Strathern JN (2002) The roles of *REV3* and *RAD57* in double-strand-break-repair-induced mutagenesis of *S. cerevisiae*. Genetics 162(3):1063–1077

Refsland EW, Harris RS (2013) The APOBEC3 family of retroelement restriction factors. Curr Top Microbiol Immunol 371:1–27

Reijns MA, Rabe B, Rigby RE, Mill P et al (2012) Enzymatic removal of ribonucleotides from DNA is essential for Mammalian genome integrity and development. Cell 149(5):1008–1022

Reijns MA, Kemp H, Ding J, de Proce SM et al (2015) Lagging-strand replication shapes the mutational landscape of the genome. Nature 518(7540):502–506

Roberts SA, Gordenin DA (2014) Hypermutation in human cancer genomes: footprints and mechanisms. Nat Rev Cancer 14(12):786–800

Roberts JD, Izuta S, Thomas DC, Kunkel TA (1994) Mispair-, site-, and strand-specific error rates during simian virus 40 origin-dependent replication in vitro with excess deoxythymidine triphosphate. J Biol Chem 269(3):1711–1717
Roberts SA, Sterling J, Thompson C, Harris S et al (2012) Clustered mutations in yeast and in human cancers can arise from damaged long single-strand DNA regions. Mol Cell 46 (4):424–435
Roberts SA, Lawrence MS, Klimczak LJ, Grimm SA et al (2013) An APOBEC cytidine deaminase mutagenesis pattern is widespread in human cancers. Nat Genet 45(9):970–976
Rogozin IB, Pavlov YI, Bebenek K, Matsuda T et al (2001) Somatic mutation hotspots correlate with DNA polymerase eta error spectrum. Nat Immunol 2(6):530–536. doi:10.1038/88732
Rogozin IB, Iyer LM, Liang L, Glazko GV et al (2007) Evolution and diversification of lamprey antigen receptors: evidence for involvement of an AID-APOBEC family cytosine deaminase. Nat Immunol 8(6):647–656
Rumbaugh JA, Henricksen LA, DeMott MS, Bambara RA (1999) Cleavage of substrates with mismatched nucleotides by Flap endonuclease-1. Implications for mammalian Okazaki fragment processing. J Biol Chem 274(21):14602–14608
Saini N, Ramakrishnan S, Elango R, Ayyar S et al (2013a) Migrating bubble during break-induced replication drives conservative DNA synthesis. Nature 502(7471):389–392
Saini N, Zhang Y, Nishida Y, Sheng Z et al (2013b) Fragile DNA motifs trigger mutagenesis at distant chromosomal loci in *Saccharomyces cerevisiae*. PLoS Genet 9(6):e1003551
Sakofsky CJ, Roberts SA, Malc E, Mieczkowski PA et al (2014) Break-induced replication is a source of mutation clusters underlying kataegis. Cell Rep 7(5):1640–1648
Sakumi K, Abolhassani N, Behmanesh M, Iyama T et al (2010) ITPA protein, an enzyme that eliminates deaminated purine nucleoside triphosphates in cells. Mutat Res 703(1):43–50
Sale JE (2013) Translesion DNA synthesis and mutagenesis in eukaryotes. Cold Spring Harb Perspect Biol 5(3):a012708
Salk JJ, Fox EJ, Loeb LA (2010) Mutational heterogeneity in human cancers: origin and consequences. Annu Rev Pathol 5:51–75
Sancar A, Lindsey-Boltz LA, Unsal-Kacmaz K, Linn S (2004) Molecular mechanisms of mammalian DNA repair and the DNA damage checkpoints. Annu Rev Biochem 73:39–85
Schmitz KM, Petersen-Mahrt SK (2012) AIDing the immune system-DIAbolic in cancer. Semin Immunol 24(4):241–245
Schopf B, Bregenhorn S, Quivy JP, Kadyrov FA et al (2012) Interplay between mismatch repair and chromatin assembly. Proc Natl Acad Sci USA 109(6):1895–1900
Sekiguchi M, Tsuzuki T (2002) Oxidative nucleotide damage: consequences and prevention. Oncogene 21(58):8895–8904
Shah KA, Shishkin AA, Voineagu I, Pavlov YI et al (2012) Role of DNA polymerases in repeat-mediated genome instability. Cell Rep 2(5):1088–1095
Shcherbakova PV, Pavlov YI (1993) Mutagenic specificity of the base analog 6-N-hydroxylaminopurine in the *URA3* gene of the yeast *S. cerevisiae*. Mutagenesis 8 (5):417–421
Shcherbakova PV, Pavlov YI (1996) 3'→5' exonucleases of DNA polymerases epsilon and delta correct base analog induced DNA replication errors on opposite DNA strands in *Saccharomyces cerevisiae*. Genetics 142(3):717–726
Shcherbakova PV, Pavlov YI, Chilkova O, Rogozin IB et al (2003) Unique error signature of the four-subunit yeast DNA polymerase epsilon. J Biol Chem 278(44):43770–43780
Shendure J, Akey JM (2015) The origins, determinants, and consequences of human mutations. Science 349(6255):1478–1483
Shishkin AA, Voineagu I, Matera R, Cherng N et al (2009) Large-scale expansions of Friedreich's ataxia GAA repeats in yeast. Mol Cell 35(1):82–92
Simandan T, Sun J, Dix TA (1998) Oxidation of DNA bases, deoxyribonucleosides and homopolymers by peroxyl radicals. Biochem J 335(Pt 2):233–240
Smith DJ, Whitehouse I (2012) Intrinsic coupling of lagging-strand synthesis to chromatin assembly. Nature 483(7390):434–438

Stillman B (2015) Reconsidering DNA polymerases at the replication fork in eukaryotes. Mol Cell 59(2):139–141

Storb U (2014) Why does somatic hypermutation by AID require transcription of its target genes? Adv Immunol 122:253–277

Strathern JN, Shafer BK, McGill CB (1995) DNA synthesis errors associated with double-strand-break repair. Genetics 140(3):965–972

Supek F, Lehner B (2015) Differential DNA mismatch repair underlies mutation rate variation across the human genome. Nature 521(7550):81–84

Tahirov TH (2012) Structure and function of eukaryotic DNA polymerase delta. Subcell Biochem 63:217–236

Tanaka S, Cao K, Niimi A, Limsirichaikul S et al (2010) Functions of base selection step in human DNA polymerase alpha. DNA Repair 9(5):534–541

Tang W, Dominska M, Gawel M, Greenwell PW et al (2013) Genomic deletions and point mutations induced in Saccharomyces cerevisiae by the trinucleotide repeats (GAA·TTC) associated with Friedreich's ataxia. DNA Repair 12(1):10–17

Taylor BJ, Nik-Zainal S, Wu YL, Stebbings LA et al (2013) DNA deaminases induce break-associated mutation showers with implication of APOBEC3B and 3A in breast cancer kataegis. Elife 2:e00534

Taylor BJ, Wu YL, Rada C (2014) Active RNAP pre-initiation sites are highly mutated by cytidine deaminases in yeast, with AID targeting small RNA genes. Elife 3:e03553

Tran HT, Degtyareva NP, Gordenin DA, Resnick MA (1999) Genetic factors affecting the impact of DNA polymerase δ proofreading activity on mutation avoidance in yeast. Genetics 152 (1):47–59

Vaisman A, Woodgate R (2015) Redundancy in ribonucleotide excision repair: Competition, compensation, and cooperation. DNA Repair 29:74–82

Veaute X, Fuchs RP (1993) Greater susceptibility to mutations in lagging strand of DNA replication in *Escherichia coli* than in leading strand. Science 261(5121):598–600

Waga S, Stillman B (1998) The DNA replication fork in eukaryotic cells. Annu Rev Biochem 67:721–751

Wagner SD, Neuberger MS (1996) Somatic hypermutation of immunoglobulin genes. Annu Rev Immunol 14:441–457

Waisertreiger IS, Liston VG, Menezes MR, Kim HM et al (2012) Modulation of mutagenesis in eukaryotes by DNA replication fork dynamics and quality of nucleotide pools. Environ Mol Mutagen 53:699–734

Walker BA, Wardell CP, Murison A, Boyle EM et al (2015) APOBEC family mutational signatures are associated with poor prognosis translocations in multiple myeloma. Nat Commun 6:6997

Wang X, Ira G, Tercero JA, Holmes AM et al (2004) Role of DNA replication proteins in double-strand break-induced recombination in *S. cerevisiae*. Mol Cell Biol 24(16):6891–6899

Watt DL, Buckland RJ, Lujan SA, Kunkel TA et al (2015) Genome-wide analysis of the specificity and mechanisms of replication infidelity driven by imbalanced dNTP pools. Nucleic Acids Res 44:1669–1680

Wu PF, Chen YS, Kuo TY, Lin HH et al (2015) APOBEC3B: a potential factor suppressing growth of human hepatocellular carcinoma cells. Anticancer Res 35(3):1521–1527

Yang W, Woodgate R (2007) What a difference a decade makes: insights into translesion DNA synthesis. Proc Natl Acad Sci USA 104(40):15591–15598

Yeeles JT, Deegan TD, Janska A, Early A et al (2015) Regulated eukaryotic DNA replication origin firing with purified proteins. Nature 519(7544):431–435

Zeng X, Winter DB, Kasmer C, Kraemer KH et al (2001) DNA polymerase eta is an A-T mutator in somatic hypermutation of immunoglobulin variable genes. Nat Immunol 2(6):537–541

Zhong X, Garg P, Stith CM, Nick McElhinny SA et al (2006) The fidelity of DNA synthesis by yeast DNA polymerase zeta alone and with accessory proteins. Nucleic Acids Res 34 (17):4731–4742

Rates of Spontaneous Mutation: Insights Gained Over the Last Half Century

John W. Drake

Introduction

Nikolai Timofeeff-Ressovsky understood the need for explicit theory and quantitation in biology. His adventures with Karl Zimmer and Max Delbrück and the somewhat romantic portrayal of those ideas by Erwin Schrödinger contributed notably to the development of population genetics and led to the modern theory of mutation. A central mystery in Timofeeff's time was the size, composition and stability of the gene, which he probed by the methods of radiation mutagenesis. A subsequent central mystery has been whether order may underlie the apparent chaos of rates of spontaneous mutation. Although the first hints of order appeared in the late 1960s, the robustness of certain formulations of mutation rates did not become apparent until the 1990s. It is now clear that each of four major groups of organisms has its own characteristic rate of spontaneous mutation. These rates and their impacts will be outlined here.

Natural selection acts on mutations arising throughout the genome. Therefore, the most interesting rates are the genomic rates. These are obtained by measuring the frequency of a mutant phenotype, converting it to a primary mutation rate, relating this rate to a mutational target of defined size, calculating the rate per average base, and then scaling up to a rate per genome, μ_g. However, many base-pair substitutions (BPSs) fail to produce a scorable phenotype, even though

Lecture on the occasion of the First Intern. Conf. dedicated to the centenary of the birth of N.W. Timoféeff- Ressovsky; Dubna, 6–9 September 2000. The paper was firstly published in the Conference Proceedings "Modern problems of radiobiology, radioecology and evolution", Dubna, JINR (2001): 35–45.

J.W. Drake (✉)
Laboratory of Molecular Genetics, National Institute of Environmental Health Sciences, Research Triangle Park, North Carolina 27709-2233, USA
e-mail: drake2@niehs.nih.gov

V.L. Korogodina et al. (eds.), *Genetics, Evolution and Radiation*,
DOI 10.1007/978-3-319-48838-7_7

they may be efficiently removed by natural selection on evolutionary time scales. Therefore, primary mutation rates must be adjusted for missed mutations. One simple way to do this is to obtain a mutational spectrum, ignore all the missense mutations (BPSs producing amino-acid substitutions), identify the chain-terminating BPSs (those producing mRNAs bearing internal UAA, UAG, UGA, or equivalent stop codons), multiply these by 64/3 or some other appropriate scalar that reflects average base composition and codon usage, and add that number of mutations to the non-BPSs.

Riboviruses (RNA viruses without a DNA phase) are among the simplest organisms. They are also genetically the least stable. This instability was surmised by early workers in RNA-virus genetics in the late 1950s because of failures to obtain stable genetic markers. Cellular and phage mutants typically exhibited revertant frequencies below 10^{-6}, whereas revertant frequencies in RNA viruses were 100-fold to 1000-fold greater. (Apparently stable mutants were later found to contain two or more mutations, each required for the mutant phenotype.) Only very recently has a robust theory been devised to characterize the relationships among riboviral mutant frequencies, population histories, and mutation rates (Drake and Holland 1999). The theory is exceedingly simple. A ribovirus enters a cell, its RNA is copied repeatedly into strands of the complementary sequence, these strands are in turn copied repeatedly into strands of the same sequence as that of the infecting particle, and these final strands are packaged and released as virus particles. This is the "stamping machine" model long ago considered and rejected for phage T2 (Luria 1951), and its topology is independent of the process of gene expression. In any round of replication, the mutation rate μ and the mutant frequency f have the same value, so that $f = \mu$. There are two such rounds of infection in a single infective cycle, and under the simplifying assumption that the mutation rate is the same in both rounds, the two frequencies are simply additive: $f = 2\mu$ per cycle. Over c consecutive cycles (as when growing a stock), $f = 2c\mu$.

The devil is in the data. The theory is applicable only when selection is insignificant and when the mutation is promptly expressed (that is, when there is little or no phenotypic masking, the viral equivalent of phenotypic lag in cellular systems). Growth must not have continued so long that the population is replaced by adaptive variants, a problem with riboviruses because of their high mutation rates and large population sizes. Even when the few available data that meet these criteria are discovered, a severe problem remains: to date, otherwise suitable mutant frequencies in riboviruses score only one or a few base pairs. It is well known that the mutation rates of different bases or base pairs in a gene vary hugely. Thus, a collection of small mutational targets will display much more variability than a collection of large mutational targets.

The useful results appear in Table 1. These rates vary by about nine-fold, which is at least the amount of variation to be expected in a random sample of nine sites in a gene made of DNA. It is therefore reasonable to assume, for the time being, that the typical mutation rate for a ribovirus is about 0.76 per genome replication or 1.5 per infection cycle. (For such a distribution, the median is a better estimator than the mean.) This is a huge rate when one considers that riboviral genomes are compact,

Table 1 Rates of spontaneous mutation per genome replication in riboviruses

Virus	μ_g
Poliovirus	0.13
Poliovirus	0.18
Poliovirus	0.18
Rhinovirus	0.67
Poliovirus	0.76
Poliovirus	0.88
Measles virus	1.00
Vesicular stomatitis virus	1.07
Vesicular stomatitis virus	1.15
Median	0.76

Data are from Drake and Holland (1999). μ_g = mutations per chromosome per replication

and that even synonymous codon changes may be deleterious because of the large role in gene expression played by RNA secondary structure. Indeed, just a tickle of mutagen, sufficient to increase the mutation rate by only a little over twofold, suffices to extinguish the population (Holland et al. 1990).

Mutator mutations increase the mutation rate, sometimes generally and sometimes only along very specific pathways such as A:T → C:G. Mutator mutations usually either disable a genomic repair function or reduce the accuracy of a replicative polymerase. Because of their very high mutation rates and great sensitivity to mutagens, ribovirus mutator mutations are expected to be weak (in order to survive to be seen in the first place) and to be under strong negative selection (because most of their progeny will bear new deleterious mutations that are expressed fully and immediately). On the other hand, the viral RNA-dependent RNA polymerase (the viral replicase) makes up a substantial fraction of the genome wherein mutator mutations might arise. Therefore, many mutations in ribovirus populations may arise in a minority of genomes carrying mutator mutations. These genomes are doomed whatever the nature of the scored mutations. Alternatively, when a mutator replicase arises during the course of infection and is expressed immediately, it may act in *trans* on non-mutator genomes. In either case, the standard rate may overestimate the rate of the majority of non-mutator genomes, but by a factor as yet unmeasured.

The next group of organisms with a characteristic mutation rate is the retroelements, including several animal retroviruses and a yeast retrotransposon. Mutations are usually scored in these organisms after a single cycle from DNA element to DNA element. This cycle consists of three rounds of replication, namely transcription by the cellular RNA polymerase followed by reverse transcription to single-stranded DNA and conversion to double-stranded DNA, both of the two final replications being conducted by the viral reverse transcriptase. Although it seems unlikely that all three steps have the same accuracy, it is not yet possible to measure differences among them reliably. Therefore, the average mutation rate per genome

Table 2 Rates of spontaneous mutation per genome replication in retroelements

Retroelement	μ_g
Bovine leukemia virus	0.03
Spleen necrosis virus	0.04
Saccharomyces cerevisiae Ty*1*	0.11
HIV-1	0.19
Murine leukemia virus	0.26
Rous sarcoma virus	0.43
Median	0.15

Data are from Drake et al. (1998)

replication is estimated by dividing the mutant frequency per cycle by 3. The data accumulated by late 1997 were tabulated in Drake et al. (1998) and are shown in Table 2.

Here the median rate is five times lower than for riboviruses. That this ratio is probably accurate is indicated by experiments showing that about 13-fold mutagenesis is required to extinguish a spleen necrosis virus population (Pathak and Temin 1992): 13/2.5 = 5.2, 0.76/0.15 = 5.1. Although this spontaneous mutation rate is still very high, it may allow mutators to make a greater impact before extinguishing a population. That such may be the case is suggested, albeit weakly, by the fact that the mutational target sizes underlying the above values are quite large, 288–173,043, so that the observed 14-fold variation in rate may reflect stochastic aspects of mutator action. Alternative, the retroelements may simply vary considerably in intrinsic mutation rate. These are experimentally distinguishable possibilities.

The accuracies of several reverse transcriptases have been measured in vitro, and conclusions have sometimes been drawn about the corresponding mutation rates in vivo. Such extrapolation is probably dangerous, because at least one study has shown that an error rate can be substantially lower in vivo than in vitro (Varela-Echavarría et al. 1992; Drake 1993).

The group of organisms with the best characterized and least variable μ_g is the DNA-based microbes. Their rates were most recently compiled by Drake et al. (1998). These rates are based on large mutational targets and excellent measurement systems. The rates, together with the genome size (G) and the rate per average base pair (μ_b), are summarized in Table 3. μ_g varies very little, while G and μ_b vary inversely by roughly 8000-fold. These data are usually taken to mean that mutation rates can evolve to finely tuned values, although why the particular value of about 1/300 is preferred, and by such diverse organisms, remains mysterious.

Because mutation rates in DNA-based microbes are much less than 1, even strong mutator mutations can be propagated. Such mutators have taught us a great deal about fidelity mechanisms. Recently, it has also become clear that such mutators can be important in nature. In experimental systems, mutators can win adaptive races if the product of their prevalence and their mutation rate exceeds the corresponding product for the wild-type (Chao and Cox 1983). However, a mutator is at a selective disadvantage in a constant environment and slowly accumulates

Table 3 Rates of spontaneous mutation per genome replication in DNA-based microbes

Organism	G	μ_b	μ_g
Bacteriophage M13	6.4×10^3	7.2×10^{-7}	0.0046
Bacteriophage λ	4.9×10^4	7.7×10^{-8}	0.0038
Bacteriophages T2 and T4	1.7×10^5	2.4×10^{-8}	0.0040
Escherichia coli	4.6×10^6	5.4×10^{-10}	0.0025
Saccharomyces cerevisiae	1.2×10^7	2.2×10^{-10}	0.0027
Neurospora crassa	4.2×10^7	7.2×10^{-11}	0.0030
Mean			0.0034

Data are from Drake et al. (1998). G = genome size in bases or base pairs
μ_b = mutations per average base per replication

mutations that reduce its mutation rate (Tröbner and Piechocki 1984). Theory predicts (Taddei et al. 1997; Tenaillon et al. 1999) and observation confirms (most recently, Oliver et al. 2000) that mutators often win adaptive races among bacteria entering new niches. However, the mutator bears a strong selective disadvantage if it remains in a less variable environment, and its later fate is likely to be either displacement by a slower but fitter nonmutator, or erasure of the mutator phenotype by subsequent mutations.

One constraint upon the interpretation of the data in Table 3 is that all of the measurements were made in microbial heaven, that is, under mild laboratory conditions using rich media. At least in the case of phage T4, neither the richness of the medium nor the degree to which it is cooked during preparation affect mutation rates (Smith and Drake 1998). However, many DNA-based organisms live in environments that are intrinsically highly mutagenic. For instance, both acid and heat degrade DNA bases in a number of ways that are either directly mutagenic or trigger the mutagenic SOS system. Does the standard mutation rate in DNA-based microbes persist in such environments? Alternatively, are cellular fidelity mechanisms compromised or overwhelmed? A partial answer is emerging. A mutation rate has been determined for the *pyrE* and *pyrF* genes of the archaeon *Sulfolobus acidocaldarius* growing at pH 3.5 at 75°C (Jacobs and Grogan 1997). Making reasonable guesses about the sizes of the target genes and of the genome, μ_g was estimated to be roughly 0.002 (Drake et al. 1998). Sequence analyses of *pyrE* mutations now tend to confirm this estimate. Thus, this extremophile has a spontaneous mutation rate close to or even smaller than the standard rate.

A comment is appropriate here about antimutator mutations, which reduce mutation rates. Kimura (1967) was the first to point out that reducing mutation rates can only be achieved at some cost, which he called the physiological cost but can also be viewed as a thermodynamic cost, namely the cost of reducing entropy. Modern organisms have highly evolved investments in maintaining replication fidelity, and this includes even the RNA viruses whose polymerases make many fewer errors than would occur if determined exclusively by hydrogen bonding.

The chance is small that one or even a few mutations could reduce mutation rates generally without detrimental side-effects such as slowed replication. This conjecture was verified for all known antimutators in bacteriophage T4. These antimutators unequivocally reduce some mutation rates (by as much as 100-fold), but primarily along the pathway A:T → G:C. At the same time, they were known to increase other mutation rates, notably transversions. When T4 antimutators were screened for average mutation rates using a mutational target of several kilobases, all were either weak mutators or without detectable impact. Thus, their narrow antimutator effect was achieved at the cost of a small reduction in overall fidelity. The idea that antimutators could be discovered that would, for instance, protect humans against cancer and heritable birth defects, is probably without merit.

The final group of organisms to consider is the higher eukaryotes, that is, plants and animals. This group of organisms differs profoundly from the microbes in genome structure. Their genomes bear a much higher fraction of DNA not directly devoted to organismal functions, including both intergenic DNA and introns. Mutations in this DNA have little impact, with minor exceptions such as the intrinsic need for spacer DNA (Kricker et al. 1992) and for short, specific boundary sequences in introns. Thus, natural selection can only act upon mutations and mutation rates in the most functional DNA, where functional means serving the needs of the organism itself and not its molecular parasites. Thus, one can consider two different genome sizes in the higher eukaryotes. One is the total genome size G as above. The other is the effective genome size G_e, which is approximately equal to the sum of the coding sequences of proteins, structural RNAs and regulatory elements. (Strictly speaking, G_e is the portion of the genome in which mutations impact upon the fitness of the organism.) In addition, mutation rates can be expressed using either of two rate units. One is the traditional value, mutations per gamete or per sexual generation. The other is mutations per germ line cell division or, in the case of plants, cell divisions between gametic generations. In many animals, there are far more cell divisions per sexual generation in the male than in the female, so that mutation rates are dominated by the male contribution and increase with paternal age.

As with the RNA viruses, few mutation rates in higher eukaryotes are based on optimal data. Although G is often well measured, Ge is not. Mutational target sizes are often unclear, and mutational spectra are rarely available. Thus, quite a bit of guesswork must enter into the calculations and the results must be interpreted with caution. The most recent compilation is in Drake et al. (1998) and is summarized in Table 4. Four mutation rates are presented: per total genome per sexual generation, per effective genome per sexual generation, per total genome per cell division, and per effective genome per cell division. The range (max/min) for each kind of rate is the ratio of the largest value to the smallest value.

Despite the uncertainty of some of the numbers, Table 4 suggests several conclusions. Numbers of mutations per total genome per sexual generation become enormous for large genomes composed mainly of spacer DNA. Perhaps more interesting, numbers of mutations per effective genome become quite large for mammalian genomes, hovering around 1 per gamete (and per zygote, because of the

Table 4 Rates of spontaneous mutation in higher eukaryotes

Organism	G	G_e	μ_{gs}	μ_{egs}	μ_g	μ_{eg}
Caenorhabditis elegans	8.0×10^7	1.8×10^7	0.16	0.036	0.018	0.004
Drosophila melanogaster	1.7×10^8	1.6×10^7	1.5	0.14	0.058	0.005
Mus musculus	2.7×10^9	8.0×10^7	30	0.9	0.5	0.014
Homo sapiens	3.2×10^9	8.0×10^7	64	1.6	0.16	0.004
Range (max/min)			400	44	27	3.5

Data are from Drake et al. (1998). G_e = effective genome size (see text). μ_{gs} = mutations per genome per sexual generation. μ_{egs} = mutations per effective genome per sexual generation. μ_g = mutations per genome per germ-line cell division. μ_{eg} = mutations per effective genome per germ-line cell division

predominance of mutations in the male). Thus, mammals share a surprising vulnerability with RNA viruses, rather small increases in their mutation rates threatening to extinguish the population. Diploidy certainly buffers the impact of this high rate, but eventually the genome must accumulate many detrimental mutations, leading to what Lynch has termed "mutational meltdown" (Lynch and Gabriel 1990).

If there exists a standard rate of mutation among the higher eukaryotes, as seen so robustly for the DNA-based microbes, Table 4 suggests where it is to be found. The four listed measures of mutation rate all show substantial variation among animals, but one of the ranges is far smaller than the others, and this range is widened by just one value. μ_{eg}, the mutation rate per effective genome per germ-line cell division, is 0.004–0.005 for three of the animals and 0.014 for the mouse. It is notable that neither 0.004–0.005 nor the average of the four values, 0.007, is much different from the standard rate (0.0034) for the DNA-based microbes. Thus, the evolution of multicellularity and a germ-cell lineage was not accompanied by the evolution of a correspondingly lower mutation rate per genome replication, with the result that modern organisms produce a large fraction of offspring carrying new deleterious mutations.

References

Chao L, Cox EC (1983) Competition between high and low mutating strains of *Escherichia coli*. Evolution 37:125–134

Drake JW (1993) Rates of spontaneous mutation among RNA viruses. Proc Natl Acad Sci USA 90:4171–4175

Drake JW, Charlesworth B, Charlesworth D, Crow JF (1998) Rates of spontaneous mutation. Genetics 148:1667–1686

Drake JW, Holland JJ (1999) Mutation rates among lytic RNA viruses. Proc Natl Acad Sci USA 96:13910–13913

Holland JJ, Domingo E, de la Torre JC, Steinhauer DA (1990) Mutation frequencies at defined single codon sites in vesicular stomatitis virus and poliovirus can be increased only slightly by chemical mutagenesis. J Virol 64:3960–3962

Jacobs KL, Grogan DW (1997) Rates of spontaneous mutation in an Archaeon from geothermal environments. J Bacteriol 179:3298–3303

Kimura M (1967) On the evolutionary adjustment of spontaneous mutation rates. Genet Res 9:23–34

Kricker MC, Drake JW, Radman M (1992) Duplication-targeted DNA methylation and mutagenesis in the evolution of eukaryotic chromosomes. Proc Natl Acad Sci USA 89:1075–1079

Luria SE (1951) The frequency distribution of spontaneous bacteriophage mutants as evidence for the exponential rate of phage reproduction. Cold Spring Harbor Symp. Quant Biol 16:463–470

Lynch M, Gabriel W (1990) Mutation load and the survival of small populations. Evolution 44:1725–1737

Oliver A, Cantón R, Campo P, Baquero F, Blázquez J (2000) High frequency of hypermutable *Pseudomonas aeruginosa* in cystic fibrosis lung infection. Science 288:1251–1253

Pathak VK, Temin HM (1992) 5-Azacytidine and RNA secondary structure increase the retrovirus mutation rate. J Virol 66:3093–3100

Smith LA, Drake JW (1998) Aspects of the ultraviolet photobiology of some T-even bacteriophages. Genetics 148:1611–1618

Taddei F, Hayakawa H, Bouton M-F, Cirinesi A-M, Matic I et al (1997) Counteraction by MutT protein of transcriptional errors caused by oxidative damage. Science 278:128–130

Tenaillon O, Toupance B, Le Nagard H, Taddei F, Godelle B (1999) Mutators, population size, adaptive landscape and the adaptation of asexual populations of bacteria. Genetics 152:485–493

Tröbner W, Piechocki R (1984) Selection against hypermutability in *Escherichia coli* during long term evolution. Mol Gen Genet 198:177–178

Varela-Echavarría A, Garvey N, Preston BD, Dougherty JP (1992) Comparison of Moloney murine leukemia virus mutation rate with the fidelity of its reverse transcriptase in vitro. J Biol Chem 267:24681–24688

Protein Assembly Disorders and Protein-Based Inheritance

Aleksander A. Rubel, Alsu F. Saifitdinova and Nina V. Romanova

Template Principle in Biology and Protein Inheritance

The template principle, which derives from the idea of N.K. Koltsov chromosome reproduction, reflects current understanding of the mechanisms of heredity and variation (Koltsov 1936). Later, the template principles became the basis for the development of the central dogma of molecular biology by Francis Crick (Crick 1970; Crick 1958). According to this principle the alternation of elements in the original polymer (template) determines the alternation of elements in the generated polymer: engagement of deoxyribonucleotides during DNA replication, of ribonucleotides during transcription and incorporation of amino acid residues in the translation. The above examples were termed by S.G. Inge-Vechtomov as the first order template processes (Inge-Vechtomov 2003). At the same time the second order template processes were described (Inge-Vechtomov 2003). Their existence is based on conformational or spatial templates, in which the spatial stacking of polypeptide chain catalyzes similar folding of newly, synthesized homologous (and not only homologous) polypeptide chains (Inge-Vechtomov et al. 2012). Prions are the protein templates that able to catalyze the joining itself monomeric molecules of the same protein and cause a change in their conformation similar to themselves. The existence of conformational templates implies the possibility of spatial variability of the protein without changing their primary structure (Inge-Vechtomov et al. 2012). This review describes some of the features of mammalian prions that

A.A. Rubel (✉) · N.V. Romanova
Laboratory of Amyloid Biology and Institute of Translational Biomedicine,
St. Petersburg State University, St. Petersburg, Russia
e-mail: arubel@mail.ru

A.F. Saifitdinova
Research Resource Center "Chromas", Research Park of St. Petersburg
State University, St. Petersburg, Russia

V.L. Korogodina et al. (eds.), *Genetics, Evolution and Radiation*,
DOI 10.1007/978-3-319-48838-7_8

associated with fatal transmissible spongiform encephalopathies (TSEs) in humans and a large variety of animals.

Transmissible Spongiform Encephalopathies

Transmissible spongiform encephalopathies (TSEs), or prion diseases, are related to spatial structure formation defects of a protein named as Prion Protein (PrP). In contrast to other neurodegenerative diseases (prion-like diseases), including Alzheimer's, Parkinson's, Huntington's and other similar diseases, TSEs are not limited to inherited and sporadic types, and may also be transmissible, i.e. introduced via food or iatrogenic (via medical appliances, drugs, blood transfusion or neural transplantation).

Prion diseases have been described for mammals, among them are chronic wasting disease of moose and deer, bovine spongiform encephalopathy, also known as mad cow disease, sheep and goat scrapie. In humans there are prion diseases, such as Creutzfeltd-Jakob's disease, kuru, etc. (Aguilar-Calvo et al. 2015). All prion diseases feature extensive latency period, spongiform vacuolation, astrocyte proliferation, PrP accumulation in cerebral tissues and neurodegeneration of the central nervous system (Parchi and Gambetti 1995). So far there is no cure that would be able to prevent or inhibit efficaciously the progress of prion diseases. All prion diseases are incurable and inevitably lethal (Aguzzi et al. 2013).

Prion Discovery Background

References to sheep scrapie endemics in Northern Europe date back to the 18th century (1732). A detailed description of spongiform encephalopathy in British sheep was made by the Scottish veterinarian J.P. McGowan in 1922 (McGowan 1922). The first descriptions of spongiform encephalopathies in humans date back to the early 20th century. Two German neuropathologists—Hans Gerhard Creutzfeldt in 1920 and Alfons Maria Jakob in 1920 independently described a sporadic neurodegenerative human disease with typical clinical symptoms. In 1928 Austrian physicians described an extremely rare lethal disease that was named Gerstmann-Streussler-Scheinker syndrome (GSS) (Vana et al. 2007). The first notable success in prion disease studies relate to the 1930s, when J. Cuille and P.L. Chele transmitted scrapie from sheep to sheep in 1936 and from sheep to goats in 1939 (Wickner et al. 2007). Twenty years later D.C. Gajdusek and V. Zigas described an incurable degenerative neural disorder among the tribes of Papua New Guinea related to funerary cannibalism now known as kuru (Gajdusek and Zigas 1957). However, the transmissible nature of this disorder was not discovered until W. Hadlow had noticed the similarity of clinical manifestations of scrapie and kuru and, consequently, the possible extensive latent stage of the disease (Hadlow 1959).

D.C. Gajdusek and his team showed that this disease developed in a chimpanzee after transplantation of neurodermal tissue taken from the brain of a human patient that had died from kuru (Gajdusek et al. 1966). Several years later the transmissible nature of another human spongiform encephalopathy, Creutzfeldt-Jakob's disease (CJD), was demonstrated by using similar approach (Gibbs et al. 1968). The following decades of prion disorder research were dedicated to description of spongiform encephalopathy manifestation in various animals and in humans: transmissible encephalopathy of mink (Burger and Hartsough 1965), feline spongiform encephalopathy (Wyatt 1990), chronic wasting disease of deer, moose and antelope (Williams and Young 1980; Jeffrey and Wells 1988; Kirkwood et al. 1990), human fatal familial insomnia (FFI) (Goldfarb et al. 1992) and, finally, bovine spongiform encephalopathy, or "mad cow disease" (Wells et al. 1987) that can be transmitted to humans, and causes the development of a variant form of Creutzfeldt-Jakob disease (vCJD) (Will et al. 1996).

The transmissible spongiform encephalopathy agent possessed some features that were uncommon for viruses. It was resistant to deep UV treatment at 254 nm wavelength, ionizing radiation and effects of nucleases and some other chemicals, that excluded the presence of nucleic acids (Alper et al. 1967). Based on this data, J. S. Griffiths offered several hypotheses, including one suggesting that the transmissible spongiform encephalopathy agent was an altered form of a cell protein that was able to reproduce its properties through an autocatalytic process (Griffith 1967).

Experimental proof of this hypothesis was suggested in the early 1980s by Stanley B Prusiner and his colleagues who were able to extract and purify the scrapie agent from cerebral tissue of infected animals, and demonstrated its relation to the accumulation of an abnormal form of the protein in the affected brain tissue (Prusiner 1982). For this infectious protein agent S. Prusiner coined the term "prion" from "*proteinaceous infectious particle*". The protein inducer of scrapie and other transmissible spongiform encephalopathies is referred to as PrP (from Prion Protein). The regular form of PrP is designated as PrP^{C} (from "Cellular"). The pathological PrP isoform inducing scrapie is designated as PrP^{Sc} (form "Scrapie") (Prusiner and Scott 1997). PrP encoding gene designated as *Prnp* (Chesebro et al. 2005; Oesch et al. 1985) was identified on the basis of PrP primary sequence. It is worth noting that *Prnp* turned out to be identical to *Sinc* gene whose ability to influence the duration of scrapie latent stage in mice had been demonstrated earlier (Dickinson et al. 1968).

PrP Structural Features and Functions

Mouse and human PrP^{C} are a di-, mono- or non- glycosylated sialoglycoproteins consisting of a 209 amino acids globular polypeptide chain with a glycophosphatidylinositol (GPI) anchor at C-terminal that attaches the complex molecule to the surface of the cell membrane (Stahl et al. 1987).

An analysis of mature PrP^C structure in humans, mice, golden hamster and bovines has revealed the following common features: an elongated unstructured N-domain containing a repeated octapeptide (PHGGGWGQ) region and a C-domain consisting of three α-helical regions and two small antiparallel β-sheets flanking the first α-helical region (Riek et al. 1997). The C-terminal is stabilized with disulfide bridges between Cys 179 and Cys 214 linking the second and third α-spiral regions (Riek et al. 1996).

PrP^C is produced in various tissues and organs: kidneys, heart, pancreas, muscles, secondary lymphoid organs, as well as the central and peripheral nervous systems, which may be indicative of a wide range of functions performed by this protein (Aguzzi et al. 2008b). For organisms deficient of PrP^C a range of biochemical changes has been reported (Brown et al. 2002). Despite the fact that *Prnp* knock-out mice generally do not feature any apparent *Prnp* related physiological changes (Bueler et al. 1992), a line with sleep pattern disorders was described, that may be indicative of a certain PrP^C impact on circadian pattern regulation (Tobler et al. 1996). In peripheral and central nervous systems PrP^C involvement in Cu^{2+} binding may be expected (Brown et al. 1997; Stockel et al. 1998; Stevens et al. 2009). Other authors support the idea that PrP may perform a metal carrier role due to the fact that metal ion content has been found different in a knock-out mouse in comparison with that of wild mice, specifically for Cu^{2+} (Choi et al. 2006) and Fe^{3+} (Singh et al. 2009). Since PrP knock-out cell cultures and PrP-null mice are more sensitive to oxidation stress, an assumption that PrP can perform an antioxidation function has been made (Brown et al. 1997; Brown et al. 2002). PrP has also been demonstrated to participate in regulation of free Ca^{2+} concentration through reacting with voltage-dependent calcium channels (Whatley et al. 1995). PrP^C has been reported to interact with nervous cell adhesion molecules and regulate intercellular cooperation at early embryogenesis stage (Santuccione et al. 2005; Malaga-Trillo et al. 2009). The very fact that PrP^C features a high yield in lymphoid tissues and operates as a signaling molecule on T-cell membrane implies a certain PrP^C role in development and functioning of the immune system (Vana et al. 2007). Since PrP^C antibodies partially inhibit T-cell proliferation, PrP^C may be assumed to play a certain role in T-cell response regulation (Bainbridge and Walker 2005). Also PrP was associated with other numerous cellular functions, such as participation in hematopoietic cell proliferation and viability regulation (Zhang et al. 2006), apoptotic and antiapoptotic activity (Aguzzi et al. 2008a), contribution to nervous tissue and neuroprotector function development (Steele et al. 2006; Zomosa-Signoret et al. 2008), axon growth influence (Hajj et al. 2007), and participation in signal transduction and synaptic transmission processes (Linden et al. 2008). With regard to Aβ peptide oligomers PrP has been demonstrated to play a membrane receptor (Lauren et al. 2009), such association on the surface of nervous cell membranes being able to produce toxic effect (Peters et al. 2015). There are many experimental indications of PrP^C ability to act as a universal receptor for amyloid oligomers (Resenberger et al. 2011). Recent research has demonstrated PrP^C neural expression and its controlled proteolysis are necessary for neural fiber myelin sheath support: mice whose neurons featured no *Prnp* expression or PrP^C

protein's inability for regular proteolysis—suffered from chronic demyelinating polyneuropathy (Bremer et al. 2009).

It is worth noting that despite the increasing amount of data on PrP association with various physiological and cytological processes, some authors do not support the idea of its important biological function which, when impaired, may result in certain disorders. They believe that the common presence of this protein gene in mammals is related to PrP structural instability, which occurs even in case of singular amino acid substitutions. This result in pathological protein development, therefore mutant PrP genes are subject to severe negative selection (Prcina and Kontsekova 2011).

Prion Concept

The two alternative forms of PrP differ from each other only in their physiochemical properties. Thus, PrP^{Sc}, in contrast to PrP^{C}, features aggregation in vivo and in vitro, heat resistance, low solubility in detergents and resistance to proteolytic agents (Prusiner 1982; Prusiner et al. 1983; Oesch et al. 1985). After PrP^{Sc} treatment with proteinase K, PrP fragment from 90 amino acids through 231 amino acids remains intact and in vivo and in vitro experiments demonstrating its ability to induce full-length PrP protein prionization (Caughey et al. 1991). The differences in PrP^{C} and PrP^{Sc} properties are attributable to structural differences between the two protein forms. According to spectral analysis data, PrP^{C} contains 42 % of α-helices and 3 % of β-sheets, while PrP^{Sc} contains 30 % α-helices and 43 % β-sheets (Pan et al. 1993). This fact has given ground to an assumption that acquisition of transmission properties by PrP is related to PrP conformational transition from a regular isoform into an abnormal form where β-folded structures develop.

In 1982 Stanley Prusiner formulated prion concept which suggested that protein content in a prion confor'mation is necessary and sufficient for infection (Prusiner 1982):

1. PrP^{Sc} can act as an infectious agent;
2. PrP^{Sc} isoform may reproduce in the absence of nucleic acid through an autocatalytic process;
3. protein transformation from PrP^{C} regular form into PrP^{Sc} infectious form is a result of conformational transition;
4. PrP^{C}—PrP^{Sc} conformational transition may be spontaneous (sporadic disorders) and induced either by PrP^{Sc} infiltration into the body (transmissible disorders), or as a result of *Prnp* gene mutations leading to PrP^{Sc} development (hereditary disorders).

Evidence supporting the prion concept was obtained experimentally on mammals in 2010 (Makarava et al. 2010). In vitro amyloid fibrils of full-length PrP protein were inoculated intracerebrally into Syrian golden hamsters which developed transmissible spongiform encephalopathy (Makarava et al. 2010). This data fully supports the hypothesis that PrP protein is the source for prion disorder development. However, nucleic acid or other polymer involvement in the prionisation process cannot be excluded completely. Nucleic acids can induce prion conversion in vivo without affecting PrP infectivity. DNA antibodies have been demonstrated to bind with PrP^{Sc} in cerebral homogenates, that suggests an affinity of PrP^{Sc} to nucleic acids (Zou et al. 2004). The authors of other in vitro research (Deleault et al. 2003) have demonstrated that some RNA fractions can increase prion replication efficiency. Most probably, the stimulating RNA impact on PrP conversion is related to RNA polyanionic nature rather than a specific RNA sequence or type (Deleault et al. 2005). Sulphated polysaccharides constitute another polyanionic class inducing prion conversion in cell-free systems. In vitro, polyanions are assumed to act as a framework for parallel PrP^{C} and PrP^{Sc} annealing due to interaction with positive N-terminal protein regions (Warner et al. 2002).

Prion Replication Mechanism

The mechanism by which prion infectivity increases is still unknown. Currently most scientists support the "nucleated polymerization" model (Lansbury and Caughey 1995). According to this model, PrP^{Sc} exists only as fibrils, and that fibril ends bind PrP^{C} and convert it into PrP^{Sc} (Fig. 1). The "nucleated polymerization" model has been supported by the findings of in vitro PrP^{Sc} amyloid aggregate in replication experiments (Kocisko et al. 1994; Saborio et al. 2001).

Infectivity of various size PrP^{Sc} particles has been assessed in vitro in the research conducted by Silveira et al. PrP^{Sc} aggregates were split into particles and fractioned by particle size. The highest infectivity was recorded for 300–600 kDa, which corresponds to 14–28 PrP molecules approximately, while oligomers with less than 5 PrP^{Sc} molecules demonstrated no infectivity (Silveira et al. 2005). It is worth mentioning that "nucleated polymerization" model describes amyloid formation without explaining why large polymers oligomerize into seeds which catalyze new rounds of prion conversion. It is recurring oligomerization that differentiates prions from typical amyloids. The factor(s) responsible for regulation of this process for PrP aggregates are still unknown.

Prion Transport

The research conducted using neuroblastoma cell cultures, has demonstrated that soluble PrP^{C} protein conversion into PrP^{Sc} pathogenic isoform occurs either on cell membrane surface or in endosomes during PrP^{C} interaction with PrP^{Sc} protein. In some cases of hereditary spongiform encephalopathy mutant PrP^{C} protein can spontaneously fold into a prion isoform in Golgi apparatus and endoplasmic

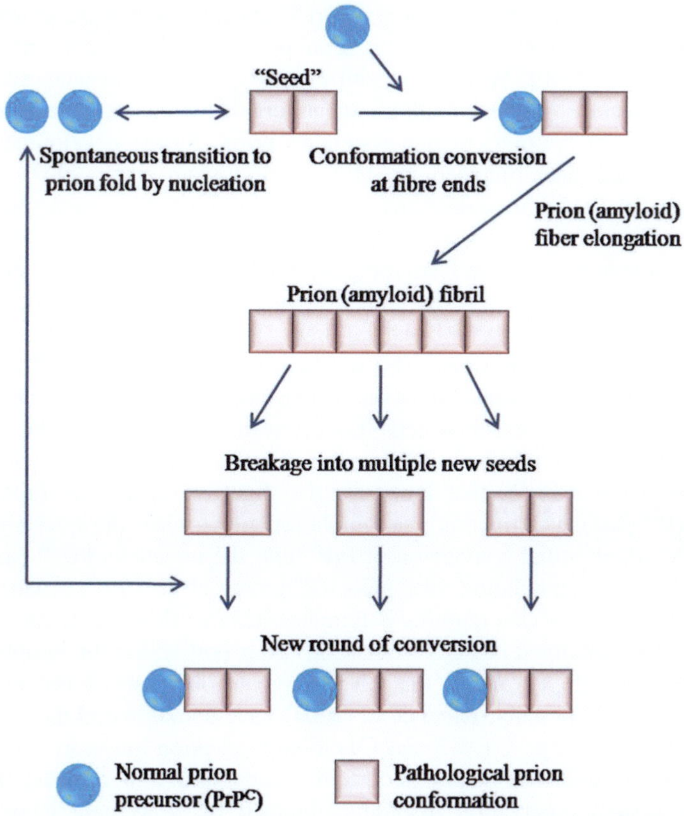

Fig. 1 Nucleated polymerization model of amyloid formation. Prions are characterized by efficient fragmentation of growing aggregates into smaller transmissible seeds

reticulum (Caughey and Baron 2006). Conflicting data is also available with regard to misfolded PrP protein conversion in the cytosol (Ma and Lindquist 2001). PrP^{Sc} aggregates may either accumulate on cell surface, in intracellular vesicles, such as lysosomes and autophagosomes, or develop extracellular depositions (Caughey and Baron 2006).

Trying to clarify the role of specific cell types in PrP^{Sc} aggregate replication and its delivery to cerebral cells we need to take in account that different prion strains can be found in different tissues and, presumably, there are cell type preferences for prion replication (Heikenwalder et al. 2007). PrP^{Sc} injection in the abdomen of model animals significantly extends the latent period of prion disease compared to direct injection of infectious agent in the cerebrum (Blattler et al. 1997). This fact evidences PrP^{Sc} accumulation in peripheral organs prior to delivery to the central nervous system (CNS). Moreover, A.F. Hill et al. have demonstrated that at initial post-infection stages PrP^{Sc} particles replicate in lymphoid organs (Hill et al. 1997). Within the lymphoreticular system follicular dendritic cells (FDC) that are mostly stored in the spleen are the main area of PrP^{Sc} concentration. FDC elimination by β-lymphotoxin receptor or dedifferentiation prevents prion disease development (Montrasio et al. 2000). Wild mice with *Prnp* expression that were subjected to abdominal injection of the infectious agent remain immune to the disease. The significant role of lymphoreticular system in prion disease development is also supported by the data obtained by M.A. Klein (Klein et al. 1997) who have demonstrated that all immunodeficiencies inhibit prion transmission suppress disease progression.

The exact path of infection from the gastrointestinal tract to CNS is still unclear. According to the data available, the following infection propagation model has been suggested. Once in the gastrointestinal tract, PrP^{Sc} remains, at least partially, resistant to proteolytic enzymes. The mechanism of infectious PrP^{Sc} particles is permeation through intestinal mucosa. An assumption has been made that their transport may involve microfold cells that act as an antigen and a pathogen portal. Upon penetration through the mucosa, prions accumulate in Peyer's patches FDC and in the nerve endings that innervate the digestive apparatus. From Peyer's patches PrP^{Sc} particles arrive into the lymphoreticular organs: the spleen, the lymph nodes, etc., where initial conversion of PrP^C into the infectious PrP^{Sc} isoform and infectious agent accumulation take place (Kitamoto et al. 1991; Montrasio et al. 2000). PrP^{Sc} transport also requires B-lymphocytes that do not replicate infectious particles yet are involved in FDC maturation. Transport within the lymphoreticular system and, probably, towards the CNS is presumably carried out by myeloid dendrite cells. Additionally, Klein et al. (2001) have demonstrated the complement component system (e.g., C1, C3 and C4) to support prion transport efficacy.

The data on extended latent period of prion disease in infected mice with lymphatic system innervation disorders obtained by Weismann laboratory has enhanced further understanding of PrP^{Sc} transport routes from lymphoreticular system cells towards the CNS (Weissmann 1999). At the same time, hyperinnervation of lymphatic organs considerably reduces prion disease latent period. Based on this data, it can be presumed that PrP^{Sc} aggregates are transported from lymphatic organs to the cerebrum directly via the sympathetic nervous system cells (Heikenwalder et al. 2007). So far the following three potential channels for infectious agent transmission are suggested: direct contact between lymphoid organ and peripheral nervous system cells, vesicular transport and interaction of PrP^C protein located on nervous cell surfaces with free-floating oligomers and PrP^{Sc} fibrils (Heikenwalder et al. 2007).

For specific prion diseases and in cases of high infectious agent concentration PrP^{Sc} can spread into the CNS cells either via the parasympathetic nervous system, or by vesicular transport directly into the cerebrum bypassing the lymphoreticular system (Bartz et al. 2005). PrP^{Sc} transport via peripheral nervous system neurons also needs further investigation. Two potential prion transmission models are considered in this regard: "domino" type and "train" type (Heikenwalder et al. 2007). In the "domino" model, infectious particle transmission occurs on the surface of nervous cells, while PrP^{Sc} converts anchored PrP^C on the surface of nervous cells into a pathogenic isoform. The "train" model suggests that inside vesicles or free floating PrP^{Sc} molecules react with nervous cells, and then, once inside the neuron, are transported to the CNS, presumably via microtubules. A similar mechanism may transport prions from neuron bodies to dendrite and axon terminals. This hypothesis is supported by the data on PrP^{Sc} transport velocity within the peripheral nervous system. It ranges from 0.7 mm (Kunzi et al. 2002) to 1–2 mm per day (Kimberlin et al. 1983), which corresponds to slow axon transport velocity (McEwen and Grafstein 1968), whereas PrP^{C} is transported by fast axon transport at the rate of 1 cm/hour approximately (Borchelt et al. 1994).

Despite the prove core role of the peripheral nervous system in PrP^{Sc} transport, the possibility of infectious particle direct transmission to the cerebrum over the hematoencephalic barrier cannot be excluded and depends on the specific properties and concentration of the individual prion strain. Furthermore, PrP^{Sc} transport can also be carried out by non-nervous system cells, e.g. Schwann cells. Once in the cerebrum, PrP^{Sc} molecules facilitate conversion of PrP^{C} monomers produced by nervous cells into a pathogenic isoform (Heikenwalder et al. 2007).

Prion Strains

Existence of multiple scrapie variants had been known long before infectious agent discovery due to the differences in their infectivity, latent period duration and pathological manifestations (Will et al. 1996). Since the characteristics of such variants can be reproduced in model animals in transmission experiments (Morales et al. 2007), they were termed "strains" similarly to viruses (Telling et al. 1996). The entire prion strain variety is defined by the properties of the same amino acid sequence. Presumably, strain variety is related to PrP^{Sc} glycosylation differences and PrP protein conformational flexibility enabling various pathological conformations (Prusiner 1998; Aguzzi et al. 2008a).

The first suggestion regarding various prion conformational statuses was made in relation to studies of Syrian golden hamsters infected with two forms of mink spongiform encephalopathy that differed in clinical manifestations and latent period duration (Bessen and Marsh 1994). After hamster brain homogenate was treated with proteinase K the molecular weight of proteinase K resistant PrP fragment in hamsters infected with different forms of the disease was demonstrated to differ by 2 kDa. This proves different structural organization of their PrP^{Sc} aggregates. Later Caughey's work demonstrated different content and nature of β-sheets structures depending on PrP^{Sc} strain (Caughey et al. 1998).

Molecular weights of proteinase K resistant PrP^{Sc} fragments were used as the basis for human prion disease classification into two major types: type I—with protease resistant core, 21 kDa, type II—19 kDa (Parchi et al. 2000; Wong et al. 2001). Sequencing has demonstrated that proteolysis takes place in the area of amino acid residues 82 and 97 respectively, which also supports the idea of strain-related PrP^{Sc} conformational differences (Parchi et al. 2000). Both strain types have featured stable reproduction in experiments in transgenic mice (Collinge et al. 1996). In addition to PrP^{Sc} structural properties, heterogeneity of human prion disease clinical manifestations also depends on the extent of PrP^{Sc} glycosylation (Sy et al. 2002). Based on a comprehensive analysis of PrP^{Sc} physical and chemical properties and clinical manifestations, human prion diseases have been classified into six sporadic Creutzfeldt–Jakob disease subtypes and a variant CJD originated from a bovine prion strain (Parchi et al. 2012).

In vitro amplified PrP^{Sc} amyloid aggregates from five different mouse prion strains retained their strain-specific properties; and after introducing into wild mice they induced the development of the disease with typical neuropathological and physico-chemical properties of the related parent strains (Sigurdson et al. 2006). Four human prion strains have also been shown to retain their biochemical properties after in vitro amplification (Castilla et al. 2008). Strain properties change neither in cases of blood transfusion related to secondary infection of humans, nor in multiple consecutive blood transfusions in experiments in model mice bearing human *Prnp* (Bishop et al. 2008).

Various prion strains may differ from each other in virulence, latent period duration and clinical manifestations of the disease, including tissue specificity of pathological changes (Prusiner 2001). The experiments conducted by Bartz et al. have shown that PrP^{Sc} structural differences may define both the latent period of the disease and the infection route (Bartz et al. 2005).

Co-existence of different types of sporadic human CJD has been demonstrated (Parchi et al. 2012). Experimental data has shown that in the situations of the same host infected by two strains, one of such strains can suppress the other strain capability to induce the disease (Manuelidis 1998). Bartz et al. suggested that this phenomenon might be related to suppression of prion replication of one strain by the other (Bartz et al. 2007).

Prion Disease Detection

PrP^{Sc} replication and aggregation happens after infection and before clinical manifestation of the disease. In the course of months and years the brain of the patient accumulates a large number of prion aggregates, that leads to the beginning of the neurodegeneration. One of the core areas of prion disease prevention researches is focusing on the development of detection methods for modified PrP conformation. These methods are based on extraordinary protease resistance of PrP^{Sc}: post-proteinase K treatment immunological detection or lack of protease resistant PrP fragment (90–231) reveals if the protein existed in the regular PrP^{C} conformation or in the protease resistant PrP^{Sc} isoform.

Timely human prion disease detection and treatment is very difficult as by the time of clinical manifestation of the disease the brain of the patient has already undergone irreversible changes which leave no time for treatment. With this regard the importance of the development and integration of early stage prion disease detection methods increases. Pathogenic strains are generally identified in experiments in transgenic mice consisting in semiquantitative analysis of vacuolation in nine areas of grey matter of the brain of infected animals and immunochemical detection of the protease resistant PrP fragment (Bueler et al. 1993). Considerable improvement in the analysis was achieved with cultivation of *Tga20* line of mice carrying 60 copies of *Prnp* gene. Due to PrP overproduction the disease development period in the brain of infected animals took 60 ($\pm$ 4) days (Karapetyan et al. 2009).

Using homogenates of the brain of both scrapie infected and uninfected hamsters, the researchers were able to reproduce PrP conversion and aggregation in vitro (Saborio et al. 2001). This method was named PMCA (protein-misfolding cyclic amplification), as it is based on a series of consecutive rounds of PrP^C conversion into the PrP^{Sc} isoform. For successful completion of PMCA reaction a very limited amount of matrix in the PrP^{Sc} conformation and substrate abundance in PrP^C conformation are required. At the first stage of co-incubation only a certain portion of PrP^C joins the aggregates and converts into the PrP^{Sc} isoform. This is followed by partial sonication and reincubation of the aggregates. It has been demonstrated that after five incubation/sonication cycles 98 % of free PrP^C is contained in the aggregates even at very low PrP^{Sc} concentration. The high sensitivity of this method allowed its adaptation for PrP^{Sc} detection in patients' blood (Castilla et al. 2005).

Prion Disease Treatment Approaches

Accumulation of data on prion nature helps to improve prion disease treatment approaches. Currently several core areas exist:

1. PrP^C structural stabilization and prevention of PrP^{Sc} binding.
2. Cell PrP production cutoff.
3. Elimination of cell protein conversion cofactors.
4. Structural rupture of PrP^{Sc} amyloid aggregates.

Systemic search for antiprion substances allowed definition of a range of agents that are able to prevent PrP^{Sc} and PrP^C reaction and, accordingly, inhibit the development of clinical manifestations of the disease (Gilch et al. 2007). Suramine and other substances from the sulfated polyanionic group inhibit the biosynthesis of cell surface proteins and oligomerization of some proteins (Stein et al. 1995). Experiments in animals have demonstrated its capability to increase scrapie latent period in hamsters (Ladogana et al. 1992). Another prion disease treatment strategy is based on PrP antibody resistance to prion conversion both in vitro and in vivo (Buchholz et al. 2006). Immunization is successfully used for prevention of pathogenic conversion of amyloid proteins, e.g. amyloid beta peptide (A_β) that induces Alzheimer's disease development (Feder'off 2009). However, infectious PrP^{Sc} particles have been shown to be unable to generate immune response

(Muller-Schiffmann and Korth 2008). This is related to constant presence of the normal PrP form in the body, including T- and . However, recent researches have demonstrated that multiple PrP epitope antibodies inhibit PrP^{Sc} reproduction in vitro and in vivo (Buchholz et al. 2006). Furthermore, mice vaccination with recombinant PrP either before or immediately after infection and passive immunization with some PrP epitope antibodies inhibited prion replication and slowed down disease progression (White and Hawke 2003). Unfortunately, as yet no evidence of immunization-related cure of prion disease after clinical manifestations is available.

Another area of prion disease treatment is based on the fact that no prion replication takes place in the absence of endogenous PrP^C, as *Prnp* knock-out mice are resistant to this infection (Bueler et al. 1993; Manson et al. 1992). In this relation attempts have been made to use in chemotherapy such substances as brefeldin A that inhibit protein synthesis and development of amyloidosis in infected cell culture (Hay et al. 1987). Extensive improvement opportunities in this area became available with RNA interference process development. Direct suppression of *Prnp* expression has been successfully achieved in experiments in cell cultures and infected mice (Tilly et al. 2003; Daude et al. 2003; Pfeifer et al. 2006). Currently methods for transport and extensive expression of small interference RNA in cerebral neurons are being developed in order to maximize the disease control efficiency (White et al. 2008).

Although suppression of PrP^C production does inhibit prion disease progression, it may induce complications related to the normal functioning of this protein. In this context efforts related to elimination of cell protein conversion cofactors picked up momentum. Prion disease chemotherapy may also include application of such compounds as mevilonin that blocks cholesterol biosynthesis and inhibits transmembrane glycoprotein transport to cell surface (Solassol et al. 2003). Special progress in this area was achieved with discovery of a 37/67 kDa receptor to laminin that is a PrP^C receptor (Gauczynski et al. 2001). Direct interaction between these proteins has been demonstrated by two-hybrid system (Hundt et al. 2001). Although this receptor is not limited to PrP metabolism, it plays an essential role in prion sustaining and development (Leucht et al. 2003). Use of laminin receptor inhibitors, such as heparin-like polysulfated glycans, prevents PrP^{Sc} binding with the receptor, which results in the latent period extension in scrapie infected mice (Farquhar et al. 1999; Adjou et al. 2003). Laminin receptor antibodies also inhibit pathogenic conformation proliferation in infected cells (Leucht et al. 2003) and block infectious particle internalization by human erythrocytes (Morel et al. 2005). However, full antibodies cannot be used in treatment not only due to problems related to overcoming the hematoencephalic barrier, but also due to the complexity of their transport, even in cases of direct injection in the brain (Zuber et al. 2008). In this context production of monochain scFvs antibodies seems to be rather promising. It was used for immunization of scrapie-infected mice, and considerable reduction of the PrP^{Sc} has been demonstrated (Zuber et al. 2008). The low molecular weight of such antibodies allows their application in treatment by immediate injection in the brain (Vana et al. 2007). Inactivation of the receptor to laminin can also be achieved with the aid of RNA interference methods (Ralph et al. 2005; Raoul et al. 2005). Therefore, a lentivirus-based RNAi gene therapy strategy using HIV-derived vectors expressing LRP-specific siRNAs represents an innovative approach in TSE treatment.

All gene therapy approaches to prion disease treatment are based use of viral transgene transport that successfully overcomes the hematoencephalic barrier and transduces terminally differentiated cells that have ceased mitotic dividing, including cerebral neurons (Crozet et al. 2004). New gene transport approaches developed for prion disease treatment are based on application of slow lentiviruses, specifically FIV (feline immunodeficiency virus) that are capable of delivering considerable amounts of genetic information even into a non-proliferating cells. Gene therapy approaches, both autonomously and combined with RNA interference, are capable of inhibiting prion disease progression, yet successful cure of such diseases would require recovery of damaged neurons. Therefore, the future of prion disease treatment lies in the development of integrated solutions for prevention of pathogenic toxic PrP^{Sc} aggregate proliferation and designing a cell therapy focused on damaged cerebral cell recovery (Relano-Gines et al. 2009).

Prion Phenomenon in Other Cases of Amyloidosis

A series of experiments performed in cell cultures and/or laboratory animals has demonstrated that amyloid transmission by mechanisms that are similar to prion is possible for most common types of amyloidoses, including Alzheimer's disease, Parkinson's disease, Huntington's disease and various forms of systemic amyloidosis (Aguilar-Calvo et al. 2015).

Prion-like infectivity may manifest at various levels, i.e. via intermolecular transmission of the modified conformation, and internal and external disease proliferation. This process may involve various cellular routes and components, while the aggregates may accumulate both extra- and intracellularly. Most forms of amyloidosis progress from one infected area into remote brain areas in a rather well structured manner. Such nature of disease development has been shown for Parkinson's disease, Alzheimer's disease, Huntington's disease and frontotemporal dementia (Moreno-Gonzalez and Soto 2011). It can be possibly explained by the fact that the development of these diseases is sustained by the prion-like transmission path. Currently the possibility of α-sinuclein (Kordower et al. 2008) and A_{β} (Braak et al. 2011) aggregates transmission in patient's brain has been verified by a number of experiments. Moreover, the majority of amyloidosis associated proteins with irregular folding have been detected circulating in blood plasma and cerebrospinal fluid. It suggests that there might be a peripheral organ involvement in the replication and proliferation of some aggregation forms of those proteins, similarly to the PrP model. For some cases infectious amyloidosis transmission between individuals has been demonstrated: even administration of a small amount of aggregates in laboratory animal brain led to development of the disease which affected the adjacent areas (Moreno-Gonzalez and Soto 2011).

Prion-lik'e intercellular transmission of amyloidosis proteins seems to be a rather promising model allowing better insight into the patterns and search for specific amyloidosis treatment approaches. However, both the verification and disproof of this model require more advanced and detailed research.

Conclusion

Understanding of prion diseases has advanced dramatically over the past half century. However, many questions are still left unanswered. Which proteins and which cell types are involved in the process of prion transmission and neuroinvasion? The molecular mechanisms of prion replication are not completely defined. The factors responsible for regulation of this process are still unknown. We do not know how we can inhibit this process. We still do not understand how prion variants capable of retaining stable phenotypic traits and we have no notion of the mechanisms that define the tropisms of prion strains. The mechanisms of neurodegeneration induced by the prion agent are not understood. We do not understand why it is less toxic to cells of the immune system, where it also undergoes live replication. Finally, the physiological function of the highly conserved, normal prion protein, PrP^C remains unknown. Clarification of the physiological function of PrP^C has the potential to help understand the mechanisms involved in prion-induced pathogenesis. The ability to answer these questions in the future can provide effective strategies geared toward defining disease etiology and dissecting molecular pathogenesis of more common neurodegenerative disorders such as Parkinson's and Alzheimer's disease. The discovery of proteins with prion-like behavior in yeast and fungi has provided some insight.

Acknowledgments This publication was supported by grant 14-50-00069 from Russian Science Foundation and by grants 14-04-08159, 15-04-06650 from Russian Foundation for Basic Research.

References

Adjou KT, Simoneau S, Sales N, Lamoury F, Dormont D, Papy-Garcia D, Barritault D, Deslys JP, Lasmezas CI (2003) A novel generation of heparan sulfate mimetics for the treatment of prion diseases. J Gen Virol 84(Pt 9):2595–2603

Aguilar-Calvo P, Garcia C, Espinosa JC, Andreoletti O, Torres JM (2015) Prion and prion-like diseases in animals. Virus Res 207:82–93. doi:10.1016/j.virusres.2014.11.026

Aguzzi A, Baumann F, Bremer J (2008a) The prion's elusive reason for being. Annu Rev Neurosci 31:439–477. doi:10.1146/annurev.neuro.31.060407.125620

Aguzzi A, Sigurdson C, Heikenwaelder M (2008b) Molecular mechanisms of prion pathogenesis. Annu Rev Pathol 3:11–40. doi:10.1146/annurev.pathmechdis.3.121806.154326

Aguzzi A, Nuvolone M, Zhu C (2013) The immunobiology of prion diseases. Nat Rev Immunol 13(12):888–902. doi:10.1038/nri3553

Alper T, Cramp WA, Haig DA, Clarke MC (1967) Does the agent of scrapie replicate without nucleic acid? Nature 214(5090):764–766

Bainbridge J, Walker KB (2005) The normal cellular form of prion protein modulates T cell responses. Immunol Lett 96(1):147–150. doi:10.1016/j.imlet.2004.08.006 S0165-2478(04)00228-7 [pii]

Bartz JC, Dejoia C, Tucker T, Kincaid AE, Bessen RA (2005) Extraneural prion neuroinvasion without lymphoreticular system infection. J Virol 79(18):11858–11863. doi:10.1128/JVI.79.18.11858-11863.2005 79/18/11858 [pii]

Bartz JC, Kramer ML, Sheehan MH, Hutter JA, Ayers JI, Bessen RA, Kincaid AE (2007) Prion interference is due to a reduction in strain-specific PrPSc levels. J Virol 81(2):689–697. doi:10.1128/JVI.01751-06 JVI.01751-06 [pii]

Bessen RA, Marsh RF (1994) Distinct PrP properties suggest the molecular basis of strain variation in transmissible mink encephalopathy. J Virol 68(12):7859–7868

Bishop MT, Ritchie DL, Will RG, Ironside JW, Head MW, Thomson V, Bruce M, Manson JC (2008) No major change in vCJD agent strain after secondary transmission via blood transfusion. PLoS ONE 3(8):e2878. doi:10.1371/journal.pone.0002878

Blattler T, Brandner S, Raeber AJ, Klein MA, Voigtlander T, Weissmann C, Aguzzi A (1997) PrP-expressing tissue required for transfer of scrapie infectivity from spleen to brain. Nature 389(6646):69–73. doi:10.1038/37981

Borchelt DR, Koliatsos VE, Guarnieri M, Pardo CA, Sisodia SS, Price DL (1994) Rapid anterograde axonal transport of the cellular prion glycoprotein in the peripheral and central nervous systems. J Biol Chem 269(20):14711–14714

Braak H, Thal DR, Ghebremedhin E, Del Tredici K (2011) Stages of the pathologic process in Alzheimer disease: age categories from 1 to 100 years. J Neuropathol Exp Neurol 70(11):960–969. doi:10.1097/NEN.0b013e318232a379

Bremer J, Heikenwalder M, Haybaeck J, Tiberi C, Krautler NJ, Kurrer MO, Aguzzi A (2009) Repetitive immunization enhances the susceptibility of mice to peripherally administered prions. PLoS ONE 4(9):e7160. doi:10.1371/journal.pone.0007160

Brown DR, Qin K, Herms JW, Madlung A, Manson J, Strome R, Fraser PE, Kruck T, von Bohlen A, Schulz-Schaeffer W, Giese A, Westaway D, Kretzschmar H (1997) The cellular prion protein binds copper in vivo. Nature 390(6661):684–687. doi:10.1038/37783

Brown DR, Nicholas RS, Canevari L (2002) Lack of prion protein expression results in a neuronal phenotype sensitive to stress. J Neurosci Res 67(2):211–224. doi:10.1002/jnr.10118

Buchholz CJ, Bach P, Nikles D, Kalinke U (2006) Prion protein-specific antibodies for therapeutic intervention of transmissible spongiform encephalopathies. Expert Opin Biol Ther 6(3):293–300. doi:10.1517/14712598.6.3.293

Bueler H, Fischer M, Lang Y, Bluethmann H, Lipp HP, DeArmond SJ, Prusiner SB, Aguet M, Weissmann C (1992) Normal development and behaviour of mice lacking the neuronal cell-surface PrP protein. Nature 356(6370):577–582. doi:10.1038/356577a0

Bueler H, Aguzzi A, Sailer A, Greiner RA, Autenried P, Aguet M, Weissmann C (1993) Mice devoid of PrP are resistant to scrapie. Cell 73(7):1339–1347 0092-8674(93)90360-3 [pii]

Burger D, Hartsough GR (1965) Encephalopathy of mink. II. Experimental and natural transmission. J Infect Dis 115(4):393–399

Castilla J, Saa P, Hetz C, Soto C (2005) In vitro generation of infectious scrapie prions. Cell 121(2):195–206. doi:10.1016/j.cell.2005.02.011 S0092-8674(05)00156-X [pii]

Castilla J, Morales R, Saa P, Barria M, Gambetti P, Soto C (2008) Cell-free propagation of prion strains. EMBO J 27(19):2557–2566. doi:10.1038/emboj.2008.181 emboj2008181 [pii]

Caughey B, Baron GS (2006) Prions and their partners in crime. Nature 443(7113):803–810. doi:10.1038/nature05294 nature05294 [pii]

Caughey BW, Dong A, Bhat KS, Ernst D, Hayes SF, Caughey WS (1991) Secondary structure analysis of the scrapie-associated protein PrP 27-30 in water by infrared spectroscopy. Biochemistry 30(31):7672–7680

Caughey B, Raymond GJ, Bessen RA (1998) Strain-dependent differences in beta-sheet conformations of abnormal prion protein. J Biol Chem 273(48):32230–32235

Chesebro B, Trifilo M, Race R, Meade-White K, Teng C, LaCasse R, Raymond L, Favara C, Baron G, Priola S, Caughey B, Masliah E, Oldstone M (2005) Anchorless prion protein results in infectious amyloid disease without clinical scrapie. Science 308(5727):1435–1439. doi:10.1126/science.1110837 308/5727/1435 [pii]

Choi CJ, Kanthasamy A, Anantharam V, Kanthasamy AG (2006) Interaction of metals with prion protein: possible role of divalent cations in the pathogenesis of prion diseases. Neurotoxicology 27(5):777–787. doi:10.1016/j.neuro.2006.06.004 S0161-813X(06)00160-4 [pii]

Collinge J, Sidle KC, Meads J, Ironside J, Hill AF (1996) Molecular analysis of prion strain variation and the aetiology of 'new variant' CJD. Nature 383(6602):685–690. doi:10.1038/383685a0

Crick FH (1958) On protein synthesis. Symp Soc Exp Biol 12:138–163

Crick F (1970) Central dogma of molecular biology. Nature 227(5258):561–563

Crozet C, Lin YL, Mettling C, Mourton-Gilles C, Corbeau P, Lehmann S, Perrier V (2004) Inhibition of PrPSc formation by lentiviral gene transfer of PrP containing dominant negative mutations. J Cell Sci 117(Pt 23):5591–5597. doi:10.1242/jcs.01484 jcs.01484 [pii]

Daude N, Marella M, Chabry J (2003) Specific inhibition of pathological prion protein accumulation by small interfering RNAs. J Cell Sci 116(Pt 13):2775–2779. doi:10.1242/jcs.00494 jcs.00494 [pii]

Deleault NR, Lucassen RW, Supattapone S (2003) RNA molecules stimulate prion protein conversion. Nature 425(6959):717–720. doi:10.1038/nature01979 nature01979 [pii]

Deleault NR, Geoghegan JC, Nishina K, Kascsak R, Williamson RA, Supattapone S (2005) Protease-resistant prion protein amplification reconstituted with partially purified substrates and synthetic polyanions. J Biol Chem 280(29):26873–26879. doi:10.1074/jbc.M503973200 M503973200 [pii]

Dickinson AG, Meikle VM, Fraser H (1968) Identification of a gene which controls the incubation period of some strains of scrapie agent in mice. J Comp Pathol 78(3):293–299 0021-9975(68)90005-4 [pii]

Farquhar C, Dickinson A, Bruce M (1999) Prophylactic potential of pentosan polysulphate in transmissible spongiform encephalopathies. Lancet 353(9147):117. doi:10.1016/S0140-6736(98)05395-1 S0140-6736(98)05395-1 [pii]

Federoff HJ (2009) Development of vaccination approaches for the treatment of neurological diseases. J Comp Neurol 515(1):4–14. doi:10.1002/cne.22034

Gajdusek DC, Zigas V (1957) Degenerative disease of the central nervous system in New Guinea; the endemic occurrence of kuru in the native population. N Engl J Med 257(20):974–978

Gajdusek DC, Gibbs CJ, Alpers M (1966) Experimental transmission of a Kuru-like syndrome to chimpanzees. Nature 209(5025):794–796

Gauczynski S, Peyrin JM, Haik S, Leucht C, Hundt C, Rieger R, Krasemann S, Deslys JP, Dormont D, Lasmezas CI, Weiss S (2001) The 37-kDa/67-kDa laminin receptor acts as the cell-surface receptor for the cellular prion protein. EMBO J 20(21):5863–5875. doi:10.1093/emboj/20.21.5863

Gibbs CJ Jr, Gajdusek DC, Asher DM, Alpers MP, Beck E, Daniel PM, Matthews WB (1968) Creutzfeldt-Jakob disease (spongiform encephalopathy): transmission to the chimpanzee. Science 161(839):388–389

Gilch S, Nunziante M, Ertmer A, Schatzl HM (2007) Strategies for eliminating PrP(c) as substrate for prion conversion and for enhancing PrP(Sc) degradation. Vet Microbiol 123(4):377–386. doi:10.1016/j.vetmic.2007.04.006 S0378-1135(07)00174-5 [pii]

Goldfarb LG, Petersen RB, Tabaton M, Brown P, LeBlanc AC, Montagna P, Cortelli P, Julien J, Vital C, Pendelbury WW et al (1992) Fatal familial insomnia and familial Creutzfeldt-Jakob disease: disease phenotype determined by a DNA polymorphism. Science 258(5083):806–808

Griffith JS (1967) Self-replication and scrapie. Nature 215(5105):1043–1044

Hadlow WJ (1959) Myopathies of livestock. Lab Invest 8:1478–1498

Hajj GN, Lopes MH, Mercadante AF, Veiga SS, da Silveira RB, Santos TG, Ribeiro KC, Juliano MA, Jacchieri SG, Zanata SM, Martins VR (2007) Cellular prion protein interaction with vitronectin supports axonal growth and is compensated by integrins. J Cell Sci 120(Pt 11):1915–1926. doi:10.1242/jcs.03459 jcs.03459 [pii]

Hay B, Prusiner SB, Lingappa VR (1987) Evidence for a secretory form of the cellular prion protein. Biochemistry 26(25):8110–8115

Heikenwalder M, Julius C, Aguzzi A (2007) Prions and peripheral nerves: a deadly rendezvous. J Neurosci Res 85(12):2714–2725. doi:10.1002/jnr.21246

Hill AF, Desbruslais M, Joiner S, Sidle KC, Gowland I, Collinge J, Doey LJ, Lantos P (1997) The same prion strain causes vCJD and BSE. Nature 389(6650):448–450, 526. doi:10.1038/38925

Hundt C, Peyrin JM, Haik S, Gauczynski S, Leucht C, Rieger R, Riley ML, Deslys JP, Dormont D, Lasmezas CI, Weiss S (2001) Identification of interaction domains of the prion protein with its 37-kDa/67-kDa laminin receptor. EMBO J 20(21):5876–5886. doi:10.1093/emboj/20.21.5876

Inge-Vechtomov SG (2003) Template principle in biology (the past, the present, the future?). Ecol Genet 1:6–15

Inge-Vechtomov SG, Galkin AP, Sopova YV, Rubel AA (2012) Prions, "Protein based inheritance" and epigenetics. In: Zakiyan SM, Vlasov VV, Dementieva EV (eds) Epigenetics, vol 21. Publishing House SB RAS, Morskoy pr. 2, Novosibirsk, Novosibirsk region, p 481

Jeffrey M, Wells GA (1988) Spongiform encephalopathy in a nyala (Tragelaphus angasi). Vet Pathol 25(5):398–399

Karapetyan YE, Saa P, Mahal SP, Sferrazza GF, Sherman A, Sales N, Weissmann C, Lasmezas CI (2009) Prion strain discrimination based on rapid in vivo amplification and analysis by the cell panel assay. PLoS ONE 4(5):e5730. doi:10.1371/journal.pone.0005730

Kimberlin RH, Hall SM, Walker CA (1983) Pathogenesis of mouse scrapie. Evidence for direct neural spread of infection to the CNS after injection of sciatic nerve. J Neurol Sci 61(3): 315–325

Kirkwood JK, Wells GA, Wilesmith JW, Cunningham AA, Jackson SI (1990) Spongiform encephalopathy in an arabian oryx (Oryx leucoryx) and a greater kudu (Tragelaphus strepsiceros). Vet Rec 127(17):418–420

Kitamoto T, Muramoto T, Mohri S, Doh-Ura K, Tateishi J (1991) Abnormal isoform of prion protein accumulates in follicular dendritic cells in mice with Creutzfeldt-Jakob disease. J Virol 65(11):6292–6295

Klein MA, Frigg R, Flechsig E, Raeber AJ, Kalinke U, Bluethmann H, Bootz F, Suter M, Zinkernagel RM, Aguzzi A (1997) A crucial role for B cells in neuroinvasive scrapie. Nature 390(6661):687–690. doi:10.1038/37789

Klein MA, Kaeser PS, Schwarz P, Weyd H, Xenarios I, Zinkernagel RM, Carroll MC, Verbeek JS, Botto M, Walport MJ, Molina H, Kalinke U, Acha-Orbea H, Aguzzi A (2001) Complement facilitates early prion pathogenesis. Nat Med 7(4):488–492. doi:10.1038/86567 86567 [pii]

Kocisko DA, Come JH, Priola SA, Chesebro B, Raymond GJ, Lansbury PT, Caughey B (1994) Cell-free formation of protease-resistant prion protein. Nature 370(6489):471–474. doi:10.1038/370471a0

Koltsov NK (1936) Hereditary molecules. In: Organization of cell. The collection experimental studies, of articles and speeches. 1903–1935. M-L Biomedgiz, pp 585–620

Kordower JH, Chu Y, Hauser RA, Freeman TB, Olanow CW (2008) Lewy body-like pathology in long-term embryonic nigral transplants in Parkinson's disease. Nat Med 14(5):504–506. doi:10.1038/nm1747

Kunzi V, Glatzel M, Nakano MY, Greber UF, Van Leuven F, Aguzzi A (2002) Unhampered prion neuroinvasion despite impaired fast axonal transport in transgenic mice overexpressing four-repeat tau. J Neurosci 22(17):7471–7477 22/17/7471 [pii]

Ladogana A, Casaccia P, Ingrosso L, Cibati M, Salvatore M, Xi YG, Masullo C, Pocchiari M (1992) Sulphate polyanions prolong the incubation period of scrapie-infected hamsters. J Gen Virol 73(Pt 3):661–665

Lansbury PT Jr, Caughey B (1995) The chemistry of scrapie infection: implications of the 'ice 9' metaphor. Chem Biol 2(1):1–5 1074-5521(95)90074-8 [pii]
Lauren J, Gimbel DA, Nygaard HB, Gilbert JW, Strittmatter SM (2009) Cellular prion protein mediates impairment of synaptic plasticity by amyloid-beta oligomers. Nature 457 (7233):1128–1132. doi:10.1038/nature07761
Leucht C, Simoneau S, Rey C, Vana K, Rieger R, Lasmezas CI, Weiss S (2003) The 37 kDa/67 kDa laminin receptor is required for PrP(Sc) propagation in scrapie-infected neuronal cells. EMBO Rep 4(3):290–295. doi:10.1038/sj.embor.embor768 embor768 [pii]
Linden R, Martins VR, Prado MA, Cammarota M, Izquierdo I, Brentani RR (2008) Physiology of the prion protein. Physiol Rev 88(2):673–728. doi:10.1152/physrev.00007.2007 88/2/673 [pii]
Ma J, Lindquist S (2001) Wild-type PrP and a mutant associated with prion disease are subject to retrograde transport and proteasome degradation. Proc Natl Acad Sci U S A 98(26):14955–14960. doi:10.1073/pnas.011578098 011578098 [pii]
Makarava N, Kovacs GG, Bocharova O, Savtchenko R, Alexeeva I, Budka H, Rohwer RG, Baskakov IV (2010) Recombinant prion protein induces a new transmissible prion disease in wild-type animals. Acta Neuropathol 119(2):177–187. doi:10.1007/s00401-009-0633-x
Malaga-Trillo E, Solis GP, Schrock Y, Geiss C, Luncz L, Thomanetz V, Stuermer CA (2009) Regulation of embryonic cell adhesion by the prion protein. PLoS Biol 7(3):e55. doi:10.1371/journal.pbio.1000055 08-PLBI-RA-3379 [pii]
Manson J, West JD, Thomson V, McBride P, Kaufman MH, Hope J (1992) The prion protein gene: a role in mouse embryogenesis? Development 115(1):117–122
Manuelidis L (1998) Vaccination with an attenuated Creutzfeldt-Jakob disease strain prevents expression of a virulent agent. Proc Natl Acad Sci U S A 95(5):2520–2525
McEwen BS, Grafstein B (1968) Fast and slow components in axonal transport of protein. J Cell Biol 38(3):494–508
McGowan J (1922) Scrapie in sheep. Scott J Agric 5:365–375
Montrasio F, Frigg R, Glatzel M, Klein MA, Mackay F, Aguzzi A, Weissmann C (2000) Impaired prion replication in spleens of mice lacking functional follicular dendritic cells. Science 288 (5469):1257–1259 8516 [pii]
Morales R, Abid K, Soto C (2007) The prion strain phenomenon: molecular basis and unprecedented features. Biochim Biophys Acta 1772(6):681–691. doi:10.1016/j.bbadis.2006.12.006 S0925-4439(06)00272-9 [pii]
Morel E, Andrieu T, Casagrande F, Gauczynski S, Weiss S, Grassi J, Rousset M, Dormont D, Chambaz J (2005) Bovine prion is endocytosed by human enterocytes via the 37 kDa/67 kDa laminin receptor. Am J Pathol 167(4):1033–1042 167/4/1033 [pii]
Moreno-Gonzalez I, Soto C (2011) Misfolded protein aggregates: mechanisms, structures and potential for disease transmission. Semin Cell Dev Biol 22(5):482–487. doi:10.1016/j.semcdb.2011.04.002
Muller-Schiffmann A, Korth C (2008) Vaccine approaches to prevent and treat prion infection: progress and challenges. BioDrugs 22(1):45–52 2215 [pii]
Oesch B, Westaway D, Walchli M, McKinley MP, Kent SB, Aebersold R, Barry RA, Tempst P, Teplow DB, Hood LE et al (1985) A cellular gene encodes scrapie PrP 27-30 protein. Cell 40 (4):735–746 0092-8674(85)90333-2 [pii]
Pan KM, Baldwin M, Nguyen J, Gasset M, Serban A, Groth D, Mehlhorn I, Huang Z, Fletterick RJ, Cohen FE et al (1993) Conversion of alpha-helices into beta-sheets features in the formation of the scrapie prion proteins. Proc Natl Acad Sci U S A 90(23):10962–10966
Parchi P, Gambetti P (1995) Human prion diseases. Curr Opin Neurol 8(4):286–293
Parchi P, Zou W, Wang W, Brown P, Capellari S, Ghetti B, Kopp N, Schulz-Schaeffer WJ, Kretzschmar HA, Head MW, Ironside JW, Gambetti P, Chen SG (2000) Genetic influence on the structural variations of the abnormal prion protein. Proc Natl Acad Sci U S A 97 (18):10168–10172 97/18/10168 [pii]
Parchi P, de Boni L, Saverioni D, Cohen ML, Ferrer I, Gambetti P, Gelpi E, Giaccone G, Hauw JJ, Hoftberger R, Ironside JW, Jansen C, Kovacs GG, Rozemuller A, Seilhean D, Tagliavini F, Giese A, Kretzschmar HA (2012) Consensus classification of human prion disease histotypes

allows reliable identification of molecular subtypes: an inter-rater study among surveillance centres in Europe and USA. Acta Neuropathol 124(4):517–529. doi:10.1007/s00401-012-1002-8

Peters C, Espinoza MP, Gallegos S, Opazo C, Aguayo LG (2015) Alzheimer's Abeta interacts with cellular prion protein inducing neuronal membrane damage and synaptotoxicity. Neurobiol Aging 36(3):1369–1377. doi:10.1016/j.neurobiolaging.2014.11.019

Pfeifer A, Eigenbrod S, Al-Khadra S, Hofmann A, Mitteregger G, Moser M, Bertsch U, Kretzschmar H (2006) Lentivector-mediated RNAi efficiently suppresses prion protein and prolongs survival of scrapie-infected mice. J Clin Invest 116(12):3204–3210. doi:10.1172/JCI29236

Prcina M, Kontsekova E (2011) Has prion protein important physiological function? Med Hypotheses 76(4):567–569. doi:10.1016/j.mehy.2011.01.002

Prusiner SB (1982) Novel proteinaceous infectious particles cause scrapie. Science 216 (4542):136–144

Prusiner SB (1998) Prions. Proc Natl Acad Sci U S A 95(23):13363–13383

Prusiner SB (2001) Shattuck lecture–neurodegenerative diseases and prions. N Engl J Med 344 (20):1516–1526

Prusiner SB, Scott MR (1997) Genetics of prions. Annu Rev Genet 31:139–175. doi:10.1146/annurev.genet.31.1.139

Prusiner SB, McKinley MP, Bowman KA, Bolton DC, Bendheim PE, Groth DF, Glenner GG (1983) Scrapie prions aggregate to form amyloid-like birefringent rods. Cell 35(2 Pt 1):349–358 0092-8674(83)90168-X [pii]

Ralph GS, Radcliffe PA, Day DM, Carthy JM, Leroux MA, Lee DC, Wong LF, Bilsland LG, Greensmith L, Kingsman SM, Mitrophanous KA, Mazarakis ND, Azzouz M (2005) Silencing mutant SOD1 using RNAi protects against neurodegeneration and extends survival in an ALS model. Nat Med 11(4):429–433. doi:10.1038/nm1205 nm1205 [pii]

Raoul C, Abbas-Terki T, Bensadoun JC, Guillot S, Haase G, Szulc J, Henderson CE, Aebischer P (2005) Lentiviral-mediated silencing of SOD1 through RNA interference retards disease onset and progression in a mouse model of ALS. Nat Med 11(4):423–428. doi:10.1038/nm1207 nm1207 [pii]

Relano-Gines A, Gabelle A, Lehmann S, Milhavet O, Crozet C (2009) Gene and cell therapy for prion diseases. Infect Disord Drug Targets 9(1):58–68

Resenberger UK, Harmeier A, Woerner AC, Goodman JL, Muller V, Krishnan R, Vabulas RM, Kretzschmar HA, Lindquist S, Hartl FU, Multhaup G, Winklhofer KF, Tatzelt J (2011) The cellular prion protein mediates neurotoxic signalling of beta-sheet-rich conformers independent of prion replication. EMBO J 30(10):2057–2070. doi:10.1038/emboj.2011.86

Riek R, Hornemann S, Wider G, Billeter M, Glockshuber R, Wuthrich K (1996) NMR structure of the mouse prion protein domain PrP(121-321). Nature 382(6587):180–182. doi:10.1038/382180a0

Riek R, Hornemann S, Wider G, Glockshuber R, Wuthrich K (1997) NMR characterization of the full-length recombinant murine prion protein, mPrP(23-231). FEBS Lett 413(2):282–288 S0014-5793(97)00920-4 [pii]

Saborio GP, Permanne B, Soto C (2001) Sensitive detection of pathological prion protein by cyclic amplification of protein misfolding. Nature 411(6839):810–813. doi:10.1038/35081095

Santuccione A, Sytnyk V, Leshchyns'ka I, Schachner M (2005) Prion protein recruits its neuronal receptor NCAM to lipid rafts to activate p59fyn and to enhance neurite outgrowth. J Cell Biol 169(2):341–354. doi:10.1083/jcb.200409127 jcb.200409127 [pii]

Sigurdson CJ, Manco G, Schwarz P, Liberski P, Hoover EA, Hornemann S, Polymenidou M, Miller MW, Glatzel M, Aguzzi A (2006) Strain fidelity of chronic wasting disease upon murine adaptation. J Virol 80(24):12303–12311. doi:10.1128/JVI.01120-06 JVI.01120-06 [pii]

Silveira JR, Raymond GJ, Hughson AG, Race RE, Sim VL, Hayes SF, Caughey B (2005) The most infectious prion protein particles. Nature 437(7056):257–261. doi:10.1038/nature03989 nature03989 [pii]

Singh A, Kong Q, Luo X, Petersen RB, Meyerson H, Singh N (2009) Prion protein (PrP) knock-out mice show altered iron metabolism: a functional role for PrP in iron uptake and transport. PLoS ONE 4(7):e6115. doi:10.1371/journal.pone.0006115

Solassol J, Crozet C, Lehmann S (2003) Prion propagation in cultured cells. Br Med Bull 66: 87–97

Stahl N, Borchelt DR, Hsiao K, Prusiner SB (1987) Scrapie prion protein contains a phosphatidylinositol glycolipid. Cell 51(2):229–240 0092-8674(87)90150-4 [pii]

Steele AD, Emsley JG, Ozdinler PH, Lindquist S, Macklis JD (2006) Prion protein (PrPc) positively regulates neural precursor proliferation during developmental and adult mammalian neurogenesis. Proc Natl Acad Sci U S A 103(9):3416–3421. doi:10.1073/pnas.0511290103 0511290103 [pii]

Stein CA, Khan TM, Khaled Z, Tonkinson JL (1995) Cell surface binding and cellular internalization properties of suramin, a novel antineoplastic agent. Clin Cancer Res 1(5):509–517

Stevens DJ, Walter ED, Rodriguez A, Draper D, Davies P, Brown DR, Millhauser GL (2009) Early onset prion disease from octarepeat expansion correlates with copper binding properties. PLoS Pathog 5(4):e1000390. doi:10.1371/journal.ppat.1000390

Stockel J, Safar J, Wallace AC, Cohen FE, Prusiner SB (1998) Prion protein selectively binds copper(II) ions. Biochemistry 37(20):7185–7193. doi:10.1021/bi972827k bi972827k [pii]

Sy MS, Gambetti P, Wong BS (2002) Human prion diseases. Med Clin North Am 86(3):551–571, vi–vii

Telling GC, Parchi P, DeArmond SJ, Cortelli P, Montagna P, Gabizon R, Mastrianni J, Lugaresi E, Gambetti P, Prusiner SB (1996) Evidence for the conformation of the pathologic isoform of the prion protein enciphering and propagating prion diversity. Science 274(5295):2079–2082

Tilly G, Chapuis J, Vilette D, Laude H, Vilotte JL (2003) Efficient and specific down-regulation of prion protein expression by RNAi. Biochem Biophys Res Commun 305(3):548–551 S0006291X03008052 [pii]

Tobler I, Gaus SE, Deboer T, Achermann P, Fischer M, Rulicke T, Moser M, Oesch B, McBride PA, Manson JC (1996) Altered circadian activity rhythms and sleep in mice devoid of prion protein. Nature 380(6575):639–642. doi:10.1038/380639a0

Vana K, Zuber C, Nikles D, Weiss S (2007) Novel aspects of prions, their receptor molecules, and innovative approaches for TSE therapy. Cell Mol Neurobiol 27(1):107–128. doi:10.1007/s10571-006-9121-1

Warner RG, Hundt C, Weiss S, Turnbull JE (2002) Identification of the heparan sulfate binding sites in the cellular prion protein. J Biol Chem 277(21):18421–18430. doi:10.1074/jbc.M110406200 M110406200 [pii]

Weissmann C (1999) Molecular genetics of transmissible spongiform encephalopathies. J Biol Chem 274(1):3–6

Wells GA, Scott AC, Johnson CT, Gunning RF, Hancock RD, Jeffrey M, Dawson M, Bradley R (1987) A novel progressive spongiform encephalopathy in cattle. Vet Rec 121(18):419–420

Whatley SA, Powell JF, Politopoulou G, Campbell IC, Brammer MJ, Percy NS (1995) Regulation of intracellular free calcium levels by the cellular prion protein. NeuroReport 6(17):2333–2337

White AR, Hawke SH (2003) Immunotherapy as a therapeutic treatment for neurodegenerative disorders. J Neurochem 87(4):801–808 2064 [pii]

White MD, Farmer M, Mirabile I, Brandner S, Collinge J, Mallucci GR (2008) Single treatment with RNAi against prion protein rescues early neuronal dysfunction and prolongs survival in mice with prion disease. Proc Natl Acad Sci U S A 105(29):10238–10243. doi:10.1073/pnas.0802759105 0802759105 [pii]

Wickner RB, Edskes HK, Shewmaker F, Nakayashiki T (2007) Prions of fungi: inherited structures and biological roles. Nat Rev Microbiol 5(8):611–618. doi:10.1038/nrmicro1708 nrmicro1708 [pii]

Will RG, Ironside JW, Zeidler M, Cousens SN, Estibeiro K, Alperovitch A, Poser S, Pocchiari M, Hofman A, Smith PG (1996) A new variant of Creutzfeldt-Jakob disease in the UK. Lancet 347 (9006):921–925

Williams ES, Young S (1980) Chronic wasting disease of captive mule deer: a spongiform encephalopathy. J Wildl Dis 16(1):89–98
Wong BS, Chen SG, Colucci M, Xie Z, Pan T, Liu T, Li R, Gambetti P, Sy MS, Brown DR (2001) Aberrant metal binding by prion protein in human prion disease. J Neurochem 78(6):1400–1408
Wyatt JM (1990) Spongiform encephalopathy in a cat. Vet Rec 126(20):513
Zhang CC, Steele AD, Lindquist S, Lodish HF (2006) Prion protein is expressed on long-term repopulating hematopoietic stem cells and is important for their self-renewal. Proc Natl Acad Sci U S A 103(7):2184–2189. doi:10.1073/pnas.0510577103 0510577103 [pii]
Zomosa-Signoret V, Arnaud JD, Fontes P, Alvarez-Martinez MT, Liautard JP (2008) Physiological role of the cellular prion protein. Vet Res 39(4):9. doi:10.1051/vetres:2007048 v07242 [pii]
Zou WQ, Zheng J, Gray DM, Gambetti P, Chen SG (2004) Antibody to DNA detects scrapie but not normal prion protein. Proc Natl Acad Sci U S A 101(5):1380–1385. doi:10.1073/pnas.0307825100 0307825100 [pii]
Zuber C, Mitteregger G, Schuhmann N, Rey C, Knackmuss S, Rupprecht W, Reusch U, Pace C, Little M, Kretzschmar HA, Hallek M, Buning H, Weiss S (2008) Delivery of single-chain antibodies (scFvs) directed against the 37/67 kDa laminin receptor into mice via recombinant adeno-associated viral vectors for prion disease gene therapy. J Gen Virol 89(Pt 8):2055–2061. doi:10.1099/vir.0.83670-0 89/8/2055 [pii]

Broadening the Genetic Diversity of Bread Wheat Using Alien Germplasm: Emphasis on Disease Resistance

Vladimir Shumny, Elena Khlestkina, Irina Leonova and Elena Salina

Introduction

Broadening the genetic diversity for economically important traits is one of the main tasks of wheat genetics and breeding. There are many evidences of low allelic variability or loss of alleles related with yield, grain quality, and resistance to biotic and abiotic stresses among modern bread wheat varieties (Fu and Somers 2009; Gultyaeva 2012; Novoselskaya-Dragovich et al. 2015). The reduction in genetic diversity was accelerated in the second half of the 20th century, when a replacement of local wheat germplasm on high-yielding commercial cultivars took place. This has been demonstrated by analysis of pedigrees, phenotypic and molecular genetic assessments of plant material (Dreisigacker et al. 2004; Khlestkina et al. 2004).

Fungal diseases are among the major constraints that affect plant growth and yield of bread wheat (*Triticum aestivum* L.). Many efforts are focused on transfer of disease resistance genes from various cereal species of Triticeae tribe to the bread wheat genome.

T. aestivum belongs to one of the four *Triticum* genus sections—the *Triticum* section consisting of hexaploid wheat species with genome BA^uD. Further section (*Dicoccoides*) is represented by tetraploid emmer wheats with genome BA^u, and includes tetraploid progenitor of bread wheat, which hybrized with *Aegilops tauschii* Coss (genome D), resulting in formation of allohexaploid wheat (BA^uD) about 10,000 years ago. Other *Triticum* sections are *Monococcum* (diploid einkorn wheat with genomes A^u and A^b) and *Timopheevi* (tetraploid and hexaploid wheat species with genomes GA^u and GA^uA^b, respectively) (Goncharov 2002). Both the B and the G genomes are related to the S genome present in *Ae. speltoides* (Kimber 1974).

V. Shumny (✉) · E. Khlestkina · I. Leonova · E. Salina
Institute of Cytology and Genetics SBRAS, Lavrentjeva Ave. 10,
Novosibirsk 630090, Russia
e-mail: shumny@bionet.nsc.ru

V.L. Korogodina et al. (eds.), *Genetics, Evolution and Radiation*,
DOI 10.1007/978-3-319-48838-7_9

High similarity between *Ae. tauschii* and *T. aestivum* D-genome and the close evolutionary relationship between the B, S and G genomes allows for the relatively straightforward transfer of genetic information from *Ae. tauschii*, *Ae. speltoides* and *T. timopheevii* to bread wheat. Further Triticeae species (from *Aegilops*, *Secale* and *Hordeum* genera) also represent an important genetic reservoir for improvement of bread wheat (Friebe et al. 1996; Schneider et al. 2008; Trubacheeva et al. 2008; Pershina et al. 2009; Molnar-Lang et al. 2014). A wide range of wheat-alien lines has been developed over the past decades (Friebe et al. 1996, 2001; Wheat Genetics Resource Center). Synthetic wheats or lines with translocations and chromosome substitutions are often used as intermediate forms for gene transfer to varieties. The development of effective and usable genetic markers has made the process of heterologous gene transfer within the Triticeae tribe more efficient (Marais et al. 2001; Adonina et al. 2012; Chen et al. 2013; Timonova et al. 2013).

A number of resistance genes to leaf, stem, and yellow rusts and powdery mildew has been transferred into the genome of common wheat from wild and cultivated wheat relatives. Successful examples include the transfer of the *Lr9* and *Lr19* genes from *Ae. umbellulata* and *Th. ponticum* species (McIntosh et al. 1995). Until recently these genes remained effective against a broad set of leaf rust races (*Puccinia triticina* Eriks.) worldwide.

However, only a small number of the resistance genes described to date are used in breeding for development of resistant wheat varieties. Mainly this is due to the negative effects of foreign genetic material on agronomic traits.

In addition to wide application of alien germplasm for improvement of bread wheat resistance to fungal diseases, genes of various Triticeae species can be useful for increasing bread wheat tolerance to unfavorable environment conditions (reviewed in Colmer et al. 2006; Nevo and Chen 2010; Inbart-Pompan et al. 2013; recent examples are given in Yudina et al. 2015a, b), improvement of bread making quality (*Aegilops* ssp. genes: Kunert et al. 2007; Rehman et al. 2008; Garg et al. 2009; Sin et al. 2011; Wang et al. 2012, 2013; Zhou et al. 2014; *T. dicoccoides* Kunert et al. 2007) and nutrition properties: various *Triticum* ssp. and *Aegilops* ssp. genes can be used for increased production of iron and zink (Rawat et al. 2009) or antioxidant compounds (Khlestkina et al. 2010; Tereshchenko et al. 2012a, b).

Triticum Species for Bread Wheat Improvement

Hexaploid Wheats T. Spelta *and* T. Macha *(BA^{u}D)*

Hexaploid wheat *T. spelta* contributed to bread wheat resistance against septoria tritici blotch via substitution of chromosomes 2D, 5A, 5D, 6B, 6D and 7D (Simon et al. 2005). Two regions of chromosome 7D from spelt wheat were associated with isolate-specific resistance expressed at the seedling (locus *QStb.ipk-7D1)* and another at the adult (*QStb.ipk-7D2*) plant stage (Simon et al. 2010). The locus *QStb.ipk-7D2* was found on the short arm of chromosome 7D in a similar position to the locus *Lr34*/*Yr18* known to provide durable resistance against multiple pathogen races (Spielmeyer et al. 2008). Substituted chromosome 6B of hexaploid wheat *T. macha* is a carrier of a major determinant of head blight resistance, chromosomes 2A, 2B, 3A, 4A, 5A and 5B have also positive effect on resistance to these pathogens (Grausgruber et al. 1998; Steed et al. 2005; Buerstmayr et al. 2011; Burt et al. 2015).

Tetraploid Wheats T. Dicoccoides and T. Durum (BA^u)

Powdery mildew resistance genes were transferred from *T. dicoccoides* to bread wheat chromosomes 6AL (Xie et al. 2003) and 5BL (*PmAS846*; Xue et al. 2012). Other genes conferring resistance to this disease were found in *T. dicoccoides* chromosomes 5BL (*Pm36*; Blanco et al. 2008), 3BL (Li et al. 2009), 2BL (Maxwell et al. 2010) and 2BS (Hua et al. 2009; Liu et al. 2012).

T. dicoccoides leaf rust (*Lr53*) and stripe rust (*Yr35*) resistance genes were transferred to bread wheat chromosome 6BS (Marais et al. 2005). Other genes for stripe rust resistance were found in *T. dicoccoides* (*YrH52* and *Yr15*, Peng et al. 2000; *Yr36*, Uauy et al. 2005).

This species is also a potential donor of Fusarium head blight resistance genes (Hartel et al. 2004; Buerstmayr et al. 2013).

T. durum is a perspective donor of stem rust resistance gene *Sr13* (Simons et al. 2011). The role of *Sr13* and the role of chromosome regions putatively harboring *Sr9*, *Sr14*, *Sr17* and *Sr28* was shown for resistance to highly virulent stem rust race Ug99 (Letta et al. 2013).

T. durum resistance genes are often transferred to bread wheat via synthetic hexaploid forms obtained by crossing *T. durum* with *Ae. tauschii*. Genome-wide association study of 181 synthetic hexaploid wheats revealed a number of stripe rust resistance loci, including that on chromosomes of the A and B genomes of *T. durum*: 1AS, 1BS, 2AS, 2BL, 3BL, 5A, 5BL, and 7AL (Zegeye et al. 2014).

Molecular analysis of hybrid lines derived from crossing of Belarusian wheat varieties with *T. durum* suggested contribution of chromosomes 4B and 5B in the formation of the complex resistance to leaf rust, powdery mildew and septoriosis (Leonova et al. 2013).

Tetraploid Wheat **T. Timopheevii** *(GA^u)*

T. timopheevii was a donor of two stem rust resistance gene: *Sr36* and *Sr37* (McIntosh et al. 2013). The gene *Sr36* (formerly *SrTt1*) was used in breeding programs for more than 40 years. *Sr36* refers to a group of the genes, providing effective resistance to the most aggressive race of stem rust, Ug99 (Yu et al. 2014). For *Sr36* gene protocols for polymerase chain reaction were developed and diagnostic markers were designed for application in marker-assisted selection (MAS) programs.

The gene *Sr37* (*SrTt2*) despite its effectiveness against Ug99 was not used in breeding to improve the resistance of wheat varieties because of the significant negative effects of T4BL-4GL translocation on the productivity traits (McIntosh et al. 1995).

Two genes determining resistance to powdery mildew, *Pm6* and *Pm27*, were transferred from *T. timopheevii* ssp. *timopheevii* into chromosomes 2B and 6B respectively (Jørgensen and Jensen 1973; Järve et al. 2000). The gene *Pm6* was widely used for creation of resistant wheat varieties and is still one of the most effective genes in many regions of the world, including Europe, China and North America (Alam et al. 2011).

Three resistance genes with different efficiency to leaf rust (*Lr18*, *LrTt1*, and *LrTt2*), and two quantitative trait loci (*QLr.icg-1A, QLr.icg-2B*) were described in the genome of different subspecies of *T. timopheevii* (Yamamori 1994; Leonova et al. 2007, 2011). The gene *Lr18*, originating from *T. timopheevii* ssp. *timopheevii* has lost its efficacy against leaf rust races worldwide and is not currently used in breeding. The presence of the genes *LrTt1* and *LrTt2* (*T. timopheevii* var. *viticulosum*) and locus *QLr.icg-2B* (unknown subspecies of *T. timopheevii*) in wheat genotypes provides resistance to the wide range of leaf rust races in West Siberian regions of Russia (Leonova et al. 2007; Timonova et al. 2013). These loci have been used for creation of breeding lines and wheat cultivar "Pamyati Maistrenko" (Laikova et al. 2013).

Additionally, three genes determining resistance to leaf rust (*Lr50*), stem rust (*Sr40*) and powdery mildew (*Pm37*) were identified in the genome of wild subspecies *T. timopheevii* ssp. *armeniacum* (Brown-Guedira et al. 2003; Perugini et al. 2008; Wu et al. 2009).

Besides the aforementioned genes, there were evidences that wheat breeding lines with translocations from *T. timopheevii* contain genetic loci responsible for resistance to spot blotch, fusarium and septoriosis (Ma and Hughes 1995; Christopher et al. 2007; Srinivasachary et al. 2008).

Diploid Wheats T. Monococcum (A^m) and T. Boeoticum (A^b)

Stem rust resistance genes *Sr21*, *Sr22*, and *Sr35* were transferred from diploid wheat *T. monococcum* to bread wheat (Kerber and Dyck 1973; Rouse and Jin 2011). *Sr21* (Chen et al. 2015) and a number of other *T. monococcum* stem rust resistance genes different from *Sr21*, *Sr22*, and *Sr35* (2 genes, Rouse and Jin 2011; gene *SrTm4*, Briggs et al. 2015) confer resistance to highly virulent stem rust race Ug99.

Two powdery mildew resistance genes (*PmTb7A.1* and *PmTb7A.2*) were identified and mapped on chromosome 7AL of another diploid wheat species, *T. boeoticum* (Chhuneja et al. 2012, 2015).

Aegilops Species for Wheat Improvement

Ae. Speltoides

Ae. speltoides, which is characterized by a high grain protein content along with fungal diseases resistance, was a donor of the leaf rust resistance genes *Lr28, Lr35, Lr36, Lr47*, and *Lr51* (reviewed in Schneider et al. 2008; Todorovska et al. 2009). Among recent examples, genes encoding leaf rust resistance have been transferred from *Ae. speltoides* into *T. aestivum* (Adonina et al. 2012). *Lr28*, *Lr36* and *Lr47* genes were shown to be highly effective against leaf rust populations in most regions of the world, although, with the exception of the *Lr28*, remaining genes were not widely used in breeding practice for development of resistant wheat cultivars due to negative effects on agronomic traits.

Ae. speltoides is also a source of *Sr32, Sr39*, *Sr47*, and *SrAes7t* genes providing resistance of common and durum wheats to stem rust including race Ug99 (McIntosh et al. 1995; Faris et al. 2008; Klindworth et al. 2012; Yu et al. 2014).

Two *Ae. speltoides* powdery mildew resistance genes (*Pm12* and *Pm32*) were introgressed into common wheat chromosomes 6BL and 1BL respectively (McIntosh et al. 1995; Hsam et al. 2003). *Pm12* is very effective against populations of powdery mildew pathogen worldwide. But this gene is not widely used in wheat breeding because of negative influence of alien chromatin on wheat productivity (Song et al. 2007).

Recently gene *Pm53* was located on the long arm of chromosome 5BL of winter wheat germplasm line and appears to be a new source of powdery mildew resistance that can be tracked with molecular markers in MAS schemes (Petersen et al. 2015).

Ae. Tauschii

Ae. tauschii, the D genome donor of bread wheat has been used extensively for the transfer of agronomic important traits to wheat, including leaf rust resistance genes *Lr41*, *Lr42*, and *Lr43* (Cox et al. 1994) as well as stem rust resistance genes *Sr33* (1DS), *Sr45* (1DS), and *Sr46* (2DS) conferring resistance to highly virulent race Ug99 (Rouse and Jin 2011; Yu et al. 2015). It was shown later that *Lr40* is the same gene as *Lr21*, while *Lr41* is a synonym of *Lr39*. *Lr43* is not a unique gene because initial germplasm lines consist of the progenies with gene combination (*Lr21* and *Lr39*) (McIntosh et al. 2013).

Other two Ug99-effective stem rust resistance genes were recently transferred to bread wheat from chromosomes 6DS (*SrTA10187*) and 7DS (*SrTA10171*) of *Ae. tauschii* (Olson et al. 2013).

Ae. tauschii is often used for introgression of resistance genes into the genome of bread wheat through the development of synthetic hexaploid forms. Association mapping with the help of a set of synthetic hexaploid wheats revealed 2 novel stripe rust resistance loci on chromosomes 3DL and 6DS originating from *Ae. tauschii* (Zegeye et al. 2014). Stripe rust resistance genes were revealed on *Ae. tauschii* chromosomes 3DS (*YrY206*, Zhang et al. 2009) and 4DS (*YrAS2388*, Huang et al. 2011). Potential of this species for improvement of stripe rust resistance was also shown by Liu et al. (2013).

Miranda et al. (2006, 2007) identified *Ae. tauschii*-derived wheat powdery mildew resistance genes *Pm34* and *Pm35* on wheat chromosome 5D. Sun et al. (2006) mapped 2 genes, *PmY201* and *PmY212*, on chromosome 5DL of *Ae. tauschii*. Maxwell et al. (2012) characterized a novel *Ae. tauschii*-derived gene (*MlNCD1*) conferring resistance to powdery mildew, mapped to the short arm of chromosome 7D in more distal position compared to the previously known *Pm38* gene.

Other Aegilops Species

Germplasm of the other *Aegilops* species (not related with hexaploid wheat origin) has been exploited to a limited extent for a search and transfer of the genes determining wheat resistance to fungal diseases. This is primarily due to methodological difficulties of introduction of foreign genetic material into the genome of common wheat by direct hybridization. Nevertheless, a number of genes for resistance to rust diseases was identified in the genomes of *Ae. ventricosa* (*Lr37, Yr17, Sr38*), *Ae. geniculata* (*Lr57, Yr40, Sr53*), *Ae. triuncialis* (*Lr58*), and *Ae. peregrina* (*Lr59*) (Seah et al. 2001; Kuraparthy et al. 2007a, b; Marais et al. 2008)

For translocation T5DL•5DS-5MgS from *Ae. geniculata*, containing leaf and stripe rust resistance genes (*Lr57, Yr40*) the absence of negative effects on endosperm texture of winter wheat cultivars has been proven (Kuraparthy et al. 2009). Positive effects on bread-making quality was detected for translocation containing cluster *Lr37-Yr17-Sr38* genes from *Ae. ventricosa* (Labuschagne et al. 2002).

A number of disease resistance genes have been transferred from other *Aegilops* species: *Ae. umbellulata* (*LrU1, LrU2, YrU1*), *Ae. caudata* (*LrAC*), *Ae. variabilis* (unknown gene), and *Ae. variabilis* (*LrV, YrV*) (Chhuneja et al. 2008; Riar Kaur et al. 2012; Spetsov et al. 2013). Apparent linkage drag was observed neither in the introgression lines obtained on the basis of *Ae. umbellulata* and *Ae. variabilis*, nor in their backcross progenies, suggesting that new genetic loci could be exploited commercially.

Rye as a Source of Disease Resistance Genes for Bread Wheat

Among the successful introduction of rye (*Secale cereale* L.) genetic material into the genome of common wheat only two translocations have practical value, and in both cases chromosome 1RS took part in these translocations.

The translocated 1RS chromosome arm of the rye cultivar 'Kavkaz' frequently occurs in wheat cultivars of European selection (Villareal et al. 1998; Purnhauser et al. 2011). It confers disease resistance against powdery mildew (*Pm8*), stem rust (*Sr31*), leaf rust (*Lr26*) and stripe rust (*Yr9*) (Lukaszewski 2000). Cultivars with this so-called 1BL.1RS translocation (Mettin et al. 1973; Zeller 1973) showed high yield potential and multiple disease resistance (Rabinovich 1998; Kim et al. 2004; Belan et al. 2010).

Translocation 1AL.1RS, obtained on the basis of rye cultivar 'Amigo' contains genes for resistance to wheat aphid (*Gb2*), powdery mildew (*Pm17*) and stem rust (unknown gene). This translocation is spread mainly in the American wheat varieties. Translocation 1AL.1RS was also found to have positive effects on drought tolerance, grain protein content and yield (McIntosh et al. 1995; Kim et al. 2004).

Other wheat-rye translocations, such as T3AS-3RS with *Sr27* gene, T4BS.4BL-5RL with *Lr25*, T6BS-6RL with *Pm20*, T2AS-2RS.2RL with *Lr45*, and T2BS-2RL containing the locus of resistance to Hessian fly are not widely used in breeding due to their negative effects on agronomic traits.

Conclusions and Prospects

Wild and cultivated wheat relatives and species from related genera are an inexhaustible source of genes providing resistance to diseases and insects. To date, more than 50 % of known genes determining resistance to rust diseases and powdery mildew originate from wheat relatives (McIntosh et al. 2013). However, only a limited number of genes were utilized in practical breeding to improve genetic basis of bread wheat resistance. Others were used only for development of introgression and isogenic lines.

One of the main reasons for limited practical application of the resistance genes is the presence of alien genetic material tightly linked with the target gene and negatively influencing agronomic traits. For example gene *Lr19* closely linked to the gene *Y*, determining the yellow color of flour is not used for development of resistance of wheat cultivars in those countries which traditionally prefer the flour of white color.

In recent years, the development of new molecular methods for tagging of disease resistance genes and technology of marker-assisted selection made it possible to eliminate excessive donor material, which may have negative effect on agronomic traits. In this way, for example, bread wheat lines with shorter introgression fragments containing *Sr37* and *Sr39* genes with effective resistance to stem rust were obtained (Yu et al. 2010; Zhang et al. 2012). The study demonstrated that the lines with reduced alien chromatin have no negative effects on productivity traits anymore.

Further difficulties can be related with insufficient level of alien gene expression due to divergence between bread wheat *cis*- or *trans*-regulatory elements and those of the donor species (Khlestkina et al. 2009; Shoeva et al. 2016) or suppression of alien genes by their orthologs in wheat (Hurni et al. 2014).

The recently developed approach for precise gene editing, CRISPR/Cas9 system (reviewed in Bortesi and Fischer 2015), has been already used successfully for editing resistance genes in wheat (Wang et al. 2014). This system allows making fine changes in plant genome and may help to overcome the difficulties mentioned above by editing regulatory sequences or direct insertion of desirable gene sequences without hybridization.

References

Adonina IG, Petrash NV, Timonova EM, Khristov YA, Salina EA (2012) Construction and study of leaf rust-resistant common wheat lines with translocations of *Aegilops speltoides* Tausch. genetic material. Russ J Genet 48:404–409

Alam MA, Xue F, Wang C, Ji W (2011) Powdery mildew resistance genes in wheat: identification and genetic analysis. J Mol Biol Res 1:20–39

Blanco A, Gadaleta A, Cenci A, Carluccio AV, Abdelbacki AM, Simeone R (2008) Molecular mapping of the novel powdery mildew resistance gene *Pm36* introgressed from *Triticum turgidum* var. *dicoccoides* in durum wheat. Theor Appl Genet 117:135–242

Belan IA, Rosseeva LP, Trubacheeva NV, Osadchaya TS, Dorogina OV, Zhmud EV, Kolmakov YV, Blokhina NP, Kravtsova LA, Pershina LA (2010) Some agronomic important features of spring wheat cultivar Omskaya 37 lines containing wheat-rye translocation 1RS.1BL. Vavilov J Genet Breed 14:632–640 (Russian)

Bortesi L, Fischer R (2015) The CRISPR/Cas9 system for plant genome editing and beyond. Biotechnol Adv 33:41–52

Brown-Guedira GL, Singh S, Fritz AK (2003) Performance and mapping of leaf rust resistance transferred to wheat from *Triticum timopheevii* subsp. *armeniacum*. Phytopathology 93:784–789

Briggs J, Chen S, Zhang W, Nelson S, Dubcovsky J, Rouse MN (2015) Mapping of *SrTm4*, a recessive stem rust resistance gene from diploid wheat effective to Ug99. Phytopathology 105:1347–1354

Buerstmayr M, Lemmens M, Steiner B, Buerstmayr H (2011) Advanced backcross QTL mapping of resistance to Fusarium head blight and plant morphological traits in a *Triticum macha* × *T. aestivum* population. Theor Appl Genet 123:293–306

Buerstmayr M, Alimari A, Steiner B, Buerstmayr H (2013) Genetic mapping of QTL for resistance to Fusarium head blight spread (type 2 resistance) in a *Triticum dicoccoides* × *Triticum durum* backcross-derived population. Theor Appl Genet 126:2825–2834

Burt C, Steed A, Gosman N, Lemmens M, Bird N, Ramirez-Gonzalez R, Holdgate S, Nicholson P (2015) Mapping a Type 1 FHB resistance on chromosome 4AS of *Triticum macha* and deployment in combination with two Type 2 resistances. Theor Appl Genet 128:1725–1738

Chen P, You C, Hu Y, Chen S, Zhou B, Cao A, Wang X (2013) Radiation-induced translocations with reduced *Haynaldia villosa* chromatin at the *Pm21* locus for powdery mildew resistance in wheat. Mol Breed 31:477–484

Chen S, Rouse MN, Zhang W, Jin Y, Akhunov E, Wei Y, Dubcovsky J (2015) Fine mapping and characterization of *Sr21*, a temperature-sensitive diploid wheat resistance gene effective against the *Puccinia graminis* f. sp. *tritici* Ug99 race group. Theor Appl Genet 128:645–656

Chhuneja P, Kaur S, Goel RK, Aghaee-Sarbarzeh M, Prashar M, Dhaliwal HS (2008) Transfer of leaf rust and stripe rust resistance from *Aegilops umbellulata* Zhuk. to bread wheat (*Triticum aestivum* L.). Genet Resour Crop Evol 55:849–859

Chhuneja P, Kumar K, Stirnweis D, Hurni S, Keller B, Dhaliwal HS, Singh K (2012) Identification and mapping of two powdery mildew resistance genes in *Triticum boeoticum* L. Theor Appl Genet 124:1051–1058

Chhuneja P, Yadav B, Stirnweis D, Hurni S, Kaur S, Elkot AF, Keller B, Wicker T, Sehgal S, Gill BS, Singh K (2015) Fine mapping of powdery mildew resistance genes *PmTb7A.1* and *PmTb7A.2* in *Triticum boeoticum* (Boiss.) using the shotgun sequence assembly of chromosome 7AL. Theor Appl Genet 128:2099–2111

Christopher MJ, Williamson PM, Michalowitz M, Jennings R, Lehmensiek A, Sheppard J, Banks P (2007) Simple sequence repeat markers associated with three quantitative trait loci for black point resistance can be used to enrich selection populations in bread wheat. Austr J Agricult Res 58:867–873

Colmer TD, Flowers TJ, Munns R (2006) Use of wild relatives to improve salt tolerance in wheat. J Exp Bot 57:1059–1078

Cox TS, Raupp WJ, Gill BS (1994) Leaf rust-resistance genes *Lr41*, *Lr42*, and *Lr43* transferred from *Triticum tauschii* to common wheat. Crop Sci 34:339–343

Dreisigacker S, Zhang P, Warburton ML, van Ginkel M, Hoisington D, Bohnband M, Melchinger AE (2004) SSR and pedigree analyses of genetic diversity among CIMMYT wheat lines targeted to different megaenvironments. Crop Sci 44:381–388

Faris JD, Xu SS, Cai X, Friesen TL, Jin Y (2008) Molecular and cytogenetic characterization of a durum wheat-*Aegilops speltoides* chromosome translocation conferring resistance to stem rust. Chromosome Res 16:1097–1105

Friebe B, Jiang J, Raupp WJ, McIntosh RA, Gill BS (1996) Characterization of wheat-alien translocations conferring resistance to diseases and pests: current status. Euphytica 91:59–87

Friebe B, Raupp WJ, Gill BS (2001) Alien genes in wheat improvement. In: Bedo Z, Lang L (eds) Wheat in a global environment. Kluver Academic Publisher, Dordrecht, pp 709–720

Fu YB, Somers DJ (2009) Genome wide reduction of genetic diversity in wheat breeding. Crop Sci 49:161–168

Garg M, Tanaka H, Ishikawa N, Takata K, Yanaka M, Tsujimoto H (2009) A novel pair of HMW glutenin subunits from *Aegilops searsii* improves quality of hexaploid wheat. Cereal Chem 86:26–32

Goncharov NP (2002) Comparative genetics of wheats and their related species. Siberian un-ty press, Novosibirsk (Russian)

Grausgruber H, Lemmens M, Bürstmayr H, Ruckenbauer P (1998) Chromosomal location of Fusarium head blight resistance and in vitro toxin tolerance in wheat using the Hobbit 'sib' (*Triticum macha*) chromosome substitution lines. J Genet Breed 52:173–180

Gultyaeva EI (2012) Genetic diversity of Russian wheat cultivars for resistance to leaf rust. Rep Russ Acad Agric Sci 2:29–32 (Russian)

Hartel KD, Berzonsky WA, Kianian SF, Ali S (2004) Expression of a *Triticum turgidum* var. *dicoccoides* source of Fusarium head blight resistance transferred to synthetic hexaploid wheat. Plant Breed 123:516–519

Hsam SLK, Lapochkina IF, Zeller FJ (2003) Chromosomal location of genes for resistance to powdery mildew in common wheat (*Triticum aestivum* L. em Thell.). 8. Gene *Pm32* in a wheat *Aegilops speltoides* translocation line. Euphytica 133:367–370

Hua W, Liu Z, Zhu J, Xie C, Yang T, Zhou Y, Duan X, Sun Q, Liu Z (2009) Identification and genetic mapping of *pm42*, a new recessive wheat powdery mildew resistance gene derived from wild emmer (*Triticum turgidum* var. *dicoccoides*). Theor Appl Genet 119:223–230

Huang L, Zhang L-Q, Liu B-L, Yan Z-H, Zhang B, Zhang H-G, Zheng Y-L, Liu D-C (2011) Molecular tagging of a stripe rust resistance gene in *Aegilops tauschii*. Euphytica 179:313–318

Hurni S, Brunner S, Stirnweis D, Herren G, Peditto D, McIntosh RA, Keller B (2014) The powdery mildew resistance gene *Pm8* derived from rye is suppressed by its wheat ortholog *Pm3*. Plant J 79:904–913

Inbart-Pompan H, Eilam T, Eshe A (2013) Searching for salt tolerance among wild relatives of wheat: what should we look for? Aust J Crop Sci 7:2116–2127

Järve K, Peusha HO, Tsymbalova J, Tamm S, Devos KM, Enno TM (2000) Chromosomal location of *Triticum timopheevii* derived powdery mildew resistance genes transferred to common wheat. Genome 43:377–381

Jørgensen JH, Jensen CJ (1973) Gene *Pm6* for resistance to powdery mildew in wheat. Euphytica 22:4–23

Kerber ER, Dyck PL (1973) Inheritance of stem rust resistance transferred from diploid wheat (*Triticum monococcum*) to tetraploid and hexaploid wheat and chromosome location of the gene involved. Canadian J Genet Cytol 15:397–409

Khlestkina EK, Salina EA, Shumny VK (2004) Genotyping of native varieties of soft wheat by the microsatellite (SSR) markers. Sel'skokhozyaistvennaya Biologiya [Agric Biol] 5:44–52 (Russian)

Khlestkina EK, Tereshchenko OY, Salina EA (2009) Anthocyanin biosynthesis genes location and expression in wheat-rye hybrids. Mol Genet Genomics 282:475–485

Khlestkina EK, Röder MS, Börner A (2010) Mapping genes controlling anthocyanin pigmentation on the glume and pericarp in tetraploid wheat (*Triticum durum* L.). Euphytica 171:65–69

Kim W, Jonson PS, Baenziger PS, Lukaszewski AJ, Gaines CS (2004) Agronomic effect of wheat-rye translocation carrying rye chromatin (1R) from different sources. Crop Sci 44:1254–1258

Kimber G (1974) A reassessment of the origin of the polyploid wheats. Genetics 78:487–492

Klindworth DL, Niu Z, Chao S, Friesen TL, Jin Y, Faris JD, Cai X, Xu SS (2012) Introgression and characterization of a Goatgrass gene for a high level of resistance to Ug99 stem rust in tetraploid wheat. G3-Genes Genomes Genet 2:665–673

Kunert A, Naz AA, Dedeck O, Pillen K, Léon J (2007) AB-QTL analysis in winter wheat: I. Synthetic hexaploid wheat (*T. turgidum* ssp. *dicoccoides* x *T. tauschii*) as a source of favorable alleles for milling and baking quality traits. Theor Appl Genet 115:683–695

Kuraparthy V, Chhuneja P, Dhaliwal HS, Kaur S, Bowden RL, Gill BS (2007a) Characterization and mapping of cryptic alien introgression from *Aegilops geniculata* with new leaf rust and stripe rust resistance genes *Lr57* and *Yr40* in wheat. Theor Appl Genet 114:1379–1389

Kuraparthy V, Sood S, Chhuneja P, Dhaliwal HS, Kaur S, Bowden RL, Gill BS (2007b) A cryptic wheat–*Aegilops triuncialis* translocation with leaf rust resistance gene *Lr58*. Crop Sci 47:1995–2003

Kuraparthy V, Sood S, See DR, Gill BS (2009) Development of a PCR assay and marker-assisted transfer of leaf rust and stripe rust resistance genes *Lr57* and *Yr40* into hard red winter wheats. Crop Sci 49:120–126

Labuschagne MT, Pretorius ZA, Grobbelaar B (2002) The influence of leaf rust resistance genes *Lr29, Lr34, Lr35* and *Lr37* on bread making quality in wheat. Euphytica 124:65–70

Laikova LI, Belan IA, Badaeva ED, Rosseeva LP, Shepelev SS, Shumny VK, Pershina LA (2013) Development and study of spring bread wheat variety Pamyati Maystrenko with introgression of genetic material from synthetic hexaploid *Triticum timopheevii* Zhuk. × *Aegilops tauschii* Coss. Rus J Genet 49:89–97

Leonova IN, Laikova LI, Popova OM, Unger O, Börner A, Röder MS (2007) Detection of quantitative trait loci for leaf rust resistance in wheat—*T. timopheevii/T. tauschii* introgression lines. Euphytica 155:79–86

Leonova IN, Budashkina EB, Kalinina NP, Röder MS, Börner A, Salina EA (2011) *Triticum aestivum—Triticum timopheevii* introgression lines as a source of pathogen resistance genes. Czech J Genet Plant Breed 47:S49–S55

Leonova IN, Badaeva ED, Orlovskaya OA, Röder MS, Khotyleva LV, Salina EA, Shumny VK (2013) Comparative characteristic of *Triticum aestivum/Triticum durum* and *Triticum aestivum/Triticum dicoccum* hybrid lines by genomic composition and resistance to fungal diseases under different environmental conditions. Rus J Genet 49:1112–1118

Letta T, Maccaferri M, Badebo A, Ammar K, Ricci A, Crossa J, Tuberosa R (2013) Searching for novel sources of field resistance to Ug99 and Ethiopian stem rust races in durum wheat via association mapping. Theor Appl Genet 126:1237–1256

Li G, Fang T, Zhang H, Xie C, Li H, Yang T, Nevo E, Fahima T, Sun Q, Liu Z (2009) Molecular identification of a new powdery mildew resistance gene *Pm41* on chromosome 3BL derived from wild emmer (*Triticum turgidum* var. *dicoccoides*). Theor Appl Genet 119:531–539

Liu Z, Zhu J, Cui Y, Liang Y, Wu H, Song W, Liu Q, Yang T, Sun Q, Liu Z (2012) Identification and comparative mapping of a powdery mildew resistance gene derived from wild emmer (*Triticum turgidum* var. *dicoccoides*) on chromosome 2BS. Theor Appl Genet 124:1041–1049

Liu M, Zhang C, Yuan C, Zhang L, Huang L, Wu J, Wang J, Zheng Y, Zhang H, Liu D, Fu D (2013) Stripe rust resistance in *Aegilops tauschii* germplasm. Crop Sci 53:2014–2020

Lukaszewski AJ (2000) Manipulation of the 1RS.1BL translocation in wheat by induced homoeologous recombination. Crop Sci 40:216–225

Ma H, Hughes GR (1995) Genetic control and chromosomal location of *Triticum timopheevii*-derived resistance to septoria nodorum blotch in durum wheat. Genome 38:332–338

Marais GF, Marais AS, Groenewald JZ (2001) Evaluation and reduction of *Lr19*-149, a recombined form of the *Lr19* translocation of wheat. Euphytica 121:289–295

Marais GF, Pretorius ZA, Wellings CR, McCallum B, Marais AS (2005) Leaf rust and stripe rust resistance genes transferred to common wheat from *Triticum dicoccoides*. Euphytica 143:115–123

Marais GF, McCallum B, Marais AS (2008) Wheat leaf resistance gene *Lr59* derived from *Aegilops peregrina*. Plant Breed 127:340–345

Maxwell JJ, Lyerly JH, Srnic G, Parks R, Cowger C, Marshall D, Brown-Guedira G, Murphy JP (2010) *MlAB10*: A *Triticum turgidum* subsp. *dicoccoides* derived powdery mildew resistance gene identified in common wheat. Crop Sci 50:2261–2267

Maxwell JJ, Lyerly JH, Srnic G, Murphy JP, Cowger C, Parks R, Marshall D, Brown-Guedira G, Miranda L (2012) *MlNCD1*: A novel *Aegilops tauschii*—derived powdery mildew resistance gene identified in common wheat. Crop Sci 52:1162–1170

McIntosh RA, Wellings CR, Park RF (1995) Wheat rust: an atlas of resistance genes. CSIRO Publishing, Australia, p 200

McIntosh RA, Yamazaki Y, Dubcovsky J, Rogers J, Morris C, Appels R, Xia XC (2013) Catalogue of gene symbols for wheat. Available at: http://www.shigen.nig.ac.jp/wheat/komugi/genes/

Mettin D, Bluthner WD, Schlegel G (1973) Additional evidence on spontaneous 1B/1R wheat-rye substitutions and translocations. In: Sears ER, Sears LMS (eds) Proceedings of the 4th international wheat genetics symposium. University of Missouri, Columbia, pp 179–184

Miranda LM, Murphy JP, Marshall D, Leath S (2006) *Pm34*: A new powdery mildew resistance gene transferred from *Aegilops tauschii* Coss. to common wheat (*Triticum aestivum* L.). Theor Appl Genet 113:1497–1504

Miranda LM, Murphy JP, Marshall D, Cowger C, Leath S (2007) Chromosomal location of *Pm35*, a novel *Aegilops tauschii* derived powdery mildew resistance gene introgressed into common wheat (*Triticum aestivum* L.). Theor Appl Genet 114:1451–1456

Molnar-Lang M, Line G, Szakacs E (2014) Wheat–barley hybridization: the last 40 years. Euphytica 195:315–329

Nevo E, Chen G (2010) Drought and salt tolerances in wild relatives for wheat and barley improvement. Plant Cell Environ 33:670–685

Novoselskaya-Dragovich AY, Bespalova LA, Shishkina AA, Melnik VA, Upelniek VP, Fisenko AV, Dedova LV, Kudryavtsev AM (2015) Genetic diversity of common wheat varieties at the gliadin-coding loci. Rus J Genet 51:262–271

Olson EL, Rouse MN, Pumphrey MO, Bowden RL, Gill BS, Poland JA (2013) Introgression of stem rust resistance genes *SrTA10187* and *SrTA10171* from *Aegilops tauschii* to wheat. Theor Appl Genet 126:2477–2484

Peng JH, Fahima T, Röder MS, Huang QY, Dahan A, Li YC, Grama A, Nevo E (2000) High-density molecular map of chromosome region harboring stripe-rust resistance genes *YrH52* and *Yr15* derived from wild emmer wheat, *Triticum dicoccoides*. Genetica 109:199–210

Pershina LA, Devyatkina EP, Belova LI, Trubacheeva NV, Arbuzova VS, Kravtsova LA (2009) Features of alloplasmic wheat-barley substitution and addition lines (*Hordeum marinum* subsp. *gussoneanum*)-*Triticum aestivum*. Russ J Genet 45:1223–1229

Perugini LD, Murphy JP, Marshall DS, Brown-Guedira G (2008) *Pm37*, a new broadly effective powdery mildew resistance gene from *Triticum timopheevii*. Theor Appl Genet 116:417–425

Petersen S, Lyerly JH, Worthington ML, Parks WR, Cowger C, Marshall DS, Brown-Guedira G, Murphy JP (2015) Mapping of powdery mildew resistance gene *Pm53* introgressed from *Aegilops speltoides* into soft red winter wheat. Theor Appl Genet 128:303–312

Purnhauser L, Bona L, Lang L (2011) Occurrence of 1BL.1RS wheat-rye chromosome translocation and of *Sr36/Pm6* resistance gene cluster in wheat cultivars registered in Hungary. Euphytica 179:287–295

Rabinovich SV (1998) Importance of wheat-rye translocations for breeding modern cultivars of *Triticum aestivum* L. Euphytica 100:323–340

Rawat N, Tiwari VK, Singh N, Randhawa GS, Singh K, Chhuneja P, Dhaliwal HS (2009) Evaluation and utilization of Aegilops and wild Triticum species for enhancing iron and zinc content in wheat. Gen Res Crop Evol 56:53–64

Rehman A, Evans N, Gianibelli MC, Rose RJ (2008) Allelic variations in high and low molecular weight glutenins at the *Glu-Dt* locus of *Aegilops tauschii* as a potential source for improving bread wheat quality. Aust J Agricult Res 59:399–405

Riar Kaur A, Kaur S, Dhaliwal YS, Singh K, Chhuneja P (2012) Introgression of a leaf rust resistance gene from *Aegilops caudata* to bread wheat. J Genet 91:155–161

Rouse MN, Jin Y (2011) Genetics of resistance to race TTKSK of *Puccinia graminis* f. sp. *tritici* in *Triticum monococcum*. Phytopathology 101:1418–1423

Schneider A, Molnár I, Molnár-Láng M (2008) Utilization of Aegilops (goatgrass) species to widen of genetic diversity of cultivated wheat. Euphytica 163:1–19

Seah S, Bariana H, Jahier J, Sivasithamparum K, Lagudah ES (2001) The introgressed segment carrying rust resistance genes *Yr17, Lr37* and *Sr38* in wheat can be assayed by a cloned disease resistance gene-like sequence. Theor Appl Genet 102:600–605

Shoeva OY, Dobrovolskaya OB, Leonova IN, Salina EA, Khlestkina EK (2016) The B-, G- and S-genomic *Chi* genes in Triticeae. Biologia Plant 60:279–284

Simon MR, Worland CA, Struik PC (2005) Chromosomal location of genes encoding for resistance to septoria tritici blotch (*Mycosphaerella graminicola*) in substitution lines of wheat. Neth J Agric Sci 53:113–129

Simon MR, Khlestkina EK, Castillo NS, Börner A (2010) Mapping quantitative resistance to septoria tritici blotch in spelt wheat. Eur J Plant Pathol 128:317–324

Simons K, Abate Z, Chao S, Zhang W, Rouse M, Jin Y, Elias E, Dubcovsky J (2011) Genetic mapping of stem rust resistance gene *Sr13* in tetraploid wheat (*Triticum turgidum* ssp. *durum* L.). Theor Appl Genet 122:649–658

Sin E, del Moral J, Hernández P, Benavente E, Rubio M, Martín-Sánchez JA, Pérez RF, López-Braña I, Delibes A (2011) Effects of the 4N(v) chromosome from *Aegilops ventricosa* on agronomic and quality traits in bread wheat. Czech J Genet Plant Breed 47:S63–S66

Song W, Xice H, Liu Q, Xice CJ, Ni ZF, Jang TM, Sun QX, Liu ZY (2007) Molecular identification of *Pm12*—carrying introgression lines in wheat using genomic and EST-SSR markers. Euphytica 158:95–102

Spetsov P, Daskalova N, Plamenov D, Moraliyski T (2013) Resistance to powdery mildew and leaf rust in wheat lines derived from a *Triticum aestivum/Aegilops variabilis* cross. Turkish J Field Crops 18:128–133

Spielmeyer W, Singh RP, McFadden H, Wellings CR, Huerta-Espino J, Kong X, Appels R, Lagudah ES (2008) Fine scale genetic and physical mapping using interstitial deletion mutants of *Lr34/Yr18*: a disease resistance locus effective against multiple pathogens in wheat. Theor Appl Genet 116:481–490

Srinivasachary Gosman N, Steed A, Faure S, Bayles R, Jennings P, Nicholson P (2008) Mapping of QTL associated with Fusarium head blight in spring wheat RL4137. Czech J Genet Plant Breed 44:147–159

Steed A, Chandler E, Thomsett M, Gosman N, Faure S, Nicholson P (2005) Identification of type I resistance to Fusarium head blight controlled by a major gene located on chromosome 4A of *Triticum macha*. Theor Appl Genet 111:521–529

Sun X-L, Liu D, Zhang H-Q, Huo N-X, Zhou R-Z, Jia J-Z (2006) Identification and mapping of two new genes conferring resistance to powdery mildew from *Aegilops tauschii* (Coss.) schmal. J Integr Plant Biol 48:1204–1209

Tereshchenko OY, Pshenichnikova TA, Salina EA, Khlestkina EK (2012a) Development and molecular characterization of a novel wheat genotype having purple grain colour. Cereal Res Commun 40:210–214

Tereshchenko OY, Gordeeva EI, Arbuzova VS, Börner A, Khlestkina EK (2012b) The D genome carries a gene determining purple grain colour in wheat. Cereal Res Commun 40:334–341

Timonova EM, Leonova IN, Röder MS, Salina E (2013) Marker-assisted development and characterization of a set of *Triticum aestivum* lines carrying different introgressions from the *T. timopheevii* genome. Mol Breed 31:123–136

Todorovska E, Christov N, Slavov S, Christova P, Vassilev D (2009) Biotic stress resistance in wheat—breeding and genomic selection implications. Biotechnol Biotec Eq 23:1417–1426

Trubacheeva NV, Badaeva ED, Adonina IG, Belova LI, Devyatkina EP, Pershina LA (2008) Construction and molecular and cytogenetic analyses of euploid ($2n$ = 42) and telocentric addition ($2n$ = 42 + 2t) alloplasmic lines (*Hordeum marinum* subsp. *gussoneanum*)-*Triticum aestivum*. Russ J Genet 44:67–73

Uauy C, Brevis JC, Chen X, Khan I, Jackson L, Chicaiza O, Distelfeld A, Fahima T, Dubcovsky J (2005) High-temperature adult-plant (HTAP) stripe rust resistance gene *Yr36* from *Triticum turgidum* ssp. *dicoccoides* is closely linked to the grain protein content locus *Gpc-B1*. Theor Appl Genet 112:97–105

Villareal RL, Banuelos O, Mujeeb-Kazi A, Rajaram S (1998) Agronomic performance of chromosome 1B and T1BL.1RS near-isolines in the spring bread wheat Seri M82. Euphytica 103:195–202

Wang K, An XL, Pan LP, Dong K, Gao LY, Wang SL, Xie ZZ, Zhang Z, Appels R, Ma W, Yan YM (2012) Molecular characterization of HMW-GS 1Dx3(t) and 1Dx4(t) genes from *Aegilops tauschii* and their potential value for wheat quality improvement. Hereditas 149:41–49

Wang S, Yu Z, Cao M, Shen X, Li N, Li X, Ma W, Weißgerber H, Zeller F, Hsam S, Yan Y (2013) Molecular mechanisms of HMW glutenin subunits from 1S(l) genome of *Aegilops longissima* positively affecting wheat breadmaking quality. PLoS ONE 8:e58947

Wang Y, Cheng X, Shan Q, Zhang Y, Liu J, Gao C, Qiu JL (2014) Simultaneous editing of three homoeoalleles in hexaploid breadwheat confers heritable resistance to powdery mildew. Nat Biotechnol 32:947–951

Wu SG, Pumphrey M, Bai GH (2009) Molecular mapping of stem-rust-resistance gene *Sr40* in wheat. Crop Sci 49:1681–1686

Xie C, Sun Q, Ni Z, Yang T, Nevo E, Fahima T (2003) Chromosomal location of a *Triticum dicoccoides*-derived powdery mildew resistance gene in common wheat by using microsatellite markers. Theor Appl Genet 106:341–345

Xue F, Ji W, Wang C, Zhang H, Yang B (2012) High-density mapping and marker development for the powdery mildew resistance gene *PmAS846* derived from wild emmer wheat (*Triticum turgidum* var. *dicoccoides*). Theor Appl Genet 124:1549–1560

Yamamori M (1994) An N-band marker for gene *Lr18* for resistance to leaf rust in wheat. Theor Appl Genet 89:643–646

Yu GT, Zhang Q, Klindworth DL, Friesen TL, Knox R, Jin Y, Zhong S, Cai X, Xu SS (2010) Molecular and cytogenetic characterization of wheat introgression lines carrying the stem rust resistance gene *Sr39*. Crop Sci 50:1393–1400

Yu LX, Barbier H, Rouse MN, Singh S, Singh RP, Bhavani S, Huerta-Espino J, Sorrells ME (2014) A consensus map for Ug99 stem rust resistance loci in wheat. Theor Appl Genet 127:1561–1581

Yu G, Zhang Q, Friesen TL, Rouse MN, Jin Y, Zhong S, Rasmussen JB, Lagudah ES, Xu SS (2015) Identification and mapping of *Sr46* from *Aegilops tauschii* accession CIae 25 conferring resistance to race TTKSK (Ug99) of wheat stem rust pathogen. Theor Appl Genet 128:431–443

Yudina RS, Leonova IN, Salina EA, Khlestkina EK (2015a) Effect of alien genomic introgressions on the osmotic stress resistance of wheat. Russ J Genet Appl Res 5:168–173

Yudina RS, Leonova IN, Salina EA, Khlestkina EK (2015b) Change of salt tolerance in common wheat after introgression of genetic material from *Aegilops speltoides* and *Triticum timopheevii*. Vavilov J Genet Breed 19:171–175

Zegeye H, Rasheed A, Makdis F, Badebo A, Ogbonnaya FC (2014) Genome-wide association mapping for seedling and adult plant resistance to stripe rust in synthetic hexaploid wheat. PLoS ONE 9:e105593

Zeller FJ (1973) 1B/1R wheat-rye chromosome substitutions and translocations. In: Proceedings of 4th international wheat genetic symposium, Mo Agric Exp Stn, Columbia, MO, p 209

Zhang H, Fu H, Ma S (2009) Intraspecific hybridization among *Aegilops tauschii* varieties and genetic analysis and molecular markers of a novel stripe rust resistance gene from *Ae. tauschii*. Chin J Appl Environ Biol 15:44–47

Zhang Q, Klindworth DL, Friesen TL, Chao S, Jin Y, Cai X, Xu SS (2012) Development and characterization of wheat lines with *Sr37* for stem rust resistance derived from wild Timopheev's wheat. In: International conference on "Plant and Animal genome XXI". San Diego, USA, p 316

Zhou JP, Yao CH, Yang EN, Yin MQ, Liu C, Ren ZL (2014) Characterization of a new wheat-*Aegilops biuncialis* addition line conferring quality-associated HMW glutenin subunits. Genet Mol Res 13:660–669

Organization and Evolution of the Duplicated Flavonoid Biosynthesis Genes in Triticeae

Elena Khlestkina and Olesya Shoeva

Introduction

Gene duplications are of a great evolutionary importance. In plant genomes, duplicated genes may result from not only segmental chromosome duplications, but also from polyploidization. The former give rise to paralogous genes, the latter—to homoeologous (orthologous) genes. Interspecific hybridization occurs frequently in monocot and dicot plant species (Masterson 1994; Wolfe 2001) . In some cases it is followed by a chromosome doubling, leading to formation of allopolyploids. Bread wheat (*Triticum aestivum* L., $2n = 6x = 42$) is derived from a complex hybridization process involving three diploid species carrying the three homoeologous genomes A, B and D (Kihara 1944, 1954). In 70–99 % homoeologous gene series of allopolyploids, all homoeologues are reportedly co-expressed (Comai et al. 2000; Kashkush et al. 2002; Bottley et al. 2006). Analysis of expression level of individual gene copies showed that expression levels and/or patterns of co-expressed homoeologues can sometimes be equal/identical (Morimoto et al. 2005; Shitsukawa et al. 2007; Khlestkina et al. 2008) or vary (Nomura et al. 2005; Appleford et al. 2006; Shitsukawa et al. 2007). Rapid changes in duplicated genes can be related with normalization of expression of their product. On the other hand duplicated gene doesn't undergo selection pressure and can evolve and acquire a new functional specialization (Marais and Rausher 2008).

Flavonoid biosynthesis pathway genes (FBGs) provide a suitable tool for studies in genetics, since they confer easily scorable phenotypes. FBGs are divided into the regulatory and the structural genes. FB enzymes are encoded by the structural genes, activated by regulatory genes encoding transcriptional factors of bHLH (or MYC), MYB or WD40-type, which form the MBW regulatory complex (Khlestkina et al. 2015).

E. Khlestkina (✉) · O. Shoeva
Institute of Cytology and Genetics SBRAS, Novosibirk, Russia
e-mail: khlest@bionet.nsc.ru

V.L. Korogodina et al. (eds.), *Genetics, Evolution and Radiation*,
DOI 10.1007/978-3-319-48838-7_10

Duplications of the Regulatory *MYC* Gene

Purple wheat grains are considered to be a useful source of dietary anthocyanins (flavonoid pigments). The trait is due to synthesis of anthocyanins in pericarp under the control of the *Pp1* and *Pp3* complementary genes. A likely candidate for *Pp3* is the *Myc1* gene encoding R/B-like bHLH transcription factor (Shoeva et al. 2014a). *Myc1* and its four duplicated copies in bread wheat genome demonstrated essential structural and functional divergence. Their transcriptional activity was tissue-specific. In particular, only the *Myc1* gene (chromosome 2A) is expressed in wheat pericarp, conferring purple color of grain. Three *Myc* genes are expressed in other parts of wheat plant, while one copy is not expressed at all. The *Myc3* (chromosome 2B) and *Myc4* (chromosome 2D) genes products are potential candidates participating in formation of the MBW regulatory complex in coleoptile and culm. The *Myc* loci mapped to chromosomes 2B and 2D have never been revealed in segregating mapping populations analyzed, suggesting missing or rare occurrence of loss-of-function alleles of these genes in wheat populations. Unlike *Myc3* and *Myc4*, the *Myc1* (*Pp3*) gene is characterized by high allelic diversity observed at the phenotypic level, detected in segregating populations (Dobrovolskaya et al. 2006; Khlestkina et al. 2010).

CHI Homoeologs

Expression of chalcone-flavanone isomerase (CHI; EC 5.5.1.6.) catalyzing the conversion of chalcones to flavanones is related with plant adaptive and protective responses to environmental stress. Three copies of this gene were sequenced and mapped to the terminal bins of the long arms of chromosomes 5A, 5B, and 5D (Shoeva et al. 2014b). Comparative analysis of the *Chi* genomic sequences in different species suggested elimination of the last intron followed by fusion of the two last exons in the common ancestor of Triticeae tribe. The coding sequences of the three *Chi* wheat copies have over 96 % identity and maintain conservative sites important for proper folding and activity of the enzyme CHI, type I. However, transcription of the three homoeologs is not always co-regulated. In particular, the three genes demonstrate different response to salinity in roots: *Chi-D1* is upregulated, *Chi-A1* responds medially, whereas *Chi-B1* is not activated at all. The observed variation in transcriptional activity between the *Chi* homoeologs is in a good agreement with structural diversification of their promoter sequences. In addition, the relation between *Chi* transcription and pigmentation in different parts of wheat plant was studied. It was shown that *Chi* is expressed in different parts of wheat plant independently on its coloration. Although in some intensively colored organs some increase of the *Chi* gene expression can be observed. Thus, the *Chi* genes are under complex regulation in wheat (Shoeva et al. 2014b).

Homoeologous and Paralogous Copies of the *F3H* Gene

Flavanone 3-hydroxylase (F3H; EC 1.14.11.9) catalyses one of the key steps of the flavonoid biosynthesis pathway (conversion of flavanones to dihydroflavonols) yielding a large family of flavonoid compounds which are involved in many biological activities. The genome of most plant species contains only a single copy of the F3h gene. In the allohexaploid bread wheat (*T. aestivum* L.), the family of genes encoding F3H is represented by four copies (3 highly similar homoeologs and 1 paralog) (Khlestkina et al. 2008). Analysis of the paralogue (*F3h-B2*) suggested this gene to encode functional enzyme that however differs from that encoded by the three homoeologous copies (*F3h-A1*, *-B1*, *-D1*) by replacement of some of the strictly conserved residues in the putative substrate-binding sites. *F3h-B2* promoter region diverged essentially from that of the *F3h-1* genes, and diversified transcriptional regulation of these two genes is observed (Khlestkina et al. 2013; Shoeva and Khlestkina 2013). Analysis of *F3h-1* and *F3h-2* coding sequence divergence suggested *F3h* to be duplicated in the common ancestor of the Triticeae tribe. Specific occurrence of the orthologous *F3h-2* genes was demonstrated in a small group of Triticeae species having one of the closely related genomes B, G or S, and in rye. *F3h-2* could acquire a new functional specialization in the common progenitor of the B, G and S genomes, and in rye (Khlestkina et al. 2013). These can be a reason for the maintenance of the duplicate *F3 h* copies in these genomes, whereas in other Triticeae genomes, *F3h-2* was likely pseudogenized.

Conclusion

Flavonoid biosynthesis pathway genes provide a suitable model for detailed studies into evolution of duplicated genes in allopolyploid genomes of Triticeae species.

References

Appleford NE, Evans DJ, Lenton JR, Gaskin P, Croker SJ, Devos KM, Phillips AL, Hedden P (2006) Function and transcript analysis of gibberellin-biosynthetic enzymes in wheat. Planta 223:568–582

Bottley A, Xia GM, Koebner RM (2006) Homoeologous gene silencing in hexaploid wheat. Plant J 47:897–906

Comai L, Tyagi AP, Winter K, Holmes-Davis R, Reynolds SH, Stevens Y, Byers B (2000) Phenotypic instability and rapid gene silencing in newly formed arabidopsis allotetraploids. Plant Cell 12:1551–1568

Dobrovolskaya OB, Arbuzova VS, Lohwasser U, Röder MS, Börner A (2006) Microsatellite mapping of complementary genes for purple grain colour in bread wheat (*Triticum aestivum* L.). Euphytica 150:355–364

Des Marais DL, Rausher MD (2008) Escape from adaptive conflict after duplication in an anthocyanin pathway gene. Nature 454:762–765

Kashkush K, Feldman M, Levy AA (2002) Gene loss, silencing and activation in a newly synthesized wheat allotetraploid. Genetics 160:1651–1659
Khlestkina EK, Röder MS, Salina EA (2008) Relationship between homoeologous regulatory and structural genes in allopolyploid genome—a case study in bread wheat. BMC Plant Biol 8:88
Khlestkina EK, Röder MS, Börner A (2010) Mapping genes controlling anthocyanin pigmentation on the glume and pericarp in tetraploid wheat (*Triticum durum* L.). Euphytica 171:65–69
Khlestkina EK, Dobrovolskaya OB, Leonova IN, Salina EA (2013) Diversification of the duplicated F3h genes in Triticeae. J Mol Evol 76:261–266
Khlestkina EK, Shoeva OY, Gordeeva EI (2015) Flavonoid biosynthesis genes in wheat. Rus J Genet Appl Res 5:268–278
Kihara H (1944) Discovery of the DD-analyser, one of the ancestors of *Triticum vulgare*. Agric Hort (Tokyo) 19:13–14
Kihara H (1954) Origin of wheat. Wheat Inf Serv 1:35–42
Masterson J (1994) Stomatal size in fossil plants: evidence for polyploidy in majority of angiosperms. Science 264:421–424
Morimoto R, Kosugi T, Nakamura C, Takumi S (2005) Intragenic diversity and functional conservation of the three homoeologous loci of the KN1-type homeobox gene *Wknox1* in common wheat. Plant Mol Biol 57:907–924
Nomura T, Ishihara A, Yanagita RC, Endo TR, Iwamura H (2005) Three genomes differentially contribute to the biosynthesis of benzoxazinones in hexaploid wheat. Proc Natl Acad Sci USA 102:16490–16495
Shitsukawa N, Tahira C, Kassai K, Hirabayashi C, Shimizu T, Takumi S, Mochida K, Kawaura K, Ogihara Y, Murai K (2007) Genetic and epigenetic alteration among three homoeologous genes of a class E MADS box gene in hexaploid wheat. Plant Cell 19:1723–1737
Shoeva OY, Khlestkina EK (2013) *F3h* gene expression in various organs of wheat. Mol Biol 47:901–903
Shoeva OY, Gordeeva EI, Khlestkina EK (2014a) The regulation of anthocyanin synthesis in the wheat pericarp. Molecules 19:20266–20279
Shoeva OY, Khlestkina EK, Berges H, Salina EA (2014b) The homoeologous genes encoding chalcone-flavanone isomerase in *Triticum aestivum* L.: structural characterization and expression in different parts of wheat plant. Gene 538:334–341
Wolfe KH (2001) Yesterday's polyploidization and the mistery of diploidization. Nat Rev Genet 2:233–241

Kinase Cascade of DNA Damage Checkpoint

Natalia Koltovaya

Introduction

Ionizing radiation induces a wide spectrum of DNA damage, including double-strand breaks (DSBs) of DNA which play an important role in cell death and genetic instability (Gaziev 1999). The response of cells to DNA damage is considered in numerous reviews (Glazer and Glazynov 1999; Korolev 2007). According to the proposed model (Rouse and Jackson 2002) DNA damage is initially detected by specific repair factors having a high affinity to specific types of primary DNA lesions. In some cases the damage can be relatively easily repaired and, thus, it is quickly restored soon after being detected. In this case the damage does not activate the checkpoint response, since the signaling system Mec1 (Tel1) has no time to detect the damage. If the DNA damage does not restore quickly and is retained for a long period of time, the Mec1/Ddc2 complex is delivered to the site of damage. It modifies the targets in the neighborhood of the damage, for example, repair enzymes and H2A histones. If this is followed by complete repair, a global response to the DNA damage is again prevented. However, if DNA repair still cannot be completed, other complexes are recruited. The Rad53 and Chk1 kinases are activated by the Mec1-dependent way, and a global response to DNA damage is initiated. It includes cell cycle arrest, further alteration of the chromatin structure, and regulation of the cell repair potential. All these reactions facilitate repair and prevent the progression of the key stages of the cell cycle. There are several checkpoints of genome integrity throughout the cell cycle.

The following problems are gaining in importance: domain structure of proteins, physical interaction between repair and checkpoint proteins, their structural changes and activation during repair and checkpoint response. An effective and widely used way to regulate the activity of many enzymes and proteins is chemical modification

N. Koltovaya (✉)
Joint Institute for Nuclear Research, Dubna, Russia
e-mail: koltovaya@jinr.ru

V.L. Korogodina et al. (eds.), *Genetics, Evolution and Radiation*,
DOI 10.1007/978-3-319-48838-7_11

by kinases (Hoeksta et al. 1991a, b). Protein kinases comprise a large family of enzymes catalyzing the transfer of the terminal phosphate residue from ATP onto different groups in the protein structure. The transfer of the negatively charged group is accompanied by large structural rearrangements of a target protein (Gusev 2000). This phosphorylation can have various consequences. For example, it can activate the enzyme, increase affinity to other proteins, or change protein localization.

Activation of Checkpoint

The signal for checkpoint is ssDNA (Garvic et al. 1995). Checkpoint activation is proportional to the degree of DSB degradation and to the length of ssDNA (Melo et al. 2001; Zou and Elledge 2003). The protein complexes MRX (Mre11/Rad50/Xrs2) and Ku bind to the DSB ends, maintain and process them. Checkpoint activation requires subsequent loading of RPA on ssDNA which is an analog of the bacterial protein ssb (*single strand binding protein*). RPA prevents the formation of a secondary structure by the ends of ssDNA. The formation of ssDNA-RPA nucleoprotein filaments permits the Mec1/Ddc2 and Mec3/Rad17/Ddc1 complexes to be recruited to the DNA damage. The structure of the Mec3/Rad17/Ddc1 complex is similar to the structure of the PCNA (*proliferating cell nuclear antigen*) homotrimer that retains DNA polymerase on DNA and increases its processivity (Kaur et al. 2001). PCNA is loaded on DNA during replication by means of the RFC (*replication factor C*) complex. The PCNA-like complex Mec3/Rad17/Ddc1 is loaded on the DSB end by using the RFC-like complex Rad24/Rfc2–5 containing the Rad24 instead of the Rfc1 subunit (Longhese et al. 1998; Lindsey-Boltz et al. 2001; Green et al. 2000). The Rad24/RFC complex recognizes DSB structures with ssDNA (Ellison and Stillman 2003) and binds to RPA that covers the ssDNA region (Zou et al. 2003). After ATP hydrolysis the Mec3/Rad17/Ddc1 complex dissociates from the Rad24/RFC complex (Majka and Burgers 2003).

Mec1 Kinase

The binding of Mec1 kinase to the DNA DSB requires the Ddc2 protein. The loading of the Mec1/Ddc2 complex depends on preloading of RPA on DNA (Lee et al. 1998, 1999) and is realized through the interaction of both proteins with RPA (Rouse and Jackson 2002; Zou and Elledge 2003). Mec1 physically interacts with the Rfa1, Rfa2, and Rfa3 subunits (Gavin et al. 2006; Nakada et al. 2005): in particular, the C–terminal fragment of Mec1 interacts with Rfa1 and Rfa2.

The Mec1 kinase belongs to the conserved family of the so-called phosphatidylinositol-3-kinases. The yeast kinase Tel1 and the mammalian kinases

ATM, ATR, and DNA-PK also belong to this family (Elledge 1996). The catalytic domain of PI-3 kinases is observed in the C-terminus sequence of the Mec1 (Fig. 1) and Tel1 kinases, but despite this fact the kinases do not phosphorylate lipids. Instead, all of the above-mentioned kinases phosphorylate proteins by the consensus SQ/TQ, which often form clusters named as SCD (SQ/TQ-rich cluster domains). Mec1 responds to different types of DNA damage and plays the primary role in checkpoint activation by DNA damage and by the replication block (Longhese et al. 1998). Although Tel1 is also involved in repair, recombination, and cell cycle regulation, it plays a minor role in response to the DSBs (Morrow et al. 1995; Nakada et al. 2003; Sanchez et al. 1996).

The Mec1/Ddc2 checkpoint complex binds to the RPA complex through the Ddc2 subunit (Robison et al. 2004), and this binding leads to its activation. Two Mec1 subunits phosphorylate each other, and the complex disintegrates releasing the active form of the Mec1 kinase. The interaction of the Mec3/Rad17/Ddc1 complex with the Mec1 results in further phosphorylation of Mec1 and enhances its activity. The active Mec1 kinase causes chemical modification of RPA. It is known that in undisturbed cells the Rfa2 subunit of RPA is phosphorylated in S and G2 (Brush et al. 1996). In response to DNA damage and replication blocking by HU two subunits (Rfa1 and Rfa2) are phosphorylated (Brush et al. 1996; Brush and Kelly 2000; Bartrand et al. 2004). Both kinds of RPA phosphorylation, during the cell cycle or in response to DNA damage and replication blocking, are controlled in *S. cerevisiae* by the Mec1 kinase. Induced phosphorylation also depends on Rad53. RPA phosphorylation is necessary for G1/S, intra S, and G2/M checkpoints and seems to provide signaling in cooperation with the Rad24/RFC complex (Longhese et al. 1996; Kim and Brill 2003).

CDK1 Kinase

CDK1/CDC28 kinase is necessary for repair and checkpoint. CDC28/CDK1 is the major protein kinase regulating the cell cycle progression and belongs to the CDK (cyclin-dependent kinase) family of cyclin-dependent serine/threonine kinases (Mendenhall and Hodge 1998). CDK kinase substrates have, as a rule, multiple sites of phosphorylation (S/T-P-x-K/R). Activation of the CDK catalytic subunit requires binding to the regulatory subunit, cyclin. Systematic analysis revealed 360 substrates for the active form of CDC28/CLB2 kinase (Ubersax et al. 2003). CDK kinase structure has two globules with an ATP located in the gap between them (Bartova et al. 2004). The gap is formed with the involvement of T- and G-loops localized in different globules. The full activation occurs as a result of cyclin joining, phosphorylation of the T169 residue in the T-loop, and dephosphorylation of the Y19 and T18 residues in the G-rich loop. The kinase structure thus changes, the gap widens, and the phosphate group of ATP becomes accessible for the substrate. The *cdc28-srm* mutation represents a substitution of conserved glycine by serine (G20S) in the G-loop (Kholmurodov et al. 2006). Computer molecular

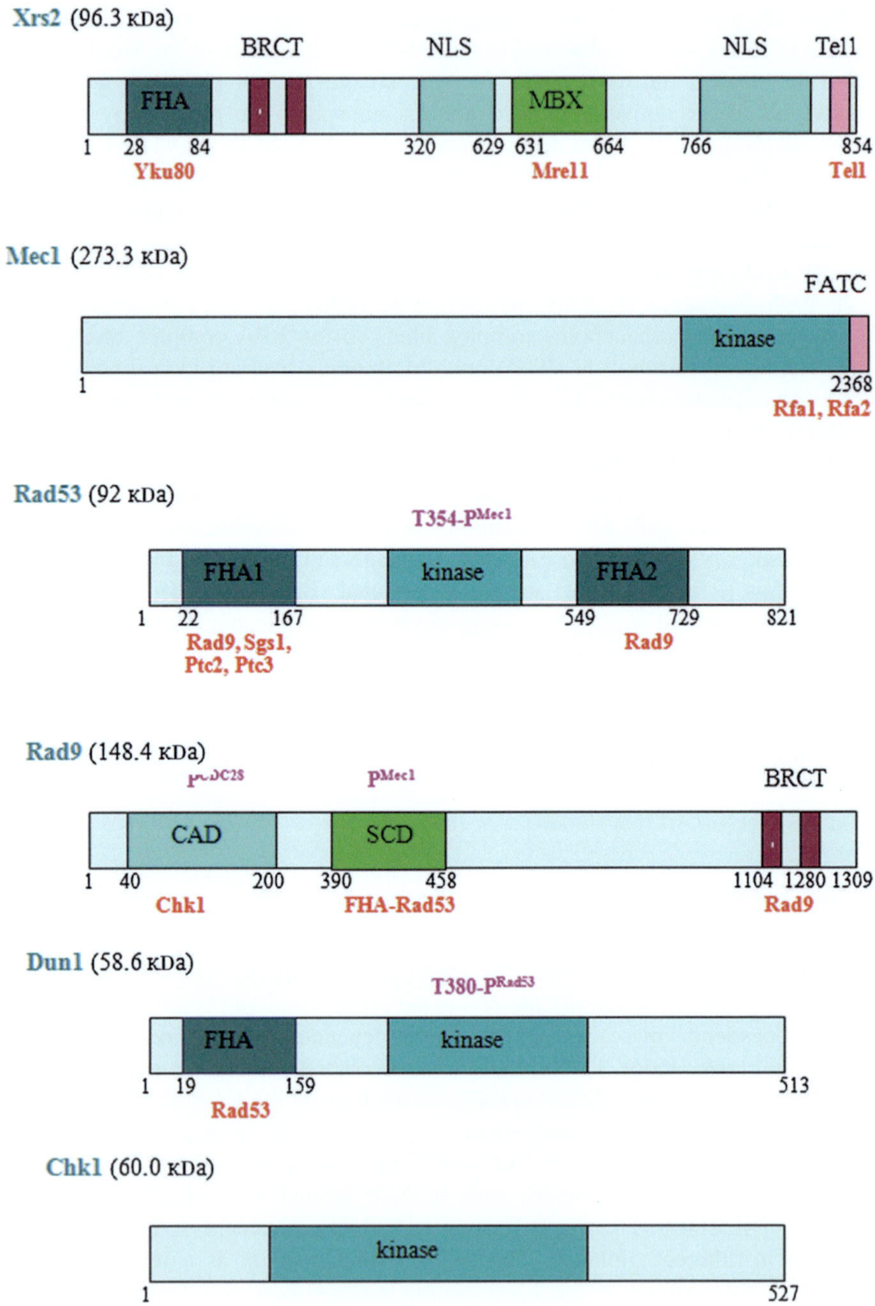
Xrs2 (96.3 кDa)
BRCT
NLS
NLS
Tel1
FHA
MBX
1
28
84
320
629
631
664
766
854
Yku80
Mre11
Tel1
Mec1 (273.3 кDa)
FATC
kinase
1
2368
Rfa1, Rfa2
Rad53 (92 кDa)
T354-P^Mec1
FHA1
kinase
FHA2
1
22
167
549
729
821
Rad9, Sgs1,
Ptc2, Ptc3
Rad9
Rad9 (148.4 кDa)
p^Mec1
BRCT
CAD
SCD
1
40
200
390
458
1104
1280
1309
Chk1
FHA-Rad53
Rad9
Dun1 (58.6 кDa)
T380-P^Rad53
FHA
kinase
1
19
159
513
Rad53
Chk1 (60.0 кDa)
kinase
1
527

◀ **Fig. 1** A structure of Xrs2, Mec1, Rad53, Rad9, Dun1 and Chk1. Indicated is the position of structural elements determining the sites of binding Xrs2-Tel1 (Tel1) and Xrs2-Mre11 (MBX, *Mre11-binding domain*); nuclear localization signal (NLS); domain of kinase activity Rad53 (kinase) and Rad53-Rad9 interaction (FHA, *forkhead-associated domain*); interactions Rad9-Chk1 (CAD, *Chk1 activation domain*); interactions Rad9-Rad53 (SCD, *[S/T]Q cluster domain*); interactions Rad9-Rad9 (BRCT, *BRCA1 C-terminus*); kinase and FHA-domain of Dun1 kinase. Proteins binding with these domains are indicated beneath the boxes. Indicated for proteins Rad53, Rad9, Dun1 are the sites of phosphorylation

dynamic modeling has shown that the substitution of one amino acid residue causes marked changes in the kinase structure and probable changes in the interaction of the kinase with cyclins, substrates, and ATP.

The CDC28 kinase provides the choice of a repair pathway (HR or NHEJ) in G1 and G2, since its activity is required for degradation of DSB ends by the MRX complex (Ira et al. 2004). Degradation of DSB depends almost completely on the MRX complex in G2 phase and only partially in G1 phase (Deide and Gottschling 2001). It is supposed that the CDC28 kinase in G2 phase regulates the phosphorylation status of the MRX complex and of other factors involved in DSB processing. Indeed, some proteins participating in DNA recombination are CDK substrates, including Srs2 helicase and two subunits (Mre11 and Rad50) of the MRX complex (Ubersax et al. 2003; Liberi et al. 2000). The CDC28/CLB2 kinase and the MRX complex were demonstrated to be functionally associated in the course of the mitotic cycle, recombination, and repair (Aylon and Kupiec 2005; Grandin and Charbonneau 2003).

Analysis of the amino acid sequence of Mre11 endonuclease revealed several potential consensus sites of phosphorylation by the CDC28 kinase. However, mutations at these sites have no influence on the activation of checkpoint control and, probably, on the degradation of the break ends (Ira et al. 2004). Mre11 endonuclease is a constituent of a large complex together with helicases Srs2 and Sgs1 in unirradiated cells. Irradiation induces phosphorylation of the Srs2 helicase by the CDC28 kinase and Mec1 kinase-dependent formation of individual complexes Sgs1-Mre11 and Srs2-Mre11 (Chiolo et al. 2005). Phosphorylation of both proteins (Srs2 and Mre11) is indispensable for the formation of the subcomplexes. Abnormalities of Srs2 helicase sites of CDC28 kinase-mediated phosphorylation block their formation. Probably, the larger complex is used as a center of storage of the enzymes before they are needed.

The Xrs2 protein is a component of the MRX complex and contains the sequence MBX (*Mre11 binding domain of Xrs2*) -binding site of Mre11 in the C-termini (Fig. 1). On each side of the site, NLS (*nuclear localization signal*) sequences are located in positions 320–629 and 766–854 (Tsukamoto et al. 2005). Xrs2 regulates movement of Mre11 from the cytoplasm to the nucleus, where it binds to Rad50. As a result, the Mre11, Rad50, and Xrs2 proteins form the MRX complex in the ratio of 2:2:1. Rad50 mediates ATP-dependent binding of the complex with DSB (Raymond and Kleckner 1993). Binding of the complex with ATP causes conformational changes of Rad50, thus stimulating Mre11 nuclease activity. The role of phosphorylation of Mre11 and Rad50 by the CDC28 kinase is unknown so far.

Rad53 and Chk1 Kinases

Signaling from the sensors to the cell division apparatus occurs as a result of phosphorylation of the adaptors of the Chk1 and Rad53 kinases by the Mec1 kinase (Blankley and Lydall 2004; Ma et al. 2006; Sweeney et al. 2005). Hyperphosphorylation of the adaptor protein Rad9 is specifically induced by DNA damage rather than by inhibition of DNA replication (Sun et al. 1998). Rad9 is required for signal transduction in G1 and G2/M phases. In S phase a signal is mediated by another adaptor, namely, by the Mrc1 protein associated with the replication fork (Alcasabas et al. 2001; Osborn and Elledge 2003).

The hypophosphorylated Rad9 is found in a cell as a component of two complexes, 850 and 560 kDa, containing Ssa1 and Ssa2 chaperones that, probably, remodel it (Bosch and Lowndes 2004). The smaller complex also contains the Rad53 kinase. The Rad9 protein has two BRCT domains in the C-terminus sequence (Fig. 1). These domains facilitate Rad9-Rad9 interaction in the case of DNA damage (Soulier and Lowndes 1999). This motif was found in many proteins involved in repair, DNA recombination, and checkpoint.

The adaptor protein Rad9 has multiple sites of Mec1 kinase-mediated phosphorylation (Naiki et al. 2004; Emili 1998). In the case of DNA damage Rad9 binds to the damage and is fully phosphorylated by the Mec1 (Naiki et al. 2004). The hyperphosphorylated Rad9 becomes a component of the smaller complex. Hyperphosphorylation of Rad9 depends on the Mec1 and Tel1 kinases, and under some conditions the complexes Mec3/Rad17/Ddc1 and Rad24p/RFC are also required (Emili 1998; Vialard et al. 1998). In asynchronous or rapidly growing cultures, in G2/M cells synchronized by nocodasole, or in the cells whose arrest is induced by HU, hyperphosphorylation of Rad9 depends on the genes *MEC1* and *TEL1*. In G1 cells synchronized by the α-factor, hyperphosphorylation of Rad9 depends, in addition to the *MEC1* gene, on the functioning of the *RAD17*, *RAD24*, *MEC3*, and *DDC1* genes, probably, due to increased processing of the MRX complex by the Mec3/Rad17/Ddc1 complex whose loading requires the Rad24p/RFC complex. Thus, the participation of checkpoint genes in the response to DNA damage is specific and depends on the stage of the cell cycle.

In UV-irradiated cells, the phosphorylated Rad9 protein interacts with the phosphorylated Rad53 kinase (an ortholog of the human kinase hChk2) (Vialard et al. 1998). The mutant Rad9 protein devoid of the key target phosphorylation residues SQ and TQ becomes unable to bind and activate the Rad53 kinase in response to the action of MMC (Schwartz et al. 2002). The interaction between the phosphorylated forms of Rad9 and Rad53 is mediated by the FHA-domains of the Rad53 kinase (Sun et al. 1998). In many mechanisms of cell control the domain FHA (*forkhead-associated*) serves as a module of protein–protein interaction, facilitates recognition and binding with phosphorylated proteins (Durocher et al. 1999). The FHA-domain is about 140 amino acid residues in size and presents a compact globular structure. Structural analysis of the FHA1-Rad53 domain has shown that the amino acid residues are packed in two antiparallel β-layers with a short α-helix at the C-termini (Liao et al. 1999). Rad53 has two FHA-domains: FHA1 (N-terminal) and FHA2 (C-terminal) (Fig. 1). Both domains bind directly to the phosphorylated Rad9 (Durocher et al. 1999; Sun et al. 1998). These domains, however, differ in binding specificity. FHA1 preferentially binds to ST(P), and FHA2 binds both to S(P)T and to ST(P).

Rad53 is an immediate target of the Mec1. Rad53 activation requires phosphorylation of multiple sites by the Mec1 kinase (Sweeney et al. 2005). Effective direct phosphorylation of Rad53 by Mec1 is observed only in the presence of the phosphorylated form of Rad9. The stimulating activity of Rad9 needs phospho- and FHA-dependent interaction with Rad53 to permit Rad53 to be recognized as a substrate by Mec1. In turn, the cell cycle is renewed as a result of inactivation of the Rad53 kinase. The phosphorylated forms of two phosphatases of the PP2C type (*protein phosphatase type 2C*), Pts2p and Ptc3p, bind to the FHA1-domain of Rad53 and inactivate the Rad53-dependent pathway due to Rad53 dephosphorylation (Leroy et al. 2003). It should be noted that Ptc3 phosphatase also dephosphorylates the regulatory residue T169 of the CDC28 kinase (Mendenhall and Hodge 1998) and is also a target for the CDC28 kinase (Ubersax et al. 2003). It may be assumed that the CDC28 kinase is involved both in activation and inactivation of checkpoint.

The Mec1 kinase plays a dual role in the activation of the Rad53 kinase. After detection of DNA damage, Mec1/Ddc2 phosphorylates the adaptor molecule of Rad53 by multiple S/T-Q motifs. Phosphorylated Rad9 recruits Rad53 to the place of damage by the FHA-dependent way. The delivery of Rad53 to the place of damage determines direct phosphorylation by Mec1 Rad53 bound to the Rad9 protein. Phosphorylation of multiple Rad53 motifs makes possible autophosphorylation, which terminates the conversion of the kinase to the active form (Vialard et al. 1998; Gilbert et al. 2001). Autophosphorylation of Rad53 leads to a nine fold increase of activity and depends on the kinase concentration (Ma et al. 2006). DNA damage causes oligomerization of part of Rad53 molecules in vivo. At a low concentration, preincubation of Rad53 with the Mec1/Ddc2 complex induces activation of the Rad53 kinase. Hyperphosphorylated active Rad53 kinase relieves itself of Rad9 releasing it for the next rounds of Rad53 activation. It has been shown that the kinase activity of Rad53 is necessary to perform a function in checkpoint control (Fay et al. 1997; Kim and Weinert 1997).

The protein Mrc1 (*mediator of the replication checkpoint*) mediates the Rad53 kinase activity under replication block conditions in response to the depletion of nucleotides or to the presence of alkylated damage (Alcasabas et al. 2001; Tanaka and Russell 2001). As Rad9, the adaptor protein Mrc1 is hyperphosphorylated, but only in response to the replication block. The protein contains a canonic sequence with several SQ/TQ motifs which are sites of Mec1 kinase-mediated phosphorylation (Kim et al. 1999). In the absence of an immediate response, the replication stress causes DNA damage in the *mrc1* mutant. The damage activates the adaptor protein Rad9 and DNA damage-induced checkpoint with subsequent activation of the Rad53 kinase. In the *mrc1* mutant, the Rad53 kinase is not activated in the absence of Rad9 (Alcasabas et al. 2001).

As in the case of Rad9, the Rad53 kinase can be activated by the phosphorylated adaptor protein Mrc1 through the delivery of the Rad53 kinase to the Mec1/Ddc2 complex, permitting direct phosphorylation or autophosphorylation. Then the activated Rad53 kinase itself phosphorylates the adaptor protein Mrc1. Probably, further activation of the Rad53 kinase is thus promoted through the formation of an activation loop. Mutations at the Mrc1 protein phosphorylation sites prevent Mrc1 phosphorylation and block Rad53 activation, but they do not affect the function of the adaptor protein Mrc1 in replication.

The adaptor protein Rad9 participates in phosphorylation and activation of two kinases, Rad53 and Chk1 (Schwartz et al. 2002; Gillbert et al. 2001; Sanchez et al. 1999). Correspondingly, Rad9 has two domains (Fig. 1) activating different signaling kinase pathways independently of one another (Blankley and Lydall 2004). The N-terminal region contains the CAD domain (*Chk1 activation domain*) of phosphorylation and activation of the Chk1 kinase required for signal transduction in the case of cell division arrest induced by damage in the telomeric region of chromosomes in the *cdc13-1* and *yku70Δ* mutants, but not upon UV light induction or inhibition of telomeric nuclease. The substitution of residues blocking DNA damage-induced phosphorylation of the SQ and TQ motifs of Rad9 disturbs only the Rad53-mediated response rather than the Chk1-dependent one (Schwartz et al. 2002). The Chk1 kinase has no FHA-domains. It seems likely that the adaptor protein Rad9 uses different mechanisms in regulating the activity of Rad53 and Chk1 kinases.

In intact cells, regulated and cell cycle stage-dependent phosphorylation of Rad9 takes place (Vialard et al. 1998), and in the case of DNA damage it is hyperphosphorylated by the Tel1 and Mec1 kinases. Rad9 has potential sites of phosphorylation by the CDC28 kinase that controls the cell cycle progression (Ubersax et al. 2003). The CAD-domain has no sites of phosphorylation by the Mec1 kinase, but has three potential sites of phosphorylation by the CDC28 kinase. It has to be elucidated if indeed CDC28-dependent phosphorylation regulates the Rad9-Chk1 signaling pathway. It should be noted that potential sites of phosphorylation are conserved, and they were discovered in orthologs of Rad9.

Phosphorylation of Chk1 and Rad53 leads to activation of the G1/S, intra S, or G2/M checkpoints and to the cell cycle arrest, as well as to transcriptional activation of repair genes, repression of cyclin transcription, and to stabilization of replication forks (Chen and Sanchez 2004; Weinert 1998). The Rad53 kinase plays a decisive role throughout the whole cell cycle in checkpoint control of DNA damage and replication blocking (Longhese et al. 1998), while the Chk1 kinase functions only in G2/M phase in the response to DNA damage (Sanchez et al. 1999). The role of Chk1 in S checkpoint was not demonstrated, but indirect evidence points to the possibility of an optional role of the Chk1 kinase in the response to the replication block: *dun1* cells without Chk1 are more sensitive to HU than *dun1* null mutants (Schollaert et al. 2004).

Dun1 Kinase

The signaling kinase cascade also involves another substrate of the Rad53 kinase, Dun1 kinase. The Dun1 kinase takes part in the activation of repair genes as a result of a replication block or DNA damage (Zhou and Elledge 1993), in G2/M checkpoint (Pati et al. 1997; Gardner et al. 1999), and in inducible phosphorylation of the Rad55 repair protein (Bashkirov et al. 2003). Rad55 is also phosphorylated by the Rad53 kinase (Herzberg et al. 2006). The dihybrid analysis showed the existence of physical interaction between Rad53 and Rad55. The Rad53 kinase directly phosphorylates S14-Rad55. The Dun1 kinase belongs to the Chk2-family of protein kinases that also includes the human kinase Chk2/hCds1. These kinases have at least one FHA-domain (Fig. 1), which specifically binds to a phosphorylated amino acid residue, predominantly to phosphothreonine T(P). The FHA-domain of the Dun1 kinase is required for direct phosphorylation of Dun1 by Rad53 in vitro and in vivo and apparently participates in the interaction between Rad53 and Dun1 (Bashkirov et al. 2003). Phosphorylation of the T380 residue by Rad53 in the activation loop of Dun1 is responsible for Dun1 activation (Chen et al. 2007). It was found that the T354 residue in the activation loop of Rad53 is necessary for its activation. Phosphorylation of threonine in the activation loop is a conserved mechanism of activation of kinases of the Chk2-family.

The Dun1 kinase provides the synthesis of deoxyribonucleotides (dNTP) induced by DNA damage or by replication stress via transcriptional activation of ribonucleotide reductase (RNR) genes (Huang et al. 1998) and inactivation of the ribonucleotide reductase inhibitor Sml1 (Zhou and Elledge 1993; Zhao and Rothstein 2002). RNR catalyzes the limiting stage of dNTP synthesis: conversion of dNDP to dNTP. The RNR activity directly influences the level and balance of the nucleotide pool and is regulated in the course of replication and repair. Transcription of *RNR* genes is regulated by the transcription corepressor Crt1-Ssn6-Tup1. When DNA replication is blocked, Dun1 takes part in phosphorylation of the transcription corepressor subunit Crt1. The result is dissociation of Crt1 from the promoter and subsequent derepression of transcription of *RNR* genes (Huang et al. 1998). Transcription of repair genes after DNA damage is induced by a different way with the involvement of the regulators of transcription Swi4 and Swi6. An immediate target of the Rad53 kinase is the transcription factor Swi6 participating in transcription of some inducible genes in response to DNA damage. This leads to a delay of the entry of cells into S phase as a result of inhibition of transcription of cyclins genes *CLN* (Sidorova and Breeden 1997). Thus, Swi6 is a target for G1/S arrest. The Dun1 kinase takes part in G2/M checkpoint by inhibiting Clb2 cyclin degradation and the exit from mitosis by blocking activation of the MEN (*mitotic exit network*) complex (Pati et al. 1997; Gardner et al. 1999).

Dun1 is involved in the regulation of postreplication repair (Hammet et al. 2002). The block of replication induces temporary binding of the Dun1 kinase to the Pan2/Pan3 complex, thus providing accessibility of the Ccr4/Caf1 complex and its activation by Dun1. The Pan2/Pan3 and Ccr4/Caf1 complexes regulate stability of *RAD5*-mRNA. Since the *RAD5* gene belongs to the *RAD6*-group of postreplication repair genes, Dun1 indirectly influences the work of this repair pathway. Rad6 has the RING-finger domain mediating the interaction of K63-specific ubiquitin ligase Ubc13 and Mms2 with Rad18 and K48-specific ubiquitin ligase Rad6 (Ulrich and Jentsch 2000). The heterocomplex generates a signal activating DNA repair through the HR pathway (i.e., error-free) instead of NHEJ (i.e., error-prone) (Ulrich and Jentsch 2000; Xiao et al. 2000; Brusky et al. 2000). The protein surface regions mediating the cooperation of Rad5 and Rad18 can also mediate homodimerization of the corresponding subunits and cause dissociation of the pentamer in subcomplexes Rad5-Ubc13-Mms2 and Rad6-Rad18, which induce a Rad6-dependent signal for activating the error-prone pathway (Ulrich and Jentsch 2000). Thus, the target process of a checkpoint response is regulation of gene expression not only through control of transcription activation (Zhou and Elledge 2000) but also through cotranscriptional regulation of posttranscriptional reactions of mRNA 3'end processing and polyadenylation (Kleiman and Manley 2001). The Dun1 kinase has several additional functions in repair of DNA damage along with the checkpoint function of cell division arrest. These include induction of transcription of repair enzyme genes (Huang et al. 1998) and direct modification of repair proteins, e.g., phosphorylation of the repair protein Rad55 (Bashkirov et al. 2000) and Nej1 component of DNA ligase IV (Ahnesorg and Jackson 2007). Nej1 was found to contain the site of phosphorylation by the Dun1 kinase. Mutation of this site reduces NHEJ efficiency.

Conclusion

The participants in checkpoint control activation are more or less known, whereas the target proteins are known only in rare cases. They require further investigation and are the subject of individual consideration.

References

Ahnesorg P, Jackson SP (2007) The non-homologous end-joining protein Nej1p is a target of the DNA damage checkpoint. DNA Repair 6:190–201 (Amst)

Alcasabas AA, Osborn AJ, Bachant J et al (2001) Mrc1 transduces signals of DNA replication stress to activate Rad53. Nat Cell Biol 3:958–965

Aylon Y, Kupiec M (2005) Cell cycle-dependent regulation of double-strand break repair: a role for the CDK. Cell Cycle 4:E61–E63

Bartova I, Otyepka M, Kriz Z, Koca J (2004) Activation and inhibition of cyclin-dependent kinase-2 by phosphorylation: a molecular dynamics study reveals the functional importance of the glycine-rich loop. Protein Sci 13:1449–1457

Bartrand AJ, Iyasu D, Brush GS (2004) DNA stimulates Mec1-mediated phosphorylation of replication protein A. J Biol Chem 279:2762–26767

Bashkirov VI, King JS, Bashkirova EV et al (2000) DNA repair protein Rad55 Is a terminal substrate of the DNA damage checkpoints. Mol Cell Biol 20:4393–4404

Bashkirov VI, Bashkirova EV, Haghnazari E, Heyer WD (2003) Direct kinase-to-kinase signaling mediated by the FHA phosphoprotein recognition domain of the Dun1 DNA damage checkpoint kinase. Mol Cell Biol 23:1441–1452

Blankley RT, Lydall D (2004) A Domain of Rad9 specifically required for activation of Chk1 in budding yeast. J Cell Sci 117:601–608

Bosch M, Lowndes NF (2004) Remodelling the Rad9 checkpoint complex: preparing Rad53 for Action. Cell Cycle 3:119–122

Brush GS, Kelly TJ (2000) Phosphorylation of the replication protein a large subunit in the saccharomyces cerevisiae checkpoint response. Nucleic Acids Res 28:3725–3732

Brush GS, Morrow DM, Hieter P, Kelly TJ (1996) The ATM homologue MEC1 is required for phosphorylation of replication protein A in yeast. Proc Natl Acad Sci USA 93:15075–15080

Brusky J, Zhu Y, Xiao W (2000) UBC13, a DNA-damage-inducible gene, is a member of the error-free postreplication repair pathway in saccharomyces cerevisiae. Curr Genet 37:168–174

Chen Y, Sanchez Y (2004) Chk1 in the DNA damage response: conserved roles from yeasts to mammals. DNA Repair 3:1025–1032 (Amst)

Chen SH, Smolka MB, Zhou H (2007) Mechanism of Dun1 activation by Rad53 phosphorylation in saccharomyces cerevisiae. J Biol Chem 282:986–995

Chiolo I, Carotenuto W, Maffioletti G et al (2005) Srs2 and Sgs1 DNA helicases associate with Mre11 in different subcomplexes following checkpoint activation and CDK1-mediated Srs2 phosphorylation. Mol Cell Biol 25:5738–5751

Deide SJ, Gottschling DE (2001) Exonuclease activity is required for sequence addition and Cdc13p loading at a de novo telomere. Curr Biol 11:1336–1340

Durocher D, Henckel J, Fersht AR, Jackson SP (1999) The FHA domain is a modular phosphopeptide recognition motif. Mol Cell 4:387–394

Elledge SJ (1996) Cell cycle checkpoints: preventing an identity crisis. Science 274:1664–1672

Ellison V, Stillman B (2003) Biochemical characterization of DNA damage checkpoint complexes: clamp loader and clamp complexes with specificity for 5' recessed DNA. PLoS Biol 1:E33

Emili A (1998) MEC1-dependent phosphorylation of Rad9p in response to DNA damage. Mol Cell 2:183–189

Fay DS, Sun Z, Stern DF (1997) Mutations in SPK1/RAD53 that specifically abolish checkpoint but not growth-related functions. Curr Genet 31:97–105

Gardner R, Putnam CW, Weinert T (1999) RAD53, DUN1 and PDS1 define two parallel G2/M checkpoint pathways in budding yeast. EMBO J 18:3173–3185

Garvik B, Carson M, Hartwell L (1995) Single-stranded dna arising at telomeres in cdc13 mutants may constitute a specific signal for the RAD9 checkpoint. Mol Cell Biol 15:6128–6138

Gavin AC, Aloy P, Grandi P et al (2006) Proteome survey reveals modularity of the yeast cell machinery. Nature 440:631–636

Gaziev AI (1999) DNA damage in cells exposed to ionizing radiation. Radiats Biol Radioekol 39:630–638

Gilbert CS, Green CM, Lowndes NF (2001) Budding yeast Rad9 is an ATP-dependent Rad53 activating machine. Mol Cell 8:129–136

Glazer VM, Glazunov AV (1999) Molecular genetic analysis of DNA double-strand break repair in yeast saccharomycetes. Russ J Genet 35:1449–1469

Grandin N, Charbonneau M (2003) Mitotic cyclins regulate telomeric recombination in telomerase-deficient yeast cells. Mol Cell Biol 23:9162–9177

Green CM, Erdjument-Bromage H, Tempst P, Lowndes NF (2000) A novel Rad24 checkpoint protein complex closely related to replication factor C. Curr Biol 10:39–42

Gusev NB (2000) Proteinkinases: structure, classification, properties and biological role. Soros Obraz Zh 6:4–12

Hammet A, Pike BL, Heierhorst J (2002) Posttranscriptional regulation of the RAD5 DNA repair gene by the Dun1 kinase and the Pan2–Pan3 poly(A)-nuclease complex contributes to survival of replication blocks. J Biol Chem 277:22469–22474

Herzberg K, Bashkirov VI, Rolfsmeier M et al (2006) Phosphorylation of Rad55 on serines 2, 8, and 14 is required for efficient homologous recombination in the recovery of stalled replication forks. Mol Cell Biol 26:8396–8409

Hoeksta MF, Demaggio AJ, Dhillon N (1991a) Genetically identified protein kinases in yeast: I. Transcription, translation, transport and mating. Trends Genet 7:256–261

Hoeksta MF, Demaggio AJ, Dhillon N (1991b) Genetically identified protein kinases in yeast: II. DNA metabolism and meiosis. Trends Genet 7:293–297

Huang M, Zhou Z, Elledge SJ (1998) The DNA replication and damage checkpoint pathways induce transcription by inhibition of the Crt1 repressor. Cell 94:595–605

Ira G, Pellicioli A, Balijja A et al (2004) DNA end resection, homologous recombination and DNA damage checkpoint activation require CDK1. Nature 431:1011–1017

Kaur R, Kostrub CF, Enoch T (2001) Structure–function analysis of fission yeast Hus1-Rad1-Rad9 checkpoint complex. Mol Biol Cell 12:3744–3758

Kholmurodov KhT, Kretov DA, Gerasimova AS, Koltovaya NA (2006) Molecular dynamics simulation of replacing of conserved glycine by serine in the G-loop in the *cdc28-srm* yeast mutant using the crystal lattice of human CDK2 kinase. Biofizika 51:679–691

Kim S, Weinert TA (1997) Characterization of the checkpoint gene *RAD53/MEC2* in *Saccharomyces cerevisiae*. Yeast 13:735–745

Kim HS, Brill SJ (2003) *MEC1*-dependent phosphorylation of yeast RPA1 in vitro. DNA Repair 2:1321–1335 (Amst)

Kim ST, Lim DS, Canman CE, Kastan MB (1999) Substrate specificities and identification of putative substrates of ATM kinase family members. J Biol Chem 274:3738–3743

Kleiman FE, Manley JL (2001) The BARD1-CstF-50 interaction links mRNA 3' end formation to DNA damage and tumor suppression. Cell 104:743–753

Korolev VG (2007) Molecular mechanisms of DNA double-strand break repair in eukaryotes. Radiats Biol Radioekol 47:389–401

Lee SE, Moore JK, Holmes A et al (1998) Ku70, Mre11/Rad50 and RPA proteins regulate adaptation to G2/M arrest after DNA damage. Cell 94:399–409

Lee SE, Paques F, Sylvan J, Haber JE (1999) Role of yeast *SIR* genes and mating type in directing DNA double-strand breaks to homologous and non-homologous repair paths. Curr Biol 9:767–770

Leroy C, Lee SE, Vaze MB et al (2003) PP2C phosphatases Ptc2 and Ptc3 are required for DNA checkpoint inactivation after a double-strand break. Mol Cell 11:827–835

Liao H, Byeon IJ, Tsai MD (1999) Structure and Function of a New Phosphopeptide-Binding Domain Containing the FHA2 of Rad53. J Mol Biol 294:1041–1049

Liberi G, Chiolo I, Pellicioli A et al (2000) Srs2 DNA helicase is involved in checkpoint response and its regulation requires a functional Mec1-dependent pathway and Cdk1 activity. EMBO J 19:5027–5038

Lindsey-Boltz LA, Bermudez VP, Hurwitz J, Sancar A (2001) Purification and characterization of human DNA damage checkpoint Rad complexes. Proc Natl Acad Sci USA 98:11236–11241

Longhese MP, Fraschini R, Plevani P, Lucchini G (1996) Yeast *pip3/mec3* mutants fail to delay entry into S phase and to slow DNA replication in response to DNA damage, and they define a functional link between Mec3 and DNA primase. Mol Cell Biol 16:3235–3244

Longhese MP, Foiani M, Muzi-Falconi M et al (1998) DNA damage checkpoint in budding yeast. EMBO J 17:5525–5528

Ma JL, Lee SJ, Duong JK, Stern DF (2006) Activation of the checkpoint kinase Rad53 by the phosphatidyl inositol kinase-like kinase Mec1. J Biol Chem 281:3954–3963

Majka J, Burgers PM (2003) Yeast Rad17/Mec3/Ddc1: a sliding clamp for the DNA damage checkpoint. Proc Natl Acad Sci USA 100:2249–2254

Melo JA, Cohen J, Toczyski DP (2001) Two checkpoint complexes are independently recruited to sites of DNA damage in vivo. Genes Dev 15:2809–2821

Mendenhall MD, Hodge AE (1998) Regulation of Cdc28 cyclin-dependent protein kinase activity during the cell cycle of the yeast. Micriobiol Mol Biol Rev 62:1191–1243

Morrow DM, Tagle DA, Shiloh Y et al (1995) *TEL1*, an *S. cerevisiae* homolog of the human gene mutated in ataxia telangiectasia, is functionally related to the yeast checkpoint gene *MEC1*. Cell 82:831–840

Naiki T, Wakayama T, Nakada D et al (2004) Association of Rad9 with double-strand breaks through a Mec1-dependent mechanism. Mol Cell Biol 24:3277–3285

Nakada D, Shimomura T, Matsumoto K, Sugimoto K (2003) The ATM-related Tel1 protein of *Saccharomyces cerevisiae* controls a checkpoint response following phleomycin treatment. Nucleic Acids Res 31:1715–1724

Nakada D, Hirano Y, Tanaka Y, Sugimoto K (2005) Role of the C-terminus of Mec1 checkpoint kinase in its localization to sites of DNA damage. Mol Biol Cell 16:5227–5235

Osborn AJ, Elledge SJ (2003) Mrc1 is a replication fork component whose phosphorylation in response to DNA replication stress activates Rad53. Genes Dev 17:1755–1767

Pati D, Keller C, Groundine M, Plon SE (1997) Reconstitution of a *MEC1*-independent checkpoint in yeast by expression of a novel human fork head cDNA. Mol Cell Biol 17:3037–3046

Raymond WE, Kleckner N (1993) RAD50 Protein of *S. cerevisiae* exhibits ATP-dependent DNA binding. Mol Gen Genet 238:390–400

Robison JG, Elliott J, Dixon K, Oakley GG (2004) Replication protein A and the Mre11-Rad50-Nbs1 complex co-localize and interact at sites of stalled replication forks. J Biol Chem 279:34802–34810

Rouse J, Jackson SP (2002) Interfaces between the detection, signaling, and repair of DNA damage. Science 297:547–551

Sanchez Y, Desany BA, Jones WJ et al (1996) Regulation of RAD53 by the ATM-like kinases MEC1 and TEL1 in yeast cell cycle checkpoint pathways. Science 271:357–360

Sanchez Y, Bachant J, Wang H et al (1999) Control of the DNA damage checkpoint by chk1 and rad53 protein kinases through distinct mechanisms. Science 286:1166–1171

Schollaert K, Poisson J, Searle JJ et al (2004) A role for *Saccharomyces cerevisiae* Chk1p in the response to replication blocks. Mol Biol Cell 15:4051–4063

Schwartz MF, Duong JF, Sun Z et al (2002) Rad9 phosphorylation sites couple Rad53 to the *Saccharomyces cerevisiae* DNA damage checkpoint. Mol Cell 9:1055–1065

Sidorova JM, Breeden LL (1997) Rad53-dependent phosphorylation of Swi6 and down-regulation of CLN1 and CLN2 transcription occur in response to DNA damage in *Saccharomyces cerevisiae*. Genes Dev 11:3032–3045

Soulier J, Lowndes NF (1999) The BRCT domain of the *S. cerevisiae* checkpoint protein Rad9 mediates a Rad9–Rad9 interaction after DNA damage. Curr Biol 9:551–554

Sun Z, Hsiao J, Fay DS, Stern DF (1998) Rad53 FHA domain associated with phosphorylated Rad9 in the DNA damage checkpoint. Science 281:272–274

Sweeney FD, Yang F, Chi A et al (2005) *Saccharomyces cerevisiae* Rad9 acts as a Mec1 adaptor to allow Rad53 activation. Curr Biol 15:1364–1375

Tanaka K, Russell P (2001) Mrc1 channels the DNA replication arrest Signal to checkpoint kinase Cds1. Nat Cell Biol 3:966–972

Tsukamoto Y, Mitsuoka C, Terasawa M et al (2005) Xrs2p regulates Mre11p translocation to the nucleus and plays a role in telomere elongation and meiotic recombination. Mol Biol Cell 16:597–608

Ubersax JA, Woodbury EL, Quang PN et al (2003) Targets of the cyclin-dependent kinase Cdk1. Nature 425:859–864

Ulrich HD, Jentsch S (2000) Two RING finger proteins mediate cooperation between ubiquitin-conjugating enzymes in DNA repair. EMBO J 19:3388–3397

Vialard JE, Gilbert CS, Green CM, Lowndes NF (1998) The budding yeast Rad9 checkpoint protein is subjected to Mec1/Tel1-dependent hyperphosphorylation and interacts with Rad53 after DNA damage. EMBO J 17:5679–5688

Weinert T (1998) DNA damage checkpoints update: getting molecular. Curr Opin Genet Dev 8:185–193

Xiao W, Chow B, Broomfield S, Hanna M (2000) The *Saccharomyces cerevisiae* RAD6 group is composed of an error-prone and two error-free postreplication repair pathways. Genetics 155:1633–1641

Zhao X, Rothstein R (2002) The Dun1 checkpoint kinase phosphorylates and regulates the ribonucleotide reductase inhibitor Sml1. Proc Natl Acad Sci USA 99:3746–3751

Zhou Z, Elledge SJ (1993) *DUN1* encodes a protein kinase that controls the DNA damage response in yeast. Cell 75:1119–1127

Zhou BB, Elledge SJ (2000) The DNA damage response: putting checkpoints in perspective. Nature 408:433–439

Zou L, Elledge SJ (2003) Sensing DNA damage through ATRIP recognition of RPA–ssDNA complexes. Science 300:1542–1548

Zou L, Liu D, Elledge SJ (2003) Replication protein A—mediated recruitment and activation of Rad17 complexes. Proc Natl Acad Sci USA 100:13827–13832

Part III
Radiobiology Effects and Mechanisms

The Evolution of Radiobiological Thought: Past History and Future Predictions

Carmel Mothersill and Colin Seymour

Early Ideas

Theories concerning the nature of the action of radiation date back to its discovery in 1895 by German physicist Wilhelm Roentgen who discovered a kind of ray that could travel through solid wood or flesh and yield radiographs of living people's bones (Lea 1962) Roentgen called them X-rays, with X standing for unknown. In 1896 French physicist Henri Becquerel reported that uranium compounds, emitted rays that would fog a photographic plate. The scientific community with the exception of the Curies[1] continued to focus its attention on Roentgen's X-rays, ignoring the weaker Becquerel rays or uranium rays. At this stage most of the research was on the physics and chemistry of the interaction between the rays and living tissue (Lea 1962). Using novel equipment the Curies characterised the electric currents emitted by uranium and suggested it was a fundamental property of the atom and not changed by the physical or chemical state of the element. This was a fundamental conceptual shift which ultimately leads to the major revolutions in the field of atomic structure and quantum physics (Lea 1962).

The fact that radiation could cause biological effects was recognised almost immediately. Burns were described by Roentgen who concluded that the rays could harm living tissue and therapeutic effects were also described early on by Tarkhanov who after irradiating frogs and insects with X-rays in early 1896, several weeks after Röntgen's discovery, concluded that these newly discovered rays not

[1]Marie Skłodowska-Curie and Pierre Curie.

C. Mothersill (✉) · C. Seymour
Department of Medical Physics and Applied Radiation Sciences, McMaster University, Hamilton, ON, Canada
e-mail: mothers@mcmaster.ca

C. Seymour
e-mail: seymouc@mcmaster.ca

V.L. Korogodina et al. (eds.), *Genetics, Evolution and Radiation*,
DOI 10.1007/978-3-319-48838-7_12

only photograph, but also "affect the living function" (Kudriashov et al. 2008). Calabrese has researched the early use of X-rays to treat a number of diseases e.g. (Calabrese and Dhawan 2013; Calabrese et al. 2014, 2015) however mechanisms were not really considered until much later and the genetic consequences were not defined until 1927 when Hermann Mueller published data showing genetic effects of radiation resulting in chromosomal damage (Muller 1927). The first hint of controversies to come is researched in exquisite detail by Calabrese (2015) who details the history of this time. He reveals that Mueller's work which is the acknowledged basis of ideas that the effect (in this case number of mutations) increased as the dose, ignored work going on in a colleague's laboratory which appears to show the first reports of low dose non-linearity seen as sparing/adaptive response/hormesis. This debate is still in full swing.

The Genetic Revolution

Despite the doubts now apparent, in the day the genetic revolution was dominant and increasing genetic damage induced by increasing radiation dose was the dominant theory of radiation action. In 1932 Mueller moved to Berlin, to work with Nikolai Timofeev-Ressovskii a Russian geneticist. Radiation genetics was driving the field of genetics at the time because of the ease of quantifying exposure and the perceived ability to control dose by using external sources of radiation. This limited the problem experienced with chemicals where the chemical remained in the body for varying times, was metabolised and targeted specific organs. None of these problems occurred with X-rays. Additionally, by using model organisms such as bacteria or fruit flies, quantitative data concerning lethal doses to populations could be reliably produced (reviewed in Timofeev-Resovsky et al. 1971). The genetic revolution which attributed virtually every characteristic of life to "genetics" to the exclusion of environmental, social, neural or behavioural science, was also extremely important at the time as it provided a "scientific" basis for some of the extreme ideologies of the time which peaked as Eugenics in Nazi Germany and Lysenkoism in Russia. Survival of scientists in this period often depended on doing research which "bought in" to the prevailing ideologies. Maybe to a less extreme extent—such political manipulation of scientific direction is one of the major limitations on scientific progress even today.

Discovery of DNA and the Evolution of Target Theory

The discovery of DNA and the formalization of target theory were occurring around the same time. Friedrich Miescher discovered nuclein in 1869 (discussed in Dahm 2007). This eventually was named nucleic acid, then deoxyribonucleic acid. In 1919 Phoebus Levene reported "there is no doubt of the polynucleotide structure of the yeast nucleic acid." (Levene 1919). In 1944 Oswald Avery demonstrated that hereditary units, or genes, are composed of DNA (Avery et al. 1944). In 1953 Watson and Crick propose a double helix structure (Watson and Crick 1953).

In 1935, Timofeev-Resovskij published the major work, *Über die Natur der Genmutation und der Genstruktur*, with Karl Zimmer, and Max Delbrück (the green book) (Timofeev-Resovskii et al. 1935). It was considered to be a major advance in understanding the nature of gene mutation and gene structure. The work was a keystone in the formation of molecular genetics and it also led to formalised target theory in radiobiology which was published by Lea in 1946 providing a mathematical framework for the theory (Lea 1946). The overlap between target theory and DNA research led to acceptance that DNA was the primary target of radiation damage, not only in the target theory model, but also models proposed by Katz in 1967 (Katz et al. 1972) known as the "Amorphous Track Structure Model".

Kellerer and Rossi (1974) proposed "The Theory of Dual Radiation Action" and Chadwick and Leenhouts (1973) proposed "The molecular (L-Q) model" all of which were modifications of target theory in that radiation hits to the cell caused damage.

Coincidentally, the adoption of the clonogenic assay post 1956 (Puck and Marcus 1956) as a major technique in radiobiology and the limits this placed on the cell lines which could be used (i.e. high PE and ability to form colonies), led to a self fulfilling prophesy.

Takuma Nomiya writing in J. Radiat. Res. in 2013 (Nomiya 2013) said "Target theory is one of the essential concepts for understanding radiation biology. Although many complex interpretations of target theory have been developed, its fundamental principle is that 'inactivation of the target(s) inside an organism by radiation results in the organism's death". Nomiya admits that the number of targets and their location within the organism is not always clear but that the target can be considered as a "fundamental biological unit". The fact that this thinking is still dominant in 2013 is quite startling given the evolution of thought in biology including omics and system approaches (Tapio 2013; Hatzi et al. 2015). It may perhaps be partly explained by scientific demographics. When radiation was discovered and ever since, a considerable knowledge of or collaboration with nuclear physicists was required to use the technology. Instrumentation is complex and precise measurement of dose and calibration of machines requires mathematics and physics knowledge at a level which is often not part of life sciences education even today. In addition, the main source of employment involving radiation use, outside of academia, was the nuclear industry or government defence departments, which heavily recruited engineers and physicists, with perhaps a token epidemiologist. This pattern is changing now due to much greater interdisciplinary collaboration and due to the accelerating pace of instrumentation and technological development in the life sciences permitting interrogation of basic mechanisms in radiation biology.

Physics Meets Biology

As a result of increased collaborations, the disciplines of physics and biology are now meeting and melding. It is important to appreciate the completely different conceptual frameworks and research approaches which have always existed in these disciplines. The mind of the physicist thinks in mathematical terms as a fundamental language, where problems are defined by equations, solutions postulated and experiments done to validate or prove that the predicted solutions are correct. When things do not fit as in the quantum revolution, it is assumed that there is another way to solve the equation or that something fundamental has been missed. The rules of mathematics allow for emergent properties of systems which were unpredictable to exist and allow for inductive thinking which is almost heresy in biology (although this is changing!).

Biology developed much more as a verbal, descriptive science where observation was critical and documentation of cause and effect relationships led to advances where so-called "breakthroughs" resulted from the sudden coming together of seemingly unrelated experimental results. To put it simply, in physics, the focus is on cause while in biology the focus is on effect. The relevance of this to radiobiology is that while physicists traditionally studied tracks and ray or particle interactions with living material, biologists became interested in how the mutations arose, what biochemistry underlies carcinogenesis what signaling pathways are turned on in cells receiving a dose and similar questions. It is interesting to note here that the discipline of radioecology is currently undergoing a major transition from cause (dose) driven analysis concerning transfer pathways and concentration ratios etc. to effect driven research exemplified by moves to an ecosystem approach (Mothersill and Seymour 2010; Bréchignac and Paquet 2013; Bradshaw et al. 2014; Aleksakhin et al. 2014)—perhaps something similar is happening in radiobiology with the move to system biology approaches discussed later.

The Age of Reductionism

Reductionism is the belief that a system is no more than the sum of its parts. This means that to completely understand a system, you must understand it at the most basic level, such as at the atomic level, and from there, you can rebuild the system. There are always intermediate levels that can be used, such as starting from DNA in a cell rather than trying to span from sub-atomic particles but such intermediate steps are temporary, until an understanding of a more basic level is obtainable. Berk in his very interesting discussion paper entitled "Reductionism and the failure of radiobiology" (Berk 2004) discusses the failure of radiobiology to make major contributions to radiotherapy. He concludes that the problem is the assumption that studying simple models (such as cell cultures or inbred laboratory animals) enables one to gain information of use in human medicine. The big problem with such reductionist thinking is that it ignores the possibility of emergent properties of systems which are not predictable from the basic understanding of parts. It also precludes consideration of chaotic mechanisms or complexity theory. Thus it is a severely limiting concept which up until recently has dominated biology. In physics the quantum revolution did much to move the field away from reductionism (Berk 2004). Reductionist approaches also prevent thinking "outside the box" which questions fundamental assumptions in a field or at least requires reexamination at regular intervals as scientific thought and available techniques advance. Such "inductive" thinking lead directly to the birth of the "new radiobiology" which came originally from questioning of fundamental assumptions in genetics (Seymour et al. 1986; Kadhim et al 1992; Mothersill and Seymour 2012).

The Birth of the "New Radiobiology"

Classic radiobiology developed around target theory which had a number of predictions. Over the years, these predictions became beliefs and were rarely questioned. They became enshrined on the flimsiest of evidence as truths (Mothersill and Seymour 2013). Examples are the concept of iso-effect per fraction which really only applies in established cell lines with high plating efficiency and no adaptive response (Elkind and Whitmore 1967; Hall and Giaccia 2012). Among the assumptions in radiobiology which in our opinion have seriously held back both radiotherapy and radiation protection are the following:

- Cells respond to radiation as individuals.
- "Elkind" repair will be complete between fractionated doses.
- The repaired cell will behave as though it were unirradiated.
- All lethal mutations will be expressed in the first mitosis post irradiation.

These ideas and many others were held as true and those finding data that did not fit were either doing the experiments incorrectly or were ignored. In fact there was a joke in radiobiology that if you wanted to find the first evidence of a new theory, you should look in a journal widely held to be the "last resort" where odd results which did not get through conventional establishment referees, often found a home.

Now modern biology has demonstrated the importance of cell signaling and in radiobiology, signaling is seen as increasingly important in the now accepted "non-targeted" phenomena of co-operative repair, abscopal and bystander effects and coordinated tissue responses to radiation exposure (Salomaa et al. 2010; Hei et al. 2011; Mothersill and Seymour 2012). The concepts of the microenvironment and stem cell niche are thought be critical to understanding tumour control by radiation (Blyth and Sykes 2011; Mannino and Chalmers 2011; Richardson 2011; Pajonk and Vlashi 2013; Yi et al. 2013; Liu et al. 2015) which is a major step forward from ideas that individual cell hypoxia or double strand break repair capacity were paramount (Berk 2004; Hall and Giaccia 2012). Other advances may have come more from the "fresh look" afforded by the entry of significant numbers of biologists into radiobiology although many of the leaders in classical radiobiology were well aware that the prevailing paradigms might not be entirely correct—to quote Mort Elkind "It is concluded that multiple mechanism may be involved in cell killing and, in addition, that the processes connected with the shoulder region of the survival curve amount to only the tip of an iceberg whose size and properties are yet to be fully appreciated" (Elkind et al. 1987).

The odd effects being described around the time of the above quotation include the evidence that repair of radiation damage may be incomplete or erroneous and that apparently normal fully repaired cells could carry the potential to express non-clonal, sporadic or unpredictable mutations at any time, many generations after the radiation exposure occurred and include mis-repair (Alper and Cramp 1989), lethal mutation (Alper et al. 1988), also known as delayed cell or reproductive death (Chang and Little 1991) and genomic instability (Wright 2000). Another related concept which was very controversial at this time was the idea that cells or organisms could have a "memory" of radiation damage. There were data suggesting that re-irradiation of a radiotherapy field resulted in adverse strong responses—the so-called "field effect" (Nieder et al. 2000; Kusunoki and Hayashi 2008; Halliday et al. 2012). On the other hand there were many datasets showing that prior exposure to low doses of radiation stimulated protective responses so that larger doses were less effective (Mitchel 2015). When these effects were systematically studied, it became apparent that dose, dose rate and timing were all important in determining outcome (Pateras et al. 2015).

Nowadays, most radiobiologists accept these phenomena as real and the term "non-targeted effects" has been coined to describe them but many consider them to be low dose effects governed by different mechanisms, the significance of which in radiotherapy or radiation protection, is unknown. However they are considered by most of the established agencies concerned with such matters, not to undermine the central role of DNA as the target determining radiobiological response (BEIR VII Report 2006; UNSCEAR Report 2010).

The Ecological Approach and the Rise of "System Radiobiology"

As mentioned previously, the debate over holism or reductionism pervades many fields in radiation science. It is particularly evident in radioecology which is grappling with the difficulty of looking at radiation effects in complex communities where multiple factors input to determining final outcome (Konradov 1994). Reductionists would claim that if we understood the effect of radiation on every species and the contributions of everything else in the ecosystem to the modulation of that effect, we would understand everything but apart from the fact that it is impossible to do that, the approach ignores complexity theory, emergence, iterative effects and the concept of critical tipping points or transition points known from chaos theory (Balzano and Sheppard 2003; Sheppard et al. 2008). These concepts are also being studied in system radiobiology although that field is dominated currently by attempts to develop simple models to describe radiobiological phenomena. This approach is likely to fail through ignoring the reality of complexity in biology although inductive reasoning can sometimes be used to develop a general model from empirical observations from particular experimental conditions (Dale and Jones 2007).

The Revival of Inductive Reasoning

Inductive reasoning is defined in Wikipedia as follows "Unlike deductive arguments, inductive reasoning allows for the possibility that the conclusion is false, even if all of the premises are true. Instead of being valid or invalid, inductive arguments are either *strong* or *weak*, which describes how *probable* it is that the conclusion is true. Another crucial difference is that deductive certainty is impossible in non-axiomatic systems, such as reality, leaving inductive reasoning as the primary route to (probabilistic) knowledge of such systems". The key points for radiobiological thought are that there is a recognition that the fundamental assumptions which lead to the design of experiments may be false. Also all results are viewed as probably supporting (or not) a particular theory. Thus inductive reasoning has led to the growth of modelling and system approaches and the acceptance of probabilistic theories. It has also been seen in the work of those who questioned the fundamental assumptions which stagnated the field in the 70s and the re-examination of which have led to the new understanding of non-targeted radiation effects. This rise in inductive and holistic thinking is not confined to radiobiology but is seen in other fields of science as well as in art (holism) and literature where it is a popular device used by modern authors to create suspense.

Where to Next?

So given this big shift from reductionism to a more holistic approach in radiobiology can we hope in the future for real contributions to therapy and risk analysis coming from basic radiobiology? Can we prove that Steele was wrong? He said in the introduction to his book *Basic Clinical Radiobiology* (1997).

> There is no doubt that radiobiology has been very fruitful in the generation of new ideas and in the identification of potentially exploitable mechanisms. A variety of new treatment strategies have been produced, but unfortunately few of these have so far led to demonstrable clinical gains … the ability of laboratory science to guide the radiotherapist in the choice of specific protocols is limited by the inadequacy of the theoretical and experimental models: It will always be necessary to rely on clinical trials for the final choice of a protocol. (Steele 1997)

We think that if the fundamental shift from certainty to probability is accepted, new approaches may become possible. A few suggested concepts which could lead to new strategies for proactive risk prevention or for development of new therapies are discussed below.

Population Level Response

Moving away from the concept of individual survival would totally change the approach to radiotherapy. Currently both chemotherapy and radiation treatments plan to deliver the highest dose to the tumour which can be tolerated without harm to normal tissue but the underlying philosophy is to kill every last tumour cell. If we think in terms of physiology and accept that cells respond within the context of their "society" then tumour infrastructure becomes a target and disruption of infrastructure may be achieved with far less toxic regimes. Similarly if cell communication and signaling is what has been compromised in a tumour, resetting that is an option which does not require sterilization of the tumour cells.

In terms of protection, population level responses mean that the linear-non-threshold (LNT) model of the dose response relationship for radiation carcinogenesis cannot be correct and research is needed to determine the spectrum of responses that are possible as well as the factors which determine the ultimate outcome.

Iterative Responses

Non-targeted radiation effects/responses in tissues also allow for iterative responses which change as the circumstances in the tissue/organ/organism change. Adaptive responses discussed below are one type of iterative response but the general process is more flexible and can include frequent subtle adjustments to cellular processes to optimize outcomes. Since these iterative processes occur, presumably, in tumour as well as normal cell populations, they could provide more novel intervention points for tumour control if they were fully understood.

Adaptive Responses

These may be thought of as a subgroup of iterative responses but in radiobiology they are extremely important because they mean that chronic exposures to low doses could, in theory, protect against a high dose. This again means that the LNT model does not really predict the radiation dose response. Evidence for reduced effect of high doses following exposure to low doses exists for wild animal species living in contaminated environments (Ulsh et al. 2004; Audette-Stuart et al. 2011), but an intriguing possibility is that nuclear medicine tests aimed at diagnosis or simulations conducted during treatment planning, might induce a systemic adaptive response which could alter response to radiotherapy. Other intriguing possibilities are that if adaptive responses are induced by radiation exposure, then the clinical assumption of iso-effect per fraction in radiotherapy should not be true. This subject was explored by the US Radiation Research Society History Symposium in 2015 and the conclusion was that a rigorous examination of the evidence did not in fact support iso-effect per fraction.

Relationship to Personalized Medicine?

Personalised medicine is a very popular idea now. It recognises that people differ in their responses to medical treatment and that preventative medicine and probiotics cannot be applied using a "one size fits all" approach. In radiobiology the issues are identifying radiosensitive individuals and tailoring treatment or preventative administration of radioprotectors to optimize benefit to individuals and not to the general population. Clearly to do this, it is essential to understand the drivers of radiation response and to have robust biomarkers capable of predicting response.

Multiple Stressor Issues

Finally a new area of concern in the field of environmental radiobiology and radioecology is how to deal with multiple stressors. We all know that there are many different chemical and physical agents in the environment including radiation but we do not know how to regulate radiation exposure to take account of other stressors which can or may modulate the radiation effect. This is one area where in the future complexity modeling may be very important.

Conclusion

To conclude, this review attempts to look at the historical evolution of radiobiological thought from the earliest days after radiation was discovered, through the years where mutations, double strand breaks in DNA and target theory dominated research, to the present day where non-targeted mechanisms have been identified. The implications of these mechanisms are still the subject of controversy and they are not included in risk estimates for radiation protection or in treatment planning for radiotherapy.

Acknowledgments We acknowledge continued support from the Canada Research Council Canada Research Chairs Programme, The Canadian Natural Sciences and Engineering Research Council (NSERC) and the National CFIDS Foundation Inc (Chronic Fatigue and Immune Deficiency Syndrome).

References

Aleksakhin RM, Udalova AA, Geras'kin SA (2014) V.I. Vernadskiĭ's theory of the biosphere and modern problems of radioecology. Radiat Biol Radioecol 54(4):432–439

Alper T, Cramp WA (1989) The role of repair in radiobiology. Experientia 45(1):21–33

Alper T, Mothersill C, Seymour CB (1988) Lethal mutations attributable to misrepair of Q-lesions. Int J Radiat Biol 54(4):525–530

Audette-Stuart M, Kim SB, McMullin D et al (2011) Adaptive response in frogs chronically exposed to low doses of ionizing radiation in the environment. J Environ Radioact 102(6):566–573

Avery OT, Macleod CM, McCarty M (1944) Studies on the chemical nature of the substance inducing transformation of pneumococcal types: induction of transformation by a desoxyribonucleic acid fraction isolated from pneumococcus type III. J Exp Med 79(2):137–158

Balzano Q, Sheppard A (2003) RF nonlinear interactions in living cells-I: nonequilibrium thermodynamic theory. Bioelectromagnetics 24(7):473–482

Berk LB (2004) Reductionism and the failure of radiobiology. JACR 1(5):304–307

BEIRVII (2006) Health Risks from Exposure to Low Levels of Ionizing Radiation Phase 2 report of Natl Acad Press. Available at http://www.nap.edu/read/11340/chapter/1

Blyth BJ, Sykes PJ (2011) Radiation-induced bystander effects: what are they, and how relevant are they to human radiation exposures? Radiat Res 176(2):139–157

Bradshaw C, Kapustka L, Barnthouse L et al (2014) Using an Ecosystem Approach to complement protection schemes based on organism-level endpoints. J Environ Radioact 136:98–104

Bréchignac F, Paquet F (2013) Radiation-induced risks at low dose: moving beyond controversy towards a new vision. Radiat Environ Biophys 52(3):299–301

Calabrese EJ, Dhawan G (2013) The historical use of radiotherapy in the treatment of sinus infections. Dose-Response Publ Int Hormesis Soc 11:469–479

Calabrese EJ, Dhawan G, Kapoor R (2014) Use of X-rays to treat shoulder tendonitis/bursitis: a historical assessment. Arch Toxicol 88(8):1503–1517

Calabrese EJ, Dhawan G, Kapoor R (2015) The use of X rays in the treatment of bronchial asthma: a historical assessment. Radiat Res 184(2):180–192

Calabrese EJ (2015) On the origins of the linear no-threshold (LNT) dogma by means of untruths, artful dodges and blind faith. Environ Res 142:432–442

Chadwick KH, Leenhouts HP (1973) A molecular theory of cell survival. Phys Med Biol 18 (1):78–87

Chang WP, Little JB (1991) Delayed reproductive death in X-irradiated Chinese hamster ovary cells. Int J Radiat Biol 60(3):483–496

Dahm R (2007) Discovering DNA: Friedrich Miescher and the early years of nucleic acid research. Hum Genet 122(6):565–581

Dale RG, Jones B (2007) Radiobiological modelling in radiation oncology. British Institute of Radiology Publications, London

Elkind MM, Whitmore GF (1967) The radiobiology of cultured mammalian cells. Gordon and Breach, New York

Elkind MM, Utsumi H, Ben-Hur E (1987) Are single or multiple mechanisms involved in radiation-induced mammalian cell killing? Br J Cancer Suppl 8:24–31

Hall EJ, Giaccia AJ (2012) Radiobiology for the radiologist. Lippincott Williams & Wilkins, Philadelphia

Halliday GM, Damian DL, Rana S, Byrne SN (2012) The suppressive effects of ultraviolet radiation on immunity in the skin and internal organs: implications for autoimmunity. J Dermatol Sci 66(3):176–182
Hatzi VI, Laskaratou DA, Mavragani IV et al (2015) Non-targeted radiation effects in vivo: a critical glance of the future in radiobiology. Cancer Lett 356(1):34–42
Hei TK, Zhou H, Chai Y et al (2011) Radiation induced non-targeted response: mechanism and potential clinical implications. Curr Mol Pharmacol 4(2):96–105
Kadhim MA, Macdonald DA, Goodhead DT et al (1992) Transmission of chromosomal instability after plutonium alpha-particle irradiation. Nature 355(6362):738–740
Katz R, Sharma SC, Homayoonfar M (1972) The structure of particle tracks, vol 1. In: Topics in radiation dosimetry: radiation dosimetry, p 568. Elsevier
Kellerer A, Rossi H (1974) The theory of dual radiation action. Curr Top Radiat Res 8:85–158
Konradov AA (1994) Statistical approaches to the analysis of multivariate heterogeneous biological systems. Radiat Biol Radioecol 34(6):877–886
Kudriashov IB, Kudryashov YB, Lomanov MF (2008) Radiat Biophys (Ionizing Radiations). Nova Publication
Kusunoki Y, Hayashi T (2008) Long-lasting alterations of the immune system by ionizing radiation exposure: implications for disease development among atomic bomb survivors. Int J Radiat Biol 84(1):1–14
Lea DE (1946) The inactivation of viruses by radiations. Br J Radiol 19:205–212
Lea D (1962) Actions of radiations on living cells, 2d edn. Cambridge University Press, Cambridge
Levene PA (1919) The structure of yeast nucleic acid. IV. Ammonia hydrolysis. J Biol Chem 40 (2):415–424
Liu Y, Kobayashi A, Fu Q et al (2015) Rescue of targeted nonstem-like cells from bystander stem-like cells in human fibrosarcoma HT1080. Radiat Res 184(3):334–340
Mannino M, Chalmers AJ (2011) Radioresistance of glioma stem cells: intrinsic characteristic or property of the "microenvironment-stem cell unit"? Mol Oncol 5(4):374–386
Mitchel REJ (2015) Adaption by low dose radiation exposure: a look at scope and limitations for radioprotection. Dose-Response: A Pub Int Hormesis Soc 13(1)
Mothersill C, Seymour C (2010) Eco-systems biology–from the gene to the stream. Mutat Res 687 (1–2):63–66
Mothersill C, Seymour C (2012) Changing paradigms in radiobiology. Mutat Res/Rev Mutat Res 750(2):85–95
Mothersill C, Seymour C (2013) Uncomfortable issues in radiation protection posed by low-dose radiobiology. Radiat Environ Biophys 52(3):293–298
Muller HJ (1927) Artificial transmutation of the gene. Science (N.Y.) 66(1699):84–87
Nieder C, Milas L, Ang KK (2000) Tissue tolerance to reirradiation. Semin Radiat Oncol 10 (3):200–209
Nomiya T (2013) Discussions on target theory: past and present. J Radiat Res 54(6):1161–1163
Pajonk F, Vlashi E (2013) Characterization of the stem cell niche and its importance in radiobiological response. Semin Radiat Oncol 23(4):237–241
Pateras IS, Havaki S, Nikitopoulou X et al (2015) The DNA damage response and immune signaling alliance: Is it good or bad? Nature decides when and where. J Pharmacol Exp Ther 154:36–56
Puck TT, Marcus PI (1956) Action of x-rays on mammalian cells. J Exp Med 103(5):653–666
Richardson RB (2011) Stem cell niches and other factors that influence the sensitivity of bone marrow to radiation-induced bone cancer and leukaemia in children and adults. Int J Radiat Biol 87(4):343–359
Salomaa SI, Wright EG, Hildebrandt G et al (2010) Editorial. Non-DNA targeted effects. Mutat Res 687(1–2):1–2
Seymour CB, Mothersill C, Alper T (1986) High yields of lethal mutations in somatic mammalian cells that survive ionizing radiation. Int J Radiat Biol Relat Stud Phys Chem Med 50(1): 167–179

Sheppard AR, Swicord ML, Balzano Q (2008) Quantitative evaluations of mechanisms of radiofrequency interactions with biological molecules and processes. Health Phys 95(4): 365–396

Steel G (1997) Basic clinical radiobiology. CRC Press, Boca Raton, FL

Tapio S (2013) Ionizing radiation effects on cells, organelles and tissues on proteome level. Adv Exp Med Biol 990:37–48

Timoféeff-Ressovskii NW, Zimmer KG, Delbrück M (1935) Über die Natur der Genmutation und der Genstruktur. In: Nachrichten von der Gesellschaft der Wissenschaften zu Göttingen. Mathematisch-Physikalische Klasse. Fachgruppe VI, Biologie S(l):189–245

Timofeev-Resovskii NV, Ginter EK, Glotov NV, Ivanov VI (1971) Genetic and somatic effects of x-rays and fast neutrons in experiments on Arabidopsis and Drosophila. Soviet Genetics 7 (4):446–453

Ulsh BA, Miller SM, Mallory FF et al (2004) Cytogenetic dose-response and adaptive response in cells of ungulate species exposed to ionizing radiation. J Environ Radioact 74(1–3):73–81

UNSCEAR 2010 Report: "*Summary of low-dose radiation effects on health*". Available at: http://www.unscear.org/unscear/publications.html

Watson JD, Crick FH (1953) Molecular structure of nucleic acids; a structure for deoxyribose nucleic acid. Nature 171(4356):737–738

Wright EG (2000) Inducible genomic instability: new insights into the biological effects of ionizing radiation. Med Confl Surviv 16(1):117–130 (discussion: 131–133)

Yi S-Y, Hao Y-B, Nan K-J, Fan T-L (2013) Cancer stem cells niche: a target for novel cancer therapeutics. Cancer Treat Rev 39(3):290–296

Strategies of Adaptation Under Prolonged Irradiation vs Chronic Exposure

Victoria L. Korogodina, Elena B. Grigorkina and Ludmila P. Osipova

Introduction

After the Second World War, Nikolay W. Timofeeff-Ressovsky has been appointed head of Biophysics laboratory in the South Urals to investigate the radiation effects on ecosystems. N.V. Luchnik (Fig. 1) worked in this laboratory and published some papers on the problem of low-dose radiation effects. Luchnik described the non-linear effects induced by low dose radiation in plants: fast increasing of a number of cells with chromosomal abnormalities and stimulation to cells' division (Luchnik 1958). He demonstrated firstly that stimulation of proliferation is not caused by an appearance of the chromosomal abnormalities.

Use of radiation technology as well nuclear accidents have stimulated the radiobiology researches: the bystander effect (Mothersill and Seymour 2000) and genomic instability (Korogodin et al. 1977; Morgan 2011; Yalkovskaya et al. 2011) continued in generations (Korogodin et al. 1977; Dubrova 2003), adaptive response (Upton 2001), and epigenetic regulation of gene expression (Kovalchuk and Baulch 2008). The consequences of radiation accidents are constantly studied (Shevchenko et al. 1992; Geras'kin et al. 2008; Grigorkina and Olenev 2009; Yablokov et al. 2009; Mousseau and Møller 2013). A process of adaptation was firstly described by Fisher (1930) who showed a correlation between fitness probability of organisms and their genetic changes.

V.L. Korogodina (✉)
Joint Institute for Nuclear Research, Dubna, Moscow Region, Russia
e-mail: korogod@jinr.ru

E.B. Grigorkina
Institute of Plant and Animal Ecology, UD RAS, Yekaterinburg, Russia

L.P. Osipova
Institute of Cytology and Genetics, SB RAS, Novosibirsk, Russia

V.L. Korogodina et al. (eds.), *Genetics, Evolution and Radiation*,
DOI 10.1007/978-3-319-48838-7_13

Fig. 1 Sitting, from *left* to *right*: N.V. Luchnik, N.A. Poryadkova, N.W. Timofeeff-Ressovsky. Miassovo, circa 1950s

Strategy of adaptation aims population survival and depends on exposure type: acute impact induces repair recovery (Korogodin and Malyutina 1959), prolonged or time-varying irradiation needs expansion of variability, and constant exposure produces radioresistance in population. There are different mechanisms induced by radiation (Mothersill and Seymour 2006; Morgan 2011; Grigorkina 2010; Korogodina et al. 2013). One of the specific radiation phenomena is radioadaptation, which occurs in generations (Shevchenko et al. 1992).

The aim of this paper is the analysis of adaptation processes induced by low-dose prolonged irradiation and chronic exposure. The consequences of both types of irradiation will be considered for plants, animals, and human populations.

Objects and Methods

Characteristics of the Objects

Here the characteristics of objects, features of sites, and effective methods, which were used to analyze general principles of adaptation in plant, animal, and human are presented. The objects are considered in the conditions of time-varying irradiation induced by the radiation sources or due to the eco-physiological features, and conversely, at exposure to the constant in time radiation factor. The principle lab experiments are described in the Table 1, and the natural tests are presented in the Table 2.

Table 1 Lab studies: objects, dose (D), dose-rate (DR), and methods of investigations

Object	Pea seeds: lines "Capital" (Luchnik 1958) and *Pisum arvense* "Nemchinovsky-817" (Korogodina et al. 1998)
Dose	D = 7 cGy, DR = 0.3 ~ 19.1 cGy/h immersion in solution of uranium fission products (β—radiator, ~ 98 %) (Luchnik 1958) Irradiation in water with ^{60}Co γ-rays (Korogodina et al. 1998)
Method	Cytogenetic studies, statistical modeling (Korogodina et al. 1998, 2013)

Table 2 Natural studies: objects, sites, radiation: dose (D), dose-rates (DR), soil contamination (C), methods

Plant	Source/site (a)	NPP/JINR facilities
	Dose	Balakovo NPP (20-km zone) (SCEP SR 2000): γ-radiation DR ~ 0.10–0.15 μSv/h; ^{137}Cs, C ~ 5–10 Bq/kg JINR facilities (1998) (Korogodina et al. 2013): Neutron D & DR: 1 mSv & 0.8 μSv/h (for two months)
	Source/site (b)	"Chernobyl" site (~ 100 km from NPP)
	Dose	γ-radiation ~ 0.10-0.15 μSv/h (SCEP SR 2000) ^{137}Cs, C ~ 30 Bq/kg (Kresson 1998)
	Object	Plantain seeds (*Plantago major*) collected in natural populations
	Method	Cytogenetic studies, statistical modeling (Korogodina et al. 1998; 2013)
Animal	Source/site (c)	Head part of Eastern Urals Radioactive Trace zone (EURT) ^{90}Sr is the main contaminant
	Dose	1957: ^{90}Sr, C = 18.5 MBq/m^2 (Nikipelov et al. 1989) Current levels of ^{90}Sr C = 5.5–15.0 MBq/m^2 (Pozolotina et al. 2008)
	Object	Pygmy wood mice (*Sylvaemus uralensis*), $LD_{50/30}$: 7.0 ± 0.4 Gy Field mice (*Apodemus agrarius)* $LD_{50/30}$: 10.0 ± 0.2 Gy Terrestrial animals, high migratory activity (Grigorkina and Olenev 2009)
	Source/site (d)	Southern Urals, the coastal zone of the Techa River
	Dose	^{90}Sr, C = 43.8 kBq/m^2; there are ^{137}Cs and $^{239–240}Pu$ (Aarkrog et al 2000)
	Object	Northern mole-voles *(Ellobius talpinus*) $LD_{50/30}$: 5.0 ± 0.7 Gy Subterranean burrowing animals (typical digger), quite low migratory activity (Grigorkina and Olenev 2009)
	Method	Adaptive response (Grigorkina 2010), statistical methods: the Student and Mann-Whitney tests
Human	Source/site (e)	Pribaikal'e—nuclear tests in the Semipalatinsk polygon, 1950s
	Dose	1950s: Total D (adult individuals) = 10–40 cSv; it is several times greater for the children (Nepomnyashchikh et al. 1999)
	Object	Russian
	Source/site (f)	North Siberia-nuclear tests in the Novaya Zemlya archipelago, 1950s
	Dose	2000: ^{137}Cs, C = 118.2 Bq/kg (lichen), and 162.1 Bq/kg (venison) (Osipova et al. 2000)
	Object	Nenets, food chain: moss—venison—man
	Method	Genetic and medical investigations (Osipova et al. 2000), statistical modeling (Korogodina et al. 2013)

Statistical Modeling

Radiation-induced process of adaptation can be described as a random appearance of abnormal cells in resistant and sensitive subpopulations presented by Poisson (P) and geometric (G) laws respectively (Florko and Korogodina 2007).

This adaptation process (P + G) originates late correlative events under environment conditions (Florko and Korogodina 2007). Accumulation of the late rare abnormalities without selection is described by the P-law. The G-law is detected when the high late damages' frequency is accompanied by selection. G-law finding is a sure sign of genetic instability. The seedlings' distribution on the number of proliferated cells is divided into three ones presented by lognormal (LN) law with different parameters (LN1 + LN2 + LN3), where LN1 and LN2 correspond to the resistant and sensitive subpopulations, and a stimulation of proliferation adds the third component LN3. Strong radiation stress transforms the LN-law into G-one.

Some methods of approximations and estimation of the distribution parameters (sample means m(P, G, LN1, LN2, LN3) and values (NP, NG, LN1, LN2, LN3) were described in (Korogodina et al. 2013).

Results

Plants

Laboratory experiments: The experiments of Luchnik (1958) and Korogodina et al. (1998) have demonstrated the radiation effects induced by prolonged internal and external impacts (Fig. 2). The seeds of Ural line (1) are more resistant than the JINR one (2) (Fig. 2a, b), and it is a reason that the percentage of abnormal mitosis falls at the appearance of multiple abnormalities in meristem (b, 2). Mitotic index has approximately increased a two-fold for both lines (a). Figure 2 reflects main genetic effects of adaptation.

The statistical modeling revealed that the appearances of DNA damages obey G-law in the dose-rate interval of 0.3–1.5 cGy/h, and the P- one at the higher intensity (2–20 cGy/h) (Korogodina et al. 2013). Analysis of the distribution parameters showed that the dose-rates 0.3–1.5 cGy/h is characterized by an appearance of late correlative CAs accompanied by seedlings' selection whereas higher ones ($\sim$10–20 cGy/h) induce primary direct DNA damages which are rare and independent.

Investigation of seeds collected in natural populations: The effects of short-prolonged impacts of the NPP fallouts (sites P2, P4, P6) and JINR facilities (P12) as well as chronic exposure in the "Chernobyl" site P11 are presented in the Table 3, and the results of statistical modeling are shown in the Table 4.

In 1998, neutron radiation increased the MA value (Table 3) resulting in stimulation of resting cells (Table 4) in plants growing in the JINR territory. In 1999,

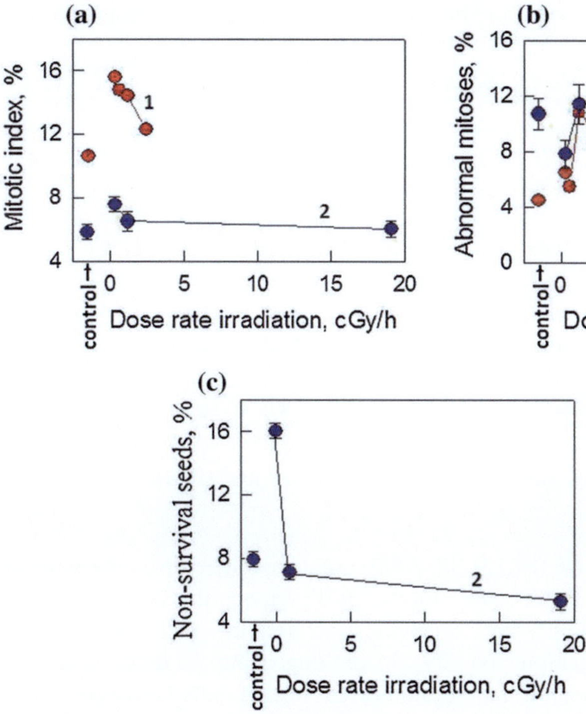

Fig. 2 Mitotic index (**a**), abnormal mitoses (**b**) and non-survived cells (**c**) versus dose-rate radiation in cells of pea seedlings in experiments of Luchnik (1958) (*1*) and of Korogodina et al. (1998) (*2*). Dose irradiation: 7 cGy; dose-rates: 0.3, 0.6, 1.2, and 2.4 cGy/h (*1*), and 0.3, 1.2, and 19.1 cGy/h (*2*)

Table 3 Non-survival of seeds (1-S), frequency of cells with abnormalities (CAs), mitotic activity (MA) in meristem

Site	1-S, %	CAs, %	MA
1998			
P2	34.2	3.1 ± 0.8	6.0 ± 0.6
P4	19.9	2.5 ± 0.6	6.3 ± 0.7
P11	55.6	4.4 ± 0.9	7.3 ± 1.0
P12	12.3	1.2 ± 0.3	14.9 ± 0.9
1999			
P2	72.6***	3.2 ± 0.4**	17.8 ± 1.2***
P4	83.6***	6.8 ± 0.9***	17.5 ± 1.4***
P6	71.6	5.1 ± 0.4	17.9 ± 1.1
P11	43.6*	5.5 ± 0.5*	9.8 ± 0.5**
P12	37.5***	5.6 ± 0.8***	8.0 ± 0.6***

Comparing 1998 and 1999 data: $^{*}p > 0.1$; $^{**}p > 0.5$; $^{***}p < 0.001$

Table 4 Parameters of the distributions of plantain seeds on the cells with abnormalities (P + G) and the proliferated cells (LN1 + LN2 + LN3) (χ^2/df < 1)

Seeds	*mG****	N_G*	*mP**	N_P**	*m(LN1)*	*m(LN2)*	*m(LN3)*	N_{LN1}	N_{LN2}	N_{LN3}
1998										
P2	0.05	0.10	0.17	0.45	5.3	–	22.0	0.61	0.00	0.04
P4	0.07	0.06	0.17	0.64	2.5	12.0	–	0.48	0.32	0.00
P11	0.70	0.09	0.45	0.21	4.0	12.0	23.0	0.32	0.13	0.03
P12	0.06	0.12	0.21	0.69	12.0	22.0	33.0	0.29	0.19	0.40
1999										
P2	**1.05**	0.13	0.36	0.11	8.0	22.5	**33.0**	0.12	0.14	7.7e-3
P4	**2.01**	0.02	1.26	0.12	12.0	22.5	**33.0**	0.11	0.03	0.02
P6	**2.45**	0.04	0.96	0.21	13.3	23.0	**37.0**	0.21	0.02	0.06
P11	0.32	0.10	0.84	0.33	9.0	18.0	3.0	0.33	0.15	0.08
P12	1.81	0.12	0.36	0.39	10.0	22.0	3.9	0.39	0.04	0.19

Standard errors of the *G* + *P* − parameters do not exceed: *20–30 %; **10–15 %; ***40–50 %

the high summer temperatures (SCEP SR 2000) combined synergistically with the NPP fallouts pronounced dramatically in the NPP region: the *mG* and *mLN3* parameters increased (Table 4), and survival seeds fell to 20–30 % (Table 3), which is the limit for population viability. So prolonged impact induces genome instability coupled with selection and stimulates the resting cells to divide.

The chronic exposure of plants growing in the soil-polluted site P11 led to high levels of CAs and 1-S both years (Table 3). Despite the hot summer the P_{P11} and G_{P11} parameters are not changed within the standard error in 1999 (Table 4) that indicates stabilization of the genetic processes. In 1998, the mP_{P11} is more than this parameter for other sites, and it is equal to them in 1999. Stable-high sample mean indicates accumulation of the late CAs without selection in the P_{P11}-subpopulation. It suggests origination of the resistant subpopulation of cells.

Animals

Adaptation of animals inhabiting the contaminated areas can be detected by testing of their adaptive response (AR). Figure 3 presents the AR values in terrestrial rodents *S. uralensis* (a) and subterranean burrowing animals *El. talpinus* (b) in their impact and control groups.

Terrestrial rodents: EURT zone is characterized by specific configuration being an extended (about 300 km) and narrow territory with a rapidly falling pollution gradient. Due to small transversal size of radioactive cloud nuclear fallout was concentrated along the axis of its movement. Mice are able to pass through significant distances that are comparable with the cross-section size of contaminated zone. Special study of mouse-like rodents' migrations in the EURT zone showed

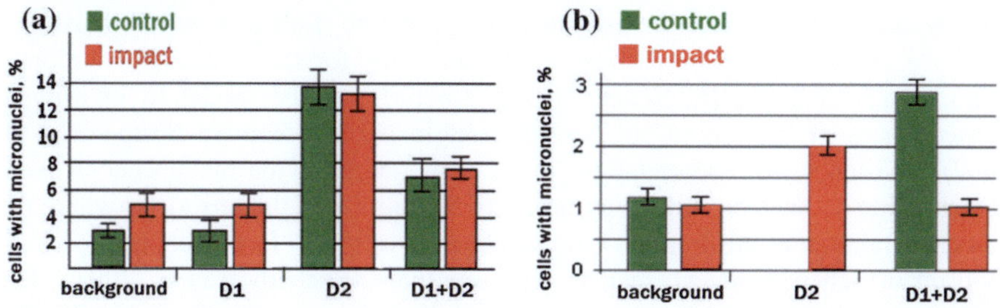

Fig. 3 Adaptive response in *S. uralensis* (**a**) and *El. talpinus* (**b**) from the EURT zone and control areas. Animals groups: background (impact and control); D1 "adapting dose", 20 cGy (mice); D2 "damaging dose", 2.0 Gy (mice and mole-voles); D1 + D2 groups (mice and mole-voles)

that migrants portion in different years varied from 5 up to 30 % (Grigorkina and Olenev 2013). Pollution gradient and migration activity contribute to radiation effect on animals and can be explanation why the control and impact mice' samples demonstrate similar adaptive response (Fig. 3a).

Analysis of the micronuclei frequency in bone marrow cells showed insignificant difference between both mice' species in impact and control D1 + D2 groups. So the radioresistance doesn't play the mayor role in the adaptive response.

Subterranean burrowing rodents: Mole-voles (*El. talpinus)* are characterized by a low ability for dispersal. Micronuclei frequency in the D1 + D2 impact group was smaller than in the control one (Fig. 3b). The burrowing rodents dwelling in the vicinities of the Techa River demonstrate the pronounced AR in comparison with the terrestrial ones that testifies to their radioadaptation (Grigorkina 2010). Both eco-physiological features and territorial insulation of their settlement in the EURT zone for a half-century history since the Kyshtym accident promoted the development of radioadaptation.

It was not find average changes of the CAs frequency in the impact and control mole-voles (Gileva 2002), that is natural take into account their high radiosensitivity (example is demonstrated in Fig. 2b). However, cells containing numerous micronuclei of various shape (spheroid, tubule-like, and comma-shapes) were detected, the frequency of these cells in the impact group was four times higher than in the control (Grigorkina 2010). The latter circumstance can indicate a radioresistance in these rodents.

Human

The consequences of nuclear tests (1950s) in the Semipalatinsk test site (Nepomnyashchikh et al. 1999; Sukhorukov et al. 2000) and Novaya Zemlya archipelago (Shcherbov et al. 2000; 2007) were studied in human populations (Osipova et al. 1999; Medvedev et al. 2009a). Radiation effects of the Semipalatinsk nuclear test fallouts were investigated in the South Baikal zone (Pribaikal'e). Table 2e presents the summarized irradiation dose for the adults and children. Population of the Tundra Nenets (North Siberia) was the model to investigate consequences of the fallouts of nuclear tests in Novaya Zemlya (Osipova et al. 2002). The long-lived radionuclides in the components of the biogeocenosis were detected that showed considerable radioactive pollution of the territories where the native population of the Russian Utmost North lives today. In this connection their food chain lichen-reindeer—man is very important (Shcherbov et al. 2001). The ^{137}Cs contamination (2000) in lichen and venison are given in the Table 2f. The investigations of blood lymphocytes of the individuals living in Pribaikal'e and North Siberia were performed (Osipova et al. 2000; Medvedev et al. 2009b).

Figure 4 shows distributions of persons of different generations on the frequency of blood cells with abnormalities (Korogodina et al. 2013). Random samples of persons were divided into four groups corresponding to the individuals who had been immediately irradiated by the fallouts, and their children, grandchildren, and great-grandchildren. The sample of the known healthy individuals was used as the control group living in the city of Novosibirsk.

To illustrate this Fig. 4, it should be noted that the types of the distributions of the individuals with normal and bad activated blood lymphocytes are the P- and G-ones, respectively (Korogodina et al. 2013). It is clear because the instability and selection processes lead to the growing number of died sensitive blood cells, depletion of the lymphocyte pool, and high percentage of fatal cases of sensitive individuals.

Figure 4 showed that genetic instability characterizes blood cells of individuals living in both territories of North Siberia and Pribaikal'e: their P-sample means differ a little, and exceed significantly the control value. The instability process continues across three generations of irradiated persons, but there is a difference between adaptation processes in tested settlements. In North, the groups of persons aged 20–55 years are G-distributed (**c**,**b**); the Poisson law dominates in the elder group (**a**). In the Pribaikal'e population, contribution of the G-distribution is observed in the groups of persons older than 40 years; it increases according their age (**e**,**f**). These data have important meaning: there is the resistant group of persons living in the Tyumen region with an increased frequency of aberrations in blood cells, even in old age (**a**).

Discussion

Prolonged impact: This type of radiation exposure was studied in lab (Table 1: plant) and nature (Table 2: plant, animal, human). Plant population growing near the NPP and in the JINR territory experience periodically prolonged radiation impacts (Table 2, a), as well as human population living in the Pribaikal'e was short-prolonged irradiated in 1953 due to nuclear tests in Semipalatinsk polygon (Table 2, e). We consider the radiation effect on mice inhabiting the EURT zone

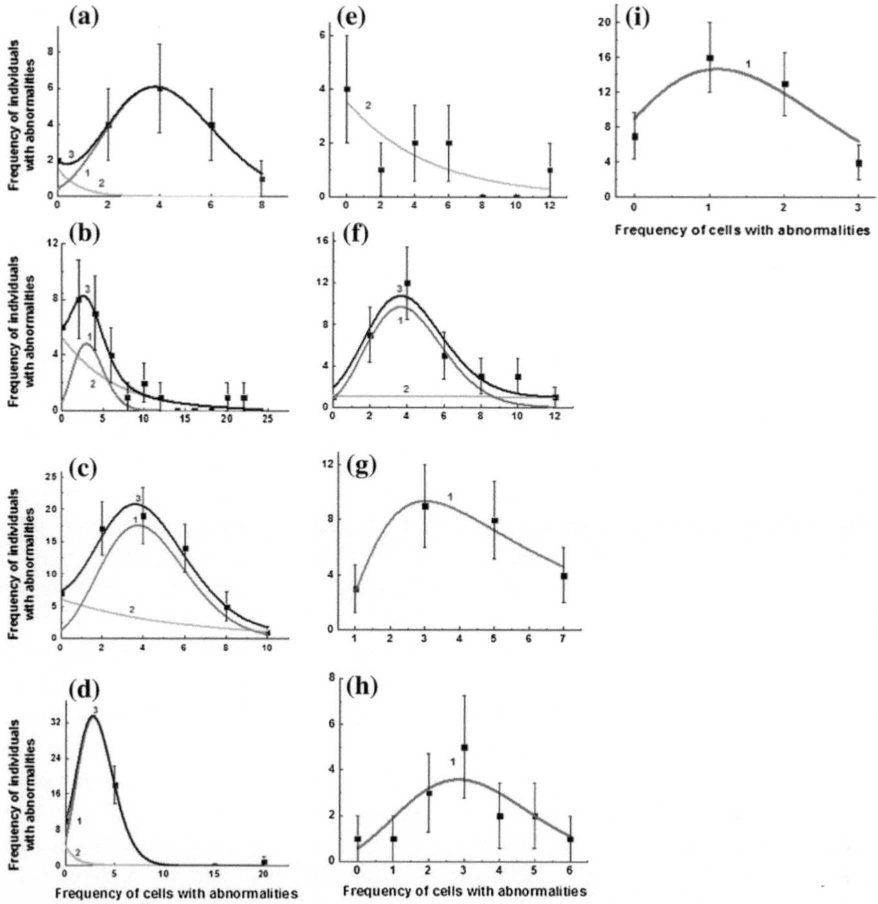

Fig. 4 Statistical modeling of distributions of persons of different generations on the frequency of cells with CAs occurrence (Korogodina et al. 2013). Sample of the persons living in the North Siberia (**a**–**d**); Pribaikal'e (**e**–**h**); the city Novosibirsk (**i**). Parents (**a**, **e**); children (**b**, **f**); grandchildren (**c**, **g**); great grandchildren (**d**, **h**). *1* Poisson component; *2* geometric one; *3* their sum. Experimental data are shown with their standard errors

(Table 2, c) also as prolonged one because genome fit is continuous (Korogodina et al. 2013), and the changing dose-rate doesn't allow selection of resistant specimens in migrating rodents. Consequences of migrations are considerably decreasing probability that certain adaptive changes may be fixed and inherited in a series of generations of mobile species' rodents.

Our studies showed cytogenetic changes, chromosomal instability and stimulation of proliferation of resting cells (Korogodina et al. 2013; Yalkovskaya et al. 2011). Investigations of other scientists demonstrated involving of the epigenetic regulation of gene expression (Kovalchuk et al. 2004) and activation of multiple mitogen-activated protein kinase (MAPK) pathways (Lyng et al. 2006) linked to growth factor-mediated regulation of cellular events. All these explain general features of low-dose prolonged irradiation to induce genome destabilization.

Genetic and epigenetic mechanisms contribute into multiple changes of morphophysiological characteristics to enhance fertility and eventually lead to increase "evolution material" (term of Timofeeff-Ressovsky 1939). The alteration of developmental timing is observed as stress response that allows the plants to complete their life cycle in a timely manner. Plants can be induced to flower by responding to stress factors such as drought (Nir et al. 1972; Hopkinson 1977), high salinity (Kolar and Senkova 2008), low temperature (Hatayama and Takeno 2003), high-intensity light (King et al. 2008) and gamma radiation (Custers and De Jong 1986; Dhillon et al. 2014). Xu et al. (2014) have found a common epigenetic mechanism regulating stress-induced flowering, related the microRNA miR169 family members.

In the case of terrestrial rodents inhabiting the EURT zone, Grigorkina and Olenev (2013) offered that mice form a flowing population; radiation stress doesn't lead to extinction of animals but to their high breeding rate and reduced fetal mortality (Grigorkina 2007). A key role in the survival of mice populations plays a bond their immune system and fertility. It is known that strong immunity means low fertility (Christiansen 2013), but stress is an immunosuppressant (Yarilin 1999). Radiation stress is the trigger of switching of ontogenetic development' pathways in cyclomorphic mammals and contributes to increase of a share of rodents breeding in the year of his birth (more radiosensitive). Another (more resistant) part of irradiated population survives the winter and reproduces the following year. Population of rodents in radiocontaminated environment is compelled to include more energy-intensive way of maintenance its ability to live (Grigorkina and Olenev 2011). Thus polyalternativeness of small mammals' ontogeny development is one of the main mechanisms of population adaptive strategy. Like strategy of a balance between survival and reproduction was observed in the wild Scottish sheep (Graham et al. 2010).

Obviously, short-prolonged irradiation induces genetic instability which provides phenotypic variability. It is expanded by epigenetic gene regulation and stimulation of resting cells that contribute into survival. Living organisms use different mechanisms to increase their fertility.

Chronic exposure: Effects of chronic irradiation were investigated in plant growing in the "Chernobyl site" (Table 2, b), and in subterranean mole-voles (genus *Ellobius*) dwelling on the banks of the Techa River (Table 2, d). The indigenous peoples of the North Siberia live in the tundra and small settlements. Nenets persons are chronically exposed due to the permafrost kept radionuclides and their short food chain moss—venison—man (Osipova et al. 2000; Shcherbov et al. 2001) (Table 2, f).

The chronic exposure of plants growing in the "Chernobyl" site resulted in stabilization of the high CAs and 1-S levels (Table 3). The statistical modeling revealed accumulation of late CAs without selection in a plant subpopulation that reflects its elevated radioresistance. The limited genetic instability in plants

(Table 4) should provide variability of their cytomorphological and physiological characteristics (Pozolotina 1996).

Adaptation processes and biological consequences of chronic low-dose exposure in small mammals are modified by eco-physiological peculiarities as well as main habits of species and configuration of radiocontaminated zone. Convincing evidence of compelled genetic radioadaptation was obtained at the example of mole-voles (*El. talpinus*) inhabiting radiocontaminated zone during a long period in the course of changing (ca. 50) generations. In spite of that average cytogenetic, morphologic and physiological characteristics do not differ in the impact and control mole-voles, it was revealed a group of cells with multiple micronuclei in the impact rodents (Grigorkina 2010), that can indicates their increased radioresistance. It is verifies by presence of AR in the impact group in comparison with the control one (Fig. 3).

Adaptation to chronic radiation exposure results in stabilization of genetic variability level and reduced survival, and appearance of a resistant subpopulation. Adaptive response is a mechanism of cell protection against ionizing radiation as well as other stress factors. There is a growing consensus which suggests different AR mechanisms acting at different levels of abiotic stress. It is true for both ionizing radiation (Mothersill and Seymour 2006) and other abiotic stress factors (for example, salinity stress and heavy metal toxicity in plants; Tereshchenko et al. 2012).

Statistical modeling of adaptation processes allows evaluating the resistant group in the human populations living in the contaminated areas (Korogodina et al. 2013). Figure 4 demonstrates statistical laws of the appearance of chromosomal abnormalities in blood lymphocytes in generations. There is a resistant group of North individuals (Fig. 4 **a**-**d**, gr.1) which is characterized by increased frequency of chromosomal abnormalities in blood lymphocytes and is described by P-law. There is no like group in the Pribaikal'e town in the tested population, because a group of individuals described by G-law increases with persons' age (Fig. 4 **e**-**h**, gr.2) and after 60 years dominates.

Table 5 presents adaptation specifics to the prolonged and chronic impacts. Both effects are manifested in the adaptation processes, and have the same compounds. Duration of stable radiation exposure determines the final result: diversity or resistance of population.

Table 5 Specifics of adaptation processes induced by prolonged and chronic exposures

Prolonged irradiation: diversity	Chronic exposure: resistance to stresses
Increasing of variability:	*Radioresistance*:
Activation of genomic instability	Involving of different ways of selection
Stimulation of resting cells	Adaptive response
Stimulation of flowering	Selection of resistant objects in generations
Increasing fertility	Resistance to synergic effects

Conclusions

- Low-dose irradiation induces processes of adaptation. Survival strategy depends on exposure type. Type of exposure is determined by radiation sources, configuration of radiocontaminated zone and its climate conditions, eco-physiological peculiarities of species and other factors.
- Prolonged low-dose impact induces adaptation processes which aim to increase population diversity. Radiation stress induces genome-wide destabilization accompanied by selection. In addition, epigenetic gene expression and stimulation to divide the resting cells expand material for adaptation. Genetic variability originates diversity of morphologic and other characteristics. Stress changes parameters of reproduction that increases material for evolution. Synergic combination stresses effect of variability and selection.
- Chronic exposure induces adaptation processes which lead to radioadaptation of population. Radioadaptation of population means an appearance of resistant subpopulations of cells and organisms, stabilization of adaptation process, and high percentage of genetic abnormalities and decreased survival.

Acknowledgments This study was partly funded by Basic Research Programs of the Ural Branch of Russian Academy of Sciences (N 15-4-2-21 and 15-3-4-49). Special thanks to our colleague Dr. Elena K. Khlestkina for helpful discussions.

References

Aarkrog A, Trapeznikov AV, Molchanova IV, Yushkov PI, Pozolotina VN, Polikarpov GG, Dahlgaard H, Nielsen SP (2000) Environmental modeling of radioactive contamination of floodplains and sorlakes along the Techa and Iset rivers. J Environ Radioact 49:243–257

Christiansen OB (2013) Reproductive immunology. Mol Immunol 55(1):8–15

Custers JBM, De Jong J (1986) Induced changes in growth and flowering of chrysanthemum after irradiation and in vitro culture of pedicels and petal epidermis. Euphytica 35:137–148

Dhillon RS, Saharan RP, Jattan M, Rani T, Sheokand RN, Dalal V, Von Wuehlisch G (2014) Molecular characterization of induced mutagenesis through gamma radiation using RAPD markers in Jatrophacurcas L. Afr J Biotechnol 13:806–813

Dubrova YE (2003) Radiation-induced transgenerational instability. Oncogene 22:7087–7093

Fisher RA (1930) The genetical theory of natural selection. Oxford University Press, Oxford

Florko BV, Korogodina VL (2007) Analysis of the distribution structure as exemplified by one cytogenetic problem. PEPAN Lett 4:331–338

Geras'kin SA, Fesenko SV, Alexakhin RM (2008) Effects of non-human species irradiation after the Chernobyl NPP accident. Environ Int 34:880–897

Gileva EA (2002) Chromosomal instability at rodents from EURT zone: interspecific comparisons. Radiat Biol Radioecol 42(6):670–673 (Russian)

Graham AL, Hayward AD, Watt KA, Pilkington JG, Pemberton JM, Nussey DH (2010) Fitness correlates of heritable variation in antibody responsiveness in a wild mammal. Science 330 (6004):662–665

Grigorkina EB (2007) The reproductive strategy and immune status of rodents in an environment altered by industrial activity. Doklady Biol Sci 412:27–29

Grigorkina EB (2010) Low-dose radiation effects: adaptive response in rodents (*Ellobius talpinus Pall.*) inhabiting radionuclide-contaminated environment. Doklady Biol Sci 430:42–44

Grigorkina EB, Olenev GV (2009) Radioadaptation of rodents in the zone of local radioactive contamination (Kyshtim accident RUSSIA): 50 years on. Radioprotection 44(5):129–134

Grigorkina EB, Olenev GV (2011) East Urals Radioactive Trace: adaptive strategy of rodents' population. Radioprotection 46(6):437–443

Grigorkina EB, Olenev GV (2013) Migration of rodents in the Eastern Urals Radioactive Trace Zone (radiobiological aspect). Radiats Biol Radioecol 53 (1): 76–83 (Russian)

Hatayama T, Takeno K (2003) The metabolic pathway of salicylic acid rather than of chlorogenic acid is involved in the stress-induced flowering of Pharbitis nil. J Plant Physiol 160:461–467

Hopkinson JM (1977) Formative factors—tropical pasture legume seed crops. In: Proceedings of the Australian Seeds Research Conference. Tamworth, pp 14–23

King RW, Hisamatsu T, Goldschmidt EE, Blundell C (2008) The nature of floral signals in Arabidopsis. I. Photosynthesis and a far-red photo-response independently regulate flowering by increasing expression of flowering locus T (FT). J Exp Bot 59:3811–3820

Kolar J, Senkova J (2008) Reduction of mineral nutrient availability accelerates flowering of *Arabidopsis thaliana*. J Plant Physiol 165:1601–1609

Korogodin VI, Malyutina TS (1959) Recovery of irradiated yeast cells. Rus J "Nature" (Priroda) 10: 82–85 (Russian)

Korogodin VI, Bliznik KM, Kapultsevich YG (1977) Regularities of radioraces formation in yeasts. Comm. 11. Facts and hypotheses. Radiobiologiya 17:492–499 (Russian)

Korogodina VL, Panteleeva A, Ganicheva I et al (1998) Influence of the low gamma-irradiation dose rate on mitosis and adaptive response in meristem cells of pea seedlings. Radiat Biol Radioecol 38:643–649 (Russian)

Korogodina VL, Florko BV, Osipova LP (2013) Radiation-induced processes of adaptation: research by statistical modeling. Springer, The Netherlands

Kovalchuk O, Baulch JE (2008) Epigenetic changes and nontargeted radiation effects—is there a link? Environ Mol Mutagen 49:16–25

Kovalchuk I, Abramov V, Pogribny I, Kovalchuk O (2004) Molecular aspects of plant adaptation to life in the Chernobyl zone. Plant Physiol 135(1):357–363

Kresson E (ed) (1998) The Atlas of caesium-137 contamination in Europe after the Chernobyl accident. European Commission, DG XII, Safety Programme for Nuclear Fission (Protection against Radiation)

Luchnik NV (1958) Influence of low dose irradiation on mitosis in pea. Newsletter of Ural Dept Moscow Soc Nat Investigators 1:37–49 (Russian)

Lyng FM, Maguire P, McClean B, Seymour C, Mothersill C (2006) The involvement of calcium and MAP kinase signaling pathways in the production of radiation-induced bystander effects. Radiat Res 165(4):400–409

Medvedev VI, Korshunov LG, Chernyago BP, Kovalenko VV, Matorova NI, Karchevsky AN, Osipova LP, Kelman MP (2009a) Radiation exposure of the Semipalatinsk nuclear test site in the Southern Siberia. Prob Biogeochem Geochem Ecol 2(10):57–65 (Russian)

Medvedev VI, Korshunov LG, Chernyago BP, Kovalenko VV, Matorova NI, Karchevsky AN, Osipova LP, Kelman MP (2009b) Radiation influence of the Semipalatinsk nuclear testing ground on South Siberia (experience of long term studies in East and middle Siberia and comparison of the results with materials on Western Siberia). In: Proceedings of the III international conference "Radioactivity and radioactive elements in environment" (Tomsk, 23–27.VI.2009), pp 348–355 (Russian)

Morgan WF (2011) Radiation induced genomic instability. Health Phys 100:280–281

Mothersill CE, Seymour CB (2000) Genomic instability, bystander effect and radiation risks: implications for development of protection strategies for man and environment. Radiat Biol Radioecol 40:615–620

Mothersill C, Seymour C (2006) Radiation-induced bystander and other non-target effects: novel intervention points in cancer therapy. Curr Cancer Drug Targets Rev 6:447–454

Mousseau TA, Møller AP (2013) Perspectives on Chernobyl and Fukushima health effects: what can be learned from Eastern European research? J Health & Pollut 3(5):2–6
Nepomnyashchikh AI, Chernyago BP, Kuznetsov AF, Medvedev VI (1999) Local fallout on the South Irkutsk region from nuclear tests. Reports, RAS 369(2):258–260 (Russian)
Nikipelov BV, Romanov GN, Buldakov LA et al (1989) Accident in the Southern Urals on 29 September 1957. IAEA Information Circulars 368. Report on a Radiological Accident in the Southern Urals on 29 September 1957. IAEA, Vienna
Nir I, Goren R, Leshem B (1972) Effects of water stress, gibberellic acid and 2-chloroethyl trimethylammonium chloride (CCC) on flower differentiation in Eureka lemon trees. J Am Soc Hortic Sci 9:774–778
Osipova LP, Posukh OL, Koutzenogii KP, Sukhorukov FV, Matveeva VG, Grafodatskii AS, Konovalova NA, Sukhovey YG, Petrov SA, Lefranc G, Lefranc M-P (1999) Epidemiological studies for the assessment of risks from environmental radiation on Tundra Nentsi population. NATO Science: Environmental Security. In: Baumstark-Khan C et al (eds) Fundamentals for the assessment of risks from environmental radiation, vol 55, pp 35–42
Osipova LP, Posukh OL, Ponomareva AV, Matveeva VG, Shcherbov BL, Strakhovenko VD, Sukhorukov FV (2000) Medical genetic studies of Tundra Nenetz population and estimation of radiation situation in their habitation region. Siberian J Ecol 7(1):61–66 (Russian)
Osipova LP, Koutsenogii KP, Shcherbov BL, Strakhovenko VD, Petrov SA (2002) Environmental radioactivity for risk assessment of health status in nature and human population of Northern Siberia. Proceedings of the 5-th international conference on environmental radioactivity in the Arctic and Antarctic (St. Petersburg 16–20.VI.2002), pp 124–127
Pozolotina VN (1996) Radiation-induced adaptation processes in plants. Ecology 2:111–116 (Russian)
Pozolotina VN, Molchanova IV, Karavaeva EN et al (2008) The current state of terrestrial ecosystems in the Eastern Ural radioactive trace: contamination levels and biological effects. Akademkniga, Ekaterinburg (Russian)
Shcherbov BL, Strakhovenko VD, Malikova IN, Osipova LP, Sukhorukov FV, Stepin AS (2000) Comparative characteristics of modern radioactive pollution of West Siberian territories adjacent to the Semipalatinsk and Novaya Zemlya testing grounds (on the examples of Altai and Pur-Taz interfluvium). Siberian J Ecol 7(1):51–60 (Russian)
Shcherbov BL, Malikova IN, Sukhorukov FV, Strakhovenko VD, Osipova LP (2001) Artificial radionuclides and heavy metals in the food chain of the indigenous population in some areas of Siberia. Siberian J Ecol 2:1 (Russian)
Shcherbov BL, Malikova IN, Osipova LP, Strakhovenko VD, Sukhorukov FV (2007) Radioecological situation on the Siberian territories where native people live on the border XX–XXI centuries. Probl Biogeochem Geochem Ecol 1(3):13–21 (Russian)
Shevchenko VA, Pechkurenkov VL, Abramov VI (1992) Radiation genetics of the native populations. Genetic consequences of the Kyshtym accident. Nauka, Moscow
State Committee on the Environment Protection (SCEP) in the Saratov Region (2000) Conditions of environment in the Saratov region in 1999. Acvarius, Saratov (Russian)
Sukhorukov FV, Malikova IN, Gavshin VM, Kovalev SI, Shcherbov BL, Melgunov MS, Strakhovenko VD, Tsybulchik VM (2000) Technogenous radionuclides in the environment of West Siberia (Radiation sources and pollution levels). Siberian J Ecol 1:31–38 (Russian)
Tereshchenko OYu, Khlestkina EK, Gordeeva EI, Arbuzova VS, Salina EA (2012) Relationship between anthocyanin biosynthesis and abiotic stress in wheat. EWAC Newsl (Proc15-th EWAC Conf 2011 Novi Sad) 15: 72–75
Timofeeff-Ressovsky NW (1939) Genetik und evolution. Zeitschrift für inductive Abstammungs and Vererbungslehre 76:158–218 (German)
Upton AC (2001) Radiation hormesis: data and interpretations. Crit Rev Toxicol 31(4–5):681–695
Xu MY, Zhang L, Li WW, Hu XL, Wang MB, Fan YL, Zhang CY, Wang L (2014) Stress-induced early flowering is mediated by miR169 in *Arabidopsis thaliana*. J Exp Bot 65:89–101
Yablokov AV, Nesterenko VB, Nesterenko AV (2009) Chernobyl: consequences of the catastrophe for people and the environment. Ann NY Acad Sci 1181: vii–xiii (1–327)

Yalkovskaya LE, Grigorkina EB, Tarasov OV (2011) Cytogenetic consequences of chronic irradiation in rodent populations inhabiting the Eastern Ural Radioactive Trace Zone. Biophysics 56(1):140–144
Yarilin AA (1999) Radiation and immunity. Interference of ionizing radiation with key immune processes. Radiats Biol Radioecol 39(1):181–189 (Russian)

Mathematical Modeling of the DNA Double-Strand Break Repair in Mammalian and Human Cells

Oleg V. Belov, Marina S. Panina, Munkhbaatar Batmunkh and Nasser Sweilam

Introduction

Radiation-induced DNA double-strand breaks (DSBs) are one of the most deleterious types of DNA lesions, since it can lead to induction of structural gene mutations, chromosome aberrations, and possible initiation of the malignant cell transformation, which can be a causal event of carcinogenesis. Until the present, most of the experimental data on regularities of DSB repair have been obtained for sparsely ionizing radiations (γ- and X-rays) with the low linear energy transfer (LET). However, the amount of knowledge on high-LET radiations also increases. Acquisition of these new data requires development of approaches which enable summarizing of newly revealed regularities. In this regard, mathematical models represent a promising tool to understand DNA repair stages hardly accessible for measurements.

Most of the existing papers on simulation of DSB rejoining in eukaryotes focus on the lesions induced mainly by sparsely ionizing radiations of low-LET (Cucinotta et al. 2008; Bastin et al. 1992; Goodhead 1985; Kiefer 1988; Cucinotta et al. 2000; Taleei et al. 2012; Taleei and Nikjoo 2013; Taleei 2013). A few papers deal with high-LET radiations (Friedland et al. 2010; 2011a, b). The latter ones, however, consider only NHEJ process as the dominant mechanism in G1 phase. In contrast to these findings, our study is aimed at developing a model that would be able to describe several DSB repair pathways after exposure to both low- and

O.V. Belov (✉) · M.S. Panina · M. Batmunkh
Laboratory of Radiation Biology, Joint Institute for Nuclear Research, Dubna, Russia
e-mail: dem@jinr.ru

M. Batmunkh
National University of Mongolia, Ulaanbaatar, Mongolia

N. Sweilam
Mathematics Department, Faculty of Science, Cairo University, Giza, Egypt

V.L. Korogodina et al. (eds.), *Genetics, Evolution and Radiation*,
DOI 10.1007/978-3-319-48838-7_14

high-LET radiations taking into consideration important regulatory factors including the damage complexity and cell cycle. The model offered in the current work takes into account most of the recently identified enzymes involved in the non-homologous end-joining (NHEJ), homologous recombination (HR), single-strand annealing (SSA), and two alternative end-joining pathways and meets the requirements mentioned above. Our calculations are validated by comparing the kinetics of different repair stages to experimentally observed time-courses of the corresponding radiation-induced fluorescent foci. The results on final DSB rejoining have been verified against the data on γ-H2AX foci detection after irradiation.

Mathematical Model

The suggested mathematical model consists of five parts where the first one represents a possible approach to estimate the initial yield of radiation-induced DSBs and other parts are referred to numerical models of NHEJ, HR, SSA, and alternative end-joining repair systems respectively. The basic scheme depicting the simulated repair stages is presented in Fig. 1. In our work we have used the mass-action kinetics approach to simulate the processing of DNA lesions by specific repair enzymes. The kinetic parameters of the model are estimated by the fitting of the calculated curves to the experimental data. The kinetics of DSB induction and remaining are calculated by the following formula:

$$dN_0/dt = \alpha(L)N_{ir}dD/dt - V_{\mathrm{NHEJ}} - V_{\mathrm{HR}} - V_{\mathrm{SSA}} - V_{\mathrm{micro\text{-}SSA}} - V_{\mathrm{Alt\text{-}NHEJ}}, \quad (1)$$

where $N_0 = N_{\mathrm{ncDSB}} + N_{\mathrm{cDSB}}$; V_{NHEJ}, V_{HR}, V_{SSA}, $V_{\mathrm{micro\text{-}SSA}}$, and $V_{\mathrm{Alt\text{-}NHEJ}}$ are the terms characterizing elimination of DSBs by the NHEJ, HR, SSA, micro-SSA and Alt-NHEJ repair pathways, respectively. In Eq. (1), we have also introduced the fraction of irreparable DSBs (N_{ir}) which results in γ-H2AX foci remained in the cell 24 h and later after irradiation. In our calculations we assume that both Ku- and MRN-initiated pathways compete for DNA lesions and the choice of the pathway is regulated by three major factors: the of DSB, cell cycle phase, and the speed of different repair mechanisms (Shibata et al. 2011).

To simulate stages of the major DSB repair pathways we used a biochemical approach based on nonlinear kinetics. A dynamic change of intracellular concentrations of main intermediate complexes are generally expressed by the following differential equations:

$$\frac{dX}{dt} = V_+(X_i, N_0) - V_-(X_i, N_0), \quad (2)$$

where X_i is the intracellular level of the ith enzymatic complex, t is time, N_0 is the total yield of DSBs, the functions V_+ and V_- describe the complex accumulation and degradation, respectively. In total, the model represents a system of 29 ordinary differential equations with the corresponding initial conditions.

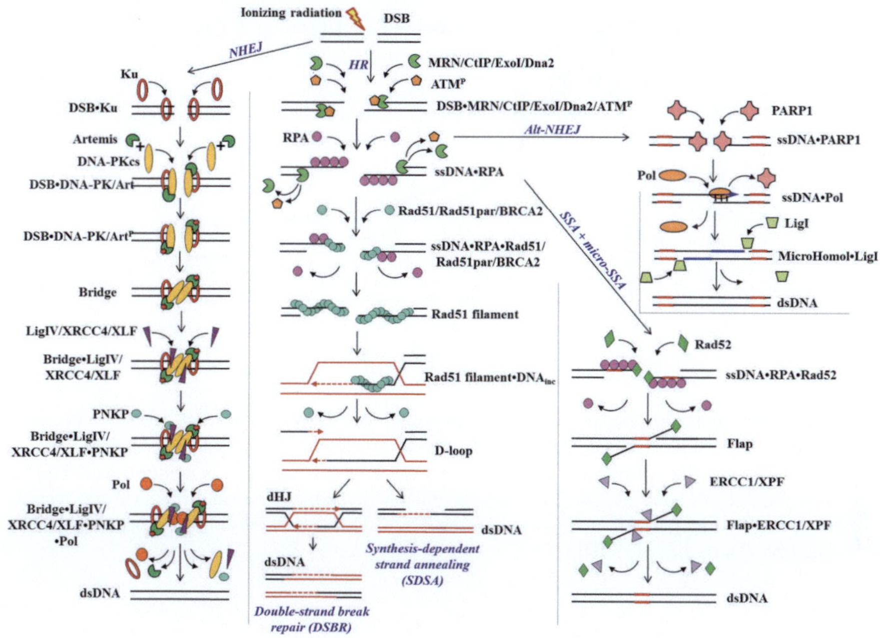

Fig. 1 The basic scheme of the major pathways for DSB repair in mammalian and human cells

Fig. 2 Kinetics of DNA-PKcs foci DNAafter exposure to γ-rays (^{137}Cs) and 1 GeV/u ^{56}Fe ions at the dose of 1 Gy. The calculated curves are compared to the experimental data on foci induction in human skin fibroblasts (HSF42) after corresponding exposures (symbols) (Asaithamby et al. 2008). The error bars represent SEM

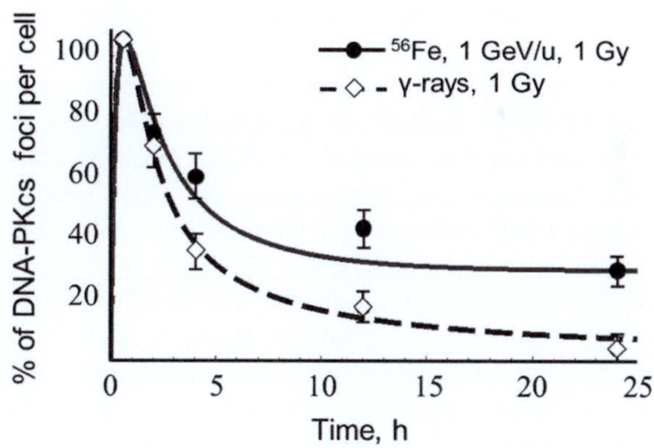

Results

Using our model, we have reconstructed time-courses of fluorescent foci specific to different DSB repair stages. Figure 2 represents the results of computation for the NHEJ stage where the DNA-dependent protein kinase, catalytic subunit (DNA-PKcs), is recruited to the site of repair. The calculated results are expressed as the percentage of pT2609 foci scaled per the number of foci half an hour after exposure, as it was done in the experiment by Asaithamby et al. (2008).

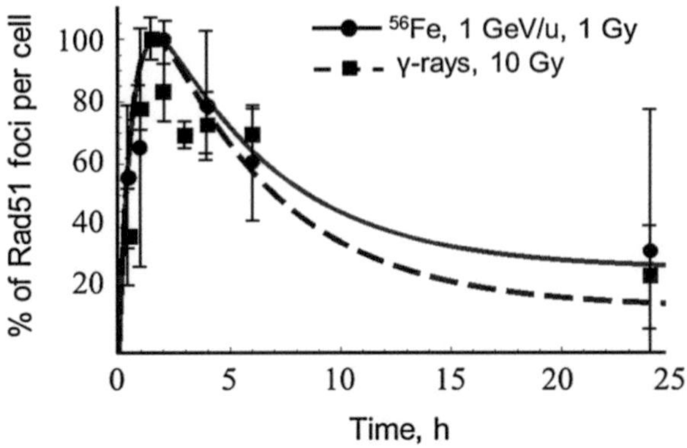

Fig. 3 Kinetics of Rad51 recruitment after exposure to 1 GeV/u ^{56}Fe ions (1 Gy) and γ-rays (10 Gy). The curves are the calculated results; *filled circle* are the scaled experimental data on relative number of cells with Rad51 foci in HF19 human fibroblasts after irradiation with 1 Gy of ^{56}Fe ions (Anderson et al. 2010); *filled square* are the scaled experimental data on the average number of Rad51 foci per cell in Chinese hamster lung fibroblast (V79-4) after exposure to 10 Gy of γ radiation (Harper et al. 2010)

Figure 3 shows the kinetics of Rad51 protein recruitment, one of the most important stages of HR. Here the calculated and measured data are scaled per the maximal values measured 2 and 1.5 h after exposure with ^{56}Fe and γ-rays, respectively.

The final yields of γ-H2AX foci calculated by combining the NHEJ, HR, SSA, and Alt-NHEJ models are presented in Fig. 4. Using our model, we have also calculated the time-course of γ-H2AX foci in cells defective in NHEJ, HR, and SSA functions (data not shown). All these results have been obtained for the radiation with different LET values starting from about 0.02 keV/μm for γ-rays up to 236 keV/μm for 0.3 GeV/u ^{56}Fe ions. The model is also able to describe the repair kinetics for LET values up to 440 keV/mm for which the experimental data on DSB induction are available. However, to prove exactly the model validity for the broader LET range, additional experimental measurements on γ-H2AX foci remaining are required. In relation to a radiation dose, our model possesses a different degree of validity for different repair stages. The kinetics of Ku recruitment is well reconstructed over the range of at least 1–137 Gy; time-courses of DNA-PKcs, RPA, Rad51, and γ-H2AX foci induction are valid for, at least, 1 Gy, 1–4 Gy, 1–10 Gy, and 1–2 Gy, respectively. The conclusion on the validity is based on the comparison of the calculated results with the experimentally assessed kinetics of the corresponding repair stages.

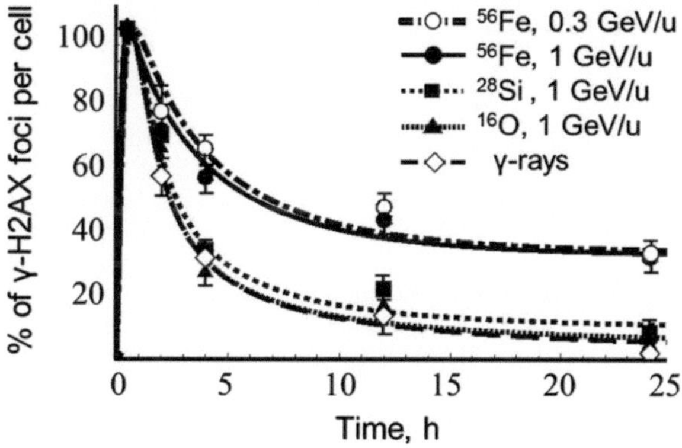

Fig. 4 Kinetics of γ-H2AX remaining after exposure to 1 Gy of iron, silicon, oxygen particles and γ-rays. The curves are the calculated results; symbols are the experimental data referred to irradiation of human skin fibroblasts (HSF42) (Asaithamby et al. 2008). The error bars represent SEM

Conclusions

The offered model is attempted to show a possible mechanistic description of major repair pathways capable to eliminate DNA DSBs in mammalian and human cells exposed to ionizing radiations of wide charge and LET ranges. Simulation of these pathways has been carried out in compliance with the concepts of a modern biology system and computation methods for studying complex biological networks. Our model provides topological views of the NHEJ, HR, SSA, and two Alt-NHEJ pathways, which contribute to clarifying their biological relations. It shows possible connections between the biochemical processes of the repair systems, some of which are still hardly accessible for experimental measurements.

By means of our model we have tested the hypothesis that at least two alternative end-joining pathways operate in response to induction of DSBs when classical NHEJ becomes unavailable. Results of calculations for NHEJ-deficient cells have shown that even in case of high-LET irradiation a limited fraction of complex DSBs may be processed by using these alternative pathways. It leads to the slower repair kinetics but the remaining DSB level appears to be higher than it might be expected if no other alternative mechanisms exist.

The simulation carried out for HR-deficient cells has confirmed one more hypothesis that the functionality of Ku- and DNA-PK-dependent NHEJ is particularly perturbed by high-LET induced clustered DSBs making this pathway less relevant to resolving these lesions (Moore et al. 2014). In our study, the evidence for this is the slower repair kinetics for ^{12}C-irradiated HR-deficient cells compared with those exposed to X-rays.

In general, the developed model (Belov et al. 2015) summarizes a lot of recent findings on regularities of DSB repair after exposure to radiation with different LET. It stimulates proposing new experiments to assess repair stages which are insufficiently studied to the date. It mainly refers to the branching of HR at its late stages, and to Ku-independent alternative end-joining pathways.

References

Anderson JA, Harper JV, Cucinotta FA, O'Neill P (2010) Participation of DNA-PKcs in DSB repair after exposure to high- and low-LET radiation. Radiat Res 174:195–205

Asaithamby A, Uematsu N, Chatterjee A, Story MD, Burma S, Chen DJ (2008) Repair of HZE-particle-induced DNA double-strand breaks in normal human fibroblasts. Radiat Res 169:437–446

Bastin M, St-Onge L, Bouchard L (1992) A mathematical model of homologous recombination in cultured cells. J Theor Biol 156:513–524

Belov O, Krasavin E, Lyashko M, Batmunkh M, Sweilam N (2015) A quantitative model of the major pathways for radiation-induced DNA double-strand break repair. J Theor Biol 366:115–130

Cucinotta FA, Nikjoo H, O'Neill P, Goodhead DT (2000) Kinetics of DSB rejoining and formation of simple chromosome exchange aberrations. Int J Radiat Biol 76:1463–1474

Cucinotta FA, Pluth JM, Anderson JA, Harper JV, O'Neill P (2008) Biochemical kinetics model of DSB repair and induction of γ-H2AX foci by non-homologous end joining. Radiat Res 169:214–222

Friedland W, Jacob P, Kundrat P (2010) Stochastic simulation of DNA double- strand break repair by non-homologous end-joining based on track structure calculations. Radiat Res 173:677–688

Friedland W, Jacob P, Kundrát P (2011a) Mechanistic simulation of radiation damage to DNA and its repair: on the track towards systems radiation biology modelling. Radiat Prot Dosim 143:542–548

Friedland W, Dingfelder M, Kundrat P, Jacob P (2011b) Track structures, DNA targets and radiation effects in the biophysical Monte Carlo simulation code PARTRAC. Mutat Res 711:28–40

Goodhead DT (1985) Saturable repair models of radiation action in mammalian cells. Radiat Res 104:58–67

Harper JV, Anderson JA, O'Neill P (2010) Radiation induced DNA DSBs: contribution from stalled replication forks? DNA Repair 9:907–913

Kiefer J (1988) A repair fixation model based on classical enzyme kinetics. In: Kiefer J (ed) Quantitative models in radiation biology. Springer, Berlin, pp 171–180

Moore S, Stanley FKT, Goodarzi AA (2014) The repair of environmentally relevant DNA double strand breaks caused by high linear energy transfer irradiation—no simple task. DNA Repair 17:64–73

Shibata A, Conrad S, Birraux J, Geuting V, Barton O, Ismail A, Kakarougkas A, Meek K, Taucher-Scholz G, Löbrich M, Jeggo PA (2011) Factors determining DNA double-strand break repair pathway choice in G2 phase. EMBO J 30:1079–1092

Taleei R (2013) Modelling and calculation of DNA damage and repair in mammalian cells induced by ionizing radiation of different quality. Doctoral thesis, Department of Oncology-Pathology, Karolinska Institutet. Stockholm

Taleei R, Nikjoo H (2013) Biochemical DSB-repair model for mammalian cells in G1 and early S phases of the cell cycle. Mutat Res 756:206–212

Taleei R, Weinfeld M, Nikjoo H (2012) Single strand annealing mathematical model for double strand break repair. J Mol Eng Syst Biol 1:1–10

Mathematical Analysis of Regulatory Networks and Damage Repair Efficiency in Bacterial Cells

Aleksandr Bugay, Maria Vasilyeva, Aleksandr Parkhomenko and Evgeny Krasavin

Introduction

The genetic stability of living organisms is largely achieved by existence of a complex hierarchy of repair systems, which maintain normal cell activity and are used to stabilize DNA damaged by physical and chemical nature factors. Defects in the DNA repair systems are the main cause of the increased mutation level or cell death. Therefore, it is important to study the influence of defects in different repair process stages on the cell's ability to withstand damaging factors.

In bacteria, such as *Escherichia coli*, cell survival and genomic stability after UV radiation depends on the repair mechanisms induced as part of the SOS response to DNA damage. The SOS response has become a paradigm for the field of DNA repair at least for the last 60 years since the pioneering work by Weigle, who found that preliminary irradiation of *Escherichia coli* cells by low UV doses before infection, significantly increases the survival of UV-irradiated λ phage, and increases the mutations frequency in phages (Weigle 1953). Several years later Witkin suggested that it was the evidence of a damage-induced DNA repair system in bacterial cells (Witkin 1976). Finally, Radman formulated the idea of "mutation-prone" replication mechanism involving the *lexA* and *recA* genes, which was named "SOS repair" (Radman 1975). The knowledge of molecular mechanisms of excision repair and postreplication repair had been established near that time (Rupp and Howard-Flanders 1968), but the processes of SOS repair and SOS mutagenesis remained unknown for a long period of time.

In the beginning of the 1970s a group of over 40 genes (SOS regulon) expressed during SOS response and regulated by RecA and LexA proteins was identified. But only in 1998–1999 it was discovered (Tang et al. 1999) that SOS repair

A. Bugay (✉) · M. Vasilyeva · A. Parkhomenko · E. Krasavin
Laboratory of Radiation Biology, Joint Institute for Nuclear Research, Dubna, Russia
e-mail: bugay@jinr.ru

V.L. Korogodina et al. (eds.), *Genetics, Evolution and Radiation*,
DOI 10.1007/978-3-319-48838-7_15

(“translesion synthesis”, TLS) was ensured by DNA polymerase V (UmuD′$_2$C), that was encoded by *umuD* and *umuC* genes. In a few years, Pol V was shown to function as a single complex with activated RecA* protein (Jiang et al. 2009) .

After the discovery of SOS repair key mechanisms, the induced mutagenesis became the subject of intensive mathematical research that systematizes the known structure of repair mechanisms and mutagenesis events based on experimental facts. First studies were related with simulation of gene expression and protein kinetics during SOS response (Aksenov 1999; Krishna et al. 2007; Ni et al. 2007), until it became possible to calculate the number of mismatches in DNA as result of TLS (Belov et al. 2009). Most of studies were focused on mathematical description of mutation process in *E. coli* wild-type cells.

The current research is focused on the problem of capability of different repair systems to handle cellular damage in case of malfunction of one or several repair pathways. This crucial step in bacterial SOS response is connected with initial lesion removal performed by nucleotide excision repair (NER). Under normal operation of NER system in microorganisms, up to 80 %, and in the cells of higher organisms up to 70 %, of the primary lesions are removed before DNA replication (Lin et al. 1997). Therefore, the main attention of this study will be focused on modeling of the NER system and its impact on mutagenesis.

Model Formulation

DNA Repair Pathways

As an external damaging factor we consider ultraviolet (UV) radiation because its effect on bacterial cells has been well studied experimentally. As it is known, typical DNA chain lesions under UV irradiation are thymine-thymine cys-syn cyclobutane photodimers and pyrimidine-pyrimidone (6–4) photoproducts (Cadet et al. 1992). *Escherichia coli* bacterial cells have several ways to remove this type of UV lesions (Fig. 1).

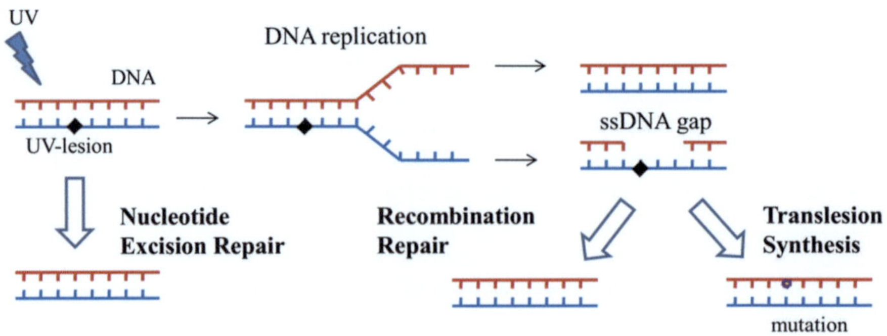

Fig. 1 A scheme of the key pathways of UV damage repair in bacterial cells

First of all, the lesions are effectively removed by the photoreactivation repair system, i.e. the process of dimer monomerization under visible light by the special photolyase enzyme (Rupert 1975). Since most of the experiments are carried out in the conditions excluding the influence of photoreactivation on the process under study, we will assume the absence of photoreactivation. Another way of lesion removal is the nucleotide excision repair (NER) system (Van Houten 1990). In *E. coli* cells special enzymes encoded by *uvr*-genes seek and cut a single-stranded DNA (ssDNA) fragment that is 12 nucleotides long and contains a dimer. The appeared gap is then sealed by DNA polymerase I and DNA ligase.

If so many lesions emerge in the DNA chain that cannot be removed by excision repair, the single-strand DNA gaps are formed in the daughter strands at the places of lesions in the mother strand during DNA replication (Rupp and Howard-Flanders 1968). Some of them can be restored in the course of postreplicative repair processes. In bacterial cells, one of the gap repair mechanisms is recombination exchange between sister duplexes (recombination repair) (Kuzminov 1999), the key regulator of which is the RecA protein. When it binds with DNA, a protease conformation is produced; also, this protein splits the repressor LexA protein, which triggers inducible SOS repair. In a complicated interaction system, products of the *umuC* and *umuD* genes form a polymerase Pol V complex, which can synthesize DNA on the damaged matrix at the cost of the appearance of errors (TLS —translesion synthesis) (Wang 2001), thereby realizing the mutagenic pathway of post-replicative repair. It should be noted that the SOS induction of the *uvrA, uvrB,* and *uvrD* genes also stimulates the functioning of the excision repair system.

DNA Replication and ssDNA Production

The replication of the *E. coli* circular chromosome is bidirectional. Two replication forks start to move at similar rates in opposite directions from the origin site around the chromosome until they meet at the replication terminus.

When a replication fork encounters a barrier (damaged DNA), it stalls or collapses until the repair systems restore the fork's movement. There are three typical pathways when a replication can be blocked. At first, if a replication fork meets a photodimer, the replication complex can resume its movement leaving an ssDNA gap on a daughter strand (Rupp and Howard-Flanders 1968). Second, a similar situation can take place when a protein complex is bound to the photodimer. It is assumed that protein blocks should be resolved by means of recombinational repair machinery. Third, when ssDNA interruption (a nick) is encountered on a template strand, it forces the replication fork to collapse (Kuzminov 1999). In this case, recovery occurs by recombinational repair involving the RecBC and RecA proteins. All three cases lead to the formation of ssDNA (the "secondary" lesions).

The kinetic equations for DNA replication at the presence of defects were derived in (Gauthier et al. 2010). Here we introduce an extended version of this model:

$$\frac{df}{dt} = 2v_0\rho, \tag{1}$$

$$\frac{d\rho}{dt} = I(1-f) - \frac{v_0\rho\sum_i n_i}{L(1-f)} + \sum_i \frac{\rho_i}{\tau_i} - \frac{v_0\rho(2\rho + \sum_i \rho_i)}{(1-f)}, \tag{2}$$

$$\frac{d\rho_i}{dt} = \frac{v_0\rho n_i}{L(1-f)} - \frac{\rho_i}{\tau_i} + \frac{v_0\rho\rho_i}{(1-f)}, \tag{3}$$

where $f(t)$ is the fraction of replicated DNA, v_0 is the replication fork velocity in non-irradiated DNA, L is the *E. coli* genome length, $\rho(t)$ is the local density of moving replication forks, $\rho_i(t)$ is the density of stalled replication forks, and $I(t)$ is the rate of replication fork production (nucleation rate). We should note that the nucleation rate is rather low and can be kept constant $I = I_0$ for prokaryotes, but not for eukaryotes. In the latter case, the time-dependent nucleation occurs non-uniformly along the chromosome, i.e. there is a large amount of progressing replication forks at the same time.

The ssDNA production rate can be calculated as the amount of bases s missed until the stalled replication fork resumes movement. Thus, the total rate of the ssDNA gap production should be proportional to the numbers $L\rho_i$ of stalled replication forks:

$$\frac{ds^+}{dt} = \sum_i v_i L\rho_i, \tag{4}$$

where $v_i = L_g/\tau_i$ is the effective rate of the ssDNA production on the lesion of the 'i' type, and L_g is the average length of the gap.

NER *System*

NER is a multistep, ATP-dependent process. The NER mechanism consists of initial damage recognition and verification, excision of the ssDNA fragment containing the damaged base, and resynthesis of the appeared gap (see Fig. 2). In *E. coli*, more than six gene products—UvrA, UvrB, UvrC, UvrD helicase, DNA polymerase and DNA lygase—are involved in these stages.

In solution, UvrA forms a dimer that binds to the damaged DNA template. UvrB interacts with the $UvrA_2$ dimer, producing the $UvrA_2B$ complex. This complex alters ATPase activity and directly recognizes the damaged DNA. Upon binding to DNA, the $UvrA_2B$-DNA complex undergoes conformational changes. The lesion remains in a close contact with UvrA and then it is transferred to UvrB. $UvrA_2$ then dissociates from the complex. A stable UvrB-DNA complex is generated. UvrC can

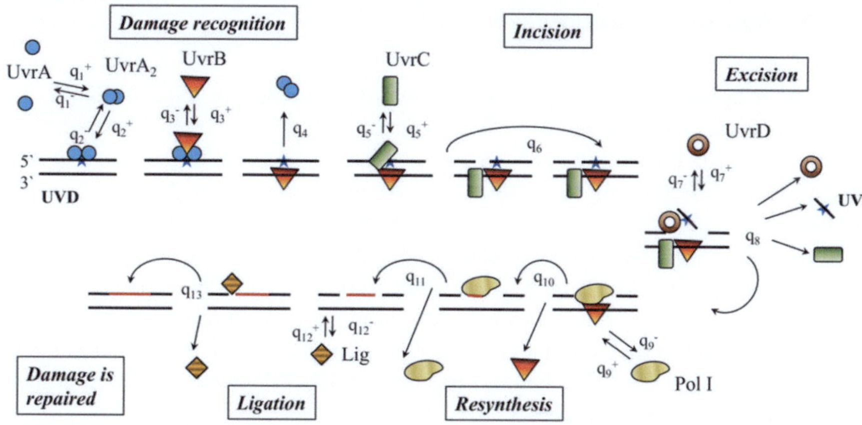

Fig. 2 A scheme of the NER mechanism in bacterial cells

bind to this complex and form a pre-incicion complex, catalyzing the 3′ and 5′ incisions which take approximately 12 nucleotides apart. After the incision events, the UvrD helicase binds to the complex and starts unwinding the resulting 12 nucleotides fragment until its excision and UvrC removal. Finally, the DNA polymerase Pol I fills the appeared ssDNA gap releasing UvrB. Repair synthesis ends with bridging the ends of a gap by the DNA ligase.

A quantitative description of the described events can be provided using the reaction scheme given in Fig. 2. According to it, we can write equations for protein concentrations following conventional chemical kinetics. Time dependence of each substance is governed by a system of nonlinear differential equations for the respective concentrations $x_i(t)$:

$$\frac{dx_i}{dt} = V_{ij}^{(+)}(x_j, \rho_j) - V_{ij}^{(-)}(x_j, \rho_j), \tag{5}$$

where, $X_i (i = 1 \ldots N)$ are the intracellular concentrations of the regulatory proteins of N species, t is time. V_{i+} and V_{i-} are the rates of the ith protein synthesis and decay, respectively. Nonlinear functions $V^{(\pm)}$ are determined by reaction schemes following standard rules of chemical kinetics.

The opportunity to distinguish all complexes and stages of NER process allows us to compare different NER-deficient mutant bacterial cells. They can be divided into two major classes. The first one is unable to repair the primary DNA lesion, and the second is unable to restore the produced gaps. In this paper, we will consider *uvrA*$^-$and *polA*$^-$ mutant cells as respective examples of these two defect cases. For this purpose, the quantity s of the produced enzimatic gaps should be subdivided into two parts: $s = s_1 + s_2$. The first part s_1 contains the photodimer opposite the gap, and the second part s_2 does not contain it (see Eq. 4).

SOS System and Mutagenesis

A typical scheme of SOS-system regulation is presented in Fig. 3. The LexA protein is a repressor of about 40 SOS genes including its own one. Repression can be removed by its proteolitic cleavage by the RecA* protein filament. The latter is formed after RecA binding to ssDNA (one RecA monomer per 3–4 nt of DNA). The RecA* protease also takes part in homologous recombination.

Inactivation of LexA repressor leads to derepression of inducible SOS system genes including *uvrA*, *uvrB*, *recX*, *dinI*, *umuC*, *umuD*, and SOS-response starting. The RecA* protease is able to cleave the UmuD protein transferring it into the active state UmuD′. The UmuD and UmuD′ proteins form the $UmuD_2$ and $UmuD'_2$ homodimers and the UmuDD′ heterodimer in solution. Moreover, the homodimers demonstrate a subunit exchange.

All three types of dimers interact with the UmuC protein. The result of the corresponding binding reactions is three protein complexes: $UmuD_2C$, UmuDD′C, and $UmuD'_2C$. The $UmuD_2C$ complex is involved in the cell cycle regulation. The UmuDD′C complex plays an inhibiting role in SOS mutagenesis by suppressing $UmuD'_2$ activity. The $UmuD'_2C$ complex plays the central role in TLS. However, the $UmuD'_2C$-RecA*-ATP complex, but not only $UmuD'_2C$ itself, represents the active Pol V core (Jiang et al. 2009). Evidently, this ensures the TLS location on the damaged DNA template, capped by the RecA* filament. An extended PolV-mut complex can also include the SSB protein and subunits of DNA polymerase III.

RecX and DinI proteins play an important role in SOS system regulation. The DinI protein binds to the RecA* filament providing a cap that prevents cleavage of the UmuD protein and, consequently, suppresses $UmuD'_2C$ complex formation. This forms the negative feedback limiting SOS mutagenesis. The function of RecX

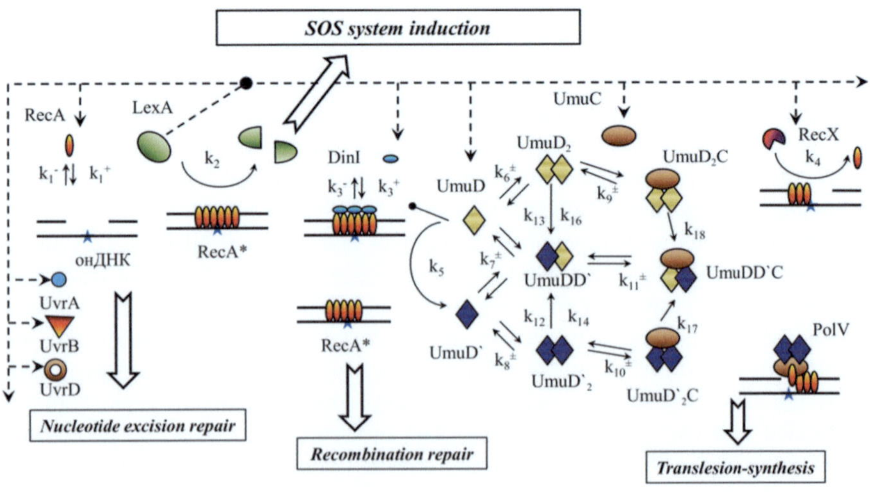

Fig. 3 The scheme of SOS regulation and translesion synthesis in bacterial cells

protein is to prevent RecA* filament growth, thus forming the negative feedback limiting the maximal level of the induced signal.

The mathematical description for kinetics of SOS proteins and protein complexes is also based on standard kinetics rules according to Eq. 5. A detailed derivation is given in (Belov et al. 2009; Bugay et al. 2015).

Let us consider the process of mutation formation. To be activated, the UmuD'$_2$C complex must first bind a RecA* containing the ssDNA gap, and then start the repair synthesis at the highest rate v_T. A detailed analysis of the mutation probability during SOS repair was performed in (Belov et al. 2009). It is known that PolV Mut can realize TLS both on damaged (containing a photodimer) and undamaged DNA templates. Using this fact, the overall number n_{mut} of errors in newly synthesized DNA during translesion synthesis depending on time t and UV energy fluence Ψ, can be calculated as follows:

$$n_{mut} = \mu_1 \int\limits_0^t n_d(t', \psi)\, dt' + \mu_2 \int\limits_0^t n_n(t', \psi)\, dt', \tag{6}$$

where the numbers n_d and n_n of the translesion synthesis events on the damaged and undamaged DNA matrices are derived from equation set 5. The respective probabilities are μ_1 and μ_2.

Results and Discussion

The formulated model makes it possible to calculate the replication kinetics, the amount of DNA lesions, protein concentration and the level of induced mutagenesis. The main result is the number of incorrect base substitutions reflecting the mutation frequency within the framework of the model presented in Fig. 4. As expected, the level of induced mutations is low in wild-type cells, slightly higher in *polA*$^-$ cells, and very high in *uvrA*$^-$ cells. All the calculated results excellently coincide with the experimental observations.

To study NER performance in the wild type cells, we have calculated the characteristic time of UV damage (UVD) elimination using the NER depending on UV fluence (Fig. 5a). Under the conditions of moderate UV fluences, UVD are eliminated during about 40–120 min after irradiation, as it is observed in vivo (Lin et al. 1997). Upon high fluences, the efficiency of the NER reduces. To make clear the strategy used by bacteria for DNA damage removing when NER doesn't work correctly, we have compared the wild type bacterial cells and the cells defective in NER genes (*uvrA*$^-$ and *polA*$^-$). It follows that the defect in the primary damage repair system (NER) leads to elevated induction of secondary repair systems (recombination and TLS) as one may expect. Keeping in mind SOS induction of key NER proteins (UvrA and UvrB) one may ask what would happen in the inverse situation, when NER functions normally, but recombination or TLS is defective.

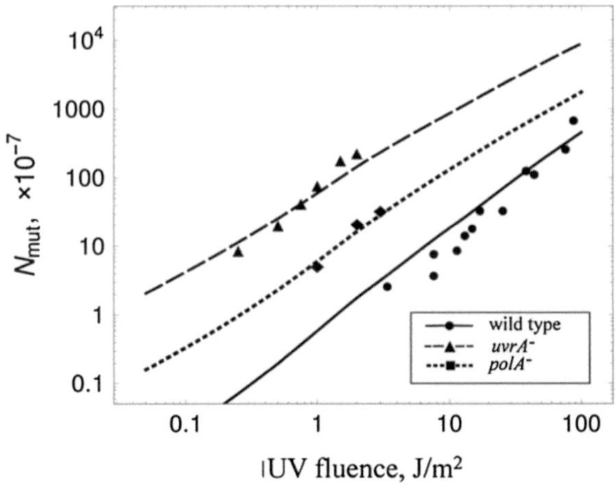

Fig. 4 Frequency of induced mutations depending on the fluence of UV energy for different strains having defects in the excision repair system. The experimental data from (Kato et al. 1974) and (Bates et al. 1989) are used

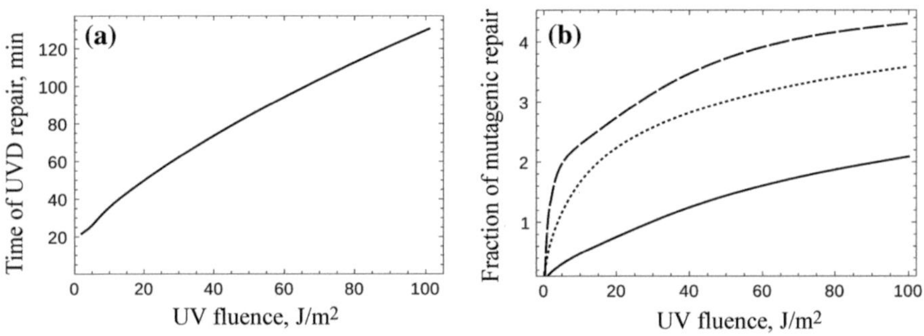

Fig. 5 Efficiency of functioning of the bacterial repair systems: **a** dependence of the characteristic time of UVD elimination by the NER on the UV fluence in wild type cells and **b** number of the ssDNA restored by the TLS with respect to the ssDNA recovered by the homologous recombination for wild type cells (*solid line*), *uvrA*⁻ mutants (*dashed line*), and *polA*⁻ mutants (*dotted line*)

Malfunction in one of secondary repair systems which does not lead to sufficient increase in the overall photodimer removal rate. Therefore we can conclude that SOS dependence of NER mainly provides the repair protein level elevation under high levels of UV damage.

We have considered two primary classes of NER mutants having defects in primary damage recognition *uvrA*⁻ or in excision gap sealing *polA*⁻. One can ask whether specific defects recruit secondary repair systems in a comparable way or not. To answer this question, we have compared the efficiency of functioning of the secondary repair systems by calculating the ratio between the number of ssDNA recovered during the TLS and that recovered by the homologous recombination at the given time and UV fluence (Fig. 5b). This comparison in a quantitative form has been made for the first time. Recombination is dominant for low UV fluences,

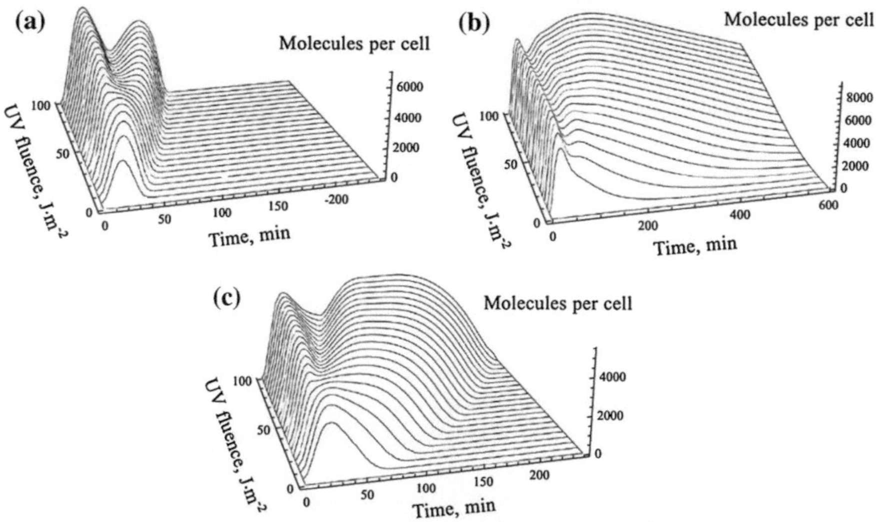

Fig. 6 Dynamics of the ssDNA concentration in the case of wild type cells (**a**), *uvrA*$^{-}$ mutants (**b**), and *polA*$^{-}$ mutants (**c**)

but in NER mutants TLS becomes twice as efficient following the ten times lower amount of the initial damage. We also note that the equal efficiency ratio is achieved approximately during 60 min after irradiation for any of the cell genotypes considered. This fact indicates the accurate timing of the cell repair systems. Less evident result is that TLS repair cannot be more efficient than by four times with respect to recombination independently of NER mutant type. Thus, the cell is restricted in its capacity to repair damages in the error-prone manner.

Another issue that requires special attention is the behaviour of the ssDNA level. Our model has demonstrated the appearance of local minima of the ssDNA levels (Fig. 6).

This phenomenon might indicate a time moment, when the error-free repair switches to the error-prone one. Maxima in the inducing signal appear when the efficiency of the TLS is about twice as high as that of the recombination. To additionally test this effect, we have performed calculations in the case of impairments in the TLS functioning. Modulation of the ssDNA level disappears in this case. This assumption correlates with single-cell fluorescence experiments (Friedman et al. 2005), where the temporal modulation of the promoter activity of SOS genes was observed. This modulation was shown to be under control of the UmuDC operon, like in our model. However the effect was stochastic and disappeared upon averaging the cell population. But in our case we deal with the averaged cell behavior, where modulation of ssDNA level might reflect switching between the error-free and error-prone DNA repair. We note that the LexA key protein is not subjected to significant modulation. Thus, an experimental test should be made to find out the origin of the phenomena found. Measurements of the DNA degradation level under UV were carried out long time ago (Tang and Ross 1985) and those data do not provide sufficient details about phenomena of interest. As an alternative, the kinetics of the RecA protein should be measured in its protease form RecA*, whose concentration curve is similar to that of ssDNA.

Conclusion

Current development of molecular biology made it possible to construct a mathematical model of UV induced mutagenesis, which reproduces a chain of the key molecular events from a primary photodamage to the appearance of point mutations. It has become possible to study the influence of defects in particular genes on the damage repair, and perform calculations of the mutation yield for specific bacterial genotypes.

Attention was mainly paid to the mathematical description of the NER system, including the detailed kinetics of the repair complexes. This level of complexity allows one to consider the repair-deficient *E. coli* cells, where one or several repair pathways are broken. Here, we have demonstrated this opportunity with two examples of *uvrA*⁻ and *polA*⁻ mutant cells. These cells have elevated levels of DNA degradation and the increased induced mutation frequency after UV irradiation. Also, the comparison of the relative efficiency of the recombination and TLS has been performed for the first time. The TLS has been found incapable of being efficient more than four times than the homologous recombination.

The model correctly reproduces the available experimental observations of protein concentration kinetics and induced mutation frequencies. The model approach was able to describe a subtle effect: the temporal modulation of SOS response. Our results have indicated that *umuDC* genes form a negative feedback that limits the induced signal (ssDNA) level. However, a detailed study of the above effects requires a more precise description of the single cell level dynamics by means of stochastic modeling.

Thus, the developed approach can provide a strong basis for the general understanding of the interplay between replication and repair as a response to external damaging factors.

References

Aksenov SV (1999) Dynamics of the inducing signal for the SOS regulatory system in *Escherichia coli* after ultraviolet irradiation. Math Biosci 157:269–286

Bates H, Randall SK, Rayssiguier C, Bridges BA, Goodman MF, Radman M (1989) Spontaneous and UV-Induced mutations in *Escherichia coli* K-12 strains with altered or absent DNA polymerase I. J Bacteriol 171:2480–2484

Belov OV, Krasavin EA, Parkhomenko AYu (2009) Model of SOS-induced mutagenesis in bacteria *Escherichia coli* under ultraviolet irradiation. J Theor Biol 261:388–395

Bugay AN, Krasavin EA, Parkhomenko AYu, Vasilyeva MA (2015) Modeling nucleotide excision repair and its impact on UV-induced mutagenesis during SOS-response in bacterial cells. J Theor Biol 364:7–20

Cadet J, Anselmino C, Douki T, Voituriez L (1992) Photochemistry of nucleic acids in cells. J Photochem Photobiol B 15:277–298

Friedman N, Vardi S, Ronen M, Alon U, Stavans J (2005) Precise temporal modulation in the response the SOS DNA repair network in individual bacteria. PLoS Biol 3:238

Gauthier MG, Herrick J, Bechhoefer J (2010) Defects and DNA replication. Phys Rev Lett 104:218104

Jiang Q, Karata K, Woodgate R, Cox MM, Goodman MF (2009) The active form of DNA polymerase V is UmuD'$_2$C RecA* ATP. Nature 460:359–363

Kato T, Rothman RH, Clark AJ (1974) Analysis of the role of recombination and repair in mutagenesis of *Escherichia coli* by UV Irradiation. Genetics 87:1–18

Krishna S, Maslov S, Sneppen K (2007) UV-induced mutagenesis in *Escherichia coli* SOS response: a quantitative model. PLoS Comput Biol 3(3):0451–0462

Kuzminov A (1999) Recombinational repair of DNA damage in *Escherichia coli* and bacteriophage λ. Microbiol Mol Biol Rev 63:751–813

Lin CG, Kovalsky O, Grossman L (1997) DNA damage-dependent recruitment of nucleotide excision repair and transcription proteins to *Escherichia coli* inner membranes. Nucleic Acids Res 25(15):3151–3158

Ni M, Wang SY, Li JK, Ouyang Q (2007) Simulating the temporal modulation of inducible DNA damage response in *Escherichia coli*. Biophys J 93:62–73

Radman M (1975) SOS repair hypothesis: phenomenology of an inducible DNA repair which is accompanied by mutagenesis. In: Hanawalt P, Setlow R (eds) Molecular mechanisms for repair of DNA, Part A. Plenum Press, pp 355–367

Rupert CS (1975) Enzymatic photoreactivation: overview. In: Hanawalt P, Setlow R (eds) Molecular mechanisms for repair of DNA, Part A. Plenum Press, pp 7387

Rupp WD, Howard-Flanders P (1968) Discontinuities in the DNA synthesized in an excision-defective strain of *Escherichia coli* following ultraviolet irradiation. J Mol Biol 31:291–304

Tang M, Ross M (1985) Single-strand breakage of DNA in UV-Irradiated *uvrA*, *uvrB*, and *uvrC* mutants of *Escherichia coli*. J Bacteriol 161:933–938

Tang M, Shen X, Frank EG, O'Donnell M, Woodgate R, Goodman MF (1999) UmuD'$_2$C is an error-prone DNA polymerase, *Escherichia coli* Pol V. Proc Natl Acad Sci 96:8919–8924

Van Houten B (1990) Nucleotide excision repair in *Escherichia coli*. Microbiol Rev 54:18–51

Wang Z (2001) Translesion synthesis by the UmuC family of DNA polymerase. Mutat Res 486:59

Weigle JJ (1953) Induction of mutation in bacterial virus. Proc Natl Acad Sci 39:628–636

Witkin EM (1976) Ultraviolet mutagenesis and inducible DNA repair in *Escherichia coli*. Bacteriol Rev 40:869

Radiation Risks and Confusions

Helmut Abel and Gudrun Erzgräber

Origin of the 'Linear No-Threshold Hypothesis'

The question, whether natural ionizing radiation can have biological consequences has interested scientists already in the first decades of the last century. There were numerous hints that high expositions are connected with biological consequences. To assume from this that there are effects also as the result of the natural ionizing radiation required first of all and at least a large amount of data materials.

After the analysis of various *Drosophila* experiments in the famous work "On the nature of gene mutations and gene structures" (Timofeeff-Ressovsky et al. 1935) the scientists Timofeeff-Ressovsky, Delbrück and Zimmer draw following conclusions:

- The radiation induced mutation rate is independent from the doserate and is directly proportional to the dose.
- no minimal radiation dose is to be expected and therefor the curve of proportionality can be extrapolated in the same form against zero.

By this a "linear no-threshold hypothesis" was formulated.

Under the supposition that every ionization causes a mutation it was calculated that at the most only 1 of 1000 spontaneous mutations in *Drosophila* can be traced back to the natural ionizing radiation. As a result of this it was pointless to look for

---

Lecture on the occasion of the First Intern. Conf. dedicated to the centenary of the birth of N.W. Timoféeff-Ressovsky; Dubna, 6–9 September 2000. The paper was firstly published in the Conference Proceedings "Modern problems of radiobiology, radioecology and evolution", Dubna, JINR, 2001, 173–177.

H. Abel (✉) · G. Erzgräber
Max Delbrück Center (MDC), Berlin-Buch, Germany
e-mail: helmut.abel@gmx.de

V.L. Korogodina et al. (eds.), *Genetics, Evolution and Radiation*,
DOI 10.1007/978-3-319-48838-7_16

possible processes which prevent that each ionization results in a mutation. That such processes exist has been meanwhile proven.

Yet in this particular work, at that time, only the question had to be answered whether the natural ionizing radiation can be neglected in attempts to explain spontaneous mutations in *Drosophila*. The clear positive answer led to the statement that genes are arranged atom-formations and spontaneous mutations are to be traced back to thermodynamic changes of the arrangements in atom formations.

Radiation Induced Malign Processes

The questioning in regard to spontaneous and radiation induced malign processes in human cells is bedded simultaneously. According to estimations (Clarke 1997) the relationship between primarily spontaneous and primarily natural radiation induced gene-defects is not in the range of 1000 to 1, but of 10 million to 1.

The fact that genes in spite of these extremely high rates of primary spontaneous defects still can maintain integrity became explainable by the discovery of the intracellular DNA-repair.

A lot of spontaneous and radiation induced defects become repaired by reducing themselves. However, for malign processes one cannot ask for the relationship between the total of primarily spontaneous defects and the total of natural radiation induced defects but only for the clearly smaller rates of critically defects for cancer induction. In addition it has to be taken into account that different types of cancer are also very different in regard to their causes (Trott and Rosemann 2000).

It is generally accepted that the process of cancerogenesis occurs in a series of disturbing effects in the DNA. None of the single effect causes cancer by itself. The number of necessary effects and the sequence can vary between different types of cancer. Translocations and other rearrangements in the DNA are typical defects in the malign cells.

In the case of radiation induced cancerogenesis these defects appear to be related to DNA double-strand breaks. In general double-strand breaks are effectively repaired; exceptions are cluster of double-strand breaks.

Therefore is it first of all necessary to consider the number of these clusters in dependence of the dose and the radiation types of the natural ionizing radiation.

An exact estimation in the exposition-range of natural radiation is only theoretically possible and requires the knowledge of the spatial DNA-organization.

DNA-repair experiments on somatic cells with different types of radiation, heavy ions, light nuclei, fast neutrons and gamma-radiation, resulted in the knowledge that the DNA in somatic-cells is nucleus-membrane-associated into more than 1000 repair-autonomous subunits per cell (Regel 1988; Erzgräber et al. 1992).

Based on these results it was possible to receive knowledge's of the spatial DNA-organization and to calculate quantenphysically the frequency of the rate of irreparable cluster of double-strand breaks in relation to the dose and to every interesting radiation type (Rosemann 1992).

The natural ionizing radiation consists (concerning the dose) of about 50 % gamma radiation respectively high energy electrons and about 50 % alpha radiation of radon and its decay-products. For the gamma and electron parts the calculated proportion between cluster and the total number of primary double-strand breaks is so small, that this number of cluster can be neglected in comparison with the number of cluster in the case of the radon-alpha part. In the case of the radon-alpha part the proportion between cluster and the total number of primary double-strand breaks remains even constant at 0.16 from the dose zero up to about some 100 mGy. This means that given the case an alpha-particle passes through a DNA-subunit and to induce double-strand breaks, than always an irreparable cluster will be formed in constant probability.

Hence, when epidemiological studies let to expect hints that natural ionizing radiation induced cancer-risks then this especially true to lung cancer as a result of the radon-part. That there is a link between lung cancer and radon-exposition was already recognized without any doubts some decades before in the case of uran-miners. However the concentration of radon in the mines was extremely higher as outside of the mines, especially in houses.

Formal calculations on the base of models for respiration and the 'linear no-threshold hypothesis' showed for natural radon-concentrations in houses lung-cancer rates from some to 10 %. The natural radon-concentrations in houses are strong varying around a factor 100 and more. Possibly this fact could be reflected in epidemiological studies.

A lot of epidemiological studies are existing (Wichmann et al. 1998; Cohen 1995) but confirmations of the formal calculations for a linear positive correlation between lung-cancer risk and radon-concentration in the range of the natural radon concentration have not been found.

Wichmann et al. formulate in their study of the years 1992–1996, which includes about 70 % of the German territory, "that no Radon-risk in houses is recognizable". A likewise extensive US-study from 1995 Cohen finds even "a negative correlation between lung-carcinoma-mortality and natural radon-exposition". In many publications about radiation risks of radon such results are ignored.

For instance the Swiss Office of Health formulates (Bundesamt für Gesundheit 1999) exclusively supporting on formal model-calculations on the base of 'linear no-threshold hypothesis', that "Radon our health is more damaging than the emissions as a result of Chernobyl and all tests with nuclear-weapons".

An interesting contra-consideration has given Schüttmann (1999). About 150 years ago was lung cancer a very seldom illness. In only 0.06 % was lung cancer the cause of death. The radon-concentration was however at that time the same as today.

Cause of the contradictions between the 'linear no-threshold hypothesis' and the results of the epidemiological studies is the ignorance of the DNA-repair-mediating reduction of both spontaneous and radiation-induced DNA-disturbances and the multiple steps character of cancerogenesis.

The ‘linear no-threshold hypothesis’ must not be wrong, but it cannot be interpreted in the sense of a dose-linear increase of the totality of all cancer-induction-processes in the range of the natural ionizing radiations.

Consequences of Confusions

The ICRP recommended 1 mSv per year as radiation-exposition-limit for the general public. The natural average is about some mSv per year and the maximum is about some 10 mSv per year.

If already the natural radon is no dangerous for health, than the gamma-and electron radiations of the natural expositions are also no source of danger.

In the general public the limit of 1 mSv per year has caused the feeling of a constant present cancer-risk as a result of the natural ionizing radiation.

So the caution, which formed the motive of the ICRP, has transformed into a risk itself. It is expressed in a growing radiation phobia with irrational reactions. The effects can be become fatally.

Unfounded fears against medical-radiological check-ups endanger diagnosis and therapy.

In cases of necessary interventions a radiation phobia and irrationality could become counterproductive. So long exist nuclear energy plants; so long exist all their risks. In case of a havarie irrationality can enormously increase the real consequences.

Confusions about radiation risks can be more dangerously than the real risk itself.

References

Bundesamt für Gesundheit (1999) Radon-Informationen zu einem strahlenden Thema 3003, 2, Bern

Clarke RH (1997) The ICRP protection policy. Int Zeitschrift für Kernenergie 8(9)

Cohen BL (1995) Test of the linear-no threshold theory of radiation carcinogenesis for inhaled radon decay products. Health Phys 68(2):157–174

Erzgräber G, Rosemann M, Abel H, Regel K (1992) DNA-structures and radiation injury. Adv Space Res 12:91–101

Regel K (1988) DNA superstructures–the key to understanding the mechanism of action of radiation in cells. Radiobiol Radiother (Berl) 29(5):529–560 (German)

Rosemann M (1992) Ein mikrodosimetrisches Modell zur Reparatur strahleninduzierter DNA-Doppelstrangbrüche in Säugerzellen. Dissertation, Humboldt-Universität Berlin

Schüttmann W (1999) Bewertung des Lungenkrebsrisikos durch Wohnungsradon. Strahlenschutzpraxis 4 (German)

Timofeeff-Ressovsky NW, Delbrück M, Zimmer KG (1935) Über die Natur der Genmutation und der Genstruktur. Nachrichten von der Gesellschaft der Wissenschaften zu Göttingen, Fachgr. 6, N.F. 1, Nr 13 (German)

Trott KR, Rosemann M (2000) Molecular mechanisms of radiation carcinogenesis and the linear, non threshold dose response model of radiation risk estimation. Radiat Environ Biophys 39:79–87

Wichmann HE, Kreienbrock L, Kreuzer M et al (1998) Lungenkrebsrisiko durch Radon in der Bundesrepublik Deutschland (West). Ecomed-Verlag ISBN 3-609-51500-7

The Significance of Chemosignaling Between Irradiated and Non-irradiated Organisms in Bystander Effect

Boris P. Surinov, Valentina G. Isaeva, Natalia N. Dukhova and Andrey D. Kaprin

Introduction

In previous studies it was described that exposure to ionizing radiation in sublethal doses reveals volatile components (VCs) in the urine of animals, which are missing under normal conditions. Exposure of these VCs to intact conspecifics reduces their immunoreactivity and the number of hemocytes (Surinov et al. 2005). It was found that, depending on the radiation dose, such excretions have the properties of chemo signals that attract (Surinov and Dukhova 2004) or repulse (Surinov 2007) intact animals.

In the published data it has been reported only about the role of signaling in natural physiological conditions (Beauchamp 2003; Hurst and Beynon 2004; Brennan and Kendrick 2006), in particular, in the regulation of animal breeding (Novikov 1988), or for limitation of contact between healthy and infected individuals (Moshkin et al. 2002; Penn and Potts 1998; Kavaliers and Colwell 1995).

In the literature we have found no information about chemosignaling induced by ionizing radiation in animals, with the exception of works by the authors of this paper. These chemosignals also influenced the behavioral responses of intact recipients, showing the attractive properties.

These substances caused a chain reaction of multiplication of violations in groups of animals, spreading them from irradiated mice to intact recipients both by direct contact and indirectly, through their urine (Surinov et al. 2004). This phenomenon illustrates the concept of the biological strengthening of the effects of disturbing factors, formulated by Timofeev-Ressovsky (1996). The described phenomenon could not explain appropriately the combination of increasing attractiveness of irradiated and intact animals, accompanied by a decline in the

B.P. Surinov (✉) · V.G. Isaeva · N.N. Dukhova · A.D. Kaprin
A. Tsyb Medical Radiological Research Centre, National Medical Research Radiological Centre of the Ministry of Health of the Russian Federation, Obninsk, Kaluga Region, Russia
e-mail: surinovboris@rambler.ru

V.L. Korogodina et al. (eds.), *Genetics, Evolution and Radiation*,
DOI 10.1007/978-3-319-48838-7_17

immune reactivity of intact animals. Further studies have shown the biological significance of the phenomenon under consideration. It turned out that intact animals produce the natural volatiles, chemosignals, which significantly increase the immune reactivity of the recipients irradiated with sublethal radiation doses (Surinov et al. 2008).

This paper presents examples of restoration of the immune system in irradiated mice and rats with the use of VCs, the biological feasibility of which has previously been considered (Surinov et al. 2007, 2008, 2012). In the present paper, this information is analyzed from the point of view of certain general biological concepts (Timofeev-Ressovsky 1996; Korogodin 2013).

Materials and Methods

Laboratory male mice of the inbred CBA strain and hybrids F1 (CBA × C57Bl/6), 2–3 months old (25–30 g) were kept in standard plastic boxes of 5–6 individuals each with free access to water and food, at room temperature (20–22 °C) and natural lighting. In some experiments the Wistar male rats (180–200 g) were used.

The whole-body irradiation of the animals was carried out with doses of 1 or 4 Gy using a ^{60}Co source with a dose-rate of 3.2 mGy/sec on the Gamma-Cell-220 equipment (Atomic Energy Canada Limited, Canada).

The donors of urine volatile components were intact and irradiated (4 Gy) individuals.

To obtain urine samples, a sheet of filter paper (paper litter) was placed at the bottom of the box. Access to this paper was limited by a perforated screen elevated above the bottom at 0.5 cm. The paper litter containing the urine of irradiated mice absorbed during third day after exposure or the urine of intact mice was transferred for one day under the mesh screen in the box with 1 Gy-irradiated or intact recipients.

To study the VCs influence on the development of the immune response, the paper litter with urine samples was placed into boxes with the immunized animals. In 24 h after exposure with bedding, the animals were injected intraperitoneally with 0.2 ml of suspension of sheep red blood cells (SRBC) in a dose of 1×10^8 cells/mouse in medium 199. In 4 days the mice were decapitated under ether anesthetic to determine the mass and cellularity of the spleen. The number of antibodies forming cells (AFC) in the spleen was determined by the Cunningham method. In each of the studied groups there were 5–7 individuals.

The effect of VCs on the phagocytic activity of peritoneal macrophages of mice-recipients was assessed by their ability to absorb latex particles of diameter 1.5 microns (DIA*M firm). The mice were decapitated under ether anesthetic. The suspension of macrophages was obtained by washing the abdominal cavity with 199 medium supplemented with 10 % embryonic calf serum. The reaction of phagocytosis was performed by mixing 1 ml of suspension of macrophages at a concentration of 10^6 cells/ml with 1 ml of latex suspension at a concentration of 10 mcl/ml. The mixture was incubated for 30 min at 37 °C. Smears were fixed for

30 min in 96 % ethyl alcohol and dyed using the Romanovsky-Gimsa method. The number of absorbed latex particles was counted in terms of phagocytic activity of macrophages (phagocytic index).

Experiments were reproduced at least three times. Statistic analysis of the results was performed using Student's t-test.

Results and Discussion

The 1 Gy ionizing radiation causes suppression of mice immune reactivity. Thus, within three days after irradiation of animals, the spleen weight, cellularity and the number of AFC induced by injection of SRBC (Table 1) are significantly reduced. As a result of exposure of these mice during the third day after irradiation to VCs of intact syngenic animals, the spleen cellularity and the number of AFC significantly exceeded these parameters in the irradiated animals not exposed to VCs. Almost the same effects were observed after exposure to VCs of the individuals irradiated with a dose of 4 Gy.

A similar effect of these VCs on the recipient mice on the 7th day after irradiation (1 Gy) was also accompanied by a significant increase in the ability to the immune response to antigen (Table 1). In this case, unlike the effect observed on the 3rd day after irradiation, no change was observed in the spleen mass and cellularity.

Table 1 Immunological parameters ($M \pm m$) in irradiated 1 Gy male mice F1 (CBA × C57Bl/6) after exposition to volatile components (VCs) of urine of intact mice in different post-irradiation periods

Groups of animals	Time after irradiation, days	Spleen		
		Mass, mg	Number of cells, 1×10^6	AFC, 1×10^3
Control	–	130 ± 6.5 (100 ± 5.0)	189 ± 12.3 (100 ± 6.5)	121 ± 9.0 (100 ± 7.4)
Irradiated 1 Gy	3	$94.7 \pm 3.8^*$ (72.8 ± 2.9)	$137 \pm 8.4^*$ (72.5 ± 4.4)	$52 \pm 3.1^*$ (42.9 ± 2.6)
Irradiated 1 Gy + VCs intact mice		$82.8 \pm 2.5^{*,**}$ (63.7 ± 1.9)	$134 \pm 14^*$ (70.9 ± 7.3)	$61.7 \pm 2.6^{*,**}$ (50.9 ± 2.2)
Irradiated 1 Gy	7	110 ± 3.8 (84.6 ± 2.9)	164 ± 14 (86.8 ± 7.4)	50 ± 2.2 (41.2 ± 1.8)
Irradiated 1 Gy + VCs intact mice		105 ± 5.2 (80.8 ± 4.0)	145 ± 13 (77.2 ± 6.3)	$66.2 \pm 2.4^{*,**}$ (54.6 ± 2.0)

Note In brackets % to control
*significantly different ($p < 0.05$) from control; **significantly different ($p < 0.05$) from 1 Gy

We also evaluated the immune response in mice irradiated with a dose of 1 Gy at different times after their exposure to VCs of intact or irradiated (4 Gy) individuals (Table 2).

As a result of exposure of these mice to VCs of intact animals during the third day after exposure to ionizing radiation, the spleen cellularity and the number of AFC were significantly higher than these parameters in the group of irradiated mice not exposed to VCs.

In more remote periods, within 7 days after exposure, effects of VCs on irradiated mice increased—the spleen cellularity and the number of AFC were more than twice of those for the animals not exposed to VCs.

During this observation period the mass and cellularity of the spleen also increased, especially after exposure to VCs produced by mice irradiated with 4 Gy. Therefore, the single exposure of the 1 Gy irradiated mice to VCs of intact or irradiated (4 Gy) animals was accompanied by stimulation of the immunity. The effect does not only persist for long, but also builds up with time.

Table 2 Dynamics of immunological parameters (M ± m) in irradiated 1 Gy male mice CBA after one-day exposition to volatile components (VCs) of urine of intact or irradiated (4 Gy) syngenic mice during the third day after irradiation

Groups of animals	Time after exposition days	Mass of spleen, mg	Number of cells at spleen, $1\text{x}10^6$	AFC in spleen, 1×10^3
Irradiated 1 Gy	1	101 ± 2.3 (100 ± 2.3)	80.0 ± 7.1 (100 ± 8.9)	104 ± 2.9 (100 ± 2.8)
Irradiated 1 Gy + VCs Intact mice		83.6 ± 4.3* (82.8 ± 4.3)	83.3 ± 3.4 (104 ± 4.3)	126 ± 6.0* (121 ± 5.8)
Irradiated 1 Gy + VCs irradiation 4 Gy mice		86.0 ± 3.3* (85.1 ± 3.3)	103 ± 3.4* (129 ± 4.3)	125 ± 15.0 (120 ± 14.4)
Irradiated 1 Gy	3	90.4 ± 1.8 (100 ± 2.0)	74.0 ± 5.1 (100 ± 6.9)	67.4 ± 9.0 (100 ± 13.4)
Irradiated 1 Gy + VCs intact mice		83.4 ± 5.2 (92.3 ± 5.8)	84.0 ± 9.8 (114 ± 13.2)	96.7 ± 6.0* (143 ± 8.9)
Irradiated 1 Gy + VCs irradiation 4 Gy mice		99.0 ± 4.0 (110 ± 4.4)	82.0 ± 12.4 (111 ± 16.8)	113 ± 6.8* (168 ± 10.0)
Irradiated 1 Gy	7	83.4 ± 2.7 (100 ± 3.3)	90.0 ± 3.2 (100 ± 3.5)	85.0 ± 6.7 (100 ± 7.9)
Irradiated 1 Gy + VCs Intac t mice		102 ± 2.8* (122 ± 3.4)	116 ± 17.5 (129 ± 19.4)	183 ± 4.5* (215 ± 5.3)
Irradiated 1 Gy + VCs irradiation 4 Gy mice		115 ± 3.8* (138 ± 4.6)	104 ± 5.1* (115 ± 5.6)	195 ± 10.4* (229 ± 12.2)

Note In brackets % to irradiated mice; *significantly different ($p < 0.05$) from irradiated 1 Gy

The above data rather differ from the previously described ones with respect to the impact of a single exposure to the same VCs, but at different times after irradiation of the mice with a dose of 1 Gy. In this case, stimulation of humoral immunity in the mice irradiated with the same dose was less pronounced. This is confirmed by the above data relating to the development of the immuno-potentiating effect in the course of time after exposure of irradiated animals to VCs.

These data are in fact opposite to the results obtained in the process of studying the immune reactivity of intact mice exposed to VCs produced by mice after exposure to 4 Gy. In the latter case, there was inhibition of humoral immunity, which lasted more than three days.

In one series of experiments, the impact of VCs produced by male mice CBA after exposure to 4 Gy radiation (third day), on intact animals and those irradiated with 1 Gy was simultaneously evaluated.

It was found that, as a result of exposure to VCs of intact mice or those subjected to 1 Gy radiation, there was a decrease or increase in the number of AFC in the spleen down to 62.0 ± 5.6 % ($p < 0.05$) and up to 95.6 ± 17.5 %, respectively.

In the mice irradiated with a dose of 1 Gy not exposed to VCs, the number of AFC in the spleen was 58.3 ± 1.9 % ($p < 0.05$) compared with the intact animals (100 ± 2.6 %).

Therefore, post-radiation VCs have different directions in the action depending on the state of recipients. It should be emphasized that the described here immunostimulatory effect was revealed at relatively weakly expressed immunodeficiency, as in this case at deficiency caused by irradiation with a dose of 1 Gy.

The presence in urine of chemosignals able to recover, under certain conditions, relatively small post-radiation damage was confirmed by experiments with rats which investigated the number of peripheral blood corpuscles after exposure to VCs (Table 3). Exposure of irradiated (1 Gy) rats during the third day of post-radiation period to VCs of intact or irradiated animals caused quantitative changes in the cellular composition of blood.

Thus, within three days after irradiation (1 Gy) the number of red blood cells in rats was 7.75 ± 0.24 1×10^{12}/liter. Exposure of such animals to the components

Table 3 Cell structure of peripheral blood (M±m) in irradiated 1 Gy (3d day) rat Wistar after exposition to volatile components (VCs) of intact or irradiated (4 Gy) animals

Groups of animals	Erythrocytes, 1×10^{12}/l	Leucocytes, 1×10^{9}/l	Lymphocytes, 1×10^{9}/l	Neurophiles, 1×10^{9}/l
Irradiated	7.75±0.24 (100±3.1)	10.7±1.4 (100±13.0)	3.2±0.27 (100±8.4)	2.4±0.45 (100±18.8)
Irradiated + VCs intact rat	9.65±0.24* (123±3.1)	14.1±1.5 (132±14.0)	5.5±0.87* (172±27.0)	4.54±0.61* (189±25.4)
Irradiated + VCs irradiated rat	9.4±0.48* (121±6.2)	9.8±0.7 (91.6±6.5)	3.4±0.3 (106±9.4)	2.8±0.18 (117±7.5)

Note *significantly different ($p < 0.05$) from irradiated rat groups not exposed to VCs of irradiated or intact rats increased this parameter 1.2 times as compared to the rats not exposed to VCs.

The influence of VCs of the intact or irradiated (4 Gy) animals on the intact or irradiated (1 Gy) rats also resulted in a 1.7 and 1.9 times increase in the absolute number of, respectively, blood lymphocytes and polymorphonuclear leukocytes (Table 3).

Among the findings of particular interest is the fact that the immuno- stimulatory effect on irradiated mice is intrinsic not only in VCs that are contained in the urine of mice, but also in the intact individuals themselves. It was found that the presence of even one intact male mouse in the group of irradiated mice (1 Gy) within three days increased their immune reactivity (Table 4), whereas the presence in the group of intact animals of only one irradiated male depressed their immune reactivity (Table 5). Hence, VCs produced by one animal are likely to be sufficient to demonstrate their chemosignal properties aimed at mobilizing the immune system of the animals in the group.

The immunostimulatory effect observed under the combined housing in one box of intact mice with individuals irradiated at a higher dose (4 Gy) was unstable and not reproduced in all series of experiments.

Table 4 Immunological parameters (M ± m, in brackets % to control) in irradiated (1 Gy) male mice CBA after co-housing for three days with intact male

Groups of animals	Spleen		
	Mass, mg	Number of cells, 1×10^6	AFC, 1×10^3
Intact	113 ± 5.6 (100 ± 5.0)	106 ± 6.8 (100 ± 6.4)	182 ± 8.3 (100 ± 4.6)
1 Gy	76.6 ± 3.8* (67.8 ± 3.4)	72.0 ± 2.0* (67.9 ± 1.9)	141 ± 19.3 (77.5 ± 10.6)
1 Gy + intact male	86.4 ± 3.8* (76.5 ± 3.4)	108 ± 8.6** (102 ± 8.1)	228 ± 32.4** (125 ± 17.8)

Note In brackets % to intact control; *significantly different ($p < 0.05$) as compared with intact animals; **significantly different ($p < 0.05$) as compared with irradiated animals

Table 5 Immunological parameters (M ± m, in brackets % to control) in intact male mice CBA after co-housing for three days with irradiated male

Version of experiment	Mass of spleen, mg	Number of cells in spleen, 1×10^6	AFC in spleen, 1×10^3
Control - housing with intact individual	104±1.7 (100±2.1)	132±9.0 (100 + 7.1)	90.1±4.9 (100±11.2)
Experiment - housing with irradiated individual	95.3±2.1 (92.1±2.0)	107±6.4 (81.1±5.0)	74.1±4.6 (82.8±6.0)*

Note In brackets % to irradiated mice; *significantly different ($p < 0.05$) from control

Similar experiments, but with mice irradiated with a dose of 4 Gy, resulted in irregular results and in case of exposure of animals to 6 Gy the immunostimulatory effect was not found.

A series of experiments studied the effect of post-radiation VCs on animals with changes in the immune reactions induced by non-radiation factors. Immune response to foreign antigen was selected as one of these models.

These antigens were sheep red blood cells (SRBC). To some extent, this is a model of response to infection. Mice recipients were exposed to VCs produced by irradiated (4 Gy) individuals one day before immunization, immediately after it, in 3 and 7 days after injection of SRBC (Table 6).

The number of AFC in the spleen was evaluated in 4 days after injection of antigen. As a result of exposure to post-radiation VCs to the antigenic stimulus the number of AFC in the spleen decreased to 30.4 ± 6.5 % relative to controls.

Such a reaction is fully consistent with the previously obtained data showing the immunosuppressive properties of post-radiation VCs excreted in the urine of irradiated animals (Surinov et al. 2004, 2005). After exposure of the recipient mice to VCs immediately after the administration of antigen, an immune response remained unchanged. In that case, when the recipients were exposed to VCs within 1, 3 or

Table 6 Immunological parameters (M ± m) at male mice F1 (CBA × C57Bl/6) after exposition to volatile components (VCs) of irradiated (4 Gy) individuals in different time relative to immunization

Groups of animals	Time exposure, days	Spleen		
		Mass, mg	Number of cells, 1×10^6	AFC, 1×10^3
Control	1 day before immunization	95.4 ± 3.3 (100 ± 3.5)	122 ± 9.8 (100 ± 8.0)	23.0 ± 0.5 (100 ± 2.2)
Exposition to VCs irradiated		102 ± 3.9 (107 ± 4.0)	141 ± 8.0 (116 ± 7.0)	7.0 ± 1.5* (30.4 ± 6.5)
Control	Immediately after immunization	119 ± 3.7 (100 ± 3.1)	143 ± 5.0 (100 ± 3.5)	27.5 ± 3.1 (100 ± 11.3)
Exposition to VCs irradiated		104 ± 6.4 (87.4 ± 5.0)	135 ± 14.0 (94.4 ± 9.8)	30.8 ± 2.9 (112 ± 10.5)
Control	1 day after immunization	103 ± 5.0 (100 ± 4.9)	142 ± 13.0 (100 ± 9.2)	46.6 ± 2.9 (100 ± 6.2)
Exposition to VCs irradiated		102 ± 4.8 (105 ± 5.0)	147 ± 10.2 (93.4 ± 6.5)	89.9 ± 5.0* (193 ± 11.0)
Control	3 days after immunization	77.4 ± 2.6 (100 ± 3.4)	106 ± 6.0 (100 ± 5.7)	18.0 ± 3.8 (100 ± 21.0)
Exposition to VCs irradiated		74.6 ± 1.9 (96.4 ± 2.5)	83.3 ± 6.8* (78.6 ± 6.4)	22.0 ± 5.0 (122 ± 27.7)
Control	7 days after immunization	77.2 ± 1.9 (100 ± 2.4)	112 ± 4.9 (100 ± 4.4)	9.5 ± 1.5 (100 ± 16.0)
Exposition to VCs irradiated		76 ± 2.1 (98.4 ± 2.7)	88.0 ± 6.7* (78.6 ± 6.0)	25.0 ± 6.0* (263 ± 60.7)

Note In brackets % to control; *significantly different ($p < 0.05$) from control

7 days after immunization, the number of AFC increased compared with the control to 193 ± 11.0, 122 ± 27.7 and 263 ± 60.7 %, respectively.

The presented here results of effects of VCs of irradiated mice on syngenic mice before and after immunization showed the dual nature of the studied chemosignals —the direction of their influence on animals depends on the state of the immune system. Before immunization a suppression was observed, whereas after immunization—stimulation of the humoral immune response.

Study of the effect of VCs on the phagocytic activity of peritoneal macrophages of irradiated mice showed (Table 7) that the phagocytic index (number of particles absorbed by one of phagocytic macrophages) in mice-recipients after exposure to VCs of intact animals increased 1.4 times in comparison with irradiated unexposed animals.

As in the case of antibody genesis, this information also differs from the previously obtained data showing the inhibitory effect of post-radiation VCs on phagocytosis.

The scientific literature on the role of chemosignalling in natural physiological conditions is quite voluminous (Beauchamp and Yamazaki 2003; Hurst and Beynon 2004). In particular, the influence of chemosignaling on the selective mating of animals was described Novikov (Novikov 1988), which is important to ensure the viability of offspring. In pathological conditions, chemosignaling restricts contacts of healthy and infected individuals (Moshkin et al. 2002; Penn and Potts 1998; Kavaliers and Colwell 1995).

The data presented here show that the combination of mutual attractiveness of irradiated and intact individuals with (Surinov et al. 2007) immunostimulating effects described here indicate the existence of a biologically appropriate complex of reactions regulated by chemosignals. In this complex are involved both the animals producing chemosignals induced by various factors and the individuals whose response to the signals depends on their physiological state.

Our data allow us to revise the approach to some biological traits of the bystander effect in the interaction between irradiated and non-irradiated organisms (Mothersill and Seymour 2001; Mothersill et al. 2007; Surinov 2007).

It was believed previously that such an effect is caused by post-radiation VCs and is aimed only at the suppression of immunity, but the above data demonstrate

Table 7 The phagocytic activity (M±m) of peritoneal macrophages of irradiated (1 Gy, three day) mice after exposure to the volatile components (VCs) of urine of intact or irradiated (4 Gy) of individuals

Animals group	Phagocytic index
Intact	9.0±0.4
Irradiated	13.2±1.0*
Irradiated + VCs of intact individuals	18.5±0.6*, **
Irradiated + VCs of irradiated individuals	12.8±1.2*

Note *significantly different ($p < 0.05$) from control; **significantly different ($p < 0.05$) from irradiated the opposite effect such as the impact the VCs of intact animals have on irradiated individuals can result in the restoration of immunity. Such a phenomenon (immunostimulation) was also observed after the exposure to VCs of animals after antigenic administration.

Hence, in our experiments we found that the orientation of the immunomodulatory effect of chemosignals depended on time of exposure to VCs in reference to the effects of ionizing radiation or antigenic stimulus, i.e. the physiological condition of the recipient. This manifests itself most clearly in respect of post-radiation effects of VCs induced by exposure to radiation with a dose of 4 Gy, which causes immunosuppression in intact animals, but stimulates the immune response in animals immunized with an antigen, or irradiated with low doses (1 Gy) of radiation.

Under certain conditions, the VCs which are immunosuppressive with respect to intact animals have a regenerating and stimulating activity in relation to the individuals immunocompromised by the radiation and non-radiation effects.

Therefore, the bystander effect in groups of animals can have both suppressive and repaired consequences. Dependence of these effects upon the state of the donors of chemosignals and individuals-recipients clearly demonstrates a biological advisability of regulation of mechanisms that allow animals through chemosignaling to recognize individuals with impaired immunity and ensure their stimulation and reparating effects.

The comparison of data on the influence of chemosignals of intact mice on the immune reactivity of the irradiated mice through the modification of behavioral responses reveals the mechanisms of their biological significance in terms of exposure to radiation. It was shown that after irradiation the VCs are produced that provide increased mutual attractiveness of irradiated and intact mice, but only in the range of sub-lethal doses of radiation (Surinov and Shpagin 2007). In a terminal period after exposure to lethal doses, VCs with repulsive (aversive) properties appear. As a result, intact mice enhance communications with animals irradiated with sublethal doses, but decrease communications with those irradiated at lethal doses (Surinov 2007). It is also important that it is preferable to increase the attractiveness of intact mice to syngenic irradiated animals (Surinov and Shpagin 2007). Therefore, after irradiation in groups of animals the mechanisms of chemosignaling are involved in the process that can be labeled as "sanitary" effect. Modification of the spectrum of chemosignals produced by the irradiated animals after exposure allows the other members of the group to evaluate the depth of lesions and the chance to activate the immune system through chemical communication between organisms. The appearance of aversive VCs in animals after exposure to lethal doses of radiation limits the communication, thereby reducing the likelihood of infection. The above phenomena quite accurately illustrate V.I. Korogodin's concept of the properties of living organisms, among which an important role is assigned to behavioral information (Korogodin 2013). It provides the link between organisms and the environment. One of the essential characteristics of behavioral information is the expediency and purposefulness of action. The experimental data presented in the article show the mechanisms of chemosignaling in terms of exposure to radiation appropriate to changing behavioral responses, guiding them to enhance the viability of the group.

Of special interest is the selectivity of the "sanitary" effect in respect of the group members, which is provided by these mechanisms.

Conclusion

Analysis of experimental data suggests an important role of chemosignaling mechanisms in mammals under exposure to ionizing radiation. Through such mechanisms, animals are able to determine the extent of damage, in order to distance themselves from the dying animals and to ensure the processes of reparation of damage in other individuals through communication with healthy group members. After exposure to radiation the chemosignaling mechanisms perform cognitive functions, motivating the biologically appropriate behavioral reactions of animals in groups. This chemosignaling activity can be designated as "sanitary" effect, which is biologically expedient for increasing the viability of animals in their groups. These views considerably extend the significance of the bystander effect in animals.

References

Beauchamp GK, Yamazaki K (2003) Chemical signaling in mice. Biochem Soc Trans 31:147–151

Brennan PA, Kendrick KM (2006) Mammalian social odours: attraction and individual recognition. Philos Trans R Soc Lond B Biol Sci 361(1476):206–278

Hurst JL, Beynon RJ (2004) Scent wars: the chemobiology of competitive signaling in mice. BioEssays 26(12):1288–1289

Kavaliers M, Colwell DD (1995) Odours of parasitized males induce aversive response in female mice. Anim Behav 50:1161–1169

Korogodin VI (2013) Information and the phenomenon of life. In: Phenomenon of life: selected works 2. Nauka Publication, Moscow, 207–402 (Russian)

Moshkin MP, Gerlinskaya L, Morosova O, Bakhvalova V, Evsikov V (2002) Behaviour, chemo signals and endocrine functions in male mice infected with tick-borne encephalitis virus. Psychoneuroendocrinology 27:603–608

Mothersill C, Seymour CB (2001) Radiation-induced bystander effects: past history future directions. Radiat Res 155:759–67. Review

Mothersill C, Bucking C, Smith RW, Agnihotri N, O'Neill A, Kilemade M, Seymour CB (2007) Communication of Radiation-Induced Stress or Bystander Signals between Fish in Vivo. Environ Sci Technol 41:3382–3387

Novikov SN (1988) Feromony i razmnozhenie mlekopitayushchikh (Pheromones and Reproduction in Mammals). Nauka, Leningrad (Russian)

Penn DJ, Potts WK (1998) Chemical signals and parasite-mediated sexual selection. Trens Ecol Evol 13:391–396

Surinov BP (2007) Mice with radiation or toxic damage or malignant tumors produce aversive chemosignals repelling intact animals. Doklady Biol Sci 414:199–201

Surinov BP, Dukhova NN (2004) Postradiation mice's volatile secretions which are attractive for intact individuals. Radiat Biol Radioecologiya 44(6):662–665 (in Russian)

Surinov BP, Shpagin DV (2007) Effect of irradiation on the olfactory ability of male mice to distinguish between intact animals chemosignal. Radiat Biol Radioecologiya 47(1):17–21

Surinov BP, Isayeva VG, Duhova NN (2004) Teaching multiplication of secondary violations of the blood and immunity in intact mice groups, mediated volatile secretions of irradiated individuals. Radiat Biol Radioecologiya 44(4):387–391

Surinov BP, Isaeva VG, Dukhova NN (2005) Post-radiation immunosupressing and attractive volatile secretions: "Bystander effect" or allelopaty in groups of animals. Doklady Biol Sci 400 (5):28–30

Surinov BP, Sharetsky AN, Shpagin DV, Isaeva VG, Dukhova NN (2007) Post-radiated and post-stressed volatile secretions: secondary immune and behavioral reactions in groups of animals. In: Mothersill et al. (eds) Multiple stressors: a challenge for the future. Springer, pp 127–138

Surinov BP, Isaeva VG, Karpova NA (2008) Immunostimulating chemosignaling in animals with secondary immunodeficiencs. Doklady Biol Sci 418:41–43

Surinov BP, Isaeva VG, Sharetsky AN (2012) Direct and reverse bystender effect between irradiated and unirradiated organisms: the modulating role of chemosignalling in ecology. In: Mothersill et al. (eds). Springer, pp 99–109

Timofeev-Ressovsky NV (1996) Selected works. Genetics. Evolution. Biosphere. (Russian) Medicine 154–171

Part IV
Radiation in Ecological Systems

Assessing Ecological Risk from Radiation Requires an Ecosystem Approach

François Bréchignac

Introduction

Today, almost 3 decades after the Chernobyl accident and 4 years after the nuclear catastrophe at Fukushima, the environmental/ecological consequences from these extensive radioactive contaminations of the environment are still much debated with contrasting conclusions from various experimental and observation studies (Beresford and Copplestone 2011). Supported by experimental results from in situ studies conducted in contaminated territories, as well as experiments in laboratory controlled conditions, some argue for the evidence of on-going ecological harm (Boratynski et al. 2014; Galvan et al. 2014; Garnier-Laplace et al. 2011; Hiyama et al. 2012, 2013; Møller and Mousseau 2007, 2009, 2013, 2015; Møller et al. 2005, 2007, 2012, 2013, 2015a, b; Mousseau et al. 2013; Mousseau and Møller 2012), whilst others do not reach such a conclusion (Baker et al. 1999; Baker and Chesser 2000; French et al. 1974; Jackson et al. 2004; Mihok 2004; Mihok et al. 1985; Murphy et al. 2011; Rodgers and Baker 2000; Smith and Beresford 2005; Strand et al. 2014; Williams 1995). One should perhaps keep in mind from past experience that demonstrating evidence for ecological harm has always proven to be slow. Global warming from anthropogenic releases of CO_2 provides a good illustration: on-going ecological impact has been suspected for several decades before it could reach wide recognition, if not demonstration, because the system's response, further associated to large spatial and temporal scales, was governed by complexities not easily amenable to full and explicit clarification (Maslin 2013). For these reasons, and for about two decades now, environmental protection against radiation has become an important issue which stimulates both research and developments

F. Bréchignac (✉)
Institute for Radioprotection and Nuclear Safety (IRSN) and International Union of Radioecology (IUR), BP 1, Center of Cadarache, St Paul-lez-Durance, Cedex 13115, France
e-mail: francois.brechignac@irsn.fr

V.L. Korogodina et al. (eds.), *Genetics, Evolution and Radiation*,
DOI 10.1007/978-3-319-48838-7_18

suitable to assess properly environmental/ecological risk, and especially to clarify the extent of ecological detriment in contaminated territories.

During the two past decades, following the initial proposal by an expert group formed by the International Union of Radioecology (IUR 1997, 2002), a first methodology for assessment of risk from radiation to fauna and flora has been conceived and further elaborated under the auspices of a suite of European projects, FASSET (2004), EPIC (Sazykina and Kryshev 2003, 2006), ERICA (2007)and ICRP (2003, 2008). In brief, the methodology stems from the concepts and methodology currently used to protect human beings and is expanded, with some adaptations, to address animals and plants; it is therefore grounded on an approach based upon "reference organisms" considered at individual organism level, or "reference animals and plants" as per the current ICRP terminology (ICRP 2003, 2008). This methodology is now further supported by a software tool, called the ERICA tool (Brown et al. 2008), which allows the calculation of the dose rates which reference organisms are exposed to when living in a contaminated environment with known concentrations of radionuclides. The dose calculations are based upon traditional radioecological parameters such as concentration ratios, simple geometries representing the organisms, and expert judgment about exposure resulting from typical life styles.

However, the widespread adoption of this first methodology beyond its developers' community is now facing difficulties as illustrated by several recent undertakings or positions especially expressed by international organizations:

1. Considering the limitations of the "reference organism approach" and especially the mismatch between the proposed method (based upon individual related data) and the objective of protection (targeted at the population and the ecosystem), continued brain storming led several authors assembled under the IUR to recommend and undertake the development of an "ecosystem approach" in order to address this mismatch (Bréchignac et al. 2011; IUR 2012; Bradshaw et al. 2014).
2. UNEP (the UN organization to whom UNSCEAR is linked) also expressed some reservations during a Technical Meeting of the Coordination Group on Radiation Protection of the Environment, held on 2–3 July 2013 in Vienna, which addressed the "Input to safety standards taking into account the BSS and relevant ICRP/International recommendations" (IAEA 2014). "*UNEP expressed the view that radiation protection of the environment, as described in ICRP Publications 91* (2003), *103* (2007) *and 108* (2008), *is too narrow. During the process of development of the Basic Safety Standards* (IAEA 2011), *UNEP suggested that the definitions given in the BSS for 'environment' and 'protection of the environment' should have a wider perspective beyond the considerations of radiological effects to flora and fauna. UNEP remarked that, when applying the radiation protection principles of 'justification' and 'optimization' to control and manage radiological impact to people and the environment, the appropriate involvement of environmental issues such as protection of flora and*

fauna, maintenance of quality of air, soil and water as well as considerations of sustainability aspects are key topics."

3. Finally, attempting to spread the consideration of environment protection among end-users such as operators, regulators and stakeholders based on the "reference organism approach", IAEA recently stressed unexpected reluctance to incorporate it in international basic safety standards. A further Technical Meeting of the Coordination Group on Radiation Protection of the Environment gathered by IAEA on 10 September 2014 in Barcelona, entitled "Radiological protection of the environment: the challenge to move from research to international regulation" has been especially motivated by "*the reluctance by its* [IAEA] *Member States to upgrade the current system of radiological protection*" (by expanding it to address environment protection). In general terms, the issue raised was about justifying how the effort of protecting the environment against radiation would be an appropriate and proportionate response to preventing the occurrence of ecological harm. This is indeed an important question to operators and regulators, as it may result in additional costs, but this is also quite a hot issue for society at large who are greatly concerned about ecological impacts and their long-term potential implications on human health, and particularly future generations.

In order to explain why this difficulty has arisen and to identify potential remedies, in the following sections we revisit the original strategic justification for the need for environment protection from radiation. We discuss in more details the view that, at variance with other fields situated outside the nuclear field (CO_2 driven climate change impact on ecosystems, chemical toxicant impact on fish populations in the seas, see more within IUR 2012), historical and cultural reasons led to weaknesses in the initial scoping of the need for environmental protection. Better justification requires first to clarify what is actually meant when one refers to "environment protection" (Carroll 2009). A number of misunderstandings emerge from poor and inaccurate definitions. Better justification requires next developing an integrated conceptual vision that must link human and environment radiation protection on fully coherent grounds with respect to protection objectives. The ecosystem approach, which has been recommended by IUR (Bréchignac et al. 2011; IUR 2012; Bradshaw et al. 2014) as a next step, provides such a conceptual vision.

Historical and Cultural Context Within the Radiation Protection Community

The issue of "protection of the environment" against radiation has evolved within a scientific community that was historically and culturally bound more to human health issues than to dealing with wildlife and ecosystems. Indeed, the radiation protection community consisted of physicians, health physicists, radiobiologists, human radiation protection specialists, and a few radioecologists also oriented towards human health ultimate goals. This was at variance of the community dealing with impacts from chemical toxicants and other stressors for which the community included environmentalists and natural sciences specialists such as agronomists, ecotoxicologists, ecologists, marine biologists, biogeochemists, limnologists, etc.… Hence, the skills necessary to wide understanding of the complexities and multiple facets of the environment, and its relationships with the human species, were therefore poorly represented initially within the radiation protection community.

The above observation is probably linked also to the widespread initial conviction, within the radiation protection community, that specific environmental protection was not generally deemed necessary (in normal operation of nuclear facilities), and the resulting reluctance of its members to reconsider what is now called the "ICRP paradigm" stating that standards in use to protecting human beings were indirectly affording sufficient "protection of the environment". Perhaps, the radiation protection specialists were not easily inclined to recognize that the system of radiological protection was potentially weakened by its inability to explicitly address the environment (i.e. non-human biota) in addition to human beings. When this community finally decided to reconsider this "ICRP paradigm" in order to better address environment protection, this was therefore prompted more by a pressure to respond to criticisms than by a profoundly rooted conviction about its usefulness. This quite specific ambient context has dominated, and to some extent also directed, the efforts of the newly created ICRP Committee 5 in charge of developing a system framework for environment protection against radiation.

As a result, environmental protection against radiological hazards has been justified initially on unclear and perhaps incorrect grounds, obscuring proper understanding of the actual need for it, and leading at the present stage to developments which do not fully match the objectives of ecological/environmental risk assessment.

Weak Strategic Scoping at Start

ICRP Publication no. 108 (ICRP 2008) stated: *"… It is probably true that the human habitat has been afforded a fairly high level of protection through the application of the Commission's system of protection. The problem is to demonstrate convincingly that the environment is, or will be, adequately protected in different circumstances, because there are no explicit sets of agreed assessment approaches, criteria or guidelines with international authority which can help…"* And further: *"… This decision* [to develop a systematic approach for radiological assessment of non-human species] *has not been driven by any particular concern over environmental radiation hazards. Rather, it has been developed to fill a conceptual gap in radiological protection…"*

The strategic justification for the need to protect the environment against radiation has therefore not been expressed in terms of assessing and/or to preventing ecological harm, but in terms of a conceptual gap to fill and a demonstration to provide. This conceptual gap referred to the lack of consideration of the environment within the system of radiological protection by the turn of the millennium. This was recognized to diverge from current environmental legislation which did not exclude radioactive toxicants, leading to the need to provide explicit demonstration that the environment was indeed afforded protection. Filling the conceptual gap has been undertaken via expanding the system of radiological protection for humans to also address non-human species inhabiting the environment.

Such a scoping of the justification was indeed a weak strategy at the start. It is clear that "filling a conceptual gap" to fulfill a need for "demonstration" does not form a strong enough justification for operators and regulators, the main potential end-users of the protection system. If there is no concern for ecological harm, then why should one bother about protecting the environment at all?

A more robust justification, certainly much easier to understand, could have been expressed from the very beginning by acknowledging the persistence of a suspicion from some scientists, and concern within society at large, about potential ecological harm due to radiation which called for due consideration of protection of the environment. Fears tend to proliferate in proportion to the lack of both an appropriate understanding and a related pertinent system to address ecological risk. Societal concern, scientific suspicion and lack of consensus about risk are sufficient justifications on their own for the need to build a convincing and appropriate system of environment protection to assess ecological risk from radiation. The still on-going debate on whether or not the Chernobyl and Fukushima accidents are promoting ecological consequences (Beresford and Copplestone 2011; Beresford et al. 2012a, b; Mousseau and Møller 2012; Strand et al. 2014) together with the most recent publications reporting evidence for ecological impacts (Garnier-Laplace et al. 2011; Hiyama et al. 2012, 2013; Møller and Mousseau 2013, 2015) just confirms the pertinence of such a line of justification.

Exclusive Focus on Biological Effects Largely Misses Ecological Effects

The historical and cultural initial contexts described above have much influenced the conceptual views, approaches and methods elaborated so far for environment protection against radiation. Reductionism has been a main driver, first by reducing the environment to an exclusive consideration of animals and plants, or fauna and flora, and second, by restricting the approach to individual organisms' endpoints, altogether putting emphasis on biological effects (Fig. 1, left). Protection of the environment, as stressed by UNEP (IAEA 2014) actually embraces many different perspectives including living beings (humans, animals, plants and

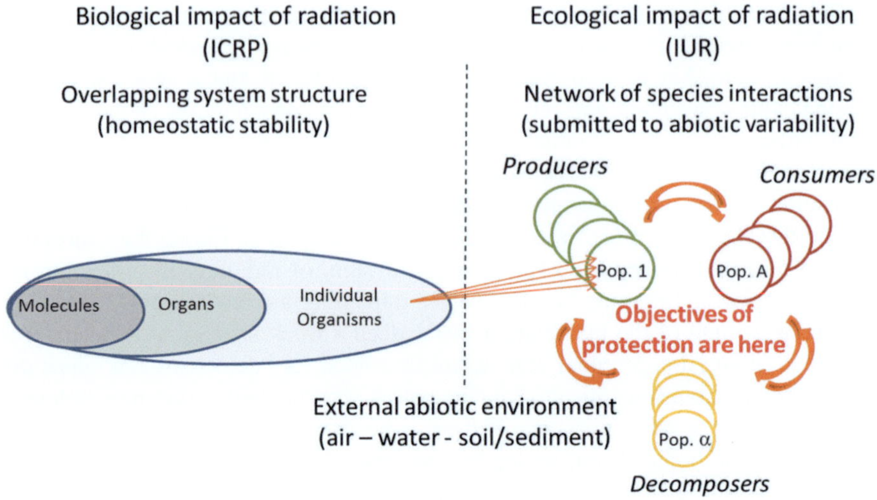

Fig. 1 Featuring biological and ecological impact of radiation within the scale of biological organization. *Left* overlapping structure of biological organization (each level encapsulates the previous one) up to individual organisms only, the relevant level to address biological impact of radiation. *Right* at higher organization levels, not currently incorporated in radiation protection, a quite different structure features a network of populations assembled in communities and ecosystems with interacting species, the relevant level to address ecological impact of radiation. Objectives of environment protection at most usually set at higher levels of biological organization

microbes/viruses), abiotic media (soils/sediments, sea and freshwater, atmosphere), landscapes and ecosystems (providing life support and many services), and a large set of sectorial legislations have been developed already to protect one or the other.

There is often an ambiguity with respect to the terminology used when mentioning "protection of the environment" when it in practice refers only to protection of "animals and plants" or "non-human biota", the environmental component sensitive to radiation. This is at variance of protection against chemical stressors where abiotic media are also considered as objects of protection, since their regulation for the protection of biota is designed upon toxicants concentrations in abiotic media. In radiation protection, abiotic media have not been afforded the same level of attention as objects of protection arguing that these were non-radiosensitive components. Abiotic media in the "reference organism approach" are only considered in terms of radionuclides transfer factors towards biota, with no reverse consideration that biota, for example, could impact the chemistry of radionuclides in abiotic media (e.g. speciation) that could in turn change the radiation responses from biota. Such an exclusive focus on living biota has inevitably led to restrictions in the capabilities of the system to assess other environmental issues such as those associated to abiotic media, or more importantly, those associated to more holistic goals or protection (integrating biotic and abiotic relationships) now expressed in terms of ecosystem services (structure/biodiversity and functions).

Simplicity and the wish to ensure immediate methodological consistency in how the system handles humans and non-human biota have dominated. This has resulted in paying limited attention to their respective objectives and related targets of protection (and related endpoints) which are different. Indeed, humans are protected at the individual level. For the environment at large, there is wide consensus recognizing that higher levels of organization are the appropriate targets of protection (populations, communities and ecosystems). As such, the proposed system appears to fulfill the demonstration goal pursued (for individual organisms), but at the expense of an in-depth consideration of what this new object of protection—the environment—actually is, and should require from the protection system. Meanwhile, the methodology does not address ecological harm proper and the demonstration goal is accordingly jeopardized. Such a conceptual and technical positioning has strongly influenced the evolving system which unsurprisingly suffers from incomplete formulation of the problem, and weaknesses as to what is meant by the environment and what is (are) the object(s) of protection (Bradshaw et al. 2014). Clarification of the object of protection is a prerequisite for a well-focused and fit-for-purpose protection system. In fact, focus upon individual organisms of non-human biota has been chosen because this was the essential content of the current data base on radiation effects but not primarily as the object of protection. The resulting mismatch prevents the ability of the system to move beyond pure biological effects and address important ecological issues (Fig. 1, right) such as interactions between species (predator/prey, competition, etc.…) and indirect cascade effects (Bréchignac 2003; Fleeger et al. 2003; Bréchignac and Doi 2009; Bradshaw et al. 2014), that is to say more holistic goals of protection.

Addressing Ecological Risk Assessment Requires an Ecosystem Approach

Understanding the biological effects of radiation in animals and plants is necessary but not enough to address ecological impacts of radiation. In order to overcome this problem, an ecosystem approach has been proposed by scientists gathered under IUR (Bréchignac and Doi 2009; Bréchignac et al. 2011; IUR 2012; Bradshaw et al. 2014) and is currently under development (Joint IUR-CERAD[1] Task Group). There is no need here to reiterate the conceptual justification for an ecosystem approach that has already been fully developed in much detail (Beresford and Copplestone 2011; IUR 2012), but it is of interest to stress how the most recent findings and developments provide further confirmation in support of this approach.

Attempts to address populations' response to radiation have been recently developed by Alonzo et al. (2008) and Lance et al. (2012) based on extrapolating

[1]CERAD: Centre (of excellence) for Environmental Radioactivity, based in Norway (www/cerad.nmbu.no).

from individual level endpoints. Leslie matrices combining life history characteristics (duration of life stages, survival and fecundity rate) and dose-rate response curves for hatching, survival and reproduction, from experimental data assembled within the FREDERICA data base, have been used to model population-level consequences of chronic external exposure of three aquatic invertebrates to gamma radiation. For two of the three populations' tested, significant reduction of population attributes were shown to happen at smaller dose rates than the lowest EDR_{10}.[2] EDR_{10} derived from the most sensitive individual endpoint, as recommended from the current methodology, therefore does not provide a safe level of protection for the species population. Safe standards for protection of the population therefore cannot be directly obtained from extrapolating the propagation of toxic effects from individual organisms to the population. Addressing the question directly at population level is made accessible by an ecosystem approach.

Another observation has been derived from a more theoretical study where ecological (species) interactions were taken into account in constructing eco-SSDs (species sensitivity distributions (SSDs) that include ecological interactions) for 1000 hypothetical toxicants that were tested against corresponding conventional SSDs (lognormal SSDs fitted to single-species chronic EC_{10S}[3] of the species present), (De Laender et al. 2008). For more than 1/4th of these tested toxicants, the mean and/or variance of the eco-SSDs were lower than for conventional SSDs, suggesting that indirect effects mediated by ecological interactions play an important role in the population response to toxicants. Here again, an ecosystem approach will take into account indirect effects of toxicants elicited by interactions between different populations.

Finally, a puzzling observation has been recently reported by Garnier-Laplace et al. (2013). These authors constructed SSDs from sets of data for individual endpoints as related to (1) biota experimentally tested in controlled conditions and (2) field observations in the contaminated territories of Chernobyl (with an appropriate method for reconstructing dose). A clear and consistent shift of the statistical curve for Chernobyl field data towards lower dose rates has been elicited (Fig. 2), almost by a factor of 10, tending to indicate that organisms of species observed in the field at Chernobyl were more radiosensitive than organisms of species tested in controlled optimum laboratory conditions. The discrepancy was interpreted in terms of the likely occurrence of confounding factors in the field, such as food availability, competition and predation, all of which are three natural and widespread interactions between species in ecosystems. Indeed, it is the rule within any ecosystem that all organisms need to cope with natural constraints arising from interactions with other species (like predation, competition) and also from limitation in food availability, all such constraints usually being relaxed when these organisms are tested in laboratory controlled conditions. This obviously contributes to the differences between biological effects (observed for individual organisms, through

---

[2] EDR_{10}: Effect Dose rate eliciting 10 % of effect.

[3] EC_{10S}: Effect concentration eliciting 10 % of effect on single-species population S.

appropriate endpoints) and ecological effects (observed at population and ecosystem levels with interacting species, through adequate endpoints). We therefore suggest that it would not be the radiosensitivity per se which is higher for in situ organisms (Chernobyl data), but rather the ubiquitous additional constraints that all individuals of a population are submitted to when they live within their ecosystem which may contribute to an apparent lower resistance to radiation stress. It is much desirable to develop an ecosystem approach by including consideration of interactions between species to account for such issues.

However, another confounding factor, perhaps more important, is that organisms observed at Chernobyl were actually exposed to a mixture of internal and external irradiation from several radiation types, at least γ and β, as arising from ^{137}Cs and ^{90}Sr respectively, whilst data obtained in controlled laboratory conditions were driven with exposure to γ external radiation only. This may also contribute significantly to the difference observed, the Chernobyl species being actually more exposed to radiation (all types, internal and external) than the laboratory ones.

It is worth at this point discussing further the issue of dosimetric estimations, an additional source of discrepancy which has recently been claimed to be the most likely cause of inconsistencies in the literature (Beresford et al. 2012a, b; Mousseau et al. 2012; Mousseau and Møller 2012). When expressed in dose rates, the dose range of concern here is that of relatively low doses (HDR_{50}in the range of hundreds of $\mu Gy\ h^{-1}$). Hence, the effects of relevance expected are essentially non-lethal at organism level. At this scale, effects are likely to be small and subtle, if detectable at all, and this leads to considering two hypotheses. (1) Rather than dose rate, the dose accumulated with time and through several successive generations would probably be a much better parameter against which trying to unravel ecological effects (paralleling the geneticists' concept of mutation accumulation). As was stressed by Garnier-Laplace and coworkers, this may explain, at least in part, the apparent radiosensitivity discrepancy between biota observed at Chernobyl and those tested in controlled laboratory conditions, as discussed above, the total accumulated dose at Chernobyl certainly exceeding that of laboratory tests. (2) One second hypothesis considers that it is not much the direct (small) effect of radiation on organisms' attributes (mortality, reproduction, mutation…) that would be of major concern with respect to ecological effects, but rather the imbalances indirectly mediated by differences in radiosensitivities amongst impacted populations of species. Then, the small and subtle (biological) effect of radiation (at organism level) would essentially act as a triggering mechanism of potentially larger (or smaller) effects at ecosystem level. An ecosystem approach is able to account for both larger time scales (succession of population generations) and indirect effects (interactions between populations of different species).

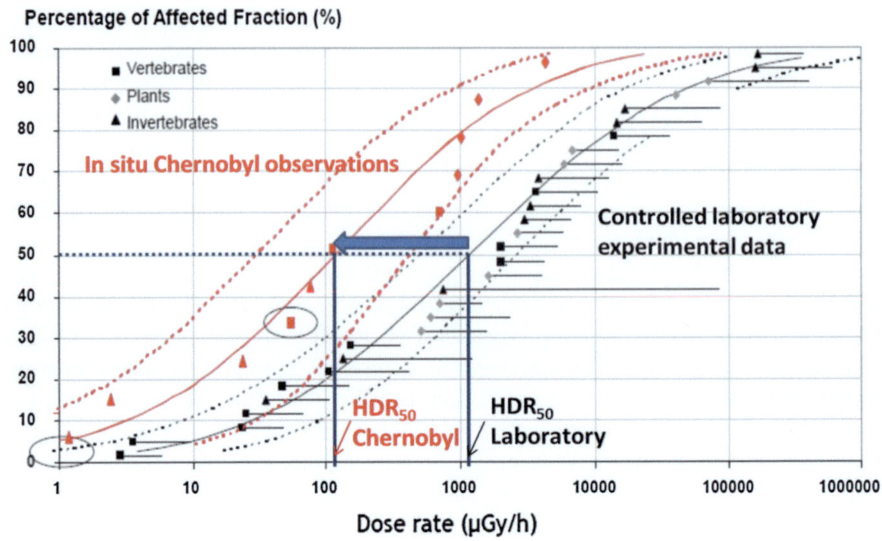

Fig. 2 Terrestrial species sensitivity distributions (SSD) of chronic responses (fitted to the set of lowest EDR_{10} values) constructed from effects data obtained in controlled conditions elicit apparent lower radiosensitivity [*blue arrow* showing lower HDR_{50} (*HDR_{50}* Hazardous Dose Rate for 50 % of the species tested.)]than when obtained from in situ observations at Chernobyl with associated dose reconstruction. (Redrawn from Garnier-Laplace et al. 2013)

Ecosystem Approach: A Conceptual Vision Integrating Humans Within the Environment

Historically, the concept started from an anthropocentric vision where the environment was only considered in so far as it mediated the transfer of radionuclides towards human beings (human exposure from radionuclides contained in food, air, water and soils) (Bréchignac 2012). Attention to living organisms other than human beings was restricted to those animals and plants of agronomical value which form the human diet, and their potential to be contaminated by radionuclides.

The concept next moved to a more biocentric vision, where attention has been extended from human beings to wildlife animal and plant species. This is the current "reference organism approach" as proposed by ICRP (2003, 2008). The misunderstandings revealed by the so-called old ICRP paradigm, stating in brief that "protecting man protects the environment", largely arose from the lack of a proper conceptual vision explaining how, and to which extent, human beings and the environment were respectively positioned and interacting. Such a vision, which had been proposed as early as in 2001 (Bréchignac 2001), then endorsed by the full Board of Council of the International Union of Radioecology (Bréchignac et al. 2003), is a prerequisite to considering if, and how, integration of human beings and the environment within a single system of radiological protection could be envisaged, a further goal currently pursued in radiation protection.

A full conceptual model integrating the *Homo sapiens* species, agronomic species, wildlife species and the abiotic media concerned in the related ecosystems is built on Fig. 3, following recommendations given earlier by Suter (1999) and Bréchignac (2016). Such an integrated model acknowledges that man and its companion agricultural species (crops, cattle, etc.…) belong to the same functional groups as have been defined for wildlife, therefore undergoing competition with wildlife species of similar hierarchical level, as in any ecosystem. What is specific here is that this ecological competition is "much more efficiently" managed by *Homo sapiens* with "smartly designed" tools, such as fertilizers and pesticides, in order to increase the biomass of agricultural products therefore out competing other species at each relevant level (weeds, phytophageous insects, etc.…).

With respect to assessing the impact of radioactive contamination of the environment through atmospheric deposition, such as in this particular example (Fig. 3), this conceptual model allows abiotic media to be put in the right context together with wildlife and the *Homo sapiens* species and their ecological interactions. Composite/complex impacts from combinations of multiple stressors are also better accounted for, as pesticides for example (used to optimize agronomic production) may not only co-act at the response level (synergistic/antagonistic responses), but also may change exposure by means of changing the ecosystem structure (some insect populations together with their feeders eradicated by insecticides, Ludwig et al. 1978; Holling 1986), or by changing the transfer characteristics (hydrophilic additives to systemic herbicides are likely to increase the penetration of leaf-deposited radionuclides).

The ecosystem approach, whilst promoting a further evolution towards an ecocentric concept, provides such an integrated vision (Bréchignac et al. 2011). In addition to biological risk it also gives access to assessment of ecological risk (IUR 2012; Bradshaw et al. 2014), the actual generic goal of environment protection. If protection of the environment has historically evolved from observed hazards to some species or categories of species or environmental threats (endangered species, desertification, pollutions impacts, etc.…) next translated into sectorial legislation, it is now acknowledged that all such detriments to the environment may alter the human species as well as other species, all being part of the ecosystem. This is because the sustainability of life is fully mediated through ecosystems' structure and functions. The human species *Homo sapiens* is but one of many different interacting species, all depending upon their ecosystem to ensure their long-term sustainability and survival. Paralleling recent environment legislation which is now expressed in more holistic terms (for a series of examples, see IUR 2012) like protection of "ecosystem integrity and services" which support all life, including that of human beings, the ecosystem approach considers that it is life survival as a whole (humans among other species) which can be jeopardized by ecological risk. Accounting for this more holistic approach is a prerequisite for conceptualizing an integrated protection system (human species, other species and the abiotic media within which they live, as well as other contaminants).

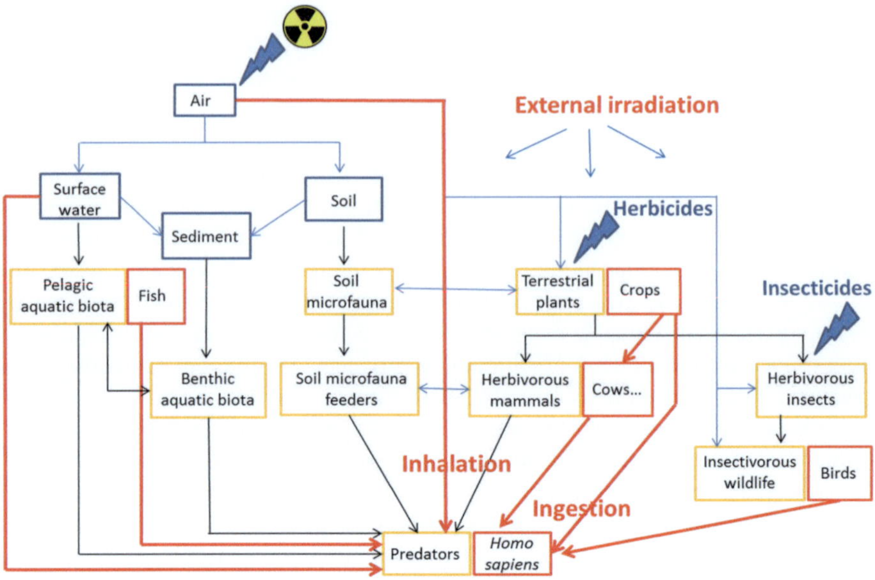

Fig. 3 Conceptual model for a scenario of release of radioactivity to the air, integrating the various ecosystem components—abiotic media (*blue boxes*), various functional categories of wildlife (*yellow boxes*)—together with the *Homo sapiens* species and its companion agricultural species (*red boxes*), in order to describe radionuclides impact on wildlife and man. *Blue arrows* represent purely physico-chemical transfers of radionuclides from abiotic compartments. *Black arrows* represent trophic transfers, highlighted in *red* when these are directed at the *Homo sapiens* species. (Redrawn from Bréchignac 2016)

Ecosystem Approach Relates Ecological Harm to Loss of Ecosystem Services

Human survival explained as being dependent on ecosystem well-being has been captured within the "ecosystem goods services" concept (Millennium Ecosystem Assessment 2005). This concept acknowledges that ecosystem components and functions are useful, if not vital, to humans, in providing them with a number of goods and services. The goods provided include food, fresh water, wood and fiber, and fuel, whilst services include, for example, supporting nutrient cycling, primary production and soil formation as a basis for regulating climate, flood, disease and water purification, and providing cultural features such as aesthetic, spiritual, educational or recreational services (Millennium Ecosystem Assessment 2005).

Just like other stressors, radiation can potentially alter the structure of ecosystems and their biotic communities, and further impact ecosystem functions which support ecosystem services useful to humans (Charpin et al. 2000). If exclusively focused on effects on individual organisms or species, the system of radiation protection of the environment cannot address ecosystem functions and thus, assessing the potential contribution of radiation to loss of ecosystem services remains out of reach. An ecosystem approach, by virtue of linking human needs with ecosystem functions, is consistent with the ecosystem services concept. In practice, relating ecological risk assessment to ecosystem services would need to be directed at identifying key drivers (and related endpoints) that are relevant to ecosystem services in radioactively contaminated ecosystems. The concept has already been used in other contexts (Nienstedt et al. 2012; Forbes and Calow 2013; Maltby 2013; van Wensem and Maltby 2013), but only a few preliminary efforts in this direction have been carried out so far with respect to radiation impact (Von Wehrden et al. 2012; Gralla et al. 2014). Ways of using the ecosystem approach in combination with the ecosystem services concept are currently being explored under the IUR-CERAD joint Task Group (terms of reference available at: www.iur-uir.org) and will hopefully provide valuable advice in support of further developments in this direction.

Conclusion

Recent controversies about the ecological impact of environmental radiation, such as originating from Chernobyl and Fukushima, point to a lack of linkage between controlled experimentation in the laboratory and in situ studies (Hinton and Bréchignac 2005; Garnier-Laplace et al. 2013) . This is primarily because they do not focus on the same endpoints and tend therefore to ignore alternative approaches to assessing effects of radiation. The former looks essentially at biological effects (interaction of radiation with molecules, cells, tissues, and whole organisms), whilst the latter looks primarily at ecological impact of radiation on populations of interacting species within ecosystems. This cultural gap between both, often substantiated by scientific disciplines and associated communities which largely operate in isolation one from the other, needs to be resolved. In fact, the situation is quite similar to, and/or originates from, the cultural gap depicted above at the start between the human radiation protection community and that dealing with environment protection against radiation.

The ecosystem approach provides a conceptual basis for assembling human and environmental radioprotection in a justified and coherent manner, therefore integrating scientific understanding gained at various levels, analytical as well as systemic, towards the ultimate goal of assessing ecological risk in situ. A further added value of importance, rooted in its ability to link ecological risk assessment to ecosystem services, will be to provide a powerful and straightforward communication tool about radiation risk.

References

Alonzo F, Hertel-Aas T, Gilek M, Gilbin R, Oughton DH, Garnier-Laplace J (2008) Modelling the propagation of effects of chronic exposure to ionising radiation from individuals to populations. J Environ Radioact 99(9):1464–1473

Baker RJ, Chesser RK (2000) The Chernobyl nuclear disaster and subsequent creation of a wildlife preserve. Environ Toxicol Chem 19:1231–1232

Baker RJ, Dewoody JA, Wright AJ, Chesser RK (1999) On the utility of heteroplasmy in genotoxicity studies: an example from Chernobyl. Ecotoxicology 8:301–309

Beresford NA, Copplestone D (2011) Effects of ionizing radiation on wildlife: what knowledge we have gained between the Chernobyl and Fukushima accidents? Integr Environ Assess Manag 3:371–373

Beresford NA, Adam-Guillermin Ch, Bonzom J-M, Garnier-Laplace J, Hinton T, Lecomte C, Copplestone D (2012a) Comment on "abundance of birds in Fukushima as judged from Chernobyl" by Møller et al (2012) Environ Poll 169:136

Beresford NA, Adam-Guillermin Ch, Bonzom J-M, Garnier-Laplace J, Hinton T, Lecomte C, Copplestone D, Della-Vedova C, Ritz C (2012b) Response to authors' reply regarding "abundance of birds in Fukushima as judged from Chernobyl" by Møller et al (2012). Environ Pollut 169:139–140

Boratynski Z, Lehmann P, Mappes T, Mousseau TA, Møller AP (2014) Increased radiation from Chernobyl decreases the expression of red colouration in natural populations of bank voles (*Myodes glareolus*). Sci Rep 4:7141; doi:10.108/srep07141

Bradshaw C, Kapustka L, Barnthouse L, Brown J, Ciffroy P, Forbes V, Geras'kin S, Kautsky U, Bréchignac F (2014) Using an ecosystem approach to complement protection schemes based on organism-level endpoints. J Environ Radioact 136:98–104

Bréchignac F (2001) Environment versus man radioprotection: the need for a new conceptual approach? Radioprotection Colloques 37-C1:161–166

Bréchignac F (2003) Protection of the environment: how to position radioprotection in an ecological risk assessment perspective. Sci Total Environ 307:37–54

Bréchignac F (2012) Environment protection: the current challenge in radioecology. In: Plastino W, Povinec PP (eds) Proceedings of conference "environmental radioactivity—new frontiers and developments". SIF, Bologna 104:3–17

Bréchignac F (2016) The need to integrate laboratory- and ecosystem-level research for assessment of the ecological impact of radiation. Integr Environ Assess Manag, October issue, in press

Bréchignac F, Doi M (2009) Challenging the current strategy of radiological protection of the environment: arguments for an ecosystem approach. J Environ Radioact 100:1125–1134

Bréchignac F, Polikarpov G, Oughton DH, Hunter G, Alexakhin R, Zhu YG, Hilton J, Strand P (2003) Protection of the environment in the 21st century: radiation protection of the biosphere including humankind—statement of the international union of radioecology. J Environ Radioact 70:155–159

Bréchignac F, Bradshaw C, Carroll S, Jaworska A, Kapustka L, Monte L, Oughton D (2011) Recommendations from the international union of radioecology to improve guidance on radiation protection. Integr Environ Assess Manag 7(3):411–413

Brown JE, Alfonso B, Avila R, Beresford NA, Copplestone D, Prohl G, Ulanovsky A (2008) The ERICA tool. J Environ Radioact 99:1371–1383

Carroll S (2009) Radiological protection of the environment from an NGO perspective. Radioprotection 44(5):439–443

Charpin FSI, Zavaleta ES, EvinerVT Naylor RL, Vitousek PM, Reynolds HL, Hooper DU, Lavorel S, Sala OE, Hobbie SE, Mack MC, Diaz S (2000) Consequences of changing biodiversity. Nature 405:234–242

De Laender F, De Schamphelaere KAC, Vanrolleghem PA, Jansen CR (2008) Do we have to incorporate ecological interactions in the sensitivity assessment of ecosystems? An examination of a theoretical assumption underlying species sensitivity distribution models. Environ Int 34:390–396

ERICA (2007) An integrated approach to the assessment and management of environmental risks from ionizing radiation. ERICA Project, European Commission, contract no. F16R-CT-2004–508847

FASSET (2004) Framework for assessment of environmental impact. Final report of EC 5th framework programme, contract FIGE-CT-2000–00102:111 p

Fleeger JW, Carman KR, Nisbet RM (2003) Indirect effects of contaminants in aquatic ecosystems. Sci Total Environ 317:207–233

Forbes VE, Calow P (2013) Use of the ecosystem services concept in ecological risk assessment of chemicals. Integr Environ Assess Manag 9(2):269–275. doi:10.1002/ieam.1368

French NR, Maza BG, Hill HO, Aschwanden AP, Kaaz HW (1974) A population study of irradiated desert rodents. Ecol Monogr 44:45–72

Galvan I, Bonisoli-Alquati A, Jenkinson S, Ghanem G, Wakamatsu K, Mousseau TA, Møller AP (2014) Chronic exposure to low-dose radiation at Chernobyl favours adaptation to oxidative stress in birds. Funct Ecol doi:10.1111/1365-2435.12283

Garnier-Laplace J, Beaugelin-Seiller K, Hinton T (2011) Fukushima wildlife dose reconstruction signals ecological consequences. Environ Sci Technol 45:5077–5078

Garnier-Laplace J, Geras'kin S, Della-Vedova C, Beaugelin-Seiller K, Hinton TG, Real A, Oudalova A (2013) Are radiosensitivity data derived from natural field conditions consistent with data from controlled exposures? A case study of Chernobyl wildlife chronically exposed to low dose rates. J Environ Radioact 121:12–21

Gralla F, Abson DJ, Møller AP, Lang DJ, von Werhden H (2014) The impact of nuclear accidents on provisioning ecosystem services. Ecol Indic 41:1–14

Hinton TG, Bréchignac F (2005) A case against biomarkers as they are currently used in radioecological risk analyses: a problem of linkage. In: Bréchignac F, Howard BJ (eds) The scientific basis for radiological protection of the environment. Tec & Doc Lavoisier, Paris, pp 123–135

Hiyama A, Nohara C, Kinjo S, Taira W, Gima S, Tanahara A, Otaki J (2012) The biological impacts of the Fukushima nuclear accident on the pale grass blue butterfly. Sci Rep 2:570; doi:10.1038/srep00570

Hiyama A, Nohara C, Taira W, Kinjo S, Iwata M, Otaki J (2013) The Fukushima nuclear accident and the pale grass butterfly: evaluating biological effects of long-term low-dose exposures. BMC Evol Biol 13:168 http://www.biomedcentral/1471-2148/13/168

Holling CS (1986) The resilience of terrestrial ecosystems: local surprise and global change. In: Clark WC, Munn RE (eds) Sustainable development of the biosphere. Cambridge University Press, Cambridge

IAEA (2011) Radiation protection and safety of radiation sources: international basic safety standards—Interim edition, IAEA safety standards series no. GSR Part 3 (Interim), IAEA, Vienna (2011)

IAEA (2014) Summary notes from the technical meeting of the coordination group on radiation protection of the environment: input to safety standards taking into account the BSS and relevant ICRP/International organizations recommendations. IAEA, Vienna, 2–3 July 2013, 8 p

ICRP (2003) A framework for assessing the impact of ionizing radiation on non-human species. Ann ICRP 33(3):266 p (publication 91)

ICRP (2007) The 2007 recommendations of the international commission on radiological protection. Ann ICRP 37(2–4):332 p (Publication 103)

ICRP (2008) Environmental protection: the concept and use of reference animals and plants. Ann ICRP 38(4–6):242 p (Publication 108)

IUR (1997) Dose and effects in non-human systems. Summary of the work of the action group of IUR. International Union of Radioecology, IUR report n° 1. Østeras, Norway, 7 p
IUR (2002) Protection of the environment: current status and future work. International Union of Radioecology, IUR report n° 3. Østeras, Norway, 23 p
IUR (2012) Towards an ecosystem approach for environment protection with emphasis on radiological hazards. International Union of Radioecology, IUR report n° 7, 82 p (978-0-9554994-4-9) (www.iur-uir.org)
Jackson D, Copplestone D, Stone DM (2004) Effects of chronic radiation exposure on small mammals in the Chernobyl exclusion zone. Nucl Energy 43:281–287
Lance E, Alonzo F, Garcia-Sanchez L, Beaugelin-Seiller K, Garnier-Laplace J (2012) Modelling population-level consequences of chronic external gamma irradiation in aquatic invertebrates under laboratory conditions. Sci Total Environ 429:206–214
Ludwig D, Jones DD, Holling CS (1978) Qualitative analysis of insect outbreak systems: the spruce budworm and forest. J Anim Ecol 47:315–332
Maltby L (2013) Ecosystem services and the protection, restoration, and management of ecosystems exposed to chemical stressors. Environ Toxicol Chem 32:974–983
Maslin M (2013) A brief history of the global warming debate. In: Maslin M (ed) Global warming, a very short introduction, 3d edn. Oxford University Press, doi:10.1093/actrade/9780199548248.001.0001
Mihok S (2004) Chronic exposure to gamma radiation of wild populations of meadow voles (*Microtus pennsylvanicus*). J Environ Radioact 75:233–266
Mihok S, Schwartz B, Iverson SL (1985) Ecology of red-backed voles (*Clethrionomys gapperi*) in a gradient of gamma radiation. Ann Zool Fenn 22:257–271
Millennium Ecosystem Assessment (2005) Ecosystems and human well-being: synthesis. Island Press, Washington, 160 p
Møller AP, Mousseau TA (2007) Species richness and abundance of forest birds in relation to radiation at Chernobyl. Biol Lett 3:483–486
Møller AP, Mousseau TA (2009) Reduced abundance of insects and spiders linked to radiation at Chernobyl 20 years after the accident. Biol Lett 5:356–359
Møller AP, Mousseau TA (2013) Low-dose radiation, scientific scrutiny, and requirements for demonstrating effects. BMC Biol 11:92. http://www.biomedcentral.com/1741-7007/11/92
Møller AP, Mousseau TA (2015) Strong effects of ionizing radiation from Chernobyl on mutation rates. Sci Rep 5:8363. doi:10.1038/srep08363
Møller AP, Surai P, Mousseau TA (2005) Antioxidants, radiation and mutation as revealed by sperm abnormality in barn swallows from Chernobyl. Proc R Soc B 272:247–252
Møller AP, Mousseau TA, de Lope F, Saino N (2007) Elevated frequency of abnormalities in barn swallows from Chernobyl. Biol Lett 3:414–417
Møller AP, Hagiwara A, Matsui S, Kasahara S, Kawatsu K, Nishiumi I, Suzuki H, Ueda K, Mousseau TA (2012) Abundance of birds in Fukushima as judged from Chernobyl. Environ Pollut 164:36–39
Møller AP, Nishiumi I, Suzuki H, Ueda K, Mousseau TA (2013) Differences in effects of radiation on abundance of animals in Fukushima and Chernobyl. Ecol Indic 24:75–81
Møller AP, Mousseau TA, Nishiumi I, Ueda K (2015a) Ecological differences in response of birds to radioactivity from Chernobyl and Fukushima. J Ornithol doi:10.1007/s10336-015-1173-x
Møller AP, Nishiumi I, Mousseau TA (2015b) Cumulative effects of radioactiveity from Fukushima on the abundance and biodiversity of birds. J Ornithol. doi:10.1007/s10336-015-1197-2
Mousseau TA, Møller AP (2012) Reply to response regarding «abundance of birds in Fukushima as judged from Chernobyl" by Møller et al (2012). Environ Pollut 169:141–142
Mousseau TA, Møller AP, Ueda K (2012) Reply to "comment on «abundance of birds in Fukushima as judged from Chernobyl" by Møller et al (2012)". Environ Pollut 169:137–138
Mousseau TA, Milinevsky G, Kenney-Hunt J, Møller AP (2013) Highly reduced mass loss rates and increased litter layer in radioactively contaminated areas. Oecologia. doi:10.1007/s00442-014-2908-8

Murphy JF, Nagorskaya LL, Smith JT (2011) Abundance and diversity of aquatic ùacro in vertebrate communities in lakes exposed to Chernobyl-derived ionizing radiation. J Environ Radioact 102:688–694

Nienstedt KM, Brock TCM, van Wensem J et al (2012) Development of a framework based on an ecosystem services approach for deriving specific protection goals for environmental risk assessment of pesticides. Sci Total Environ 415(SI):31–38

Rodgers BE, Baker RJ (2000) Frequencies of micronuclei in bank voles from zones of high radiation at Chernobyl, Ukraine. Environ Toxicol Chem 19:1644–1648

Sazykina TG, Kryshev AI (2003) EPIC database on the effects of chronic radiation in fish: Russian/FSU data. J Environ Radioact 68:65–87

Sazykina TG, Kryshev AI (2006) Radiation effects in wild terrestrial vertebrates—the EPIC collection. J Environ Radioact 88:11–48

Smith J, Beresford NA (2005) Chernobyl—catastrophe and consequences. Springer, Berlin. Praxix Publishing Ltd, Chichester, 310 p

Strand P, Aono A, Garnier-Laplace J, Hosseini A, Sazykina T, Steenhuisen F, Vivesi Battle J (2014) Assessment of Fukushima-derived radiation doses and effects on wildlife in Japan. Environ Sci Technol Lett 1:198–203

Suter GW II (1999) Developing conceptual models for complex ecological risk assessments. Hum Ecol Risk Assess 5(2):375–396

Van Wensem J, Maltby L (2013) Ecosystem services: from policy to practice. Integr Environ Assess Manag 9(2):211–213

Von Wehrden H, Fischer J, Brandt P, Wagner V, Kümmerer K, Kuemmerle T, Nagel A, Olsson O, Hostert P (2012) Consequences of nuclear accidents for biodiversity and ecosystem services. Conserv Lett 0:1–9

Williams N (1995) Chernobyl: life abounds without people. Science 269:304

Fukushima-1 and Chernobyl: Comparison of Radioactivity Release and Contamination

Tetsuji Imanaka

Introduction

A large amount of radioactivity inevitably accumulates inside the reactor core of nuclear power plant (NPP) with its operation. When NPP of 1 GWe(electricity) operates for one day, about 3 kg of nuclear fuel is consumed. Considering that about 1 kg of uranium/plutonium caused fission at the time of Hiroshima/Nagasaki atomic bombing, fission products three times larger than those of Hiroshima/ Nagasaki bombs accumulates every day in the reactor core of NPP. Therefore, about 1000 kg of fission products exists inside the reactor after one year of operation. The worst situation considered at NPP is an accident that would release radionuclides accumulated in the reactor core directly into the environment. From the beginning of NPP development, the following two accidents have been concerned to result in such situation:

- Power surge accident as a result of failing to control fission chain reaction
- Cooling failure accident as a result of losing coolant to remove decay heat from the core

The accident on April 26, 1986 at the fourth unit (RBMK, 1 GWe) of the Chernobyl NPP, USSR was a power surge accident that resulted in the immediate destruction of the reactor core and the building. With subsequent fire of the graphite blocks used as neutron moderator in the core, a large amount of radioactivity release continued about 10 days (USSR 1986).

The Fukushima-1 accident was a cooling failure accident triggered by the tsunami after the Magnitude-9 earthquake on March 11, 2011 (Japanese Government 2011; NAIIC 2012; TEPCO 2012). When the earthquake occurred at

T. Imanaka (✉)
Research Reactor Institute, Kyoto University, Kumatori-Cho, Osaka 590-0494, Japan
e-mail: imanaka@rri.kyoto-u.ac.jp

V.L. Korogodina et al. (eds.), *Genetics, Evolution and Radiation*,
DOI 10.1007/978-3-319-48838-7_19

14:46 on March 11, three units (BWR, Unit-1; 460 MWe, Unit-2 & -3; 784 MWe) out of six units at the Fukushima-1 NPP were in operation. All three reactors were automatically shut down when the seismic waves arrived at the site. Then about 40 min later, a series of tsunami wave hit the Fukushima-1. As a results of the earthquake and tsunami, all AC power, including offsite power lines and emergency diesel generator (EDG), were lost from Unit-1 to Unit-4. A long-time blackout was stipulated as unnecessary to be considered in the safety regulation documents, but it really happened. Decay heat from accumulated radionuclides could not be removed from the cores, which led to meltdown of nuclear fuels and melt-through of reactor pressure vessel (RPV) of Unit-1 to Unit-3. The hydrogen explosion occurred at Unit-1 in the afternoon of March 12 and in the morning of March 14 at Unit-3. On March 15, the largest radioactivity release during the Fukushima-1 accident occurred at Unit-2 due to failure of the containment.

Both Chernobyl and the Fukushima-1 are classified as Level-7, the worst accident according to the INES (International Nuclear Event Scale) by IAEA.

Accident Process

Chernobyl

In the early morning of April 25 1986, operators in the control room of the Unit-4 of Chernobyl NPP started the procedures to stop the reactor for the first time since the reactor began operating in December 1983. An experiment of an electric power generator using a free-wheeling turbine was scheduled at the time of the reactor shutdown. The aim of this experiment was to get emergency power for a short period before the diesel generator could provide emergency power in the case of a power failure accident. Table 1 summarizes the chronology at the initial stage of the Chernobyl accident (USSR 1986, 1991).

An emergency event started at 01:23 40′ on April 26 when the operator turned on the AZ-5 button to shut down the reactor by inserting all control rods into the core. Contrary to the intension of the operator, a positive scram phenomenon led to a small power surge at the lower part of the core, damaging nuclear fuels and channel tubes. Following the rupture of channel tubes, a large amount of vapor appeared at the core. Then, a bigger scale power surge was caused by the effect of positive void coefficient of reactivity, which led to explode the reactor and destroyed the building. According to the analysis after the accident, the explosion was supposed to occur 6–7 s after turning on AZ-5. Eyewitnesses outside the reactor building said that there was a series of explosion like fireworks up into the night sky.

Graphite in the reactor core began to burn after the initial explosion. This fire continued about ten days releasing radioactivity accumulated in the core.

Table 1 Chronology at the initial stage of the Chernobyl

Time	Event
01:00 April 25	Beginning of power reduction from the nominal power of 3.2 GWt
13:05	One of two turbines (No. 7) was separated at 1.6 GWt
14:00	Lifting of ECCS system. Following the request from the central grid controller in Kyiv, power reduction was stopped and kept in operation at half a nominal power
23:10	Restart of power reduction
00:00 April 26	Change of shift member at the control room
00:28	The operator failed to control the power and it reduced to almost zero, although the experiment was planned at the power level of 700 MWt
~01:00	After the efforts to increase power by withdrawing almost all manual control rods, the power seemed stabilized at 200 MWt. Operators decided to conduct the generator test under the planned power
01:03, 01:07	In addition to six primary pumps in operation, two more pumps were switched on. In total eight pumps began working
01:23	At this moment, the reactor was in an extremely unstable condition due to a small value of reactivity control margin as well as a big 'positive void coefficient' intensified at low power level. Operators, however, could not understand the situation
01:23 04′	The generator test began by closing the steam valve to No.8 turbine. Coolant flow of four pumps connected the test generator gradually decreased and steam production increased a little, the effect of which were compensated by a small increase of pressure and automatic insertion of control rods
01:23 40′	Chief of the shift pushed AZ-5 button (all control rod insertion)
01:23 43′	Alarms for 'power increase rate high' and 'power level high'
01:23 46–47′	Alarms for no power supply to pumps, flow rate reduction, high pressure and high water level in the steam-water separator, and failure of power control system
01:23 49′	Alarms for high pressure in the reactor vault, which meant rupture of technical channels. Alarm for power loss of control rod drive and failure of automatic control rod drive
01:24	It was written in the operator log, "01:24 Strong explosion. Control rods stopped before reaching the bottom. Power was lost for control rod drive"

According to the USSR report (USSR 1986) presented at the IAEA (International Atomic Energy Agency) conference held in August 1986 in Vienna, the main cause of the Chernobyl accident was described as an incredible combination of serious violations of regulation by the reactor operators. Five years later, a special committee of the USSR parliament reinvestigated the Chernobyl accident and found that the principal reasons of the accident were the design defects of RBMK reactor (USSR 1991). It also concluded that responsibility for the accident lay with the authorities who knew the defects of the reactor and did not take effective countermeasures.

Fukushima-1

Chronology at the initial stage of the Fukushima-1 accident is shown in Table 2 (Japanese Government 2011; NAIIC 2012; TEPCO 2012). AC power was lost because of offsite power loss due to the earthquake and emergency power loss due to the tsunami. In order to cope with such situation, emergency core cooling systems working without AC power were equipped at the Fukushima-1 NPP. Isolation condenser (IC) and steam-turbine driven high pressure core injection system (HPCI) were at Unit-1. Steam-turbine driven reactor core isolation cooling system (RCIC) and HPCI were at Unit-2 and Unit-3. However, they were designed to work only for short period, and AC power to control them was lost at Unit-1 and Unit-2. Capability to cool the reactor core was lost in the order of Unit-1, Unit-3 and Unit-2. When decay heat cannot be removed from the core of BWR, the sequence will proceed as follows:

1. Increase of temperature and pressure of the primary coolant water,
2. Safety release valve (SRV) open and leakage of coolant water from the primary loop,
3. Decrease of water level in reactor pressure vessel (RPV),
4. Exposure and temperature increase of fuel rods, beginning fuel damages and radionuclide release into coolant,
5. Zirconium-water reaction, producing hydrogen and more temperature increase,
6. Meltdown of nuclear fuels and accumulation of molten materials at the bottom of RPV,
7. Melt-through of the RPV bottom wall and fall of molten materials on the floor of drywell containment (D/W),
8. Increase of D/W pressure, possibility of containment rupture or Chinese syndrome.

At Unit-1, both IC and HPCI did not work effectively because AC power to control them was not available after the tsunami. Meltdown and melt-through of Unit-1 occurred in the evening of March 11. High pressure of containment was found in the midnight of March 11 and continued up to the afternoon of March 12. After venting of the containment pressure, hydrogen explosion occurred at the upper floor of the reactor building.

At Unit-3, DC power survived the tsunami attack. Decay heat removal was maintained by RCIC up to 11:36 March 12. Then HPCI automatically began to work. Operator tried to switch the core cooling from HPCI to fire engines. But it took several hours to open SRV before the water injection from engines and the core damage at Unit-3 began in the morning of March 13. Hydrogen explosion at Unit-3 occurred at 11:01 March 14.

At Unit-2, RCIC was working at the time of tsunami and continued to work up to 13:25 March 14. After RCIC stopped, the situation worsened rapidly as the water level decreased and pressure increased in RPV. In the evening of March 14, pressure of drywell containment (D/W) began to increase to the same level as RPV,

Table 2 Chronology at the initial stage of the Fukushima-1 accident

Time	Event
14:46 March 11	Earthquake occurred at 180 km from the Fukushima-1 NPP
14:47	Three reactors were automatically tripped by the seismic signal of large acceleration. Loss of offsite power. Actuation of all emergency diesel generators (EDG). Main stream isolation valves (MSIV) were closed at three reactors
14:50	Isolation condenser (IC) at Unit-1 and reactor core isolation cooling system (RCIC) at Unit-2 and Unit-3 were started to remove decay heat from reactor core
15:38	Arrival of tsunami. EDG located in the basement of turbine buildings were flooded by seawater. AC power was lost at Unit-1 to -4. DC batteries at Unit-1 and -2 were also damaged
23:00	Increase of radiation dose (1.2 mSv/h) in the turbine building of Unit-1, which meant a significant damage of the reactor core
02:30 March 12	Pressure of 840 kPa was observed in drywell containment (D/W) of Unit-1, about twice of the maximum design pressure of 427 kPa
05:14	Increase of radiation level in the NPP site due to radionuclide leakage from Unit-1
11:36	RCIC of Unit-3 stopped. Then at 12:35, high pressure core injection system (HPCI) of Unit-3 was automatically actuated
14:30	Vent success at Unit-1. D/W pressure decreased to 430 kPa
15:36	Hydrogen explosion at Unit-1
19:04	Start of seawater injection to the reactor core of Unit-1, using fire engines
02:42 March 13	HPCI of Unit-3 stopped. Rapid increase of reactor pressure vessel (RPV) pressure
09:08	Opening of safety release valve (SRV) at Unit-3. Rapid decrease of RPV pressure. Containment venting. Beginning of core damage at Unit-3
14:31	High radiation dose in the reactor building of Unit-3
11:01 March 14	Hydrogen explosion at Unit-3
13:25	RCIC of Unit-2 stopped. Increase of RPV pressure
18:00	Opening of SRV of Unit-2. Decrease of RPV pressure. Beginning of core damage of Unit-2
18:54	Start of seawater injection to the reactor core of Unit-2, using fire engines
22:50	High D/W pressure of Unit-2 exceeded the maximum design value. Vent was unsuccessful and D/W pressure of 750 kPa continued
06:00 March 15	Hydrogen explosion at Unit-4
7:00 ~	Decrease of D/W pressure of Unit-2, reflecting a failure of containment. A large radioactivity release from Unit-2

which meant the melt-through of RPV. Vent was unsuccessful and D/W pressure of 700 kPa continued during the midnight of March 14. Then, at 11:25 March 15 it was found that the D/W pressure of Unit-2 dropped to 150 kPa. During this period, failure of the Unit-2 containment caused the most serious radioactive release during the accident. At 09:00 March 15 radiation dose of 12 mSv/h was recorded at the gate of the Fukushima-1 site.

Meltdown and melt-through occurred at three units of the Fukushima-1 NPP. The final measure that protected these units from Chinese syndrome was the water injection by fire engines. The molten fuels and materials are considered to disperse and form debris on the floor of the containments. The exact situation of the debris is unknown even after five years from the accident.

Radioactivity Release

Daily discharge of radioactivity by the Chernobyl accident is shown in Fig. 1a (USSR 1986). Daily discharge of ^{131}I and ^{134}Cs + ^{137}Cs by the Fukushima-1 accident is Fig. 1b based on estimates by UNSCEAR (2014).

Total radioactivity released into the atmosphere is estimated by various researchers and organizations for both Chernobyl and Fukushima-1. Several estimations for major radionuclides are summarized in Table 3.

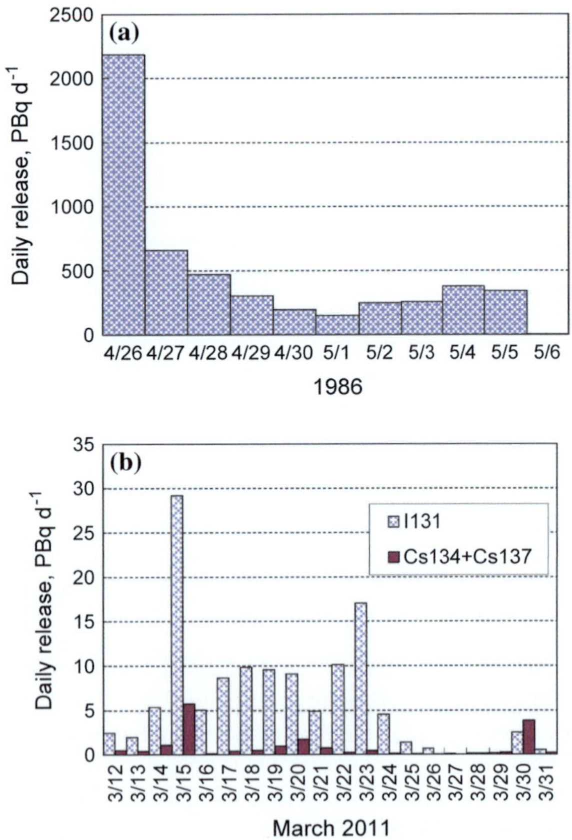

Fig. 1 Daily radioactivity discharge. **a** Chernobyl. Values are for all radionuclides except rare gases and decay-corrected to April 26, 1986. **b** Fukushima-1. Daily release is calculated by the current authors based on the hourly data in UNSCEAR (2014)

Table 3 Estimates of released radioactivity of major radionuclides into the atmosphere: Chernobyl and Fukushima. PBq

	Chernobyl			Fukushima-1		
	USSR (1986)	Seo et al. (1988)	Chernobyl Forum (2005)	NISA (2011)	Stohl et al. (2012)	UNSCEAR (2014)
^{133}Xe	9000	n.a.	6500	11,000	15,300	7300
^{131}I	760	2600	1760	160	n.a.	120
^{132}Te	640	3100	1150	0.76	n.a.	29
^{134}Cs	21	110	47	18	n.a.	9.0
^{137}Cs	37	160	85	15	36.6	8.8
^{90}Sr	8.1	20	10	0.14	n.a.	n.a.
^{95}Zr	160	240	84	0.017	n.a.	n.a.
^{103}Ru	150	470	168	7.5×10^{-6}	n.a.	n.a.
^{106}Ru	60-	200	73	2.1×10^{-6}	n.a.	n.a.
^{140}Ba	300	520	240	3.2	n.a.	n.a.
^{141}Ce	130	310	84	0.018	n.a.	n.a.
^{239}Np	1900	5900	400	0.076	n.a.	n.a.
^{239}Pu	5.2	9.3	0.013	3.2×10^{-6}	n.a.	n.a.

n.a.: not assessed

With regard to Chernobyl, USSR (1986) values are obtained mainly by estimating total depositions within the European part of USSR territory. Seo et al. (1988) analyzed all available deposition data in the northern hemisphere and evaluated total deposition within 3000 km from Chernobyl. Estimates by Chernobyl Forum (2005) are a summary from various studies.

Regarding Fukushima-1, NISA (Nuclear and Industrial Safety Agency) (2011) estimated radioactivity release based on a computer model simulating meltdown process of each unit, without referring monitoring data in the environment. So large uncertainties are considered with NISA estimates. Stohl et al. (2012) uses an inversion technique that estimates strength of radioactivity source by combining monitoring data in the environment and results of atmospheric transport simulation of released radioactivity. They used worldwide monitoring data including CTBT network. UNSCEAR (2014) is mainly based on Terada et al. (2012) which is also an inversion study using data obtained in Japan.

Regarding ^{133}Xe, rare gases were easily released from both Chernobyl and Fukushima-1. A larger release of ^{133}Xe from Fukushima-1 than Chernobyl reflected simply the difference of the reactor power between Fukushima-1 and Chernobyl: about 2000 MWe (sum of Unit-1 to Unit-3) vs. 1000 MWe.

It is clear that ^{131}I and ^{137}Cs release from Fukushima-1 were significantly smaller than from Chernobyl. Comparison of UNSCEAR (2014) with Chernobyl Forum (2005) indicates that ^{131}I and ^{137}Cs releases from Fukushima- 1 are 7 and 10 % of those from Chernobyl, respectively. Comparison of ^{137}Cs from Fukushima-1 by Stohl et al. (2012) with that from Chernobyl by Chernobyl Forum (2005) is 43 %.

Regarding ^{90}Sr, ^{239}Pu and other radionuclides, far less amounts of radioactivity release are estimated from Fukushima-1 than from Chernobyl, which reflects the difference of the accident process. In the case of Chernobyl accident, the explosion occurred inside the reactor core and the reactor materials such as nuclear fuels, graphite blocks were dispersed into the atmosphere. Therefore, the composition of discharged radionuclides from Chernobyl was similar to that in the reactor core. Meanwhile, the reactor cores did not explode in Fukushima-1 and radioactivity discharges were mostly gas and volatile radionuclides emitted from the damaged and melted reactor cores. Two hydrogen explosions occurred at Fukushima-1 under the roof of the reactor building of Unit-1 and Unit-3. But they were not inside containment vessels.

It is also noted here that liquid radioactivity directly released into Pacific Ocean by the Fukushima-1 accident is not considered in this study.

Ground Contamination

Radionuclide Composition

Reflecting the difference of radionuclide composition released into the atmosphere, the composition of ground contamination was also different between Chernobyl and Fukushima-1. Izrael et al. (1987) reported the radionuclide composition of the ground contamination within 100 km from the Chernobyl NPP, which indicates that the composition changed depending on direction and distance from Chernobyl NPP. In Table 4 the ground contamination of major radionuclides contributing gamma-ray exposure above ground is shown as relative deposition ratios to ^{137}Cs for the near western area of the Chernobyl NPP where the initial plume passed over on the first day of the accident. Relative deposition ratios around Fukushima-1 are also shown in Table 4, values of which are taken from UNSCEAR (2014) as representative for all Japan except the southern trace from Fukushima-1.

Far less ground contamination of ^{90}Sr and ^{239,240}Pu around Fukushima-1 compared with Chernobyl was confirmed by the measurement of soil samples. Table 5 shows ^{90}Sr, ^{239,240}Pu and ^{137}Cs contamination in soil taken in Iitate village located 30–40 km from the Fukushima-1 site (Imanaka et al. 2012) together with those in Kyiv about 110 km from the Chernobyl site (Garger et al. 1996). Deposition of ^{90}Sr and ^{238,240}Pu in Kyiv are 23 and 0.6 % of ^{137}Cs, respectively, while they are 0.04–0.05 % and 10^{-5}–10^{-6} % of ^{137}Cs in Iitate.

From the point of long-term effects of radioactive contamination in the environment, ^{137}Cs is the most important radionuclide for both Chernobyl and Fukushima-1. The area severely contaminated by ^{137}Cs by two accidents is compared in Table 6 (Imanaka and Kawano 2009; Sawano et al. 2013). Roughly speaking the contaminated area around Chernobyl is more than 10 times larger than Fukushima-1. It is noted, however, although Chernobyl is surrounded by land, the eastern half around Fukushima-1 is Pacific Ocean and most of discharged radioactivity is considered to stream toward the ocean by prevailing westerlies over Japan.

Table 4 Deposition ratios of major radionuclides to ^{137}Cs, contributing gamma-ray exposure at 1 m above ground

Radionuclide	Half life	Exposure rate conversion factor (nGy h^{-1})/(kBq m^{-2})	Relative deposition ratio to ^{137}Cs	
			Chernobyl (Izrael et al. 1987)	Fukushima-1 (UNSCEAR 2014)
^{95}Zr	65.5 d	2.82	20	–
^{95}Nb	35.0 d	2.92	20	–
^{103}Ru	39.3 d	1.85	16	–
^{131}I	8.04 d	1.49	18	11.5
^{132}Te	3.25 d	0.79	28	8
^{132}I	(2.30 h)[a]	8.61	28	8
^{134}Cs	2.07 y	5.97	0.4	1
^{137}Cs	30.1 y	2.18	1	1
^{140}Ba	12.8 d	0.57	22	–
^{140}La	(1.68 d)[a]	7.83	11	–
^{239}Np	2.36 d	0.60	120	–

Exposure rate conversion factors are taken from Beck (1980) for a case of 0.16 g cm^{-2} of relaxation length for depth distribution. Values for Chernobyl are on the day of the accident, April 26 1986. Values for Fukushima-1 are on March 15, 2011 when the most severe ground contamination occurred

[a]These radionuclides are treated at radioactive equilibrium with parent radionuclides

Table 5 Comparison of ^{137}Cs, ^{90}Sr and $^{239,240}Pu$ contamination in soil between Iitate village and Kyiv city

	Contamination density, Bq m^{-2}		
	^{137}Cs	^{90}Sr	$^{239,240}Pu$
Iitate village: 30–40 km North-West from Fukushima-1 NPP (Imanaka 2012)			
Sample 1	1,000,000	390[a]	0.01
Sample 2	590,000	300[a]	0.07
Sample 3	2,200,000	790[a]	0.2
Kyiv city: 110 km South from Chernobyl NPP (Garger et al. 1996)			
Average of six samples	25,000	5800	160

[a]Measured by Kyusyu Environmental Evaluation Association. Values include global fallout

Table 6 Area of severely contaminated land around Chernobyl and Fukushima-1

	^{137}Cs contamination level	
	555–1480 kBq m^{-2}	>1480 kBq m^{-2}
Chernobyl (Imanaka 2009) (km^2)	7200	3100
Fukushima-1 (Sawano et al. 2013) (km^2)	495	272

^{137}Cs contamination level of 555 and 1480 kBq m^{-2} corresponds to criteria for compulsory resettlement and alienation, respectively, around Chernobyl

Radiation Exposure from Radionuclides Deposited on the Ground

Based on the radionuclide composition in Table 4, gamma-ray exposure rate at 1 m above ground is calculated and plotted in Fig. 2a per the initial ^{137}Cs deposition of 1 MBq m^{-2} for both Chernobyl and Fukushima-1. Radiation exposure conversion factors from radionuclides on the ground are taken from Beck (1980) for a case of 0.16 g cm^{-2} of relaxation length as depth distribution and shown in the third column of Table 4. The initial exposure rate one day after deposition is 700 μGy h^{-1}for Chernobyl, while it is 80 μGy h^{-1} for Fukushima-1, nine times smaller compared with Chernobyl. As can be understandable from the data in Table 4, such radionuclides as ^{95}Z, ^{95}Nb, ^{103}Ru and ^{140}La give significant contribution to radiation exposure during the first year after the accident in Chernobyl. In Fukushima-1 on the 60th day after deposition 99 % of radiation exposure rate is due to radiocesiums (^{134}Cs + ^{137}Cs), while the contribution of radiocesiums in Chernobyl is only 4 % of exposure rate on the 60th day. Interestingly, exposure rate

Fig. 2 **a** Radiation exposure rate, μGy h^{-1} and **b** cumulative exposure, mGy at 1 m above ground normalized to the initial ^{137}Cs deposition of 1 MBq m^{-2} and relative deposition composition in Table 4

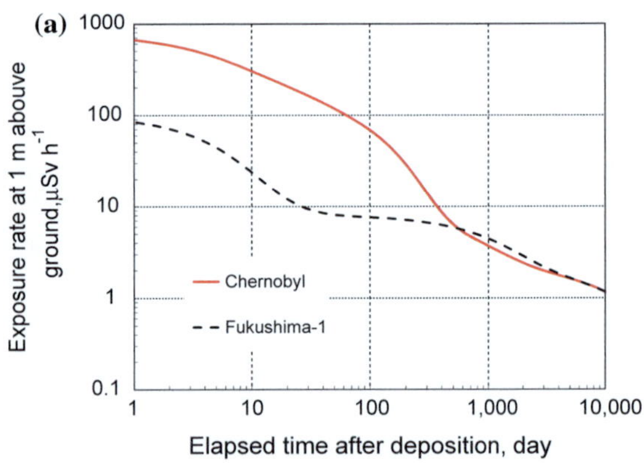

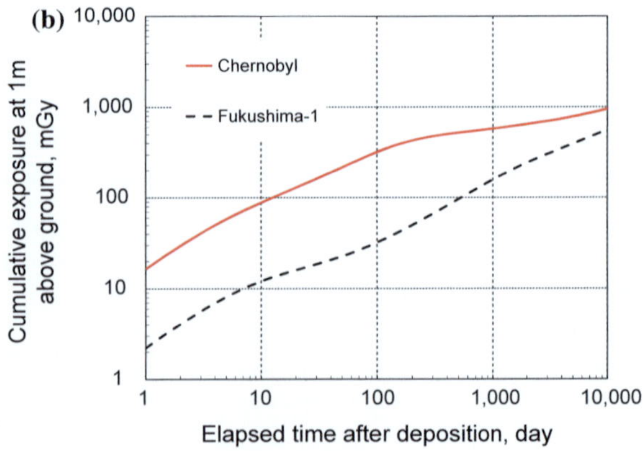

for Fukushima-1 exceeds Chernobyl at about 1.5 year after deposition and becomes almost the same after 10 year. This is due to the different deposition ratio of ^{134}Cs (half-life; 2.07 year) to ^{137}Cs (30.1 year) between Fukushima-1 (^{134}Cs:^{137}Cs = 1:1) and Chernobyl (0.5:1). The relative contribution of ^{134}Cs is larger for Fukushima-1 than for Chernobyl.

Cumulative exposures at 1 m above ground normalized to the initial ^{137}Cs deposition of 1 MBq m^{-2} is shown in Fig. 2b both for Chernobyl and Fukushima-1. Cumulative exposure for the first year (365 day) are 500 and 63 mGy for Chernobyl and Fukushima-1, respectively. Cumulative exposure for 30 year (10,960 day) are 970 and 500 mGy for Chernobyl and Fukushima-1, respectively. The contribution of radiocesiums to the cumulative exposure for the first year is 7.4 and 83 % for Chernobyl and Fukushima-1, respectively. For 30 year, they are 49 and 98 % for Chernobyl and Fukushima-1.

From the analysis above, we can say that different composition of ground contamination between Chernobyl and Fukushima-1 gives different pattern of radiation exposure mainly during the first year.

Conclusion

Chernobyl was a power surge accident caused by failing to control fission chain reaction, while Fukushima-1 was a cooling failure accident caused by failing to remove decay heat from the reactor core triggered by blackout due to earthquake and tsunami. Reflecting the difference of the accident process, radionuclide composition of ground contamination was different between the two accidents. Roughly speaking, the scale of ground contamination around Chernobyl was more than ten times larger than around Fukushima-1. On the first day of contamination, radiation exposure rate per initial ^{137}Cs deposition of 1 MBq m^{-2} was 700 μGy h^{-1} in Chernobyl, while it was 80 μGy h^{-1} in Fukushima-1, reflecting the difference of radionuclide composition of ground contamination. Cumulative exposure for the first year was 500 and 63 mGy in Chernobyl and Fukushima-1, respectively. For 30 year, they were 970 and 500 mGy, respectively.

References

Beck HL (1980) Exposure rate conversion factors for radionuclides deposited on the ground. EML-378. July 1980

Chernobyl Forum (2005) Chernobyl's legacy: health, environmental and socio-economic impacts and recommendations to the governments of Belarus, the Russian Federation and Ukraine, IAEA. https://www.iaea.org/sites/default/files/chernobyl.pdf

Garger EK, Hoffman FO, Miller CW (1996) Model testing using Chernobyl data: III. Atmospheric resuspension of radionuclides in Ukrainian regions impacted by Chernobyl fallout. Health Phys 70:18–24

Imanaka T, Kawano N (2009) Radioactive contamination and social consequences caused by the Chernobyl nuclear accident. Hiroshima Peace Sci 31:65–86

Imanaka T, Endo S, Sugai M et al (2012) Early radiation survey of Iitate village, which was heavily contaminated by the Fukushima Daiichi accident, conducted on 28 and 29 March 2011. Health Phys 102:680–686

Izrael YA, Imshennik EV, Kvasnikova EV et al (1987) Radioactive contamination of the environment in the zone of Chernobyl atomic station. Meteorol Hydrol 1987(2):5–18 (Russian)

Japanese Government (2011) Report of Japanese Government to the IAEA ministerial conference on nuclear safety—the accident at TEPCO's Fukushima Nuclear Power Stations—June 2011. http://www.kantei.go.jp/jp/topics/2011/iaea_houkokusho.html

NAIIC (2012) Report of the national diet of Japan Fukushima Nuclear Accident Independent Investigation Commission. July 5, 2012. http://warp.da.ndl.go.jp/info:ndljp/pid/3856371/naiic.go.jp/en

NISA (2011) About the status of reactor core of Unit-1, Unit-2 and Unit-3 of TEPCO Fukusima-1 NPS. 6 June 2011 (Japanese). http://donjon.rulez.jp/refeqsum/hoanRoshinhyouka-20110606008-2.pdf

Sawano S, Imanaka T, Hayashi G et al (2013) Cesium deposition mapping of Iitate village using NNSA released data. In: Proceedings of 14th workshop on environmental radioactivity, KEK Proc 2013-7, pp 136–144 (Japanese)

Seo T, Imanaka T, Koide H (1988) Radioactivity release by the Chernobyl accident. Kagaku 58:108–117 (Japanese)

Stohl A, Siebert P, Wotawa G et al (2012) Xenon-133 and caesium-137 releases into the atmosphere from the Fukushima Dai-ichi nuclear power plant: determination of the source term, atmospheric dispersion, and deposition. Atmos Chem Phys 12:2313–2343

TEPCO (2012) Investigation report on Fukushima nuclear accident. Tokyo Electric Power Company. June 20 2012 (Japanese). http://www.tepco.co.jp/nu/fukushima-np/interim/index-j.html

Terada H, Katata G, Chino H et al (2012) Atmospheric discharge and dispersion of radionuclides during the Fukushima Dai-ichi nuclear power plant accident. Part II: verification of the source term and analysis of regional-scale atmospheric dispersion. J Environ Radioact 112:141–154

UNSCEAR (2014) Levels and effects of radiation exposure due to the nuclear accident after the 2011 great east-Japan earthquake and tsunami. Annex A UNSCEAR 2013 report. http://www.unscear.org/docs/reports/2013/13-85418_Report_2013_Annex_A.pdf

USSR Government Committee for Control of Safety Works at Industries and Nuclear Energy (1991) About causes and conditions of the accident at 4-th unit of Chernobyl NPP on April 26, 1986. USSR parliament committee, 7.01.1991 (Russian)

USSR State Committee (1986) The accident at the Chernobyl nuclear power plant and its consequences. August 1986

Effects of Ionizing Radiation on Populations and Ecosystems

Stanislav A. Geras'kin, Rudolf M. Alexakhin and Alla A. Oudalova

Introduction

For an assessment of radiation impact on the environment it is essential to understand clearly: which ecological effects could be observed under different scenarios of radiation exposure? It is well known that effects of radiation exposure initially appear at the molecular level. However, this knowledge cannot be directly extrapolated to populations and ecosystems, since much more complex underlying mechanisms are involved in their response to stress (Clements and Rohr 2009). Recent discoveries (Bradshaw et al. 2014) have made an ecosystem-level perspective on radioecology more important than ever. Radioecology should clearly consider consequences of both present and impending heavy radiation accidents and shifts from a purely descriptive to a more predictive science. Greater consideration of the ecological effects of ionizing radiation should improve our ability to predict how and when populations and ecosystems will respond to radiation exposure and recover after that. A better understanding of the radiation-induced population tolerance and its costs should facilitate the identification of impacted communities and thus forecast of the ecological consequences of radiation exposure and determination of the restoration effectiveness.

In this context, a question arises: what are the patterns of radiation effects on the structure, function, and development of biological communities? Although species interactions are not the major structuring force in all ecosystems in those situations where factors such as competition and predation are important, the relative strength

S.A. Geras'kin (✉) · R.M. Alexakhin
Russian Institute of Radiology and Agroecology,
Kievskoe Shosse, 109 Km, 249020 Obninsk, Russia
e-mail: stgeraskin@gmail.com

A.A. Oudalova
Obninsk Institute of Nuclear Power Engineering, National Research
Nuclear University "MEPhI", Obninsk, Russia

V.L. Korogodina et al. (eds.), *Genetics, Evolution and Radiation*,
DOI 10.1007/978-3-319-48838-7_20

of these interactions probably influences the respond of ecosystems to radiation exposure. At present knowledge about the effects of radiation exposure at higher levels of biological organization is quite comprehensive and worth of being pondered. There are three main sources of such information: nuclear weapons testing, field studies of effects in ecosystems exposed to powerful sources of ionizing radiation, and heavy radiation accidents.

Large-Scale Studies of Ecosystems Externally Exposed to Radiation

The extremely high radiosensitivity of forest ecosystems (particularly, a coniferous forest) was discovered in the 1950s when yellowed pine needles were noted on the edge of a pine forest surrounding the experimental γ-field of Brookhaven National Laboratory, the USA, where the investigations of radiobiological effects in plants were carried out (Sparrow 1966). Simultaneously with these observations, manifestation of radiation damage of a pine forest (up to destruction) was observed after the accident in the South Urals in 1957 (Alexakhin et al. 2004). Experimental studies which were carried out in the USA, Canada, France and the USSR have given information about the dynamics of the radiation damage and recovery processes in different types of ecosystems. These studies have also enabled estimating the values of radiation dose, leading to shifts in ecosystems (Alexakhin et al. 1994; Fabries et al. 1972; Guthrie and Duole 1983; McCormick and Colley 1966; Odum and Pigeon 1970; Sparrow 1966; Woodwell 1967; Zavitovski 1977). It was found that a structure of ecosystem changed as a function of dose. The higher dose, the simpler structure of an ecosystem was observed. Another consistent observation is the loss of larger, longer-lived species and a switch to smaller, more opportunistic taxa (Woodwell 1967). Thus, the typical reactions of phytocenosis to irradiation at high doses are reduction in species diversity, changes in species dominance, reduction in productivity and changes in a community structure.

A second important conclusion from these experiments is that radiation is a form of stress which elicits community responses frequently similar to those ones resulting from other forms of stress (Woodwell 1967). As a result, the changes observed in an ecosystem after irradiation are caused not only by radiation per se, but also by the inherent nature of the ecosystem. Therefore, considerable insight into the basic nature of ecosystems and their ability to withstand or to recover from stress can be obtained through observations of the irradiated ecosystems. General outcomes of these experiments form the basis for classifying the main types of ecosystems by radiosensitivity (Table 1). It is evident that the most radiosensitive ecosystem is a coniferous forest, while the most resistant is a meadow plant community.

Table 1 Classification of ecosystems by radiosensitivity (Alexakhin 1992)

Type of ecosystem	The extent of damage at a dose, Gy		
	Low	Medium	Heavy
Agricultural crops	2	>2	–
Pine forest	2	2–20	>20
Deciduous forest	2	2–100	>100
Meadow phytocenosis	20	20–200	>200
Abandoned grass field	40	40–70	>70

Heavy Radiation Accidents

Patterns that we discussed above have been established in the radioecological studies with external exposure. However, the situation is very different in case of radioactive fallout, and knowledge gained from previous studies is insufficient to explain a biota response in this case. So, why cannot the results of experiments with irradiation of biocenoses be used without corrections for explanation of situations related to the accidental release of radionuclides?

Radionuclides uptake to natural ecosystems during radiation accidents with the release of radioactive substances occurs through different pathways, the aerial contamination being the prevailing one. Besides, there may be radionuclides discharge to the water environment, as was the case in the accident at the Japanese NPP Fukushima Daiichi in 2011. The radionuclides that escaped to the environment during an accidental release enter the biotic and abiotic chains of migration, leading to a dynamic in time and space distribution of radionuclides in ecosystems. Finally it results in a complicated space-time dynamics of exposure dose distribution in ecosystems (Table 2). The most critical situation arises when the maximum doses receive the most radiosensitive components of ecosystem. Aerial contamination of coniferous forest is a bright illustration of such radioecological situation. Coniferous trees generally have a high retention capacity and low turn-over rate for atmospherically dispersed contaminants (Tikhomirov and Shcheglov 1994). As a result, the highest radionuclide concentrations and, correspondingly, exposure doses are received by the aboveground part of plants, in particular, by needles (these are renewal once every 3–4 years). The combination of increased radionuclide concentration (exposure doses) and low radioresistance of coniferous is responsible for the maximum radiation damage observed in this case (the "red" forests in the Kyshtym and Chernobyl affected regions are most striking examples). Another example comes from the enhanced radionuclides accumulation in the forest litter. This ecological niche is the habitat of forest invertebrates, whose juvenile stages are highly radiosensitive. Ultimately, it predetermines an increased vulnerability of this component of a forest ecosystem to radiation damages. Indeed, dwellers of forest litter were severely affected at a distance of 3–7 km from the Chernobyl NPP (Krivolutsky and Pokarzhevsky 1992). In aquatic habitats, bottom deposits are characteristic of highest radionuclide concentrations, thus resulting in higher irradiation of benthic organisms.

Table 2 The comparison of the absorbed dose distribution in ecosystems in case of radiation accident and external irradiation

	External exposure of ecosystem	Radiation accident
Source of exposure	Point source	Distributed source
Type of radiation	γ or neutrons	α, β, γ in different combinations
Type of exposure	External	External and internal
Distribution of absorbed doses in ecosystem	Relatively uniform. Dose is decreased with the distance from the source	Extremely heterogeneous

Radiation effects in ecosystems depend on the species radiosensitivity and a distribution of absorbed doses within an ecosystem. However, dose distributions between components of an ecosystem are fundamentally different for situations of external exposure and radioactive material fallout (Table 2). As a result of these patterns in dose distribution there are considerable differences in doses absorbed by biota species, even when they all are presented in the same environment at the same time. Indeed, in the first year after the Chernobyl accident near the Borschevka settlement doses to biota species varied significantly, up to 250 times (Fesenko et al. 2005). It is important to point out that doses to a man in this situation would have been about 50 times lower than the maximum doses got by biota species. So, in this radioecological situation a man obviously cannot be considered as the most exposed component of the ecosystem. In addition, different types of radiation (α-, β- and γ-radiation) also result in different consequences for biota and man. It is especially evident in case of α-emitting radionuclides. For example, in case of contamination with radionuclides of ^{235}U decay chains, the soil biota species receive doses 480 times greater than man (Spirin et al. 2013).

A specific feature of heavy radiation accidents is the presence of two periods—an intensive short-term radiation impact and the subsequent long period with a slow decline in the dose rate. In radiation-impacted ecosystems two groups of effects are identified. Primary effects depend on the radiosensitivity of species that comprise an ecosystem. These effects, depending on the dose, can vary from a slight inhibition of growth to the death of organisms or even the ecosystem collapse. The ranges of lethal doses for major groups of organisms are extremely wide and cover several orders of magnitude (Whicker and Schultz 1982). It is also important that the lethal dose ranges for different groups of organisms are only partially overlapped. Therefore, a part of radiosensitive species dies after the irradiation with the sufficiently high dose. In some other species vital functions are suppressed. For radioresistant species the same dose is harmless or even stimulating. This creates backgrounds for the formation of secondary effects in the exposed ecosystems.

Irradiation of plants and animals with lethal and sublethal doses results in the disruption of ecological relationships between the components of ecosystems. Such effects may act as a trigger of perturbation and lead to consequences that may differ radically from those ones expected based on effects observed at the organismal level. Some typical examples of secondary radiation effects are given below.

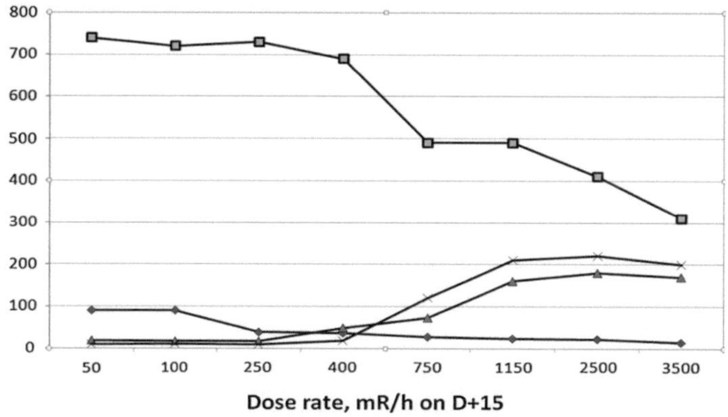

Fig. 1 Radiation effects in the meadow phytocenoses, Yanov, near the Chernobyl NPP, 1987 (Smirnov and Suvorova 1996). *Filled square* total number of plants per m^2; *cross* number of *Agrostis syreistschikowii* per m^2; *filled triangle* number of *Calamagrostis epigeios* per m^2; *filled diamond* number of plant species per 100 m^2; D + 15—dose rates were measured on 15th day after the accident

Suppression of the radiosensitive species and intensive development of the radioresistant ones. In radiation-impacted forest around the Chernobyl NPP, coniferous died completely because of their high radiosensitivity (Geras'kin et al. 2008). In contrast, a birch and grasses survived. Moreover, more radioresistant species were actively developing due to an improvement of light, temperature and nutrition conditions in the dead coniferous forest. In the meadow phytocenoses near the Chernobyl NPP due to the same reason the relative contributions of the radioresistant species increased significantly while the total number of plants and species diversity decreased sharply with the level of radiation exposure (Fig. 1). So, exclusion of radiosensitive species attenuates the competition for others.

Disruption of trophic relations as a result of the selective death of animals. The next example came from a survey of gypsy moth populations and its parasite tahinid flies inhabiting the contaminated with ^{90}Sr birch forest in the South Urals. From Fig. 2 it becomes evident that radioactive contamination can have both positive and negative effects on the consecutive trophic levels. Indeed, the infestation of caterpillars with parasites decreases, while their survival rate increases with the level of radioactive contamination. These effects become evident at levels of radioactive contamination over 1 MBq/m^2. The observed paradoxical effects are connected with ecology of these related species as well as with features of the radionuclides distribution between the components of the forest ecosystem. The caterpillars and the pupae of gypsy moth inhabit the upper tiers of the forest, where the radiation doses are negligible. In contrast, the pupae of tahinid flies occupied the litter, where the dose reached up to 70 Gy. So, in this case, the observed indirect effect, i.e. the increased survival of caterpillars and butterflies with increasing level of radioactive contamination, caused due to decrease in the number of parasites affected by radiation.

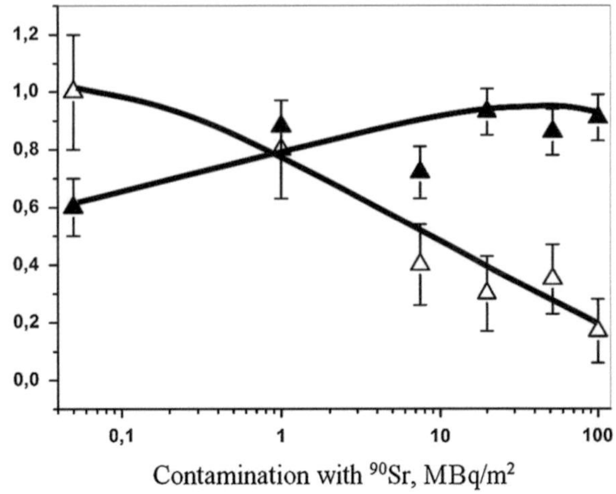

Fig. 2 Effect of radioactive contamination on the gypsy moth populations and its parasite tahinid flies (Krivolutsky et al. 1988). *Filled triangle* survival rate of gypsy moth caterpillars, relative units; *open triangle* infestation of gypsy moth caterpillars by tahinid flies, relative units

A large amount of organic residues leads to increase in the number of insect pests. This phenomenon was observed in the South Urals, where a sharp increase in the number of bark beetles was found in the irradiated forest (Alexakhin et al. 2004). In a similar way, in the 30-km ChNPP zone the forest and fruit trees were seriously injured by pests (Krivolutsky 2000).

Loss of immunity and a nidus of infectious diseases. Over longer periods another factors come into play. Radiation exposure can have substantial effects on host-pathogen relationships. These effects are predominantly associated with the serious damage to the hosts' immune system caused by ionizing radiation. In addition, chronic radiation exposure can act as a mutagenic factor that leads to the emergence of new clones in the populations of pathogenic microorganisms. It should be borne in mind that the biological and ecological requirements of individual host species will have a major influence on both the risk of radiation exposure and the subsequent degree of altered infection levels with pathogens (Morley 2012). Loss of immunity in plants and animals in the affected area has activated a nidus of infectious diseases such as tularemia, encephalitis and fungal pathogens. Indeed, within the 30-km ChNPP zone decrease in the plants resistance to disease as well as an accelerated development of new phytopathogenic forms and races with enhanced virulence were revealed (Dmitriev et al. 2011). In contrast, in a large area on the border between Belarus and Russia including the heavily contaminated Bryansk Region, the higher the contamination with ^{137}Cs was, the lower the incidence of rabies in wild Canidae was observed, with no rabid animals in regions contaminated 740–1480 kBq/m^2 ^{137}Cs or higher (Adamovich 1998). Although it is not clear what factors are impacting disease epidemiology, it seems likely that radiation is either directly affecting viral viability or more likely indirectly through a reduction in the numbers of animals acting as a source of infection. Indeed, in this region wolf litter numbers show a significant negative correlation with the density of superficial soil contamination (Adamovich 1998).

Therefore, the type and magnitude of indirect effects of radionuclides can depend on an ecosystem composition, and many other ecological factors can be much more important than radiation. Overall, the disturbances of ecological interrelations are induced by the following factors: (1) changes in microclimatic and edaphic conditions; (2) disturbances in the synchronism of seasonal phases in the development of ecologically connected groups of organisms; (3) an imbalance in food interrelations between consumers and producers; (4) changes in the biological pressure as a result of species differences in radioresistance; (5) loss of immunity and the accelerated development of new pathogenic forms and races with enhanced virulence. In addition, radiation-induced alterations in affected ecosystems are able of making open ecological niches for immigration of new species. These indirect effects cannot be deduced solely from the effects on individuals. Therefore, the use of ecological knowledge is essential for understanding the responses of populations and ecosystems to radiation.

Radioadaptation

Populations inhabiting a contaminated site may be subjected to selective pressure, and this may result in their improved resistance to pollution (Geras'kin et al. 2013). A comparison of radiosensitivity of populations among sites can thus provide the evidence of adaptation, if it is possible to show that individuals from a contaminated site display elevated resistance and that differences in resistance between populations have a genetic basis. Laboratory studies of repair inhibitions, dose-effect relationships for low- and high-LET radiations, measurements of an unscheduled DNA synthesis and an efficacy of the single strand breaks recovery suggest that the divergence of populations in terms of radioresistance is based on selection for the increased effectiveness of repair systems (Shevchenko et al. 1992). Another study of the possible mechanisms of adaptations to radioactive contamination (Kovalchuk et al. 2004) showed extremely low (more than 10-fold) recombination level and a higher level of the global genome methylation in chronically irradiated plants that may have prevented extensive genome rearrangements.

Evidence for the rapid evolutionary changes in plants in response to an environmental stress is widespread in the literature. For example, seeds from the Scots pine populations growing in the vicinity of the 'Radon' radioactive waste storage facility showed (Fig. 3a) a higher resistance to the additional γ-ray exposure than seeds from the reference population, although the impacted populations are characterized by the increased frequency of aberrant cells (Geras'kin et al. 2005). Kalchenko and Fedotov (2001) got similar results in studies of the acute γ-radiation resistance of Scots pine seeds collected in the 30-km Chernobyl NPP zone in 1997. Rapid evolution of mountain birch towards the heavy metal resistance was also observed around two subarctic copper-nickel smelters (Eranen 2008).

Evolutionary adaptation can be rapid (Levinton et al. 2003; Hendry et al. 2008) and potentially helps species withstand the stressful conditions or realize

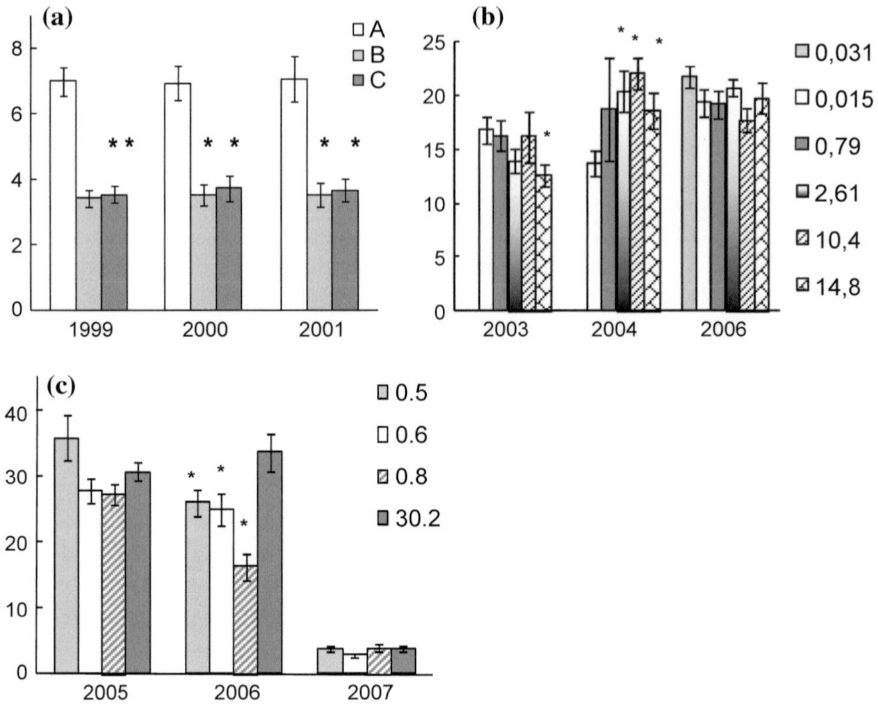

Fig. 3 Frequency of aberrant cells in seedling meristems after the acute γ-exposure of seeds collected from the chronically exposed plant populations: **a** a root meristem of Scots pine seedlings, a vicinity of the 'Radon' radioactive waste storage facility, 15 Gy (Geras'kin et al. 2005); **b** a root meristem of Scots pine seedlings, the Bryansk Region, 15 Gy (Geras'kin et al. 2011); **c** a coleoptile of crested hairgrass, the Semipalatinsk Test Site, 68.8 Gy at 2970 Gy/h in 2005 and 2006, 50 Gy at 39 Gy/h in 2007 (Geras'kin et al. 2012). *C* and *B* the technogenically impacted populations, *A* the reference population. In legends estimated dose rates at the plots are given in μGy/h. *Asterisk* difference from the reference population is significant, $p < 0.05$

opportunities arising from the change of ecological situation. However, this rule does not apply to all the cases. The challenges are to understand when evolution will occur and to identify potential winners as well as losers, such as radiosensitive species lacking the adaptive capacity in the particular radioecological situation. Contrary to the increased radioresistance of seeds from plant populations inhabiting radioactively contaminated territories described in Shevchenko et al. (1992) as well as Kalchenko and Fedotov (2001), no significant and reproducible in time difference in resistance to subsequent γ-ray exposure was found (Fig. 3b) between seeds collected from the reference and exposed Scots pine populations from the Bryansk Region affected by the Chernobyl accident (Geras'kin et al. 2011). Similarly, the seeds from the crested hairgrass populations from the Semipalatinsk Test Site that have been experiencing radiation exposure for more than a half century and are bearing the increased levels of cytogenetic abnormalities did not show (Fig. 3c) any reliable increase in resistance to the additional γ-ray exposure (Geras'kin et al. 2012).

A considerable variation in the dose rate of acute irradiation (more than 70 times, from 2970 Gy/h to 39 Gy/h) did not change the conclusion.

The absence of adaptation to chronic radiation exposure documented in some plant populations raises questions about the conditions under which radioadaptation might be expected. Which factors might constrain or promote adaptive responses? If they occur, will these responses be sufficient to keep up under the radioactive contamination conditions?

An adaptation to the radioactive contamination is equivalent to an increase in fitness under chronic exposure to radionuclides, however, the adapted population's fitness may be decreased in normal conditions (Hickey and McNeilly 1975; Diaz-Ravina and Baath 2001; Levinton et al. 2003; Eranen 2008). With such a negative correlation between the resistance and a fitness-related trait, selection for an increased resistance automatically decreases the fitness. This additional cost of increased radioresistance may be a consequence of antagonistic pleiotropic effects (Shirley and Sibly 1999) or may arise at the physiological level since the metabolic energy has to be used to produce the adaptive trait even when it is not required (Levinton et al. 2003). As a result, there are situations when resistance to environmental changes does not evolve or does not persist. Moreover, an adaptation is often observed in one species but not found in others, despite an equivalent opportunity and exposure conditions (Bradshaw 1991). Indeed, in the South Urals radioactive trace, the radioresistance increased 3–4 times in radiosensitive plants, but remained practically unchanged in radioresistant species (Shevchenko et al. 1992).

The response of populations to radiation exposure depends on many factors: a type of an organism and its associated ecological niche, the radioresistance level and the way of reproduction as well as the biophysical properties of radiation (the relative biological effectiveness, the linear energy transfer, dose rates, etc.). An improved DNA repair capacity and an ability to germinate under the abiotic stress (salinity and accelerated ageing) were shown in seeds embryos of evening primrose growing near the Chernobyl NPP on sites contaminated with γ- and β-emitters, while on the α-, β- and γ-contaminated site such adaptation was not found (Boubriak et al. 2008). This corresponds to findings about the successful adaptation of wild vetch populations on sites highly contaminated with β-emitters, but not with α-emitters (Syomov et al. 1992). It seems that if we examine radioadaptation in nature, we find as much evidence of its existence, as we do for its absence. Consequently, there are good theoretical and practical reasons for more attention being paid to the mechanisms by which populations become more radioresistant, and to those situations where radioadaptation does not appear.

Ecosystem Approach in Radioecology

The current approach for protecting environment against radiation relies on organism-level endpoints (ICRP 2008), and this approach is not matched with the stated protection goals of the international agencies: preserving life sustainability through protection of an ecosystem structure and functioning (Bradshaw et al. 2014). In addition, as described above, exposure to radionuclides can trigger non-linear changes in an ecosystem structure and function that cannot be predicted from effects on the individual organisms. These phenomena are particularly relevant when considering the potential long-term ecological effects of chronic radiation exposure. Such impacts may not be manifested as the results of direct radiobiological effects on individual organisms, but rather than the consequence of indirect effects resulting from differences in sensitivity of different species, potentially leading to changes in the habitat structure or alteration of the trophic relationships. Although radioecologists discovered the potential importance of indirect effects long time ago (Woodwell 1967; Brechignac and Doi 2009), these effects have remained understudied and not thoroughly included into the radioecological risk assessment. The development of the Ecosystem Approach to radiation protection would eliminate this inconsistency.

To effectively use the Ecosystem Approach proposed by the International Union of Radioecology (Brechignac et al. 2012), we should focus on a community of interacting species exposed to radiation rather than on a small set of species considered in isolation. Either the reference organisms or the Ecosystem Approach taken alone would not be able to cover all issues relevant to the environment protection. Each of them has its own advantages and limitations, but they reciprocally complement each other when used together, that allow meeting the overall protection goals. The Ecosystem Approach can complement the reference organism conception by enhancing its ecological contextualization and in this way facilitates an identification of the key species whose loss or decline could trigger secondary extinction. Also the Ecosystem Approach can be used for an identification of the fragile communities where the loss of species could cause a considerable reduction in the system stability. For example, keystone species that modify the resource dynamics, the trophic structure, and the disturbance regimes have the greatest potential to affect the ecosystem processes. Thus, protection of these species is essential for the ecosystem preservation. If an ecologically important species belongs to the most sensitive to radiation exposure, like coniferous, its elimination will have disproportional effects on the ecosystem function (Alexakhin et al. 2004; Geras'kin et al. 2008). Methods for managing indirect effects and for detecting adverse changes in ecosystems before they become severely affected have been developed now (Scheffer et al. 2009; Clements and Rohr 2009; Knights et al. 2013).

There seem to be no visible harmful ecosystem-level effects that can be attributed to radiation in the conditions of normal operation of nuclear facilities. However, indirect responses at population and ecosystem levels are routinely observed at dose rates occurring in case of the heavy radiation accidents (Alexakhin et al. 2004; Geras'kin et al. 2008). In general, the complexity and non-linearity of the ecosystem structure and functioning could lead to unexpected consequences from stressors' effect which otherwise appear harmless when assessed within a narrower context of the traditional organism-based approach. Regulatory structures and approaches to protecting the environment from the ionizing radiation still basically are related on outcomes which cannot completely cover the risks, which are prevalent in the real ecosystems. The ecological risk assessment is a system-level problem that cannot be simplified to reductionism. Supplement of the upward idea of the reference species approach by the top-down Ecosystem Approach therefore would enhance quite seriously the scientific relevance of the methodology of radioecological risk assessment.

Conclusion

Many of the questions in the radioecology and the techniques applied to address these questions remain solidly entrenched in a reductionist philosophy. An improved understanding of the basic ecological concepts such as the species interactions, stability-diversity relationships, and the food-web structure has enhanced the ability to predict effects of radionuclides in the populations and ecosystems. One of the reasons why the ecological effects of radiation exposure might be slow to attract a great attention is the extreme level of complexity associated with the Ecosystem Approach in radioecology. Therefore, considerable advances in our understanding of fundamental ecology could still be necessary for this approach to have widespread and consistent predictive abilities.

The role of the microevolutionary processes in a population's response to the low-level chronic exposure is still not clearly understood. Natural populations can respond to the radioactive contamination not only by the enhanced level of genetic alterations and reduction in the reproduction ability but also by radioadaptation, which means the physiological acclimation or changes in the sexual, age or genetic structure of a population. As a result, man-made pollution may result in the improved resistance to pollution. However, there are radioecological situations where enhanced radioresistance does not evolve or does not persist. Therefore, a clear understanding of the exposure history has a fundamental importance to predict how populations and ecosystems will recover from the radiation exposure.

Since radioadaptation plays an important role in response of populations to radiation exposure in natural setting, this information is directly applicable for predicting a radionuclide's impacts at the population level and for ecological risk assessment. This process needs to be considered at developing management programmes designed to minimize the biodiversity loss under the conditions of chronic exposure to radionuclides. However, with few exceptions, the importance of evolution tends to be ignored both in a broader discussion about the radiation protection criteria for the environment and in a development of the new approaches for the environmental risk assessment.

An integration of basic ecological principles into the design and implementation of the radioecological research is essential for predicting radiation effects within the context of rapidly changing environmental conditions. We currently face more variable environment with greater uncertainty about how ecosystems will respond to inevitable increases in levels of human use. Human activity has resulted in simplified ecosystems that rapidly respond to external influences in an unpredictable fashion (Folke et al. 2004). Nowadays any release of radionuclides or a radiation accident, like at the Fukushima NPP, takes place at the time when many natural ecosystems have already been under pressure from the habitat destruction, invading species, and the chemical pollution. In such a situation the effects of radiation exposure may become more harmful and lead to an appearance of adverse effects at the lower levels of radioactive contamination. Indeed, abrupt changes to alternative stable states more likely occur in the contaminated ecosystem that are simultaneously subjected to effects of the global change (Clements and Rohr 2009; Noyes et al. 2009).

When disturbances are moderate, many populations and ecosystems may be able to persist through phenotypic plasticity or genetic changes. As anthropogenic influences intensify, plasticity and genetic adaptation may be pushed to their limits. As a result, loss of natural resistance caused by long-term chronic exposure may trigger a threshold response and shift the ecosystem to an alternative state (Scheffer et al. 2001). Therefore, efforts to reduce the risk of undesired shifts between ecosystem states should address the gradual changes that affect resilience rather than focus all efforts into trying to control disturbance and fluctuations. If a state shift occurs after a threshold is exceeded, recovery may not take place until radionuclides concentrations are reduced significantly below the levels that triggered the initial shift. In light of possibility of such changes and their implications for human well-being, the capacities for self-repair of ecosystems can no longer be taken for granted. Active adaptive management will be required to help sustain or create desired states of ecosystems.

Acknowledgments The present work was supported by the Russian Scientific Foundation (grant 14-14-00666).

References

Adamovich VL (1998) Rabies in animals inhabiting radioactively contaminated territories. Rus J Ecol 29:205–208

Alexakhin RM (1992) Nuclear energy and biosphere. Energoatomizdat, Moscow (Russian)

Alexakhin RM, Karaban RT, Prister BS, Spirin DA, Romanov GN, Mishenkov NN, Spiridonov SI, Fesenko SV, YeA Fyodorov, Tikhomirov FA (1994) The effects of acute irradiation on a forest biogeocenosis; experimental data, model and practical applications for accidental cases. Sci Total Environ 157:357–369

Alexakhin RM, Buldakov LA, Gubanov VA, Drozhko YeG, Ilyin LA, Kryshev II, Linge II, Romanov GN, Savkin MN, Saurov MM, Tikhomirov FA, Kholina YuB (2004) /Ilyin LA, Gubanov VA (eds) Large radiation accidents: consequences and protective countermeasures. IzdAT Publisher, Moscow

Boubriak II, Grodzinsky DM, Polischuk VP, Naumenko VD, Gushcha NP, Micheev AN, McCready SJ, Osborne DJ (2008) Adaptation and impairment of DNA repair function in pollen of *Betula verrucosa* and seeds of *Oenothera biennis* from differently radionuclide-contaminated sites of Chernobyl. Ann Bot 101:267–276

Bradshaw AD (1991) Genostasis and the limits to evolution. Phil Trans R Soc Lond 333:289–305

Bradshaw C, Kapustka L, Barnthouse L, Brown J, Ciffroy P, Forbes V, Geras'kin S, Kautsky U, Brechignac F (2014) Using an ecosystem approach to complement protection schemes based on organism-level endpoints. J Environ Radioact 136:98–104

Brechignac F, Doi M (2009) Challenging the current strategy of radiological protection of the environment: arguments for an ecosystem approach. J Environ Radioact 100:1125–1134

Bréchignac F, Bradshaw C, Carroll S, Jaworska A, Kapustka L, Monte L, Oughton D (2012) Towards an ecosystem approach for environment protection with emphasis on radiological hazards. International Union of Radioecology, Report no 7. Cadarache, France

Clements WH, Rohr JR (2009) Community responses to contaminants: using basic ecological principles to predict ecotoxicological effects. Environ Toxicol Chem 28:1789–1800

Diaz-Ravina M, Baath E (2001) Response of soil bacterial communities pre-exposed to different metals and reinoculated in an unpolluted soil. Soil Biol Biochem 33:241–248

Dmitriev AP, Grodzinskii DM, Gushcha NI, Kryzhanoskaya MS (2011) Effect of chronic irradiation on plant resistance to biotic stress in 30-km Chernobyl nuclear power plant exclusion zone. Rus J Plant Physiol 58:1062–1068

Eranen JK (2008) Rapid evolution towards heavy metal resistance by mountain birch around two subarctic copper-nickel smelters. J Evol Biol 21:492–501

Fabries M, Grauby A, Trochain JL (1972) Study of Mediterranean type of phytocenose subjected to chronic gamma radiation. Radiat Bot 12:125–135

Fesenko SV, Alexakhin RM, Geras'kin SA, Sanzharova NI, YeV Spirin, Spiridonov SI, Gontarenko IA, Strand P (2005) Comparative radiation impact on biota and man in the area affected by the accident at the Chernobyl nuclear power plant. J Environ Radioact 80:1–25

Folke C, Carpenter S, Walker B, Scheffer M, Elmqvist T, Gunderson L, Holling CS (2004) Regime shifts, resilience, and biodiversity in ecosystem management. Ann Rev Ecol Evol Syst 35:557–581

Geras'kin SA, Kim J, Oudalova AA, Vasiliyev DV, Dikareva NS, Zimin VL, Dikarev VG (2005) Bio-monitoring the genotoxicity of populations of Scots pine in the vicinity of a radioactive waste storage facility. Mutat Res 583:55–66

Geras'kin SA, Fesenko SV, Alexakhin RM (2008) Effects of non-human species irradiation after the Chernobyl NPP accident. Environ Int 34:880–897

Geras'kin SA, Oudalova AA, Dikareva NS, Spiridonov SI, Hinton T, Chernonog EV, Garnier-Laplace J (2011) Effects of radioactive contamination on Scots pines in the remote period after the Chernobyl accident. Ecotoxicol 20:1195–1208

Geras'kin SA, Oudalova AA, Dikarev VG, Dikareva NS, Mozolin EM, Hinton T, Spiridonov SI, Copplestone D, Garnier-Laplace J (2012) Effects of chronic exposure in populations of *Koeleria gracilis* Pers. from the Semipalatinsk test site, Kazakhstan. J Environ Radioact 104:55–63

Geras'kin SA, Evseeva TI, Oudalova AA (2013) Effects of long-term chronic exposure to radionuclides in plant populations. J Environ Radioact 121:22–32

Guthrie JE, Duole JR (1983) Gamma-ray irradiation of a boreal forest ecosystem: the field irradiator-gamma (FIG) facility and research programs. Can Field Nat 97:120–128

Hendry AP, Farrugia TJ, Kinnison MT (2008) Human influences on rates of phenotypic change in wild animal populations. Mol Ecol 17:20–29

Hickey DA, McNeilly T (1975) Competition between metal tolerant and normal plant populations; a field experiment on normal soil. Evolution 29:458–464

ICRP (2008) Environmental protection: the concept and use of reference animals and plants. Publication 108 Annals of ICRP Elsevier, Amsterdam, The Netherlands

Kalchenko VA, Fedotov IS (2001) Genetic effects of acute and chronic ionizing radiation on *Pinus sylvestris* L. inhabiting the Chernobyl meltdown area. Rus J Genet 37:427–447

Knights AM, Koss RS, Robinson LA (2013) Identifying common pressure pathways from a complex network of human activities to support ecosystem-based management. Ecol Appl 23:755–765

Kovalchuk I, Abramov V, Pogribny I, Kovalchuk O (2004) Molecular aspects of plant adaptation to life in the Chernobyl zone. Plant Physiol 135:357–363

Krivolutsky DA, Tikhomirov FA, Fedorov EA, Pokargevsky AD, Taskaev AI (1988) Effect of ionizing radiation on biogeocenosis. Nauka, Moscow (Russian)

Krivolutsky DA, Pokarzhevsky AD (1992) Effects of radioactive fallout on soil animal populations in the 30 km zone of the Chernobyl atomic power plant. Sci Total Environ 112:69–77

Krivolutsky DA (2000) Problems of sustainable development and ecological indication in radioactively contaminated areas. Rus J Ecol 31:257–262

Levinton JS, Suatoni E, Wallace W, Junkins R, Kelaher B, Allen BJ (2003) Rapid loss of genetically based resistance to metals after the cleanup of a Superfund site. Proc Nat Acad Sci USA 100:9889–9891

McCormick JF, Colley FB (1966) Irradiation of natural vegetation. An experimental facility procedures and dosimetry. Health Phys 12:1467–1474

Morley NJ (2012) The effects of radioactive pollution on the dynamics of infectious diseases in wildlife. J Environ Radioact 106:81–97

Noyes PD, McElwee MK, Miller HD, Clark BW, Van Tiem LA, Walcott KC, Erwin KN, Levin ED (2009) The toxicology of climate change: environmental contaminants in a warming world. Environ Int 35:971–986

Odum HT, Pigeon RF (1970) A tropical rain forest: a study of irradiation and ecology at El Verde, Puerto Rico. TID24270. United States Atomic Energy Commission, Division of Technical Information Extension, Oak Ridge, TN

Scheffer M, Carpenter S, Foley JA, Folke C, Walker B (2001) Catastrophic shifts in ecosystems. Nature 413:591–596

Scheffer M, Bascompte J, Brock WA, Brovkin V, Carpenter SR, Dakos V, Held H, van Nes EH, Rietkerk M, Sugihara G (2009) Early-warning signals for critical transitions. Nature 461:53–59

Shevchenko VA, Pechkurenkov VL, Abramov VI (1992) Radiation genetics of natural populations: genetic consequences of the Kyshtym accident. Nauka, Moscow (Russian)

Shirley MDF, Sibly RM (1999) Genetic basis of a between-environment trade-off involving resistance to cadmium in *Drosophila melanogaster*. Evolution 53:826–836

Smirnov EG, Suvorova LI (1996) Estimation and prediction of biological effects of radioactive contamination on the plant cover in the Chernobyl affected area. In: Effects of radioactive contamination on terrestrial ecosystems in the Chernobyl affected areas. Proc. Komi Res. Center Ural Branch of RAS, Syktivkar: 27–37 (Russian)

Sparrow AH (1966) Research uses of the γ-field and related radiation facilities at Brookhaven National Laboratory. Radiat Bot 6:377–405

Spirin EV, Alexakhin RM, Panchenko SV (2013) Radioecological assessment of the potential hazard of the uranium ore deposit for biota. At Energ 115:279–284

Syomov AB, Ptitsyna SN, Sergeeva SA (1992) Analysis of DNA strand break induction and repair in plants from the vicinity of Chernobyl. Sci Total Environ 112:1–8

Tikhomirov FA, Shcheglov AI (1994) Main investigation results on the forest radioecology in the Kyshtym and Chernobyl accident zones. Sci Total Environ 157:45–57

Whicker FW, Schultz V (1982) Radioecology: nuclear energy and the environment. CRC Press Ink, Boca Raton, Florida

Woodwell GM (1967) Radiation and the patterns of nature. Science 156:461–470

Zavitovski J (ed) (1977) The enterprise, Wisconsin radiation forest. Radioecological studies. Technical information for Energy Research and Development Administration. TID-26113-p 2

The Animals of Chernobyl and Fukushima

Timothy A. Mousseau and Anders P. Møller

Introduction

There have been surprisingly few rigorous tests for radiation effects on natural, wild animal populations anywhere in the world despite tremendous public interest in such questions. There are many reasons for this vacuum in knowledge but perhaps the most significant comes from the largely unsubstantiated statements by influential intergovernmental reports (e.g. IAEA 2006; UNSCEAR 2013) that the consequences of both the Chernobyl and Fukushima disasters for natural populations have been small and inconsequential. This perspective has had the effect of minimizing broader investigator (and public) interest in the topic and largely eliminating resources that might have been available to conduct the basic research needed to rigorously address such questions. In general, resources for research are only made available following shifts in societal perceptions of the potential risks and hazards of an environmental threat. An example of such shifts can be seen for the threats posed by climate change related to CO_2 emissions. Investments in research related to climate change have increased dramatically (and rightly so) over the past two decades in response to strong lobbying by both the public and the scientific communities with funding virtually absent prior to 1990 and current funding levels at more than $2.5B annually for research alone in the United States (US Government 2013; U.S. Global Change Research Program 1989). One objective of this report is to help raise awareness of the significant threats to the environment posed by nuclear accidents and to formally call for greater investment

T.A. Mousseau (✉)
Department of Biological Sciences, and the Environment and Sustainability Program, University of South Carolina, Columbia, SC 29208, USA
e-mail: mousseau@sc.edu

A.P. Møller
Laboratoire d'Ecologie, Systématique et Evolution, CNRS UMR 8079, Université Paris-Sud, Bâtiment 362, 91405 Orsay Cedex, France

V.L. Korogodina et al. (eds.), *Genetics, Evolution and Radiation*,
DOI 10.1007/978-3-319-48838-7_21

in the basic science that is desperately needed before predictions concerning the longer term risks for flora and fauna, including humans, can be generated.

Until recently, the environmental threats posed by nuclear accidents were believed to be small and the probability of occurrence of low risk. However, following the Fukushima disaster of 2011, and recent detailed analyses of overall accident frequencies (e.g. Wheatley et al. 2015), it is now believed that the chances of large accidents are far more likely than previously assumed, with another Fukushima-scale accident or larger predicted by 2065 and a Chernobyl-scale accident by 2042. If one adds to the discussion the possibility of terrorist actions involving radiological devices (e.g. atomic or simple "dirty" bombs) (Allison 2010), there are clearly very significant incentives to develop a framework for threat assessment and the scientific knowledge needed for remediation following any such event.

Despite the relative lack of funding for ecological research related to nuclear accidents, the past decade has seen significant developments concerning the understanding of the genetic and ecological consequences of radiation in the environment. Here we summarize a few of the key developments as related to genetic and ecological consequences of the Chernobyl and Fukushima disasters on wild animal populations.

Genetic Studies in Chernobyl and Fukushima

Perhaps the first test for radiation effects on nuclear DNA mutation rates for a Chernobyl population used microsatellite markers (i.e. DNA Fingerprints) to examine de novo mutation rates in barn swallows, *Hirundo rustica*, by comparing microsatellite DNA fingerprints for parents and their offspring (Ellegren et al. 1997). This study found de novo mutation rates for this marker in a model bird species to be two- to ten-fold higher in Chernobyl when compared to control populations in Ukraine and Italy, a finding that was paralleled by a study of the offspring of Chernobyl accident liquidators (Weinberg et al. 2001). Surprisingly, there have been no other similar studies to assess heritable genetic mutations related to the Chernobyl accident. However, there have been many other studies that have employed indirect techniques to assess genetic damage (for reviews see Møller and Mousseau 2006; Zakharov and Krysanov 1996; Yaklokov et al. 2009) and when taken collectively, there is little doubt that the radioactive contaminants associated with the Chernobyl disaster have generated genetic damage and increased mutation rates, with many studies also finding phenotypic effects that were correlated to the levels of genetic damage reported. For example, Table 1 of Møller and Mousseau (2006) showed 33 studies that investigated mutations or cytogenetic effects of increased radiation around Chernobyl compared with control areas in a variety of plant and animal species. There is considerable heterogeneity in the results, with 25 of the studies showing an increase in mutations or cytogenetic abnormalities. Several studies showed an increase in mutation rates for some loci, but not for others. However, many studies were based on small sample sizes, with a resulting low statistical power and thus being unable to show differences of 25 % as being statistically significant. Only four of these studies investigated germ-line mutations and these all found significant increases.

Møller and Mousseau (2015) have recently extended their studies of mutation rates in Chernobyl populations and used a meta-analysis to examine the effects of radiation in Chernobyl across 45 published studies, covering 30 species. Overall effect size, estimated as Pearson's product-moment correlation coefficient, was very large ($E = 0.67$; 95 % confidence intervals (CI) 0.59–0.73), accounting for 44.3 % of the total variance in an unstructured random-effects model (Fig. 1). Fail-safe calculations reflecting the number of unpublished null results needed to eliminate this average effect size showed the extreme robustness of this finding (Rosenberg's method: 4135 at $p = 0.05$). Indirect tests did not provide any evidence of publication bias. The effect of radiation on mutations varied among taxa, with plants showing a larger effect than animals. Humans were shown to have intermediate sensitivity of mutations to radiation compared to other species. Effect size did not decrease over time, providing no evidence for an improvement in environmental conditions. The surprisingly high mean effect size suggests a strong impact of radioactive contamination on individual fitness in current and future generations, with potentially significant population-level consequences, even beyond the area contaminated with radioactive material.

To date, there have been relatively few studies of genetic effects related to the Fukushima disaster. A recent seminal study of butterflies exposed to radioactive contaminants associated with the Fukushima disaster found strong evidence for increased mutation rates as a direct consequence of exposure to radionuclides (Hiyama et al. 2012). The study by Hiyama et al. (2012) was greatly strengthened by laboratory experiments that used both internal and external radiation sources, and these unambiguously validated observations of the elevated mutation rates and phenotypic effects observed in the field (Møller and Mousseau 2013a). Later studies by this same group (e.g. Taira et al. 2015; Hiyama et al. 2015) provided additional

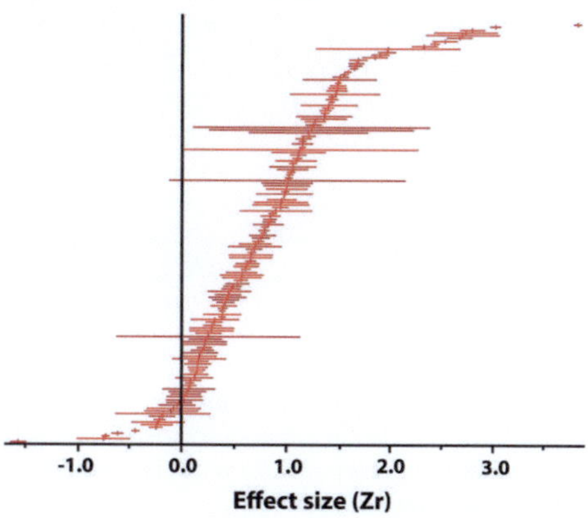

Fig. 1 Plot of the 151 effect sizes of the relationship between mutation rates and radiation from Chernobyl, ordered by increasing effect. Effect sizes are z-transformed Pearson product-moment correlation coefficient estimates, shown with the 95 % confidence intervals. The *vertical line* represents an effect size of zero. Adapted from Møller and Mousseau (2015)

support for acute and chronic effects of radiation effects, with effects decreasing over time, possibly due to reduced dose rates after several years. Of particular note was the suggestion that acquired mutations were in some cases transmitted to offspring. Collectively, these studies of butterflies provide some of the most rigorous and comprehensive analyses of chronic radiation effects in natural populations.

A recent study of barn swallows living in the Fukushima region measured dose rates to chicks in nests using TLD dosimeters and radioactivity levels of nesting materials and related this to genetic damage as measured by the frequency of single and double strand breaks in DNA of chick red blood cells (Bonisoli-Alquati et al. 2015). Radioactivity of nest samples was in the range 479 to 143,349 Bq/kg, while total external exposure varied between 0.15 and 4.9 mGy to in nests. This preliminary study found no evidence of greater genetic damage to chicks living in areas of high dose rates. Unfortunately, the range of exposure to chicks was relatively low with only two chicks found in areas of high dose rates, making this study inconclusive. There were two reasons for the lack of biological samples from areas of high radioactivity: First, barn swallow densities were very low in regions of high radiation following the disaster, and drops in bird numbers were related to ambient radiation levels. Second, during the 2012 breeding season, most researchers were not able to access the most contaminated areas because of restrictions on travel through this region.

Additional support for the hypothesis that low-dose-rate exposures can lead to elevated mutation rates comes from a recent meta-analysis of the effects of naturally occurring radioactive materials (NORM) on plant and animal populations around the world (Møller and Mousseau 2013b). This study surveyed the results from more than 5000 publications to arrive at 46 studies conducted with sufficient rigor to be included in the meta-analysis. The observed effect sizes for the 66 characters that could be extracted from these studies are presented in Fig. 2.

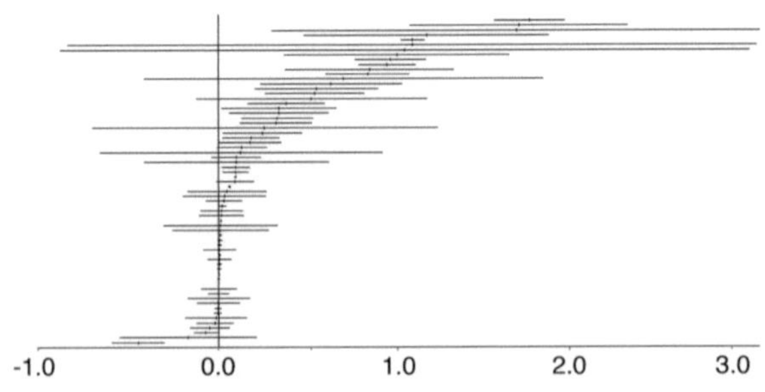

Fig. 2 Plot of the 66 effect size estimates of the relationship between level of natural background radiation and biological response variables, ordered by increasing effect size. Effect sizes are z-transformed Pearson product-moment correlation coefficient estimates (Zr), shown here with 95 % confidence intervals. *Vertical line* indicates overall mean effect size of 0.093. Adapted from Møller and Mousseau (2013b)

Although many of the individual effects were small and statistically insignificant on their own, overall there were many more that were greater than zero than expected by chance, with an overall average effect size of 0.093 (95 % CI = 0.039–0.171) indicating that exposure to naturally occurring radiation accounted for about 1 % of the variance in the traits examined. Albeit a small effect, this could still prove significant on an evolutionary time scale. The principal conclusion from this analysis was that there is extensive evidence for small, but significant negative effects of natural variation in background radiation on immune systems, mutation rates, and disease expression across a range of different animals and plants (Møller and Mousseau 2013b). Studies of naturally radioactive areas may also provide opportunities to investigate evolutionary processes of adaptation although no such studies have been conducted to our knowledge.

Developmental Effects: Albinism, Asymmetry, Brain Size, Cataracts, Sperm, and Tumors

There is an increasing array of empirical studies in Chernobyl, and now Fukushima, which document a wide range of physiological, developmental, morphological and behavioral consequences of exposure to radioactive contaminants. It is presumed that most of these effects have an underlying genetic basis although in some cases direct toxicity cannot be ruled out. Among the first visible signs of exposure were the appearance of wide spots on feathers of birds and perhaps the fur of mammals (i.e. cattle in Fukushima). These "partial albinos" (also sometimes referred to as partial leucism) have been well documented for barn swallows in Chernobyl (Ellegren et al. 1997; Møller and Mousseau 2001), and for a number of other bird species as well (Møller et al. 2013a). Barn swallows with aberrant white feathers were first detected in Fukushima in 2012 (Wild Bird Society of Japan 2013) and were observed in apparently increasing frequencies in 2013. However, such a trend could be in part related to a "screening effect" due to higher levels of scrutiny for this trait following the disaster and further investigation is needed (Mousseau et al., unpublished data; Wild Bird Society of Japan 2013). Although such partial albinos are believed to have reduced probabilities of survival (Ellegren et al. 1997; Møller et al. 2013a), there are sufficient data to suggest that this character can be inherited and may at least in part result from a mutation(s) in the germ line, based on parent-offspring resemblance (Ellegren et al. 1997).

Analysis of gametes has served as a proxy for estimates of germ line mutation rates for several species of birds in Chernobyl. Møller et al. (2004) reported that the frequency of abnormal sperm in barn swallows was up to ten times higher for Chernobyl birds as compared to sperm from males living in control areas. They also found that abnormality rates were correlated with reduced levels of antioxidants in the blood, liver and eggs of these birds, supporting the hypothesis that antioxidants likely play a significant role in protecting DNA from the direct and indirect consequences of exposure to radionuclides. And a more recent analysis of Chernobyl birds found that in nine out of 10 species examined, sperm abnormality rates were much larger for birds living in Chernobyl than those living in control areas across Europe, with the highest damage levels observed for species with longer sperm (Hermosell et al. 2013), suggesting that sperm abnormalities are likely common for birds living in radioactive areas. Similar effects on sperm morphology of small rodents have recently been reported (Kivisaari et al. 2016). Møller et al. (2008) found that barn swallow sperm swimming ability was negatively related to radiation levels while Bonisoli-Alquati et al. (2011) found that plasma oxidative status could predict sperm performance, further supporting the role antioxidants are known to play in protecting spermatogenesis from the effects of ionizing radiation. Overall these studies provide convincing evidence that spermatogenesis can be significantly impacted by low-dose radiation and the resulting male infertility may in part explain the smaller population sizes of many species that have been documented for the region (see below).

A recent study of bull sperm and testis from the Fukushima region found no evidence for significant histological changes in the testes or sperm morphology (Yamashiro et al. 2013) although this study was very preliminary with only two bulls from a relatively uncontaminated part of Fukushima represented for the analysis of sperm.

Many other cell types and tissues have been shown to be affected by Chernobyl contaminants. Møller et al. (2013a) demonstrated that the frequency of visible tumors on birds was significantly higher in radioactive areas, presumably reflecting elevated mutation rates in somatic tissues. Visible tumor rates in birds from Chernobyl were in excess of 15/1000 birds while tumors have never been reported for Danish populations despite extensive surveys (0/35,000 birds observed) (Møller et al. 2013a). In a recent survey of rodents from Chernobyl (Mappes et al., unpublished data), the frequency of liver tumors appeared related to radiation dose as measured by whole body burdens of ^{137}Cesium.

Similarly, the frequency and magnitude of cataract expression in eyes was related to radiation exposure: birds from areas with high background radiation were more likely to display opacities in one or both eyes (Mousseau and Møller 2013). As with radiation-related cataract in humans (e.g. Worgul et al. 2007), there was no relationship with the age of the birds, supporting the hypothesis that radiation was the underlying cause of cataract expression. Lehman et al. (2016) recently reported significantly increased rates of cataracts in rodents living in radioactively contaminated regions of Ukraine providing additional support for the use of cataract incidence as a reliable biomarker for exposure to ionizing radiation.

Other developmental effects that have been reported include reduced brain size in birds, which is believed to reflect the effects of ionizing radiation on developing neural tissues (Møller et al. 2011), unusual feather shapes and sizes (Møller and Mousseau 2003), and abnormal growth formations on feet and beaks (Møller et al. 2007, 2013a). Brain size effects have also been suggested for rodents living in both Chernobyl and Fukushima regions of high radioactivity (Mappes et al., unpublished). Brain size effects are thought to be related to the relatively high sensitivity of neural tissue to oxidative stress.

Population Abundances and Biodiversity in Regions of High Radiation

A key issue for conservation biologists concerns the fitness consequences of mutation accumulation and resulting developmental effects that have been observed for wild populations living in Chernobyl and Fukushima. To this end, we have conducted demographic studies aimed at documenting population sizes, numbers of species (i.e. biodiversity), sex ratios, survival and reproductive rates, and patterns of immigration for animals in both Chernobyl and Fukushima. Because of the highly heterogeneous nature of radionuclide deposition inside the contaminated regions of Chernobyl and Fukushima, it is possible to identify areas that represent the full spectrum of radiation levels, from relatively "clean" uncontaminated habitats all the way to large areas of very high radiation levels, all within short geographical distances. This heterogeneity makes it possible to conduct highly replicated tests for the effects of radiation on biological populations and communities for a single large scale event. In effect, the distribution of radioactive contaminants, especially in Chernobyl, is more akin to a mosaic or quilt work than diffusion from a point, allowing the uncoupling of radiation levels from distance from the source. It is this lack of geographic structure for radiation levels when combined with multiple tests for radiation effects across multiple habitat types that permits a sensitive analysis of radiation effects independent of other biotic and abiotic factors.

Abundance and Diversity of Birds, Butterflies and Other Invertebrates

Comprehensive surveys of animal abundance and diversity in Chernobyl were conducted by Møller and Mousseau starting in the mid-2000s. The basic sampling protocol was a "massively replicated biotic inventory" design whereby point counts of birds and invertebrates (chiefly, butterflies, dragonflies bees, grasshoppers and spiders were conducted at about 300 locations across northern Ukraine and southeastern Belarus in 2006–2008. An identical protocol was used to conduct surveys in Fukushima at 400 distinct locations in 2011–2014. To date, a total of 896 and 1500 biotic inventories have been generated for the Chernobyl and Fukushima regions, respectively. In addition to quantitative estimates of animal abundances and species diversity at each site, a large number of additional biotic and abiotic factors were measured or estimated including the type of vegetation, the distance to open water, soil type, ambient meteorological conditions, latitude, longitude, elevation, and time of day. All of these variables were included in a multivariate model and used to generate predictions for expected numbers of organisms of each species or group for each location. This model was then used to provide estimates of the variation in numbers explained by radiation independent of all the other potentially contributing factors, in essence, a partial relationship between abundance and radiation. To our knowledge, this approach has not been used in this way before to assess radioactively contaminated areas although it has been used for monitoring bird populations in Europe and North America since the 1960s. This approach is perhaps the only solution for complex ecological questions of this type, short of large scale experimental manipulations, which are generally not possible for testing the effects of nuclear fission products at a landscape scale (although see Odum and Pigeon (1970) for an overview of a large-scale experimental study of radiation effects). This approach has the added advantage of permitting assessment of ecological effects even in the absence of pre-disaster baseline data as it uses contemporary observations of distribution and abundance from unaffected areas to infer expected patterns in contaminated areas.

Contrary to popular notions, in 2006–2009 the abundance and diversity of forest and grassland birds in Chernobyl were dramatically lower in contaminated areas, showing a dose-response-like relationship, with about one-third as many birds and half as many species present in high contamination areas relative to that predicted by the models and abundances found in relatively "clean" parts of the same general region (Møller and Mousseau 2007a, b, 2011a, b). Although not every species showed declines with radiation levels, and a few even appeared to be unaffected (Galvan et al. 2014) and perhaps showed evolutionary adaptation to radiation, the overall patterns of decline were very apparent and the analyses were statistically robust. Even birds of prey showed patterns of reduced numbers in contaminated regions of Chernobyl, although it was not apparent if reduced numbers were a consequence of direct exposure to radionuclides via ingestion or via indirect effects on behavior mediated by reduced prey (Møller and Mousseau 2008; Fig. 3). In addition to population censuses, there are other lines of evidence supporting the observed declines in population sizes of birds in Chernobyl including changes in adult sex ratios (more males than females), and reductions by half in the number of older birds relative to juveniles and one-year-olds (Møller et al. 2012b). In addition, there was evidence from analyses of stable isotopes in feathers that the Chernobyl region is acting like a population sink with a higher proportion of immigrants present than in control areas or when compared to birds in historical museum collections from the same area (Møller et al. 2006).

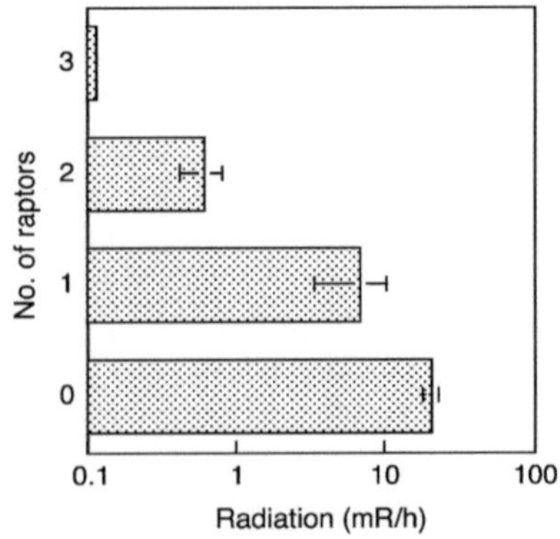

Fig. 3 Abundance of birds of prey during point counts in relation to ambient levels of radiation [mR/h, mean (SE)] at point of observation. Adapted from Møller and Mousseau (2008)

The overall pattern was very similar for birds in Fukushima in July of 2011, with the strength of the negative relationship between abundance and radiation significantly stronger in Fukushima when comparing the 14 bird species that were common to both regions (Møller et al. 2012a, 2013b). The observed stronger relationship in Fukushima could reflect the difference between acute and chronic exposures, with Chernobyl bird populations showing a response to 20 + years of selection for resistance, or this could reflect the effects of other radionuclides (e.g. I-131 and Cs-134) that were present at high levels in Fukushima during the spring of 2011 that are no longer present in Chernobyl.

Field studies in Fukushima were also conducted in 2012–2014, and the initial analyses showed a strengthening of the negative relationship between ambient radiation levels and abundance and species richness at a given site over time (Møller et al. 2015a, b; Fig. 4). Although no comprehensive surveys of raptors in Japan have yet been conducted, a recent study of goshawk (*Accipiter gentilis Fujiyama*) has reported significant declines in reproduction for this bird of prey in

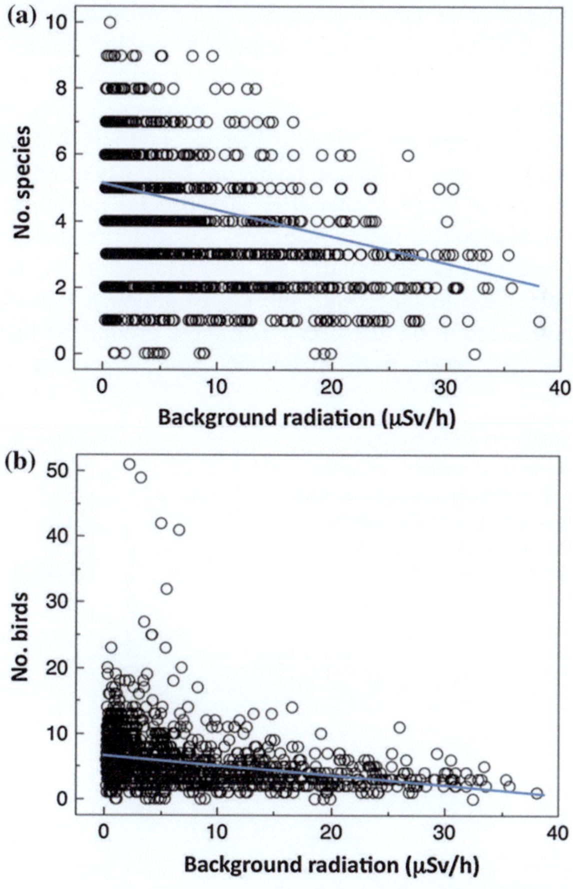

Fig. 4 The relationship between ambient background radiation and (**a**) species richness and (**b**) abundance at a given site in Fukushima 2011–2014. Both species richness and total abundances show significant drop-offs in areas of high radiation. Adapted from Møller et al. (2015b)

Fukushima following the disaster (Murase et al. 2015) although only three study areas were included in the analysis and thus attribution of the observed effect to radionuclide exposure is preliminary. Surveys of barn swallows (*Hirundo rustica*) showed significant drop-offs in abundance in the more radioactive regions of Fukushima although preliminary analyses did not indicate any relationship with genetic damage to blood cells in nestlings (Bonisoli-Alquati et al. 2015).

In the first study of its sort, Garnier-Laplace et al. (2015) calculated dose rates for 57 species of birds (almost 7000 individuals) living in Fukushima following the nuclear disaster of March 11, 2011 that were surveyed by Møller et al. (2015a, b). Doses were calculated based on radiological conditions at the point of observation and corrected for by including ecological and life history attributes of the species in the model. Dose was used to predict total number of birds while statistically controlling for potentially confounding environmental variables (e.g. habitat type, elevation, presence of water bodies; ambient meteorological conditions, time of day). Total dose was found to be a strong predictor of abundances ($P < 0.0001$), which showed a proportional decline with increasing doses with no indications of a threshold or intermediate optimum (Fig. 5). Overall, the $ED_{50\ \%}$ (the total absorbed dose causing a 50 % reduction in the total number of birds) was estimated to only be 0.55 Gy. Additional important studies of the agricultural importance of radionuclide distribution in Fukushima can be found in Nakanishi and Tanoi (2013).

It is interesting to note that as a group, butterflies also showed significant declines with radiation levels in both Chernobyl and Fukushima (Møller and Mousseau 2009; Møller et al. 2013b). One might speculate that there is something peculiar about the female ZW sex determination system shared by birds and Lepidoptera (i.e. females are heterogametic) that make these groups particularly vulnerable to mutagenic substances although this idea remains to be tested. One notion might be that because in these groups the heterogametic sex is responsible for egg production, mutational load effects on reproduction stemming from

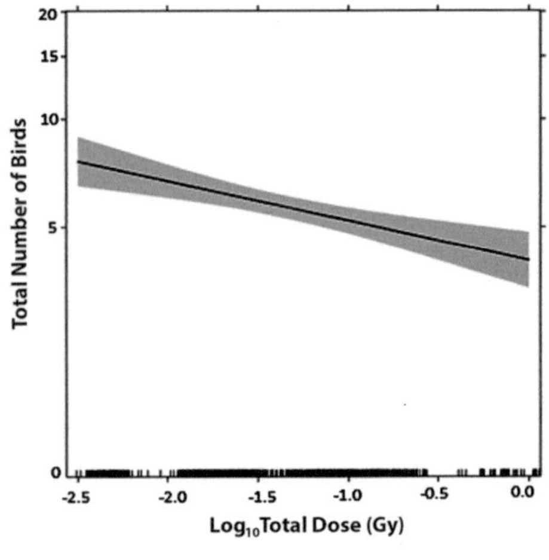

Fig. 5 The relationship between abundance of birds and reconstructed dose at 300 locations across Fukushima Prefecture in Japan, 2011–2014. Adapted from Garnier-Laplace et al. (2015)

mutation accumulation on the Z chromosome are likely to be expressed immediately following exposure as opposed to species where females are homogametic, as is the case for most sexually reproducing organisms. This might be particularly important given the apparent lack of dosage compensation in birds and Lepidoptera (Parsh and Ellegren 2013). In addition, slightly deleterious mutations may accumulate faster on sex chromosomes than on autosomes (Parsh and Ellegren 2013), and this could be at least in part responsible for the observed greater sensitivity to radiation observed in birds and Lepidoptera.

In most other invertebrate groups examined (e.g. grasshoppers, dragonflies, bees, spiders), population sizes were significantly reduced in areas of high contamination in Chernobyl 20 + years after the disaster (Fig. 6) while there was no evidence for similar declines in Fukushima and in fact spiders showed significant increases in numbers, at least during the first summer following the disaster (Møller and

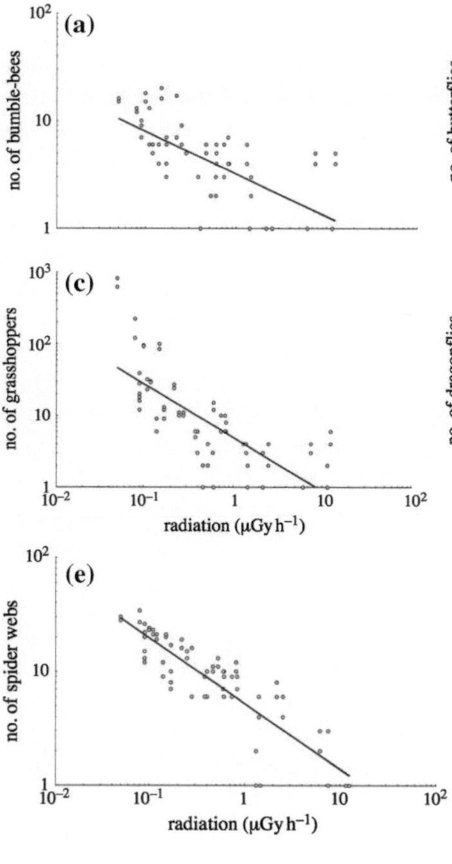

Fig. 6 Abundance of (**a**) bumble-bees, (**b**) butterflies, (**c**) grasshoppers, (**d**) dragonflies and (**e**) spider webs estimated during line transects in July 2008 in relation to background radiation (mGy h) around Chernobyl, Ukraine and Belarus. The lines are the linear regression lines. Adapted from Møller and Mousseau (2009)

Mousseau 2009; Møller et al. 2013b). It has been proposed that such differences in the time course for population effects might reflect the consequences of multi-generational mutation accumulation of recessive deleterious mutations in Chernobyl (Møller and Mousseau 2011a), which is also consistent with the immediate effects on birds and butterflies observed in Fukushima. Alternatively, increases in spider numbers could simply reflect a reduction in predation pressure (e.g. birds), a finding similar to that reported for large mammals living in the Chernobyl zone (Deryabina et al. 2015) where the lack of hunting pressure has been associated with increased population sizes in some species.

Predicting Population Responses from Physiology and Historical Base-Pair Substitution Rates

Although there is an overall pattern of decline in animal numbers in direct relation to contamination levels, there is tremendous variability among species in their apparent sensitivity to radionuclides. Many species show no significant relationship between contamination and abundance and a few species even increased with rising radiation levels. Some of the variance in apparent sensitivity can be explained by differences in life history, physiology, and behavior. For example, for birds in Chernobyl, long distance migrants, brightly colored species, and species feeding on invertebrates in the soil, showed the strongest negative responses to radiation (Møller and Mousseau 2007b). A recent analysis also found that bird species that used pheomelanin- and carotenoid-based coloration were also particularly sensitive to radionuclides (Galván et al. 2011). These findings point to a critical role played by antioxidants in defending against oxidative stress, and a likely physiological trade-off between biochemical precursors that are used for both coloration and as antioxidants. It remains to be seen if interspecific variability in population declines in Fukushima will follow a similar pattern. But overall, the interactions between antioxidant availability and allocation, and oxidative stress, appear to shape sensitivity and vulnerability of many organisms to the effects of ionizing radiation (Einor et al. 2016).

Recent evidence suggests that DNA repair may be involved in determining sensitivity to the mutagenic properties of radionuclides. A recent analysis found a significant relationship between the strength of population declines of a given species with radiation and historical mtDNA substitution rates for 32 species of birds in Chernobyl (Møller et al. 2010). Species with higher substitution rates showed the greatest declines with radiation levels suggesting that variation in DNA repair capability may be influencing population success although this hypothesis remains to be tested experimentally.

In conclusion, the radiological disasters at Chernobyl and Fukushima provide a unique opportunity to investigate genetic, ecological and evolutionary consequences of acute and chronic exposures to mutagenic sources in natural populations at regional and landscape scales. Recent advances suggest many small and large effects on biological systems from molecules to ecosystems that will likely influence ecosystem form and function for decades to centuries to come. Recent surveys of population effects in Chernobyl (Garnier-Laplace et al. 2012) suggest that populations living under the full range of natural stressors (biotic and abiotic) are almost 10 times more sensitive to ionizing radiation than predicted by conventional approaches used by some regulatory and governmental agencies (e.g. UNSCEAR 2013), thus providing some potential insights to the cause of the apparent discrepancy between empirical ecological studies and predictions from conventional radio-ecological models of radiation effects. The opportunity to compare and contrast organisms from both Chernobyl and Fukushima provides for a possible level of scientific rigor (i.e. replication) not previously available for studies of this sort, as well as analysis of the timeframe over which responses may occur and the development of predictive models to aid the management and conservation of biological systems following future nuclear accidents. Given recent advances in molecular genetic technologies, it seems likely that much new knowledge could be gained from a sustained and expansive investment in basic research related to the biological effects of radioactive mutagens within an ecosystem context that could extend far beyond the disasters at Fukushima and Chernobyl.

Acknowledgments We gratefully acknowledge the help and support of V. Bezrukov, O. Bondarenko, A. Bonisoli-Alquati, Z. Boratynski, I. Chizhevsky, N. Evangeliou, MA. Fitzpatrick, E. Flensted-Jensen, I. Galvan, S. Gaschak, D. Grodzinsky, H. Hagiwara, S. Jenkinson, S. Kasahara, K. Kawai, H. Kawai, K. Kawatsu, K. Koyama, I. Kozeretska, A. Kravets, A. Litvinchuk, T. Mappes, W. Mardal, L. Milinevski, S. Mitsui, B. Murray, M. Naboka, P. Nagarkatti, I. Nishiumi, J. Palms, H. Pastides, A. Peklo, E. Pysanets, A. Rozhok, G. Rudolfsen, S. Ruskovski, N. Saino, D.V. Shestopolov, H. Smith, H. Suzuki, K. Ueda, M. Wyatt, and many others. A very special thank you goes to T. Kanagawa and G. Milinevsky for their commitment and support to our research in Fukushima and Chernobyl.

References

Allison GT (2010) Nuclear terrorism: the ultimate preventable catastrophe. Defense Against Terrorism Rev 3:97–106

Bonisoli-Alquati A, Møller AP, Rudolfsen G, Saino N, Caprioli M, Ostermiller S, Mousseau TA (2011) The effects of radiation on sperm swimming behavior depend on plasma oxidative status in the barn swallow (Hirundo rustica). Comp Biochem Phys Part A Mol Integr Physiol 159(2):105–112. doi:10.1016/j.cbpa.2011.01.018

Bonisoli-Alquati A, Koyama K, Tedeschi DJ, Kitamura W, Suzuki H, Ostermiller S, Arai E, Møller AP, Mousseau TA (2015) Abundance and genetic damage of barn swallows from Fukushima. Sci Rep 5:9432. doi:10.1038/srep09432

Deryabina TG, Kuchmel SV, Nagorskaya LL, Hinton TG, Beasley JC, Lerebours A, Smith JT (2015) Long-term census data reveal abundant wildlife populations at Chernobyl. Curr Biol 25: R824–R826. doi:10.1016/j.cub.2015.08.017

Einor D, Bonisoli-Alquati A, Costantini D, Mousseau TA, Møller AP (2016) Ionizing radiation, antioxidant response and oxidative damage: a meta-analysis. Sci Total Environ 548–549:463–471

Ellegren H, Lindgren G, Primmer CR, Møller AP (1997) Fitness loss and germline mutations in barn swallows breeding in Chernobyl. Nature 389:593–596

Galvan I, Mousseau TA, Møller AP (2011) Bird population declines due to radiation exposure at Chernobyl are stronger in species with pheomelanin-based coloration. Oecologia 165(4):827–835. doi:10.1007/s00422-010-1860-5

Galvan I, Bonisoli-Alquati A, Jenkinson S, Ghanem G, Wakamatsu K, Mousseau TA, Møller AP (2014) Chronic exposure to low-dose radiation at Chernobyl favours adaptation to oxidative stress in birds. Funct Ecol 28:1387–1403. doi:10.111/1365-2435.12283

Garnier-Laplace J, Geras'kin S, Della-Vedova C, Beaugelin-Seiller K, Hinton TG, Real A, Oudalova A (2012) Are radiosensitivity data derived from natural field conditions consistent with data from controlled exposures? A case study of Chernobyl wildlife chronically exposed to low dose rates. J Environ Radioact doi:10.1016/j.jenvrad.2012.01.013

Garnier-Laplace J, Beaugelin-Seiller K, Della-Vedova C, Métivier JM, Ritz C, Mousseau TA, Møller AP (2015) Radiological dose reconstruction for birds reconciles outcomes of Fukushima with knowledge of dose-effect relationships. Sci Rep 5

Hermosell IG, Laskemoen T, Rowe M, Møller AP, Mousseau TA, Albrecht T, Lifjeld JT (2013) Patterns of sperm damage in Chernobyl passerine birds suggest a trade-off between sperm length and integrity. Biol Lett 9(5):20130530. doi:10.1098/rsbl.2013.0530

Hiyama A, Nohara C, Kinjo S, Taira W, Gima S, Tanahara A, Otaki JM (2012) The biological impacts of the Fukushima nuclear accident on the pale grass blue butterfly. Sci Rep 2:570. doi:10.1038/srep00570

Hiyama A, Taira W, Nohara C, Iwasaki M, Kinjo S, Iwata M, Otaki JM (2015) Spatiotemporal abnormality dynamics of the pale grass blue butterfly: three years of monitoring (2011–2013) after the Fukushima nuclear accident. BMC Evol Biol 15:15. doi:10.1186/s12862-015-0297-1

International Atomic Energy Agency (IAEA) (2006) Chernobyl's Legacy: Health, Environmental and Socio-Economic Impacts and Recommendations to the Governments of Belarus, the Russian Federation and Ukraine. The Chernobyl Forum: 2003–2005, 2nd revised version, Vienna, available at: http://www.iaea.org/Publications/Booklets/Chernobyl/chernobyl.pdf

Kivisaari K, Boratynski Z, Calhim S, Lehmann P, Mappes T, Mousseau TA, Møller AP (2016) Cut to the chase: radiation effects on sperm structure at Chernobyl (In review)

Lehmann P, Boratynski Z, Mappes T, Mousseau TA, Møller AP (2016) Cataract frequency, fecundity and background radiation in natural mammalian populations from Chernobyl. Sci Rep 6:19974

Møller AP, Mousseau TA (2001) Albinism and phenotype of barn swallows *Hirundo rustica* from Chernobyl. Evolution 55(10):2097–2104

Møller AP, Mousseau TA (2003) Mutation and sexual selection: a test using barn swallows from Chernobyl. Evolution 57:2139–2146

Møller AP, Mousseau TA (2006) Biological consequences of Chernobyl: 20 years after the disaster. Trends Ecol Evol 21:200–207

Møller AP, Mousseau TA (2007a) Species richness and abundance of forest birds in relation to radiation at Chernobyl. Biol Lett 3:483–486

Møller AP, Mousseau TA (2007b) Determinants of interspecific variation in population declines of birds after exposure to radiation at Chernobyl. J Appl Ecol 44:909–919

Møller AP, Mousseau TA (2008) Reduced abundance of raptors in radioactively contaminated areas near Chernobyl. J Ornithol 150(1):239–246

Møller AP, Mousseau TA (2009) Reduced abundance of insects and spiders linked to radiation at Chernobyl 20 years after the accident. Biol Lett R Soc 5(3):356–359

Møller AP, Mousseau TA (2011a) Conservation consequences of Chernobyl and other nuclear accidents. Biol Conserv 144:2787–2798

Møller AP, Mousseau TA (2011b) Efficiency of bio-indicators for low-level radiation under field conditions. Ecol Indic 11(2):424–430. doi:10.1016/j.ecolind.2010.06.013

Møller AP, Mousseau TA (2013a) Low-dose radiation, scientific scrutiny, and requirements for demonstrating effects. BMC Biol 11(92). doi:10.1186/1741-7007-11-92

Møller AP, Mousseau TA (2013b) The effects of natural variation in background radioactivity on humans, animals and other organisms. Biol Rev 88:226–254

Møller AP, Mousseau TA (2015) Strong effects of ionizing radiation from Chernobyl on mutation rates. Sci Rep 5:8363. doi:10.1038/srep08363

Møller AP, Surai P, Mousseau TA (2004) Antioxidants, radiation and mutations in barn swallows from Chernobyl. Proc R Soc Lond B 272:247–252

Møller AP, Hobson KA, Mousseau TA, Peklo AM (2006) Chernobyl as a population sink for barn swallows: tracking dispersal using stable isotope profiles. Ecol Appl 16:1696–1705

Møller AP, Mousseau TA, de Lope F, Saino N (2007) Elevated frequency of abnormalities in barn swallows from Chernobyl. Biol Lett 3:414–417

Møller AP, Mousseau TA, Lynnn C, Ostermiller S, Rudolfsen G (2008) Impaired swimming behavior and morphology of sperm from barn swallows *Hirundo rustica* in Chernobyl. Mutat Res, Genet Toxicol Environ Mutagen 650:210–216

Møller AP, Erritzøe J, Karadas F, Mousseau TA (2010) Historical mutation rates predict susceptibility to radiation in Chernobyl birds. J Evol Biol 23(10):2132–2142. doi:10.1111/j.1420-9101.2010.02074.x

Møller AP, Bonisoli-Alquati A, Rudolfsen G, Mousseau TA (2011) Chernobyl birds have smaller brains. PLOS One 6(2): Art. No. e16862. doi:10.1371/journal.pone.0016862

Møller AP, Hagiwara A, Matsui S, Kasahara S, Kawatsu K, Nishiumi I, Suzuki H, Ueda K, Mousseau TA (2012a) Abundance of birds in Fukushima as judged from Chernobyl. Environ Pollut 164:36–39

Møller AP, Bonisoli-Alquati A, Rudolfsen G, Mousseau TA (2012b) Elevated mortality among birds in Chernobyl as judged from biased sex and age ratios. PLoS ONE 7(4):e35223

Møller AP, Bonisoli-Alquati A, Mousseau TA (2013a) High frequencies of albinism and tumours in free-living birds at Chernobyl. Mutat Res 757:52–59

Møller AP, Nishiumi I, Suzuki H, Ueda K, Mousseau TA (2013b) Differences in effects of radiation on abundance of animals in Fukushima and Chernobyl. Ecol Indic 14:75–81

Møller AP, Mousseau TA, Nishiumi I, Ueda K (2015a) Ecological differences in response of bird species to radioactivity from Chernobyl and Fukushima. J Ornithol. doi:10.107/s10336-015-1173-x

Møller AP, Nishiumi I, Mousseau TA (2015b) Cumulative effects on interspecific differences in response of birds to radioactivity from Fukushima. J Ornithol 156:S287–S296. doi:10.1007/s10336-015-1197-2

Mousseau TA, Møller AP (2013) Elevated frequencies of cataracts in birds from Chernobyl. PLoS ONE 8(7):e66939. doi:10.1371/journal.pone

Murase K, Murase J, Horie R, Endo K (2015) Effects of the Fukushima Daiishi nuclear accident on goshawk reproduction. Sci Rep 5:9405. doi:10.1038/srep09405

Nakanishi TM, Tanoi K (2013) Agricultural implications of the Fukushima nuclear accident. Springer, Tokyo, Japan

Odum HT, Pigeon RF (1970) A tropical rain forest: a study of irradiation and ecology at El Verde, Puerto Rico. U.S. Atomic Energy Commission, Division of Technical Information, Oak Ridge, TN

Parsh J, Ellegren H (2013) The evolutionary causes and consequences of sex-biased gene expression. Nat Rev Genet 14:83–87

Taira W, Hiyama A, Nohara C, Sakauchi K, Otakai JM (2015) Ingestional and transgenerational effects of the Fukushima nuclear accident on the pale grass blue butterfly. J Radiat Res 2015:1–17. doi:10.1093/jrr/rrv068

United States (US) Government (2013) Fiscal year 2014 report to congress on federal climate change expenditures. Washington D.C.: US Government. Available from: http://www.whitehouse.gov/sites/default/files/omb/assets/legislative_reports/fcce-report-to-congress.pdf

UNSCEAR (2013) United Nations Scientific Committee on the Effects of Atomic Radiation. Rep 1: sources, effects, and risks of ionizing radiation. Report to the General Assembly, Scientific Annex A

U.S. Global Change Research Program (USGCRP) (1989) Our Changing Planet: The FY 1990 Research Plan. Executive Summary. Washington, DC, USA

Weinberg HS, Korol AB, Kirzhner VM, Avivi A, Fahima T, Nevo E, Shapiro S, Rennert G, Piatak O, Stepanova EI, Skvarskaja E (2001) Very high mutation rate in offspring of Chernobyl accident liquidators. Proc R Soc Lond B Biol Sci 268:1001–1005

Wheatley S, Sovacool B, Sornette D (2015) Of disasters and dragon dings: a statistical analysis of nuclear power incidents and accidents. arXiv preprint arXiv:1504.02380

Wildbird Society of Japan (2013) Yachō, 777: August, 2013. ISSN 0910-4488, pp 40–34

Worgul BV, Kundiyev YI, Sergiyenko NM, Chumak VV, Vitte PM, Medvedovsky C, Bakhanova EV, Junk AK, Kyrychenko OY, Musijachenko NV, Shylo SA, Vitte OP, Xu S, Xue X, Shore RE (2007) Cataracts among Chernobyl clean-up workers: implications regarding permissible eye exposures. Radiat Res 167:233–243

Yablokov AV, Nesterenko VB, Nesterenko AV (2009) Chernobyl: consequences of the catastrophe for people and the environment. Ann NY Acad Sci 1181: 327

Yamashiro H, Abe Y, Fukuda T, Kino Y, Kawaguchi I, Kuwahara Y, Fukumoto M, Takahashi S, Suzuki M, Kobayashi J, Uematsu E, Tong B, Yamada T, Yoshida S, Sato E, Shinoda H, Sekine T, Isogai E, Fukumoto M (2013) Effects of radioactive caesium on bull testes after the Fukushima nuclear plant accident. Sci Rep 3:2850. doi:10.1038/srep02850

Zakharov V, Krysanov E (1996) Consequences of the Chernobyl catastrophe: environmental health. Center for Russian Environmental Policy. Moscow, 160p

Viability of Plant Seed Progeny from the East-Ural Radioactive Trace: Radiation and Weather Conditions

Elena V. Antonova, Vera N. Pozolotina and Elina M. Karimullina

Introduction

Nikolay Timofeev-Ressovsky (1962) has formulated a proposition: the properties of living systems are determined by the integration of processes and phenomena in the three organization levels: at the molecular and cellular levels, the genetic information is recorded and can be changed by irradiation; at the ontogenetic level, it is implemented in the phenotypes; at the population level, it can be transformed by means of natural selection.

Our knowledge in this field has been expanded significantly in recent years but there is a disjoint between the molecular damage and concomitant effects at the individual and population levels of biological organization (Hinton and Brechignac 2005). New mechanisms of regulation at every level of the organization were revealed. In addition to genetic damages, there are mechanisms of phenotypic and genotypic regulations (Boyko and Kovalchuk 2011). At the ontogenetic level, there are many physiological mechanisms of regulation that could change the final results (Calow and Forbes 1998). At the population level, we must take into account some important regulations: (i) species traits; (ii) all types of interpopulation variability: individual, age, sexual variations; (iii) environmental factors that can change the radiation effects of biological systems. These factors are divided into biotic ones like predation, pathogens or parasites (Qin et al. 2011) and abiotic factors such as temperature, precipitations and so on (Holmstrup et al. 2010). Environmental factors often exert a physiological background stress that can modify the effects of pollutants, resulting in so-called "multiple stressor effects" (Crain et al. 2008; Fischer et al. 2013).

E.V. Antonova (✉) · V.N. Pozolotina · E.M. Karimullina
Institute of Plant and Animal Ecology UB RAS, Ekaterinburg, Russia
e-mail: selena@ipae.uran.ru

V.L. Korogodina et al. (eds.), *Genetics, Evolution and Radiation*,
DOI 10.1007/978-3-319-48838-7_22

There are many aspects in the problem of detection of the chronic low-level radiation exposure effects in natural populations. Yet in this paper we shall look at just one aspect which is the evaluation of the multi effects of radiation and weather conditions within certain years (time-dependent variability) on the reproductive function of several plant species.

The purpose of this investigation was the long-term study of variability of the seed progeny quality of some herbaceous plants from the East-Ural Radioactive Trace and background populations taking into account the fluctuation of weather conditions.

Sites of Investigation

The East-Ural Radioactive Trace (EURT) was formed in 1957 as a result of the accident at the PA "Mayak" (Russia). About 74 PBq of radioactive substances were released into the atmosphere. The ^{90}Sr isotope was main pollutant. In 1967, the EURT area was contaminated again with airborne radioactive sediments from the Lake Karachay shores. Lake Karachay was used by the PA "Mayak" for dumping liquid radioactive waste. In this case, ^{137}Cs was the main contaminant (Romanov et al. 1990; Aarkrog et al. 1997). At the present time, the density of contamination decreases within the central axis of EURT at a distance of 4–35 km from epicenter: for ^{90}Sr—from 69,000 to 400 kBq m^{-2}; for ^{137}Cs—from 1500 to 20 kBq m^{-2}; for ^{239,240}Pu—from 200 to 0.7 kBq m^{-2} (Pozolotina et al. 2012; Molchanova et al. 2014). The contamination levels decrease in the directions from the central axis to the peripheries by two to three orders of magnitudes (Mikhailovskaia et al. 2011). The present state of terrestrial ecosystems within the contaminated area is highly appreciated (Pozolotina et al. 2012). Plant communities of the EURT passed the main stages regenerative succession, species richness is high. For our investigations, we have chosen some impact plots located within the central axis (6–18 km from epicenter) and some buffer plots adjacent to the impact area from the western and northern sides (15–50 km).

Objects and Sample Collection

Populations exist both in space and in time, therefore, it is necessary to assess not only the survival of organisms, but also their fertility. Seed progeny of herbaceous plants is a convenient model for assessing the consequences of radiation effects on plant populations within areas of radioactive contamination. We investigated populations of brome (*Bromopsis inermis* (Leyss.) Holub), motherwort (*Leonurus quinquelobatus* Gilib.), starweed (*Stellaria graminea* L.), campion (*Melandrium album* (Mill.) Garcke) and plantain (*Plantago major* L.) from the most contaminated area of the EURT and background territories. *M. album* is an annual or biennial plant, monocarpic, cryptophyte. The other species are perennial, polycarpic plants, cryptophytes, mesophytes. The seeds of these species were collected from 40–50 plants within each impact, buffer and background populations in period from 2002 to 2014.

Dose Assessment

We assessed the absorbed doses for herbaceous plants from EURT by Erica Tool (Tier 2) using our own data of ^{90}Sr and ^{137}Cs activity concentrations in plants and soils. The total dose rates for plants from buffer area exceed the background values 1.5–22.2 (maternal plants) and 1.1–9.6 times (seeds), and from impact one 11.7–287.6 and 4.7–149.3 times accordingly (Table 1). These values are considered to be within the low levels of radiation exposure for plant species.

The detailed description of the models and transfer parameters of assessing radiation doses for different plant species cited in these articles (Karimullina et al. 2013; Antonova et al. 2014).

Evaluation of Seed Progeny Viability and Weather Conditions

Seeds of different species were germinated in roll culture at +24 °C using 12 h light/dark cycle for a period of 3 weeks. The variations of the seed progeny viability were evaluated by means of seedlings survival. All experiments were repeated 3 or 4 times, and there were a total of 49,772 seeds (from 25 to 100 seeds per each experimental container).

Weather conditions of different seasons were evaluated using monthly average temperatures, sum of temperatures of month, sum of effective temperatures, sum of precipitation, sum of precipitation at effective temperatures and aridity index (*I*) by Selyaninov (Oliver 2005). The effective temperature is the daily temperature above +10 °C. Meteorological data were obtained from two weather stations located close to background populations (#28440, 56°48′N, 60°38′E) and EURT areas (#28541, 56°05′N, 60°18′E).

Table 1 Total dose rates (μGy h^{-1}) for investigated plant species taking into account the natural background radiation (0.10 μGy h^{-1})

Populations	Maternal plants	Seed embryos
Background	0.102–0.105	0.1003–0.1005
Buffer	0.153–2.33	0.112–0.96
Impact	1.19–30.20	0.467–15.0

Statistical Analysis

MANOVA was used for statistical hypothesis testing, with Fisher's (*F*) multiple criteria, and the linear correlation analysis (*R*) in the software package STATISTICA 8.0 (Weiß 2007). In addition, the method by (Newcombe 1998) and (Wilson 1927) was used for independent samples.

Results

We have summarized our long-term observations indicating trends of the seed progeny quality. Table 2 shows the data of comparison of quality of seeds from the EURT and background samples. Down arrow indicates that the survival rate of seedlings from EURT was lower than that from background populations. Up arrow indicates that this parameter of the EURT seedlings was higher than that in the background samples. Horizontal arrow indicates no significant difference between impact and background samples. We considered the seed quality of the investigated species in different years.

Table 2 Trends of seedlings survival rate for five herbaceous species growing at the East-Ural Radioactive Trace

Species	Photo	Years of investigation												
		2002	2003	2004	2005	2006	2007	2008	2009	2010	2011	2012	2013	2014
Bromopsis inermis					↓		↑	↔	↔	↔	↓	↓		
Leonurus quinquelobatus						↑		↔	↓	↓			↔	↓
Stellaria graminea					↓		↑	↔	↔		↓	↔	↓	↑
Melandrium album					↓		↑	↑		↔	↔	↓	↓	↔
Plantago major		↔	↓	↓	↓					↑	↑		↑	

Photo of *B. inermis* is by V. Prokhorov © 2008, *L. quinquelobatus* is by V. Gumenyuk © 2012, *S. graminea* is by V. Gumenyuk © 2009, *M. album* is by V. Gumenyuk © 2013, *P. major* is by V. Gumenyuk © 2014 (all photos are from http://plantarium.ru/)

Bromopsis inermis. In 2007 we revealed that most test parameters of brome seedling from EURT area were higher than in background samples. In 2005, 2011 and 2012 the highest quality seeds were matured in background populations. Detailed characteristics of *B. inermis* seed quality from the EURT depending on the weather conditions were described earlier (Antonova et al. 2014).

Leonurus quinquelobatus. The motherwort seedlings survival varied under different weather conditions. For example, in 2009, 2010 and 2014 the seedling viability in the pollution gradient was reduced comparing to those in background populations. In other seasons this parameter was increased (2006) or remained unchanged (2008, 2013).

Stellaria graminea. Some data of the interaction between weather conditions and starweed seed progeny quality were described earlier (Pozolotina et al. 2010). In 2005, the starweed seed progeny from impact populations were less viable than those from background one. In the period 2011–2014 this effect was observed twice. In 2007 and 2014, the opposite dependence was observed: the values of most tested parameters in background samples were low. In 2008, 2009 and 2012, the results showed that the average survival rates were similar in all populations (see Table 2). In 2010, the weather conditions were extremely hot and dry. Plants of *S. graminea* from impact populations were not able to reach the flowering phase. Therefore seed collection of starweed was not sampled.

Melandrium album. Similarly to *S. graminea*, white campion belongs to Caryophyllaceae family. The investigation of two species showed that in 2005 and 2013, the survival rates of both plants from the EURT area were less than background populations, and vice versa in 2007. The offspring survival trends of these plants mismatched in other years. In 2009, we failed to collect enough *M. album* seeds for experiments. The seed capsules were damaged by the moth larva *Hadena bicruris*. *H. bicruris* is a pollinator of *M. album*, it feeds on immature seeds and thereby drastically reduces seed production (Wolfe 2002). Weather conditions are very important for *H. bicruris* oviposition. The rise of the average day temperature increases percentage of *M. album* plants oviposited by *H. bicruris*, by almost 3 % for every degree (Biere and Honders 2006). In 2009 the weather conditions contributed to raising the number of moths. From 2007 to 2009 we observed increase in the average day and sum temperature of June ($R = 0.99, p = 0.026$) and the tendency to decrease in precipitation of June ($R = -0.925, p = 0.248$) at the EURT area.

In 2010 (hot and dry summer), *S. graminea* had no seeds, while *M. album* had seeds of high quality (Table 2). This could be explained by ecological differences (resistance to drought and high temperatures) and peculiarities of reproduction system of these two species. *S. graminea* has two types of reproduction (vegetative and sexual) and perennial cycle of life, while *M. album* is mainly germiniparous and annual or biennial species. Since the formation of seeds is the basic mode of reproduction of this species, this process seems to be more successful under fluctuating weather conditions. The results of this study were presented earlier (Antonova et al. 2013).

Plantago major. Over a period of seven years, we identified the decrease in the plantain seed progeny viability for three times at the EURT area; also opposite result was observed for three times. In 2002, the quality of seed progeny from the radioactive contamination areas and background populations was similar to each other.

The question arises as to what exactly is the cause of interannual variability of plant viability? The activity concentration of radionuclides in the soil and plants has not changed significantly since 2002. Geobotanical observations showed that the state of phytocenoses was also stable. We have analyzed the weather conditions in background and impact areas from 2002 to 2014 (Table 3).

Let us now consider the effect of weather conditions on the viability of the seed progeny of plants during spring and summer months (see Table 3). The significant impact of weather conditions on the seed quality has been established only in the background populations of *B. inermis*. It may be due to the fact that the interannual variability of seedling survival from chronically exposed samples of *B. inermis* was high, exceeding background levels 2–3 times. Seeds of good quality formed with a small amount of precipitation in May and warm temperatures in June.

Seed progeny of *L. quinquelobatus* had opposite patterns: the significant effect of weather conditions on seedling survival has not been ascertained for background populations. At the same time, high-quality seeds of *L. quinquelobatus* matured in the EURT area in warm temperature in July, a sufficient amount of precipitation in May and a small amount of precipitation in August. The typical dependences of seed progeny quality on weather conditions (temperature and precipitation) in maturation period are shown in Fig. 1.

The warm temperature and low amount of precipitation was needed for seeds of *S. graminea* (see Table 3). However, the temperature in May and the sum of precipitation at effective temperature for the summer were significant for the

Table 3 Significant relationships between meteorological factors and survival of seedlings of five herbaceous plants at the EURT area and background populations

Species	Meteorological parameters			
	Background populations		Impact populations	
B. inermis	$T_{(6)}$, $\Sigma T_{(6)}$, $\Sigma T_{ef.(6)}$	+	no	
	$\Sigma P_{(5)}$	−		
L. quinquelobatus	no		$T_{(7)}$, $\Sigma T_{(7)}$, $\Sigma T_{ef.(7)}$, $\Sigma P_{(5)}$, $I_{(5)}$	+
			$\Sigma P_{(8)}$, $\Sigma P\text{-}T_{ef.(8)}$	−
S. graminea	$T_{(5)}$, $\Sigma T_{(5)}$	+	$T_{(8)}$, $\Sigma T_{(8)}$, $\Sigma T_{ef.(5;8)}$, $\Sigma P\text{-}T_{ef.(5)}$	+
	$\Sigma P_{(7)}$, $\Sigma P\text{-}T_{ef.(7;6\text{-}8)}$, $I_{(7;5\text{-}8;6\text{-}8)}$	−	$\Sigma P\text{-}T_{ef.(7)}$	−
M. album	$T_{(5;6)}$, $\Sigma T_{(5;6)}$, $\Sigma T_{ef.(6)}$	+	$\Sigma P\text{-}T_{(7)}$	+
	$\Sigma P_{(5)}$	−	$\Sigma P_{(8)}$, $\Sigma T_{ef.(8)}$, $\Sigma P\text{-}T_{ef.(8)}$	−
P. major	$\Sigma T_{ef.(5\text{-}8;6\text{-}8)}$, $I_{(5\text{-}8;6\text{-}8;7)}$	−	$T_{(7)}$, $\Sigma T_{(7)}$, $\Sigma T_{ef.(7)}$	−
	$\Sigma P_{(7;8)}$, $\Sigma P\text{-}T_{ef.(7;8;5\text{-}8;6\text{-}8)}$	−	$\Sigma P_{(7)}$, $\Sigma P\text{-}T_{ef.(6;7)}$, $I_{(8)}$	+

Positive (+) and negative (−) correlations ($R^2 = 0.12$–0.66; $p = 0.0003$–0.039); T—average air temperatures (number in brackets is symbol of month); ΣT—sum of air temperatures; $\Sigma T_{ef.}$—sum of effective temperatures; ΣP—sum of precipitation; $\Sigma P\text{-}T_{ef.}$—sum of precipitation at effective temperatures; *I*—Selyaninov's aridity index

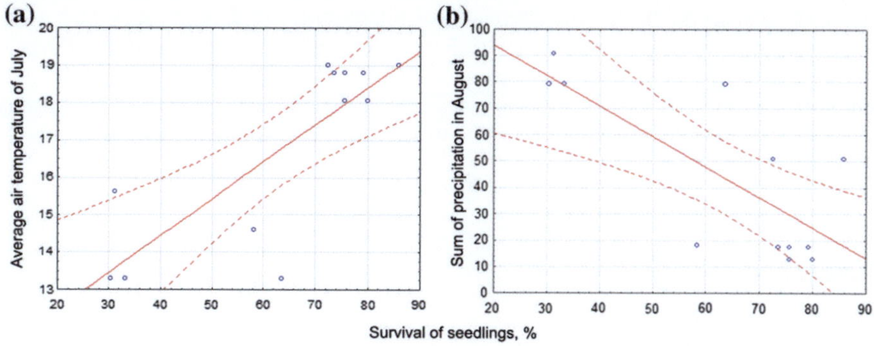

Fig. 1 Dependence of the survival seedlings in *L. quinquelobatus* on average air temperatures of July (**a**) and sum of precipitation in August (**b**) from impact populations

background samples. The temperature in May and August, as well as the sum of precipitation at effective temperature in May and July, was significant for the impact populations. Meanwhile the spring precipitation positively affected the seedling survival and the summer one had a negative influence.

The survival rate of *M. album* seedlings as well as *S. graminea* from background sites was positively correlated with temperature in May and June and had a negative correlation with precipitation in May. At the same time, the effects for *M. album* were opposite in sign to *S. graminea* at the impact sites. Sufficient precipitation during the warm days of July and small amount precipitation and fresh temperature in August were required for *M. album* populations at the EURT.

An interesting feature of *P. major* is that the opposite effect of precipitation in July and August have been observed in the background and the impact areas, as well as the amount of precipitation at effective temperature for the spring-summer period. The seeds of high quality were also formed in the *P. major* background samples at low spring and summer temperatures and in impact samples at low temperatures in June. In all cases, the factor of "precipitation" was dominant; therefore, the impact of the effects of aridity index was similar.

Resume

The results showed that the seed quality of five herbaceous plant species in different seasons depended on weather conditions. The differences of effects in the background and EURT populations indicate the specificity of the response to environmental factors in the presence of radioactive contamination. Interaction effects may be synergistic, antagonistic or additive.

All the studied species have a certain capacity for adaptation to natural and technogenic factors and respond to a complex of environmental impact within the limits of this capacity. The mechanisms of adaptation are diverse, and each population living for a long time under technogenic stress acquires specific features.

Acknowledgment This work was done with financial support from Program of the Russian Foundation for Basic Research (projects 15-04-01023 and 15-34-20639), partly from Program of UB RAS (project 15-2-4-21).

References

Aarkrog A, Dahlgaard H, Nielsen SP, Trapeznikov AV, Molchanova IV, Pozolotina VN, Karavaeva EN, Yushkov PI, Polikarpov GG (1997) Radioactive inventories from the Kyshtym and Karachay accidents: estimates based on soil samples collected in the South Urals (1990–1995). Sci Total Environ 201(2):137–154. doi:10.1016/S0048-9697(97)00098-3

Antonova EV, Karimullina EM, Pozolotina VN (2013) Intraspecific variation in *Melandrium album* along a radioactive contamination gradient at the Eastern Ural Radioactive Trace. Russ J Ecol 44(1):18–27. doi:10.1134/s1067413613010025

Antonova EV, Pozolotina VN, Karimullina EM (2014) Variation in the seed progeny of smooth brome grass, *Bromus inermis* Leyss., under conditions of chronic irradiation in the zone of the Eastern Ural Radioactive Trace. Russ J Ecol 45(6):508–516. doi:10.1134/s1067413614060034

Biere A, Honders SC (2006) Coping with third parties in a nursery pollination mutualism: *Hadena bicruris* avoids oviposition on pathogen-infected, less rewarding *Silene latifolia*. New Phytol 169(4):719–727. doi:10.1111/j.1469-8137.2005.01511.x

Boyko A, Kovalchuk I (2011) Genome instability and epigenetic modification—heritable responses to environmental stress? Curr Opin Plant Biol 14(3):260–266. doi:10.1016/j.pbi.2011.03.003

Calow P, Forbes VE (1998) How do physiological responses to stress translate into ecological and evolutionary processes? Comp Biochem Phys A 120(1):11–16. doi:10.1016/S1095-6433(98)10003-x

Crain CM, Kroeker K, Halpern BS (2008) Interactive and cumulative effects of multiple human stressors in marine systems. Ecol Lett 11(12):1304–1315. doi:10.1111/j.1461-0248.2008.01253.x

Fischer BB, Pomati F, Eggen RIL (2013) The toxicity of chemical pollutants in dynamic natural systems: the challenge of integrating environmental factors and biological complexity. Sci Total Environ 449:253–259. doi:10.1016/j.scitotenv.2013.01.066

Hinton T, Brechignac F (2005) A case against biomarkers as indicators of ecological risks: a problem of linkage. In: Bréchignac F, Howard BJ (eds) Scientific trends in radiological protection of the environment—ECORAD 2004. Aix-en-Provence, France, 2005. Tec & Doc, Lavoisier, Paris, France, 123–135

Holmstrup M, Bindesbol A-M, Oostingh GJ, Duschl A, Scheil V, Kohler H-R, Loureiro S, Soares AMVM, Ferreira ALG, Kienle C, Gerhardt A, Laskowski R, Kramarz PE, Bayley M, Svendsen C, Spurgeon DJ (2010) Interactions between effects of environmental chemicals and natural stressors: a review. Sci Total Environ 408(18):3746–3762. doi:10.1016/j.scitotenv.2009.10.067

Karimullina E, Antonova E, Pozolotina V (2013) Assessing radiation exposure of herbaceous plant species at the East-Ural Radioactive Trace. J Environ Radioact 124:113–120. doi:10.1016/j.jenvrad.2013.04.003

Mikhailovskaia LN, Molchanova IV, Karavaeva EN, Pozolotina VN, Tarasov OV (2011) Radioecological investigation of the soil cover of Eastern Urals State Radioactive Reserve and neighboring areas. Radiat Biol Radioecol 51(4):476–482

Molchanova I, Mikhailovskaya L, Antonov K, Pozolotina V, Antonova E (2014) Current assessment of integrated content of long-lived radionuclides in soils of the head part of the East Ural Radioactive Trace. J Environ Radioact 138(6):238–248. doi:10.1016/j.jenvrad.2014.09.004

Newcombe RG (1998) Interval estimation for the difference between independent proportions: comparison of eleven methods. Stat Med 17(8):873–890. doi:10.1002/(SICI)1097-0258(19980430)

Oliver EJ (2005) The encyclopedia of world climatology (Encyclopedia of Earth Sciences Series). The Springer, Dordrecht

Pozolotina VN, Antonova EV, Karimullina EM (2010) Assessment of radiation impact on *Stellaria graminea* cenopopulations in the zone of the Eastern Ural Radioactive Trace. Russ J Ecol 41(6):459–468. doi:10.1134/S1067413610060019

Pozolotina VN, Molchanova IV, Mikhaylovskaya LN, Antonova EV, Karavaeva EN (2012) The current state of terrestrial ecosystems in the Eastern Ural Radioactive Trace. In: Gerada JG (ed) Radionuclides: sources, properties and hazards. Nova Science, New York, pp 1–22

Qin GQ, Presley SM, Anderson TA, Gao WM, Maul JD (2011) Effects of predator cues on pesticide toxicity: toward an understanding of the mechanism of the interaction. Environ Toxicol Chem 30(8):1926–1934. doi:10.1002/etc.575

Romanov GN, Nikipelov BV, Drozhko EG The Kyshtym accident: Causes, scale and radiation characteristics. In: Seminar on comparative assessment of the environmental impact of radionuclides released during three major nuclear accidents: Kyshtym, Windscale, Chernobyl. Commission of the European Communities, Luxemburg, 1–5 Oct 1990, pp 25–40

Timofeeff-Ressovsky NW (1962) Some problems of radiation biogeocenology. Doctoral Thesis. Institute of Biology UB AS USSR, Sverdlovsk, 1–53

Weiß CH (2007) StatSoft Inc, Tulsa, OK.: STATISTICA, Version 8. AStA. Adv Statist Anal 91(3):339–341. doi:10.1007/s10182-007-0038-x

Wilson EB (1927) Probable inference, the law of succession and statistical inference. J Am Statist Assoc 22:209–212. doi:10.2307/2276774

Wolfe LM (2002) Why alien invaders succeed: support for the escape-from-enemy hypothesis. Am Nat 160(6):705–711. doi:10.1086/343872

Microevolution Processes in Antropogenic Radionuclide Anomalies

Dmitry M. Grodzinsky

There are territories of different size contaminated with fission products of uranium due to nuclear accidents and nuclear weapons testing in the continent and water area. Radioactive contamination in these areas is responsible for the existence of permanent ionizing radiation, and all living beings for many generations are exposed to chronic exposure in these contaminated areas. There are many examples of vast territories contaminated with radioactive substances related to the products of uranium fission. Suffice it to mention the Chernobyl Exclusion Zone, part of the land and waters of the ocean near the Fukushima Nuclear Power Plant, Bay of the Kara Sea at the confluence of the Ob River, the eastern part of the Barents Sea near Novaya Zemlya, places of atomic bomb tests, for example, test site at Semipalatinsk, south of the Sahara, some Pacific Islands, East Urals radioactive trace. There are other places heavily contaminated with radioactive substances (Las Barrios in Spain, Simi Valley, Church Rock and Denver in the USA, Sellafield in the UK, Mailuu Suu in Kyrgtzstan, etc.). Such territories are named as artificial anthropogenic radionuclide anomalies.

In the territory of the Exclusion Zone of the Chernobyl Nuclear Power Station, the dose rate of ionizing radiation is up to 5 mR/h 25 years after the accident in some places. In the most polluted areas of the Eastern Ural trace 50 years after the catastrophe the dose varies from 20 to 157 μR/h (Pozolotina et al. 2008). It is not too hard to calculate the dose values for the first years after the accident for these artificial radionuclide anomalies.

The accident of nuclear reactors in the Fukushima nuclear power plant is the cause of severe marine pollution. The sea water samples show activity up to 10,000 Bq/m^2. In the soil plot of land near the wrecked power station the contamination with ^{137}Cs reaches 1.5×10^6 Bq/m^2.

D.M. Grodzinsky (✉)
Institute of Cell Biology and Genetic Engineering,
National Academy of Sciences of Ukraine, Kiev, Ukraine
e-mail: dmgrod@gmail.com

V.L. Korogodina et al. (eds.), *Genetics, Evolution and Radiation*,
DOI 10.1007/978-3-319-48838-7_23

Living organisms in areas of nuclear accidents accumulate often very significant radiation dose due to the accumulation of radionuclides in their tissues.

Chronic exposure is an "alien" factor for autochthonic species of animals, plants and microorganisms inhabiting these areas because its impact lasts period of time which is clearly insufficient to adaptation of species populations to this factor.

On our planet, there are areas where the level of ionizing radiation dose rate is not less than in artificial anthropogenic anomaly. Elevated levels of the exposure dose of ionizing radiation are observed in many places of the Earth. It has long been known that water in many mineral springs contain elevated concentrations of radium and radon. The most famous natural radionuclide anomalies are located in all continents such as South America, Australia, Europe, and Asia. These areas are named as natural radionuclide anomalies. They are confined to the deposits of uranium and thorium located close to the surface. For example, in the state of Kerala in the municipality of Karanagapalli the coastal exposure dose reaches 3 mR/h. In Brazil, there are deposits of monazite sands from which in the states of Espirito Santos and Rio de Janeiro the external radiation levels range up to 5 mR/h (Eisebud 1987). In the Republic of Saha (Yakutia) in Russia on the dumps of uranium deposit in Elkonsky horst the dose rate reaches 3 mR/h (Zhuravskaya and Artamonova 2014). Autochthonous flora and fauna existed for many thousands of years in the natural radioactive anomalies and therefore increased exposure to ionizing radiation may not be an "alien" factor for them. It is possible that in exposure to that factor as sufficiently strongly negative the cytogenetic action of populations of species could somehow adapt. Confirmation of the admissibility of this idea is the fact that the plants found in the natural radionuclide anomaly do not form morphological abnormalities while in newly emerging artificial radionuclide anomaly occurrence of radiation-induced malformations is a mass phenomenon.

There is one very significant difference between artificial and natural radionuclide anomalies. Radioactivity in the artificial radionuclide anomaly presented uranium fission products which are easily digestible elements such as ^{137}Cs, ^{90}Sr, radioisotopes of iodine, cerium and other. In the natural radionuclide anomaly radionuclides from the series of ^{235}U and ^{238}U and thorium are substances which are poorly absorbed by plants and animals. Therefore, the internal exposure dose is considerably higher for organisms in artificial radionuclide anomalies.

Obviously we have every reason to believe that artificial radionuclide anomalies are hot spots for the flora and fauna in ecosystems because all kinds of animals, plants, fungi, bacteria and viruses are affected by chronic exposure to ionizing radiation throughout their life cycles. Above all things, the increasing doses are accompanied by an augmentation in the incidence of mutations and this phenomenon has no threshold effect.

The higher yield of mutations is causing the increasing rates of microevolution processes in areas of artificial radionuclide abnormalities. However, the effect of ionizing radiation is not limited only by mutagenesis - irradiation accompanied by the development of a number of additional responses of organisms to radiation which is associated with multiple increased variability for all living creatures in the fields of radiation.

The fact that the response of cells and multicell systems to the action of ionizing radiation is reflected by the development of two parallel processes. The first process is associated with damage of the unique cell ultra structures; the second process reflects the cell response as an active answer to the perception of radiation as the alarm signal.

The first process of cell radiation injury is associated with the damages of molecular targets the most important of which are the chromatin of the cell nucleus, mitochondrial DNA, cell membrane systems and molecular associates performing the functions of molecular machines. This way of formation of radiation damage is very thoroughly investigated as well as and the effects that result from damage to the molecular targets known as target effects. The second way is the reaction to irradiation as complex active responses of cells and multicultural systems on the perception of radiation as a signal of cytogenetic threat. It is investigated less thoroughly. Briefly described ways of the formation responses of cells and multicellular organisms to irradiation are shown in Fig. 1.

Targeted effects are revealed in the appearance of chromosome aberrations, point mutations, impaired cell membranes, inhibiting the synthesis of proteins and other substances, distortion of bioenergetic processes. These damages to a certain extent can be loosened by means of processes of reparation. In particular, single-strand and double-strand DNA breaks and other types of damages in this molecule are exposed to reparation. Reparation of membranes occurs by accelerating the membrane flows and increasing the intensity of lipid synthesis. Repair of "molecular machines" is provided by the self-assembly processes using macromolecules de novo synthesized. Obviously, differences in the radioresistance of species are due to both affection targets and the ability to repair all types of radiation damage to the targets.

Apparently the radioresistance in species populations can be increased as a result of a long natural selection for improved ability of repair systems to perform their functions. However, in order to be able to implement adaptation of the population to high levels of radiation many generations should be replaced and therefore a lot of time will be required for these results, especially for species with long ontogenesis.

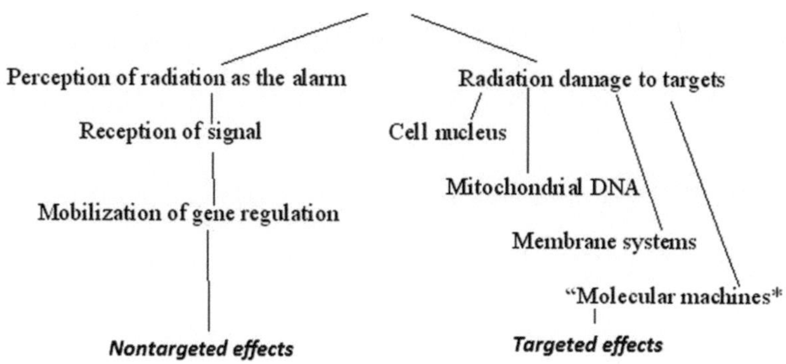

Fig. 1 Two ways of radiation effects formation

Nontargeted effects are of an entirely different nature compared with the responses of the target effects in cells and multicellular organisms. The appearance of damaged molecules in irradiated cells is perceived by cells as a signal of cytogenetic threat arising in the environment. The organism cells actively respond to this signal, and a complex process whose aim is to reduce the risk of wreck of species from the negative impact of radiation begins in cells and multicellular systems.

We see the fact that this process is really happening in the results of experiments in which there is a substantial acceleration of the exchange of ribonucleic acids. This exchange is determined by the inclusion of labeled on carbon uridine in the composition of the molecules. Protein biosynthesis accelerates after the irradiation that can be seen by observing incorporation of labeled precursors into the protein molecules (Kolomiets 2010). The results of experiments with labeled precursor of RNA are shown in Table 1.

These data indicate that active non-targeted reaction is developed in response to irradiation. Really, amplification of RNA biosynthesis occurs in parts of the plant that are not directly exposed to irradiation.

Active response of cells and multicellular systems to irradiation appears mainly in the formation of two strategies to adapt to high levels of radiation as an alien and harmful stress factors. One strategy is to improve the radiation stability of the individual. This is so-called ontogenetic adaptation. In particular, it is shown in radioadaptation. The essential point of radioadaptation is to improve the radiation stability through a variety of processes including synthesis of DNA repair enzymes, increasing the concentration of antioxidants and other substances that together form a natural background radioresistance. In addition to these biochemical mechanisms of radioresistance that increase cellular processes participate in radioadaptation namely stretch presynthetic phase in the cell cycle that improves DNA reparation. They increase a potency of repopulation recovery and intensification of cell selection as a way to reduce the genetic load in the cells of the organism.

A fundamentally different type of adaptive adaptation called phylogenetic plays an important role. Its aim is to increase the genetic and epigenetic diversity within species that is needed to create a wealth of material for natural selection to increased radioresistance. The diagram which illustrates two adaptation strategies in populations is shown in Fig. 2.

The ratio between the two paths of the organism's reaction to the action of radiation essentially depends on the dose and the method of irradiation. The first

Table 1 The changes of the specific activity of total RNA in pea roots after irradiation aboveground plant parts with UVB

Exposure	Specific radioactivity (counts per min/mg of RNA)	
	Control	Irradiated with UVB
45 min	1500	1700
60 min	1500	2050
180 min	1500	2450
26 h	1550	3250

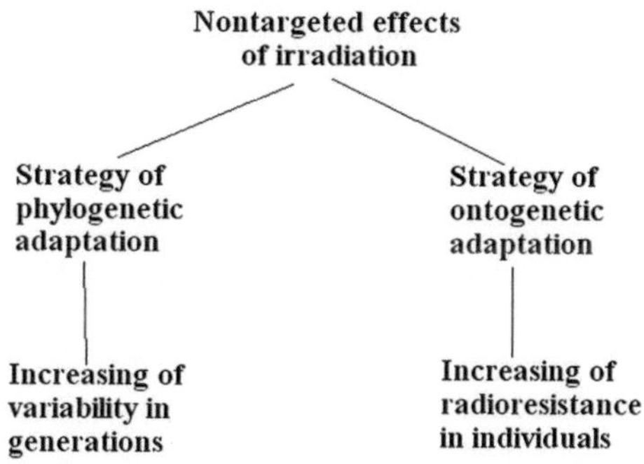

Fig. 2 Two strategies of adaptation to chronic irradiation in artificial radionuclide anomalies

path is related to the response through the radiation injuries of molecular targets within the cell and it is characteristic for the high doses and acute exposures. The second path predominates at low doses of ionizing radiation and chronic exposure.

Many radiobiological reactions are associated with the mechanisms that are involved in the regulation of genetic and epigenetic variability. For example, induction of genome instability and apparently bystander effect are associated with these mechanisms of formation of the organism's response to radiation.

The fact that the individual variation in a population exposed to radiation in small doses increases greatly can be seen in various experiments on the value of the variance (standard deviation) of quantitative traits. The increase in biodiversity is observed at different levels of organization of the organism. In plants it can be seen on the sizes of the whole plant or its individual organs and on the size of the cells in a particular tissue. For example, the following data on evaluation of the variance of quantitative values in the initial cells sizes in root meristem of pea were obtained (in percentage) (Kravets 2010):

Variance for non-irradiated plant and irradiated plants				
Dose (Gy)	0	4	6	10
Variance	100	285	290	297

It should be noted that the increase in the size variability of initials in the apical meristem is not dependent on the radiation dose. That can be attributed to a regulatory nature of this phenomenon. Here is another example of the increase in volatility during the action of chronic exposure. The coefficient of variation at the St. John's wort (*Hypericum perforatum*) growing in the exclusion zone is higher than in the population of this species in the uncontaminated area (Kripka 2010):

Coefficient of variation vs levels of radionuclide contamination				
Levels of radionuclide contamination, kBq/m^2	18.5 (control)	1.85×10^2	1.85×10^3	2.22×10^4
Coefficient of variation, %	28	36	37	43

It is shown that in the zone of radioactive contamination of Scots pine variability of growth and reproductive processes increase sharply (Skok 2005).

There are various mechanisms to improve the genetic and epigenetic diversity. Among them the next events play a greater role in plants:

- augmentation of the frequency of meiotic recombination (this process is accelerated under the influence of low-dose radiation);
- increase in the activity of transposons (the dependence of the process of irradiation is confirmed (Zaynullin 1997);
- weakening of DNA repair (it was shown to pollen of birch growing on the territory contaminated with radionuclides (Grodzinsky 2012);
- reduced role of parthenocarpic reproduction (the effect of irradiation in small doses on the increase of the frequency of sexual reproduction has been shown in plants (Grodzinsky 2012);
- radiation induction of genomic instability (Lorimore et al. 2003).

As you can see all mechanisms leading to increased genetic diversity in species populations are activated significantly by the action of radiation. It should be noted that the impact of radiation on the mechanisms to ensure an increase of heterozygosis is very noticeable. This can be seen from Table 2 which shows the dependence of the number of plants reproduced sexually on the territory contaminated with radionuclides (Kripka 2010).

Plants capable of apomixis increase the frequency of sexual reproduction and, therefore, decrease the contribution of apomictic seed formation (Kripka 2010). A similar phenomenon is observed in some polychaetes in which an increase in the role of sexual reproduction and weakening the role of vegetative propagation take place with increasing dose chronic exposure. It is interesting to note that the coefficients of variation of the quantitative values of morphological features are much higher in plants with sexual rather than vegetative propagation. That is seen from the data in Table 3.

The ratio of hermaphrodite and unisexual flowers under the influence of chronic exposure is changed so that the heterozygosis in seed generation would be increased.

Table 2 The frequency of plant St. John's wort (*Hypericum perforatum*), resulting from sexual (2n) and apomictic (4n) breeding in the Chernobyl exclusion zone

Place of records	Number of plants (in percentage)	
	Sexual origin	Apomicts
Control 10–14 μR/h	41.1	58.9
Chernobyl (low level of radionuclide pollution: 60–120 μR/h)	81.2	18.8
Chistogalovka (high level of radioactive contamination: 590–2600 μR/h)	95.0	5.0

Table 3 The ratio of hermaphrodite and unisexual flowers under the influence of chronic exposure

Place of records	Number of flowers		
	Hermaphrodite	Unisexual	
		Pistillate	Pistillate
Control	61	5	34
Kopatchi (middle level of radioactive contamination)	30	18	52
Chistogalovka (high level of radioactive contamination)	36	14	50

Table 4 The manifestation frequency of various polyploid pollen cells in mouse-ear cress (*Arabidopsis thaliana*) (the first vegetation) under chronic irradiation, %

Ploidy of pollen grain	Control	Dose, Gy	
	Unirradiated plants	0.5	5
Diploid cells	91	77	65
Tetraploid cells	1	9	15
Chimaeric tissue	8	14	20

Chronic exposure also causes a significant increase in the cell ploidy. This phenomenon is shown in Table 4 in the example of pollen grains ploidy.

Data presented in these tables point to the fact of strong influence of chronic irradiation on cytogenetical distortions in plant cells. The effects of chronic irradiation embrace various genetical events from genome to point mutations.

Formation of the organism's response to irradiation along the path of cell signaling systems explains one of the most important conflicts in radiobiology, namely, why the effects of radiation are so big if the number of radiation damages in the case of small doses is too small in comparison with the same spontaneously occurring defects in biological molecules. Apparently a principle of amplifier is implemented in this signaling pathway of response formation under the influence of small doses of ionizing radiation.

Active response to the alarm signal is carried out through the participation of epigenetic phenomena. This is the result of experiments in which there is a significant change in the cytosine methylation profiles in the DNA of the plants when they are irradiated with gamma rays or ultraviolet radiation. Figure 3 shows a general signaling diagram illustrating a process of forming a biological response to irradiation.

Accordingly chronic irradiation in a small dose-rate is perceived by the organism as a stress effect. Ontogenetic and phylogenetic adaptations should be formed in response to this stress. The first path of adaptation is associated with the activation of genes controlling DNA reparation, membrane currents, and the synthesis of antioxidants as well as a repopulation in meristem tissues. The second path of adaptation is manifested in the processes that lead to an increase in phenotypic diversity in populations, thus increases the efficiency of natural selection for

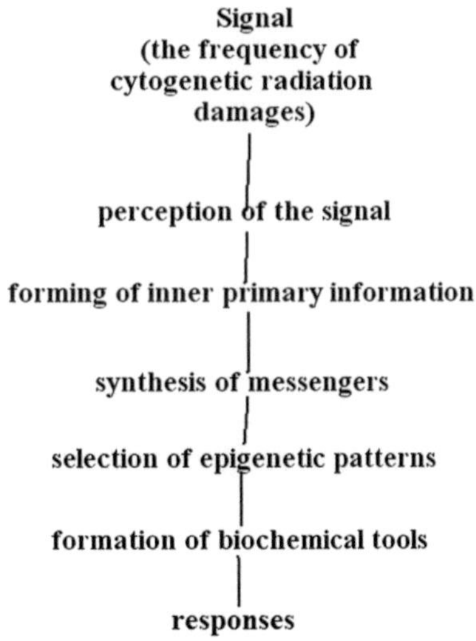

Fig. 3 Circuit response to irradiation by signal transduction pathway

enhanced radioresistance in species. Phenotypic polymorphism of population is also increasing due to the epigenetic mechanisms that lead to the appearance of pseudo mutations.

The above data convincingly demonstrate that under the conditions of chronic exposure to low dose rate a wide range of processes leading to the emergence of a population with genetically modified forms and epigenetic pseudomutations is added.

Accelerating the pace of adaptation in hot spots can be very dangerous phenomenon for the purpose of biota. In fact, spreading of mutant forms can occur that may break the biological balance in ecosystems and lead to the impoverishment of biodiversity. Since interpopulation homeostasis in biocommunities is associated with the coevolution of many species in particular in the system "parasite-host", the violation rate of evolution can disrupt the equilibrium state of the biota in ecosystems not only in hotspots but also in adjacent areas where individuals can migrate mutated.

Reported events constituting microevolution cause accelerated evolutionary changes on the intra-specific level. As some of the species populations are composed of a relatively small number of individuals in hotspots ecosystems, the probability of genetic drift due to declining fertility of some species population is replenished by immigration from areas adjacent to anthropogenic radionuclide anomaly (Møller and Mousseau 2007). This process is accompanied by an increase in the intensity of gene flow. Naturally, the high rate of microevolution processes can be revealed only in those species that are characterized by short periods of life cycle and where they have a sufficiently large number of generations under conditions of chronic exposure. The examples of stem rust pathogen of cereals fungus *Puccinia graminis* Pers. and mildew *Erysiphe* sp. demonstrated an increase in the development of these pathogens due to the emergence of new highly virulent races in the Chernobyl exclusion zone (Dmitriev et al. 2012). Results obtained both in greenhouse and in the field trials in the Chernobyl exclusion zone demonstrate the decrease in disease resistance of wheat, rye and corn cultivars under low doses chronic irradiation. Seeds of plants grown under conditions of chronic exposure are damaged greatly by fungal infection. The hotspots are gradually becoming the nidus of new virulent forms of pathogens.

Risks to biota due to an increase in micro-evolutionary processes in the anthropogenic radionuclide anomalies can result in the appearance of rare mutations, increase of the genetic and epigenetic polymorphism of populations and loss of balance of species in ecosystems. Of course, it is necessary to establish monitoring of microevolution processes in hot spots as a helpful security measure. In natural radionuclide anomalies associated with uranium deposits the range of deviation in genetic variation is not as significant as in the conditions of anthropogenic anomalies. Apparently, mechanisms stabilizing the variability in populations evolved over a long period of time.

References

Dmitriev AP, Shevchenko OI, Polishuk VP, Guscha NI (2012) Effects of low dose chronic radiation and heavy metals on plant—pathogen interactions station. In: Burlakova EB, Naydich VI (eds) The lessons of Chernobyl: 25 years later. Nova Science Publishers Inc, NY, pp 229–244

Eisenbud M (1987) Environmental radioactivity from natural, industrial, and military sources, 3rd edn. Academic Press, NY, p 475

Grodzinsky DM (2012) Response of the plant world on chronic irradiation in the zone of the Chernobyl nuclear power station. In: Burlakova EB, Naydich VI (eds) The lessons of Chernobyl: 25 years later. Nova Science Publishers Inc, NY, pp 209–228

Kolomiets OD (2010) Report of the department of biophysics and radiobiology. Institute of Cell Biology and Genetic Engineering, National Academy of Sciences of Ukraine, Kiev

Kravets EA (2010) Report of the department of biophysics and radiobiology. Institute of Cell Biology and Genetic Engineering, National Academy of Sciences of Ukraine, Kiev

Kripka AV (2010) Report of the department of biophysics and radiobiology. Institute of Cell Biology and Genetic Engineering, National Academy of Sciences of Ukraine, Kiev

Lorimore SA, Coates PhJ, Wright EG (2003) Radiation-induced genomic instability and bystander effects: interrelated nontargeted effects of exposure to ionizing radiation. Oncogene 22:7058–7069

Møller P, Mousseau TA (2007) Determinants of interspecific variation in population declines of birds after exposure to radiation at Chernobyl. J Appl Ecol 44(5):909–919

Pozolotina VN, Molchanova IV, Karavaeva EV, Mikhailovskaya LN, Antonova EV (2008) The current state of terrestrial ecosystems of the East Ural radioactive trace: pollution levels, biological effects. "Goschitskiy" Publishers, Yekaterinburg, p 204

Skok AV (2005) The variability of reproductive and growth processes of pine *Pínus sylvéstris* in different areas of chronic radioactive contamination Chernobyl of Nechernozemie Southern RF. PhD thesis, Bryansk, p 26

Zaynullin VG (1997) Genetic effects of chronic exposure to low doses of ionizing radiation. Dr Sci Thesis, Moscow, p 48

Zhuravskaya AN, Artamonova Yu S (2014) Increased radiation background and plants. Publisher SB RAS, Novosibirsk, p 152

Aquatic Plants and Animals in the Chernobyl Exclusion Zone: Effects of Long-Term Radiation Exposure on Different Levels of Biological Organization

Dmitri Gudkov, Natalia Shevtsova, Natalia Pomortseva, Elena Dzyubenko, Andrian Yavnyuk, Alexander Kaglyan and Alexander Nazarov

Introduction

Currently the radioecological situation in the Chernobyl exclusion zone (EZ) is determined primarily by long-lived radionuclides ^{90}Sr, ^{137}Cs, ^{238}Pu, ^{239}Pu, ^{240}Pu and ^{241}Am. Along with natural decontamination processes in aquatic ecosystems such as physical decay of radionuclides and their hydrological transport outside the EZ, there is a change of physical and chemical forms of radioactive substances in soils of catchment areas, their transformation and transition in the mobile and bioavailable state, washout to the closed aquatic ecosystems and accumulation by hydrobionts. This essentially deteriorates the radiation situation in lake ecosystems, which are some kind of "storage system" of radioactive substances in the EZ, and results in increase of radiation dose to aquatic species and manifests itself in a variety of radiation effects on different levels of biological systems.

In connection with slow water cycle, lakes are relatively closed systems allowing one, with a minimum uncertainty, to estimate general balance of energy and matter in ecosystem and also to analyse the dynamics of biogeochemical processes and their influence on radionuclide distribution and migration in the conditions of

D. Gudkov (✉) · N. Shevtsova · N. Pomortseva · A. Kaglyan
Institute of Hydrobiology, Kiev, Ukraine
e-mail: digudkov@gmail.com

E. Dzyubenko
G. Skovoroda State Teacher Training University, Pereyaslav-Khmelnitsky, Ukraine

A. Yavnyuk
National Aviation University, Kiev, Ukraine

A. Nazarov
State Specialized Enterprise Ecocentre, Chernobyl, Ukraine

V.L. Korogodina et al. (eds.), *Genetics, Evolution and Radiation*,
DOI 10.1007/978-3-319-48838-7_24

changing biotic and abiotic factors. The aforesaid, in our view, determines one of the key values for freshwater radioecology of investigations of lake ecosystems located in radiation contaminated territories such as areas of the main long-lived, anthropogenic radionuclides accumulation, areas of the increased long-term radiation dose rate on aquatic biota, and also areas with high probability of radiation effects realization.

The basic problems of radiation safety in the EZ concern radionuclides wash-off with surface drainage water to the river systems, their export outside the EZ and affection of the water quality in the Dnieper River, the main waterway of Ukraine. Undoubtedly, one of the most important problems is research into the long-term impact of ionizing radiation on non-human biota within the EZ. There have been several reviews on biological effects observed after the ChNPP accident and published at the international level (Sokolov et al. 1993; UNSCEAR 1996; IAEA 2006; Moller and Mousseau 2006; Hinton et al. 2007; Geras'kin et al. 2008). Unfortunately, the effects of long-term radiation exposure of aquatic ecosystems on different levels of biological organisation within the EZ are still insufficiently studied.

Thus, the main tasks of our research within the EZ were: radiation dose rate estimation due to external and internal sources of irradiation for different groups and species of hydrobionts; evaluation of dose-dependent cytogenetic, haematologic, morphometric, productional and parasitological effects due to long-term radiation impact on aquatic species.

Materials and Methods

Water bodies of studies. Our investigations were carried out during 1998–2014. The water bodies under research were the flood plain water bodies of the Pripyat River within 10-km area around the destroyed unit of the ChNPP—Azbuchin Lake, Yanovsky Crawl, Dalyokoye Lake, Glubokoye Lake as well as the cooling pond of the ChNPP and the rivers Uzh and Pripyat. The results of the cytogenetic and haematologic analyses were compared to the data obtained for hydrobionts from the reference lakes, located in the neighborhood of the Kiev City: Vyrlitsa, Opechen', Pidbirna, Goloseevo as well as Kiev Storage Reservoir (near vlgs Lyutizh and Strakholesye, Kiev Region) and the Alta River (near the town of Pereyaslav-Khmelnytsky, Kiev Region) with background levels of radioactive contamination. The hydrochemical regime of the studied water bodies is typical for natural reservoirs in the region of Kiev Polesye (woodlands) and varied in insignificant range both for water bodies within the EZ and for reference ones.

Radionuclide measurements and dose rate assessment. The ^{137}Cs concentration was measured by γ-spectrometry complex: PGT IGC-25 detector (France), "Nokia LP 4900 B" analyser ("Nokia", Finland), low-volt feeding source—crate NIM BIN, amplifier NU-8210 ("Elektronicus Merokeszulekek Gyara", Hungary) and 100-mm-thick leaden protection. The ^{90}Sr content was measured with low-background NRR-610 β-radiometer (“Tesla”, Czechia). Minimal detectable activity was 0.04 Bq under 1000 s sample exposition. ^{238}Pu and $^{239+240}$Pu content in electrolytic samples was determined by α-spectrometric tract by NUC-8192 impulse analyser (“Elektronicus Merokeszulekek Gyara”, Hungary). The ^{241}Am content was measured by x-ray-spectrometric line including x-ray detector EG&G Ortec LOAX-51370/20 CFG-SU-GMX (“EG&G Ortec”, USA) and analyser “Nokia LP 4900 B”. The results were measured in Bq/kg at natural humidity and the mistake of estimated radionuclide concentration fell within 15–25 %.

External gamma irradiation dose rate was measured by DRG-01T dosimeter (“Mekhanichesky zavod” (Mechanical Plant), Russia) and by Na-I field radiometer SRP-68-03 (“Promptibor”, Russia). The estimation of the radiation dose rate for aquatic species, due to main dose-forming radionuclides in aquatic environment and hydrobiont’s tissues, was carried out with the use of ERICA Assessment Tool (2012). All dose rate calculations were performed on the basis of our own data.

Chromosomal aberration rate analysis. The chromosomal aberration rate was measured in embryo cells of the gastropod pond snails (*Lymnaea stagnalis*) by the standard anaphase method (Pausheva 1974) and in the apical root meristems of the eight species of higher aquatic plants: common reed (*Phragmites australis*) arrowhead (*Sagittaria sagittifolia*) flowering rush (*Butomus umbellatus*) , fresh-water soldier (*Stratiotes aloides*) narrow-leaved cat’s-tail (*Typha angustifolia*) broad-leaved cat’s-tail (*Typha latifolia*) branched bur-reed (*Sparganium erectum*) and manna (*Glyceria maxima*) by the modified method (Shevtsova et al. 2005). The egg mass of the pond snails and root meristems of higher aquatic plants was preserved in Carnoy’s fluid. For staining of cytological preparations 1 g orcein was dissolved in 45 ml of boiling acetic acid and added to 55 ml of distilled water. We placed 10 eggs of the pond snail in a watch glass, added 5 ml of orcein and left at 4 °C for 24 h. As regards higher aquatic plants, we placed 5–6 tops of roots to the small melting pot, which was filled with 2/3 of the stain. The melting pot was covered with a glass slide. The stain with roots was heated over a flame of alcohol until “undercover” boiling (fogging of the cover glass) and left for 24 h. We made squash preparations from roots and removed from the egg capsules embryos in 60 % lactic acid. Analysis of chromosomal aberrations rate in cytological preparations was performed in cells at the stage of anaphase and telophase of mitosis. We studied 3547 eggs and 782 roots as well as analysed 307,540 and 153,520 anaphase and telophase cells of the pond snails and higher aquatic plants, respectively. The following types of chromosome aberration were scored: single fragment; twin fragment; single bridge; single bridge with fragment; single bridge with two fragments; twin bridge; twin bridge with single fragment; twin bridge with two fragments and multiple aberrations (more than four abnormalities per cell).

Methods of haematological studies. The leukograms of fishes and rate of abnormal red cells as different type of invaginations, ramifications, micronuclei, amitosis, etc. were analysed in peripheral blood of the perch (*Perca fluviatilis*), the (*Scardinius erythropthalmus*) and the Prussian carp (*Carassius gibelio*). Fish blood cells and their pathological changes were identified by (Ivanova 1983; Zhyteneva et al. 1989).

For leukocyte counting and determining the rate of erythrocytes with morphological and cytogenetic abnormalities, blood was collected from the tail vein of fish and film preparations were made on microscope slides. Then they were dried and fixed in methanol for 10 min. Staining was carried out by Pappenheim: with 0.3 % solution of May-Grünwald stain, 1 part of which was added to 4 parts of a phosphate buffer (pH = 6.8 − 7.2) for 10 min, and staining finished by Romanovsky solution for 30–40 min. In the film preparations, leukogram and rate of red blood cells with various abnormalities were counted. A total of 200 different types of white blood cell with abnormalities were counted in film preparation and then their percentage was calculated. The number of red blood cells with abnormalities was analyzed for 3000 erythrocytes on each film preparation. Three film preparations were made from each individual. We examined 90–100 individuals with similar size-age parameters of each fish species from water bodies within the EZ and 70–80 individuals from the reference reservoirs.

Statistical analyses. To compare the probability of increased levels of observed radiation effects in hydrobionts inhabiting the reference and contaminated areas, repeated measures models for binary data were used. The probability of increased levels of chromosomal aberrations, hemolymph structure and blood cells abnormalities was estimated for all water bodies within the EZ compared with the reference reservoirs. A P-value of less than 0.05 was considered statistically significant. Pearson correlation analysis was used to evaluate possible correlations between dose rate and observed effects in water bodies within the EZ and reference ones. For processing the obtained results, database of all cells with chromosomal aberrations, cells of molluscs' hemolymph and peripheral blood of fishes (including abnormalities) was created in the program "Microsoft Excel 97" (Microsoft, Inc.) and software package "Statistica 5.0" (Stat Soft, Inc.) was used for statistical analysis.

Results and Discussion

The territories of the EZ are characterised by significant heterogeneity of radionuclide contamination, which is significantly reflected by the radioactive substance contents in aquatic ecosystem components. This is primarily due to the composition and the dynamics of radionuclide emissions into the environment as a result of accident in 1986, as well as to the subsequent processes of radioactive substances transformation and biogeochemical migration in the soils of catchment basin and bottom sediments of reservoirs. Relatively low contents of radioactive substances are found in the river ecosystems. Due to high water change rate, the river bottom sediments have undergone decontamination processes (especially during floods and periods of high water) and over the years that passed since the accident have ceased to play the essential role as a secondary source of water contamination. The main sources of radionuclides in rivers are currently the washout from the catchment basin, the inflow from more contaminated water bodies, as well as the groundwater. On the other hand, the closed reservoirs, and in particular the lakes in the inner EZ, have considerably higher levels of radioactive contamination caused by limited water change and by relatively high concentration of radionuclides deposited in the bottom sediments. Therefore, for the majority of standing reservoirs the level of radionuclide content is determined mainly by the rates of mobile radionuclide forms exchange between bottom sediments and water, as well as by the external washout from the catchment basin.

Chromosomal aberration rate. We evaluated the cytogenetical effects level in embryo tissue of gastropod snail (*L. stagnalis*) as chromosomal aberration rate, considering it as reaction of snails to radiotoxicological condition of environment. The absorbed dose rate for snails from lakes within the EZ was registered in the range from 0.8 to 3.4 Gy/year. The highest rate was found in snails from lakes Dalyokoye and Glubokoye located within the dammed territory on the left-bank flood lands of the Pripyat River, the lowest—from the Pripyat River and Uzh River. Molluscs from the control lakes were characterised by an absorbed dose rate of about 0.3 mGy/yr. The rate of chromosomal aberrations was found in snails from lakes Dalyokoye and Glubokoye—21–22 %. In the embryo tissue of snails from Yanovsky Crawl the chromosomal aberration rate was about 18 % and in Azbuchin Lake—about 23 %. About 3.3 % of aberrant cells were registered in snail's embryo from the Pripyat River, and in snails of the Uzh River the aberration rate was about 2.3 %. The rate of chromosomal aberrations for snails from the control lakes was about 1.1–2.0 % (Fig. 1).

During 1998–2014 a tendency to decrease of chromosomal aberration level in molluscs from all the lakes of the EZ was registered.

The probabilistic prediction of the chromosomal aberration rate for gastropod snails in the lakes of the EZ has shown that spontaneous mutagenesis level (2.0–2.5 %) (Tsytsugina 1998) can be reached in Azbuchin Lake and Yanovsky Crawl in the 2020s–2030s and in lakes Daloykoye and Glubokoye—in the 2060s–2070s.

The spontaneous rate of aberrant cells for aquatic plants from our control lakes does not exceed 2.0–2.5 %. Upon the average we found 4.3 % aberrant cells in the common reed of Yanovsky Crawl during 2006–2013. The highest average rate of chromosomal aberrations in plants from Azbuchin, Glubokoye and Daloykoye lakes was registered to be respectively 5.9, 6.8 and 7.3 %. For comparison, the data obtained for reed from Goloseevo Lake amount to 1.0 %. The rate of chromosomal aberrations in reed from closed water bodies within the left-bank flood plain of the Pripyat River was 2–3 times higher than the spontaneous mutagenesis level (Fig. 2).

The single fragments were the frequently occurring aberrations in meristem cells of the common reed—on the average 57.0 % of all aberrations. Rate of single bridges was 40.7 %, multiple aberrations, including different variation of fragments and/or bridges (pair bridges or fragments, bridge and fragment, bridge and two or three fragments) was 2.3 %.

Haematological effects. The comparative leukogram of peripheral blood analysis of fishes from the stagnant water bodies of the EZ shows the presence of increased part of lymphocytes, presence of pseudobasophiles and foamy cells, as

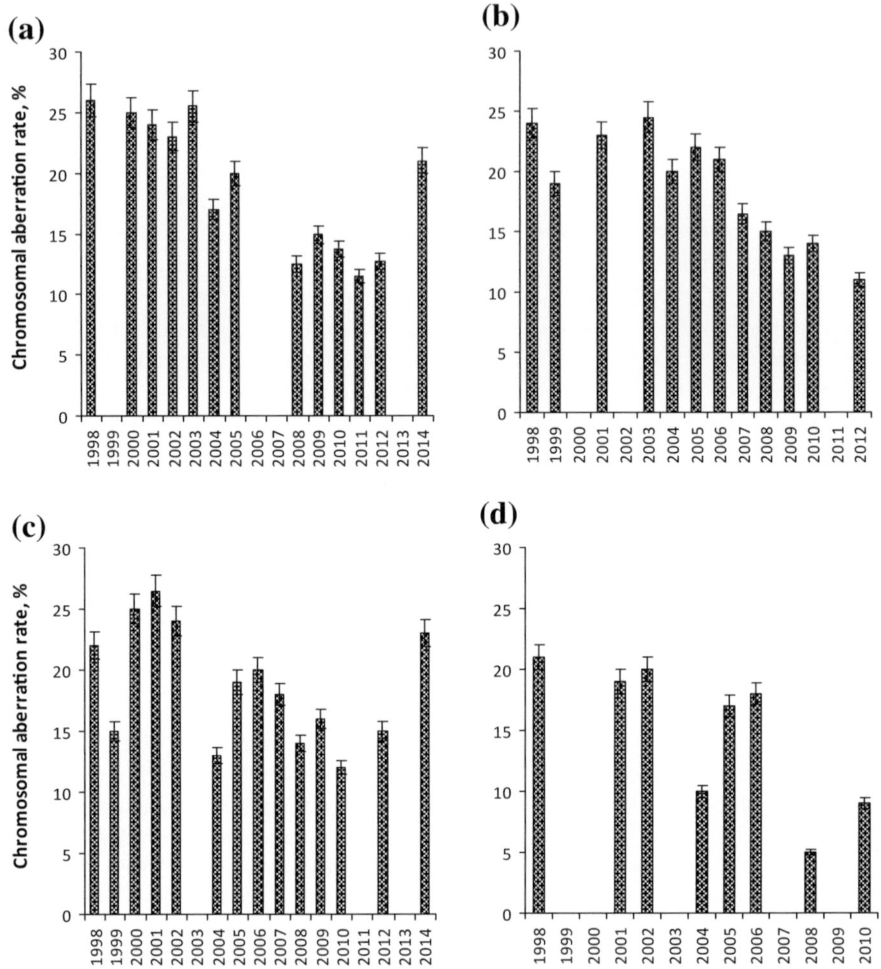

Fig. 1 **a–d** Radiation dose and chromosomal aberration rate in the pond snail (*L. stagnalis*) embryos in water bodies within the EZ and in the reference lakes during 1998–2014: **a** Glubokoye Lake; **b** Dalyokoye Lake; **c** Azbuchin Lake; **d** Yanovsky Crawl. X-axis: time of observations, years; Y-axis: chromosomal aberration rate, %. **e–h** Radiation dose and chromosomal aberration rate in the pond snail (*L. stagnalis*) embryos in water bodies within the EZ and in the reference lakes during 1998–2012: **e** Pripyat River; **f** Uzh River; **g** Opechen' Lake (reference lake) and **h** Vyrlitsa Lake (reference lake). X-axis: time of observations, years; Y-axis: chromosomal aberration rate, %

well as low maintenance of mature neutrophils, due to young granulocytes (mainly myelocytes and promyelocytes), which can testify to probable adaptation of white blood cells to the long-term radiation exposure in low doses. Correlation of red blood cells to leucocytes and thrombocytes of fishes from the EZ is higher than in the reference water body.

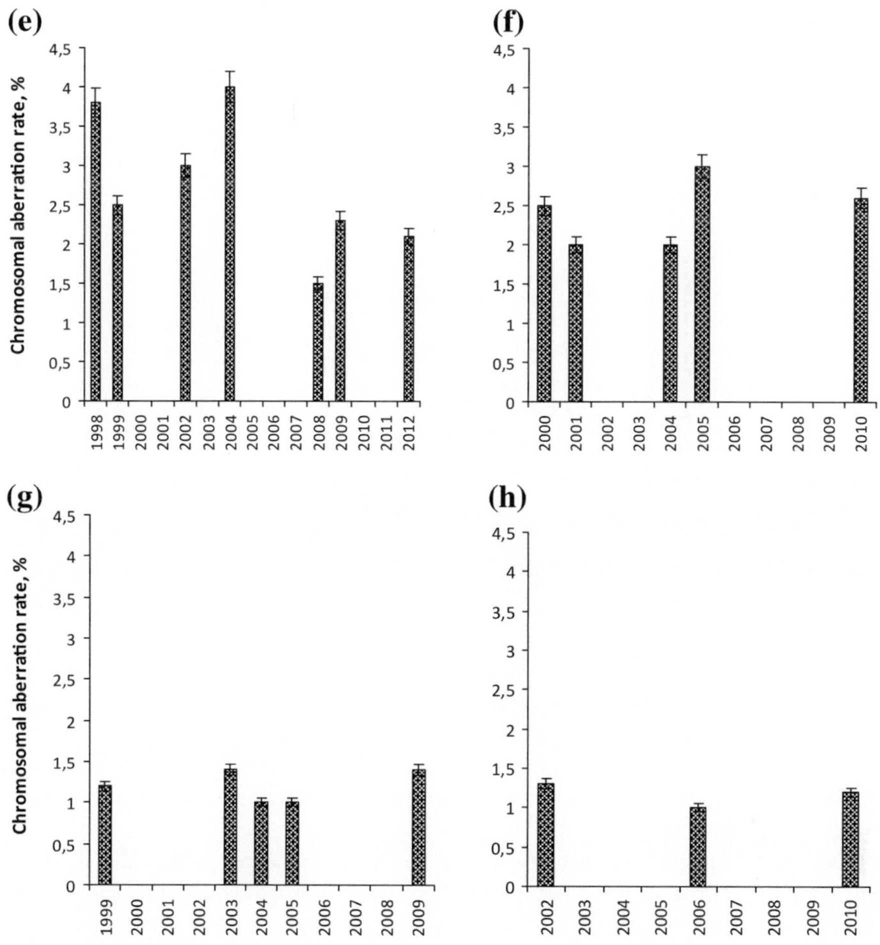

Fig. 1 (continued)

We call attention to the high rate of red cells aberrations and abnormalities in peripheral blood of fish from the stagnant water bodies within the EZ, where the absorbed dose rate in the fish organisms due to internal and external sources of irradiation has reached 269 μGy h^{-1} which is more than three orders higher in comparison with water bodies with background level of radioactive contamination. It can testify to certain mutagenous action of environment and possible display of radiation-induced genetic instability of fishes in the conditions of chronic radiation impact. Make a point of the amount of red cells with deformed shape of nucleus as different type of invaginations and ramifications as well as formation of double-nucleus cell, schistocytes (cells without nucleus), parietal nucleus, micro-cytes, nucleus and cytoplasm vacuolization, micronuclei, etc. (Fig. 3).

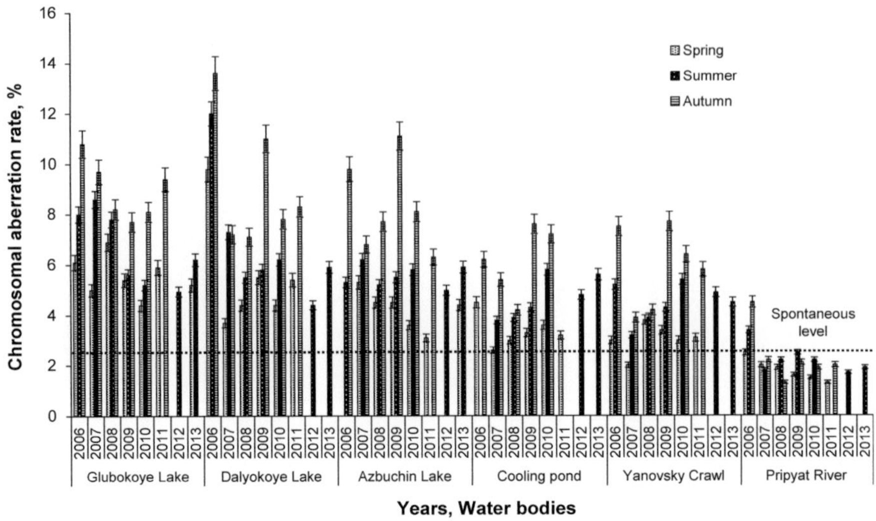

Fig. 2 Seasonal dynamics of the cytogenetical damages in cells of root meristems of the common reed (*P. australis*) in water bodies within the EZ

The atypical shape of red cell nucleus in the blood of healthy fishes, from data of row of authors, is encountered with a frequency of 0.4 ‰. The increase in the frequency of red cells with the deformed nucleus (invagination of nuclear envelope) is estimated by different authors as degenerative changes of red cells, arising from the negative impact of environmental factors on the organism of fishes (Kalinina 2002; Lugas'kova 2003).

Our studies in peripheral blood of fish from water bodies within the EZ have shown an increased level of the above-mentioned morphological damages of erythrocytes, which was generally for pray fish 4–16 times and for predatory fish 7–15 times higher in comparison with the fish from the reference water bodies with background levels of radioactive contamination (Fig. 3). The rate of micronuclei in red cells was very low and did not exceed 0.3–0.6 ‰. For the fish from reference water bodies this type of abnormality is not discovered.

Morphological indexes and teratogenic damages of molluscs. During 2009–2012 we analyzed morphological parameters and the presence of anomalies of egg capsules and egg mass of the pond snails *L. stagnalis*: despiralization or weak spiralization strand with egg capsules; multilane organization of egg capsules in egg mass; loose location of egg capsules; twin egg capsules; polyzygosity of egg capsules; egg capsules without zygotes; zygote outside the egg mass; egg capsules of larger or smaller sizes, as well as irregular shape Table 1.

In the lakes of the EZ we observed an increased percentage of abnormal molluscs shells with various forms of curvature of the last curl, often in the form of step (up to 0.5 cm) deformation occurring usually in the second year of the snail. In Yanovsky Crawl the share of abnormal shells was maximal—58.3 %, in

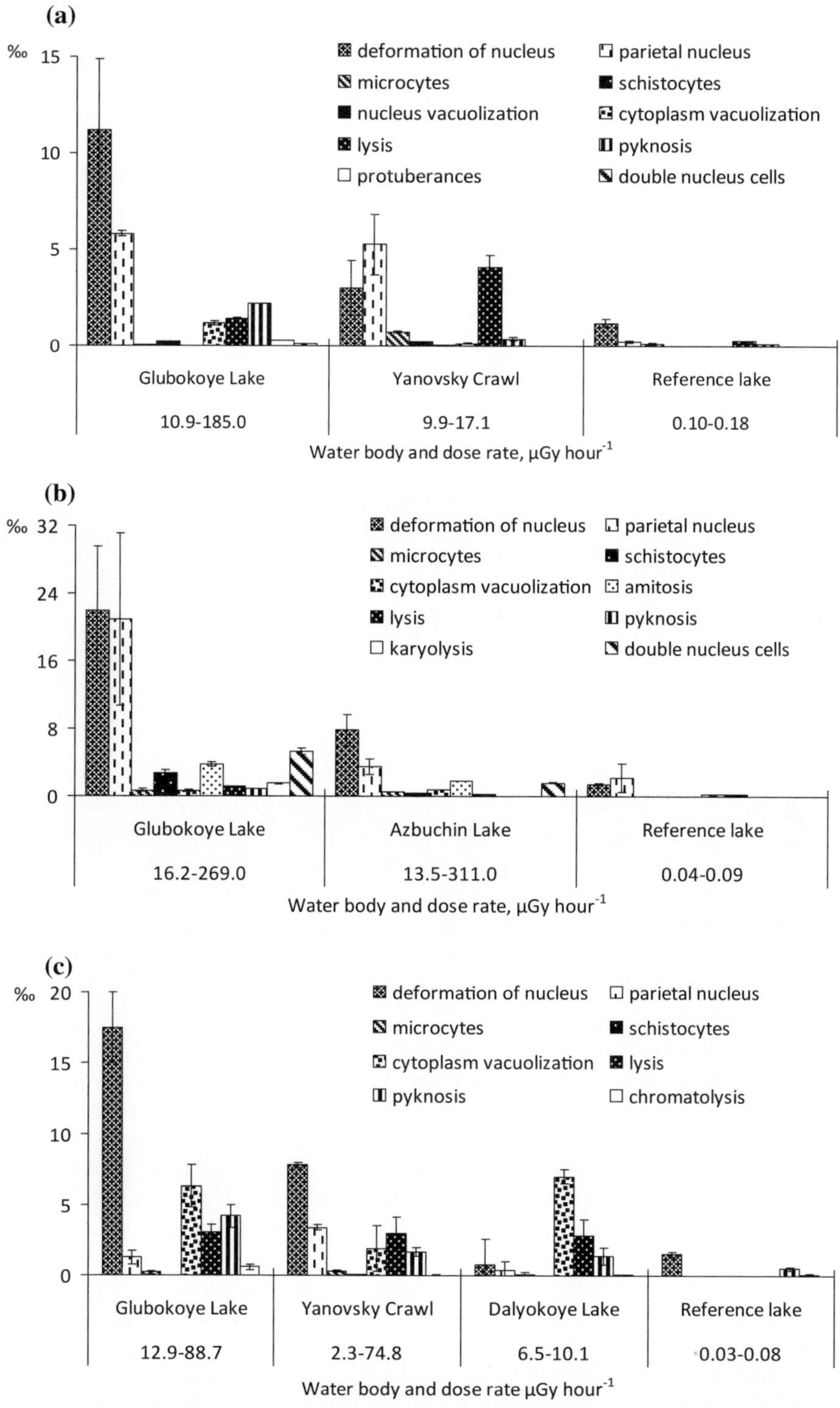
(a)
‰
15
10
5
0
deformation of nucleus
parietal nucleus
microcytes
schistocytes
nucleus vacuolization
cytoplasm vacuolization
lysis
pyknosis
protuberances
double nucleus cells
Glubokoye Lake
10.9-185.0
Yanovsky Crawl
9.9-17.1
Reference lake
0.10-0.18
Water body and dose rate, µGy hour-1
(b)
‰
32
24
16
8
0
deformation of nucleus
parietal nucleus
microcytes
schistocytes
cytoplasm vacuolization
amitosis
lysis
pyknosis
karyolysis
double nucleus cells
Glubokoye Lake
16.2-269.0
Azbuchin Lake
13.5-311.0
Reference lake
0.04-0.09
Water body and dose rate, µGy hour-1
(c)
‰
20
15
10
5
0
deformation of nucleus
parietal nucleus
microcytes
schistocytes
cytoplasm vacuolization
lysis
pyknosis
chromatolysis
Glubokoye Lake
12.9-88.7
Yanovsky Crawl
2.3-74.8
Dalyokoye Lake
6.5-10.1
Reference lake
0.03-0.08
Water body and dose rate µGy hour-1

◀ **Fig. 3** Radiation dose rate and frequency of erythrocyte abnormalities in peripheral blood of the perch *Perca fluviatilis* (**a**), the Prussian carp *Carassius gibelio* (**b**) and the common rudd *Scardinius erythropthalmus* (**c**) in water bodies within the EZ and in reference lakes during 2011–2013

Table 1 Morphological indexes and teratogenic damages of egg mass of the pond snails *L. stagnalis* in water bodies within the EZ in 2009–2012

Index	Reference lake	Yanovsky Crawl	Glubokoye Lake	Dalyokoye Lake	Azbuchin Lake	Pripyat River
Length of egg mass (mm)	33.5	26.3	30.0	27.8	31.3	31.8
Length of egg capsules (mm)	1.35	1.16	1.29	1.22	1.22	1.31
Quantity of egg capsules in egg mass	106	75	93	89	101	99
Damages of egg capsules (%)	0.8	23.6	9.2	2.4	1.6	–[a]

[a]no data

Glubokoye Lake—48.9 %, in Krasnensky old river bed (in the dammed area)—25.0 %, in Dalyokoye Lake—10 %, in Azbuchin Lake—2.8 %, in the Pripyat River (near Chernobyl town)—1.1 %. In 5 reference lakes similar anomalies were absent or did not exceed 0.7 %. There is a good correlation between the above-mentioned abnormalities and density of ^{137}Cs contamination of bottom sediments in the studied reservoirs (Fig. 4).

Parasitological effects. In reservoirs with the increased level of radioactive contamination (lakes of the dammed territory of the left-bank flood plain of the Pripyat River) the high level of the common reed damage by gall-producing arthropods, in particular by mites *Steneotarsonemus phragmitidis,* has been observed since 2000. This phenomenon was registered for the first time in the territory of Ukraine within the EZ (Gudkov et al. 2006) . We suppose the scales and speed of distribution of this phenomenon in the reservoirs of the EZ deserve special attention. As the common reed is almost cosmopolitan, it is quite logical to predict wide moving of mites in other reservoirs, in which the Polesye region (woodlands) is so rich. Thus, in 2000 and 2001 the damaged individuals were registered only in one of the seven reservoirs of sampling in the EZ—Dalyokoye Lake, whereas during 2002–2009 the damaged individuals of the common reed began to occur in all the other studied water objects. In some reservoirs we found a single affected individual only, in Yanovsky Crawl and Azbuchin Lake the described phenomenon has quickly received distribution and in 2005 the share of damaged plants was accordingly 74 and 32 %.

The highest percent of affected plants was observed in Dalyokoye Lake (at present 100 % of shoots), located in the dammed territory of Krasnensky flood-lands (Fig. 5). This territory is characterized by the maximal density of radioactive contamination within the EZ. The specific activity of radionuclides in

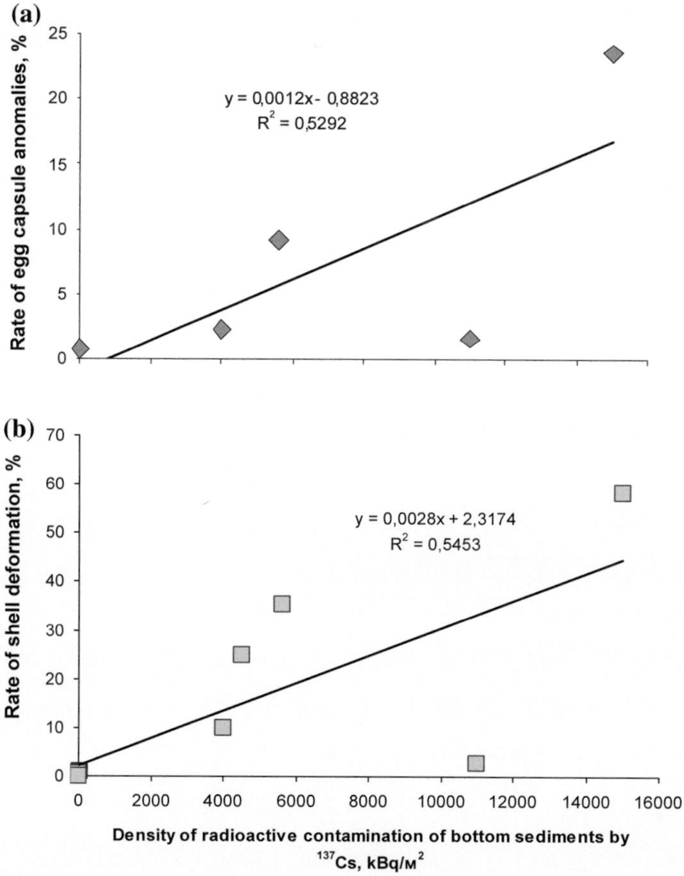

Fig. 4 Correlation between density of radioactive contamination of bottom sediments by ^{137}Cs and rate of egg capsule anomalies (**a**) and rate of morphological deformation of shells (**b**) of the pond snail *L. stagnalis*

Fig. 5 Dynamics of damage of the common reed's shoots by the cork-screw gall mites in Dalyokoye Lake

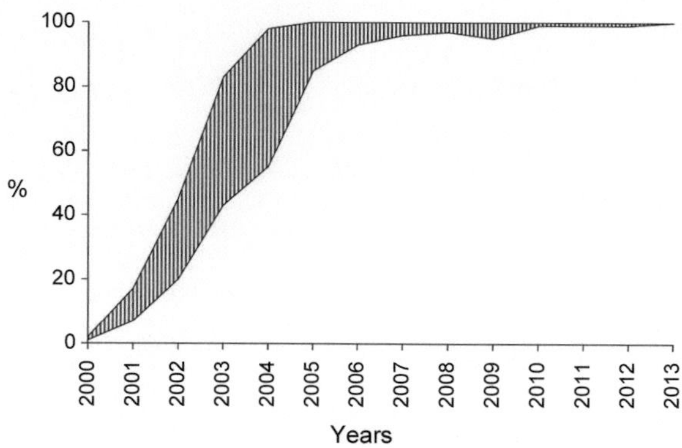

the reed's tissue during the period under consideration is 10,000 Bq/kg for ^{137}Cs and 2000 Bq/kg for ^{90}Sr (Gudkov et al. 2005). The absorbed dose caused by external gamma radiation and radionuclides incorporated in meristem tissue of plants reached more than 4.0 Gy/year during the 10–15 years after the Chernobyl accident. The damage events of common reed by larvae of gall fly of family *Chloropidae*, genus *Lipara* have been registered as well.

In 2013 the state of the common reed population in Dalyokoye Lake became critical—it has virtually disappeared from the lake, leaving a depressed individual shoots in some coastal areas. It disappeared as well in the western part of Yanovsky Crawl where the curtain of the common reed in recent years was also heavily damaged by gall mites (Fig. 6).

The influence of parasitic fungi *Claviceps purpurea* (ergot) lesions on seed production of the common reed was determined. The positive correlation between parasitic fungi lesions of plants and levels of radiation exposure was registered. High level of fungi lesions ratio for the common reed sampled from the lakes of the

Fig. 6 Damage of the common reed by the cork-screw gall mites in water bodies within the EZ: **a** normal (*left*) and damaged panicles of the common reed; **b** population of the common reed in Dalyokoye Lake at the end of vegetation in 2007; **c** population of the common reed in Dalyokoye Lake at the end of vegetation in 2013; **d** population of the common reed in Yanovsky Crawl at the end of vegetation in 2013

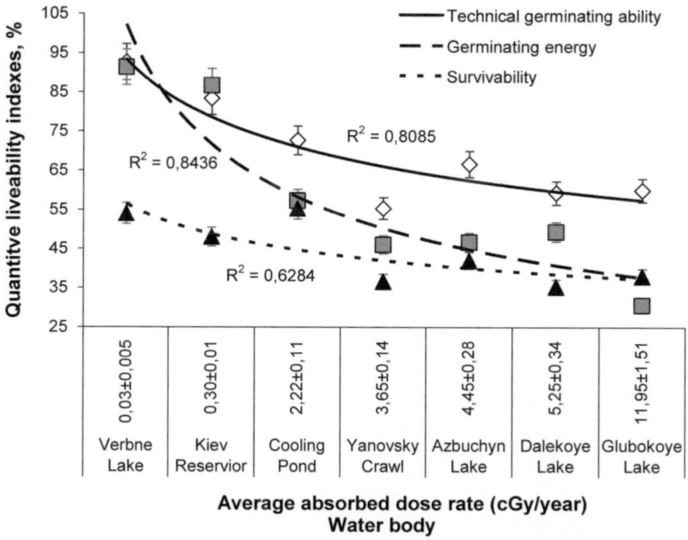

Fig. 7 Dose-dependent liveability indexes of the common reed's (*P. australis*) seeds at different levels of radioactive contamination of lakes within the EZ

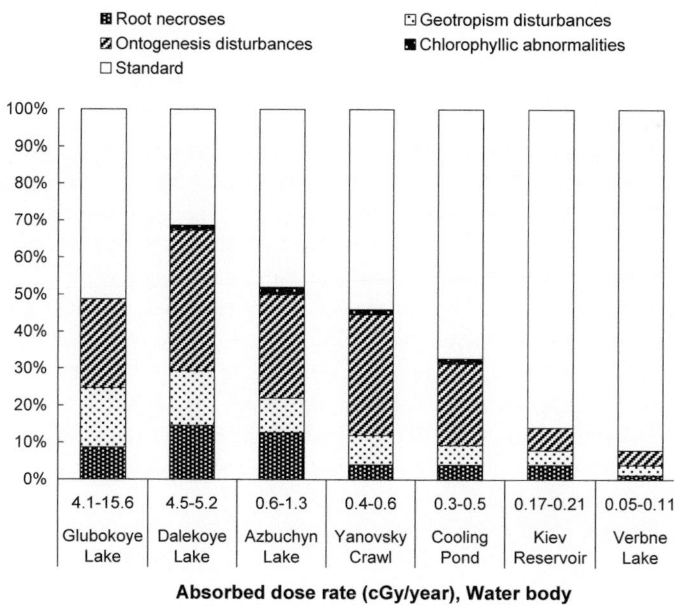

Fig. 8 Dose-dependent quantity of abnormalities of the common reed's (*P. australis*) germs in water bodies with different levels of radioactive contamination within the EZ

left-bank dammed flood-plain of the Pripyat River, which are the most radioactive contaminated within the EZ, was marked. Such significant damage of the common reed's panicles by the parasitic fungi in the lakes of the Krasnensky flood land was observed on the background of the lowest parameters of seed production in water bodies of the EZ.

Viability and abnormalities of seed progeny. Analysis of the viability of the seed progeny of the common reed at germination in the laboratory showed that in gradient of absorbed dose rate from 0.03 to 11.95 cGy/year for parental plants in lakes, there is a reduction in technical germination from 93 to 60 %, germination energy—from 91 to 30 % and seed viability—from 54 to 38 % (Fig. 7).

At the same time the number of abnormalities of seed seedlings increased significantly: necrosis of roots—from 1.3 to 14.7 %; disturbance of gravitropism—from 2.6 to 17.0 %; damages of organogenesis—from 4 to 24 %, and disturbance of chlorophyll synthesis—from 0 to 2 % (Fig. 8).

Conclusions

Self-purification of closed water bodies within the EZ is an extremely slow process. Therefore, the ecosystems of the majority of lakes, dead channels and crawls have a high level of radionuclide contamination of all components.

The established dose-related effects in hydrobionts of the lakes within the EZ indicates a damage of biological systems at sub-cellular, cellular, tissue, organ, organism and population levels as a result of chronic exposure to low doses of ionizing radiation. The rate of chromosomal aberrations in cells of aquatic species exceeds many-fold the level of spontaneous mutagenesis level to aquatic biota. The increased levels of chromosome damages may be a manifestation of radiation-induced genetic instability, which is one of the main mechanisms for the protection of living organisms from exposure to stressors with subsequent implementation at higher levels of organization of biological systems.

Haematological research of fish in the EZ shows essential changes in leukogram structure and high amount of red cells with atypical shape of nucleus as different type of invaginations, ramifications, etc. as well as cells without nucleus in the blood of fishes from the water bodies with high levels of radioactive contamination which allows one to assume that the qualitative indexes of red cells of peripheral blood of fishes are more sensible to chronic radiation influence as compared with the white blood elements.

For the common reed in the lakes of the inner EZ, the high level of affection by gall-producing mites *Steneotarsonemus phragmitidis* (up to 100 % of a vegetative population of lake) was discovered. During 2002–2013 affection by mites quickly propagated in other closed reservoirs of the EZ, essentially reducing rates of growth, seed efficiency, bioweight of plants and in some water bodies it was a reason of disappearance of the common reed population. There is no reason to finally approve that the damage of a reed is due to the impact of ionizing radiation; however, we are anxious about the fact that this mite species is registered for the first time in Ukraine within the EZ, in the territory which is the most contaminated by radionuclides. In this connection it is supposed that one of the possible reasons of a total plants disease can be a loss of parasitical stability from them in conditions of chronic radiation influence.

Acknowledgments This study was supported by the National Academy of Sciences of Ukraine and by the State Agency of Ukraine on the Exclusion Zone Management. The authors are grateful to the personnel of the state specialized enterprises "Ecocentre" and the Chernobyl NPP for promoting research within the EZ.

References

ERICA Assessment Tool 1.0 (Version November 2012) The integrated approach seeks to combine exposure/dose/effect assessment with risk characterisation and managerial considerations (http://www.erica-tool.com)

Geras'kin SA, Fesenko SV, Alexakhin RM (2008) Effects of non-human species irradiation after the Chernobyl NPP accident. Environ Int 34:880–897

Gudkov DI, Kuzmenko MI, Derevets VV, Nazarov AB (2005) Aquatic ecosystems within the Chernobyl NPP exclusion zone: The latest data on radionuclide contamination and absorbed dose for hydrobionts. In: Brechignac F, Desmet G (eds) Equidosimetry—ecological standartization and equidosimetry for radioecology and environmental ecology, vol 2. Series C: Environmental Security, Springer, Dordrecht, pp 333–342

Gudkov DI, Uzhevskaya CF, Nazarov AB, Kolodochka LA, Dyachenko TN, Shevtsova NL (2006) Lesion in common reed by gall-producing arthropods in water bodies of the Chernobyl NPP exclusion zone. Hydrobiol J 42(1):82–88

Hinton TG, Alexakhin R, Balonov M et al (2007) Radiation-induced effects on plants and animals: Findings of the United Nations Chernobyl Forum. Health Phys 93:427–440

IAEA (2006) Environmental consequences of the Chernobyl accident and their remediation: Twenty years of experience. Report of the UN Chernobyl Forum Expert Group "Environment" (EGE). IAEA, Vienna

Ivanova NT (1983) Atlas of the fish blood cells. Moscow (Russian)

Kalinina MV (2002) Feature of young chum salmon blood as indicator of water bodies contamination by heavy metals. In: Proceedings of International Conference on "New technologies for the protection of biodiversity in aquatic ecosystems", Moscow, 123 (Russian), 27–29 May 2002

Lugas'kova NV (2003) Species specific of cytogenetic stability of fish in condition of eutrophic water body. Ecology 3:235–240 (Russian)

Moller AP, Mousseau TA (2006) Biological consequences of Chernobyl: 20 years on. Trends Ecol Evol 21:200–207

Pausheva ZP (1974) Practical work on cytology of plants. Kolos, Moscow (Russian)

Shevtsova NL, Gudkov DI, Stoyko YA, Syvak EV (2005) To the method of determination of chromosome damages of higher aquatic plants at the example of common reed and arrowhead. Sc Acta of the Ternopil State Teacher's Training Univ 26(3):478–479 (Russian)
Sokolov VE, Ryabov IN, Ryabtsev IA et al (1993) Ecological and genetic consequences of the Chernobyl atomic power plant accident. Vegetation 109:91–99
Tsytsugina VG (1998) An indicator of radiation effects in natural populations of aquatic organisms. Radiat Protect Dosim 75(1–4):171–173
UNSCEAR (1996) Report to the general assembly. United Nations, New York
Zhyteneva LD, Poltavtseva TG, Rubnitskaya OA (1989) Atlas of the normal and pathological changes of the fish blood cells. Rostov-na-Donu (Russian)

Radioactive Tracers in the Black Sea: A Tool for Environmental Assessment and Ecological Regulation

Sergey B. Gulin and Victor N. Egorov

Introduction

Before the 1950–1960s, the Black Sea was known as one of the most productive seas having abundant marine life and favorable habitats. However, during the 1970s to early 1990s the Black Sea changed from a highly diversified ecosystem into eutrophic and contaminated basin with dramatic declining stocks of the living resources (Zaitsev and Mamaev 1997). Owing to its geographical location and limited water exchange with the rest of the world's oceans, the Black Sea has been also one of the marine basins most contaminated with artificial radioactivity. It is originated primarily from the atmospheric nuclear weapon tests, which were conducted mainly in 1954–1958 and in 1961–1962, and after the Chernobyl NPP accident in April 1986 (Egorov et al. 2010). Besides the direct atmospheric deposition, the Black Sea received and continues to receive substantial further input of the Chernobyl-derived radionuclides by river runoff, particularly to its north-western area from the Danube and Dnepr Rivers. As a result, the Black Sea ranks second after the Irish Sea and third after the Baltic Sea with respect to the ^{90}Sr and ^{137}Cs concentration in seawater, respectively (Nielsen et al. 2010).

It is generally difficult to evaluate the relative importance of numerous harmful factors, both natural and anthropogenic, on the marine ecosystems health. Many of these factors act interactively so that, for example, the overall effect of changes in the inputs of nutrients, organic wastes, heavy metals and artificial radionuclides would be difficult to predict (GESAMP 1990; Gulin and Egorov 2013). Therefore, environmental assessments require estimation of an integrated value of the anthropogenic impact. One of the promising approaches is to evaluate capability of marine environments for self-purification that is a numeric measure of the upper limit of the

S.B. Gulin (✉) · V.N. Egorov
The A.O. Kovalevsky Institute of Marine Biological Research,
Russian Academy of Sciences, Moscow, Russia
e-mail: s.b.gulin@imbr-ras.ru

V.L. Korogodina et al. (eds.), *Genetics, Evolution and Radiation*,
DOI 10.1007/978-3-319-48838-7_25

pollutant discharge into the basins. This value is widely defined as the environmental capacity of the water bodies, i.e. their ability to take up an impact without unacceptable changes (GESAMP 1990; Gulin and Egorov 2013; Izrael and Tsiban 1983; Polikarpov and Egorov 1986). Such an approach is particularly well developed regarding the radioactive contamination of marine basins due to the possibility to identify the origin and source of the radionuclides and to trace their migration pathways between the different compartments. This allows applying the artificial radionuclides as tracers of processes governing the self-purification of aquatic ecosystems against radioactive and chemical pollutants (Timofeeff-Ressovsky 1957).

The purpose of this chapter is to overview radiotracer applications for assessments of the self-purification rate of the Black Sea against nuclear and non-nuclear pollutants as a basis for environmental regulation of the human impact upon this unique marine basin.

Background and Data Sources

During the last years, a number of oceanographic and biogeochemical processes were successfully studied in the Black Sea using the naturally occurring and man-made radiotracers, first of all: vertical water exchange between oxic and anoxic waters in the abyssal Black Sea basin (Buesseler et al. 1991; Egorov et al. 1999, 2010; Sanchez et al. 1991), migration and transformation of riverine waters over the NW Black Sea shelf (Stokozov and Buesseler 1999), the capacity of the Black Sea catchment basins to retain pollutants (Gulin et al. 2002, 2003), the sediment accumulation rates in the abyssal and shallower areas (Anderson et al. 1989; Buesseler and Benitez 1994; Calvert et al. 1991; Crusius and Anderson 1991; Gulin et al. 2002; Yücel et al. 2012), particle fluxes in the water column including the scavenging of the particle-reactive pollutants (Wei and Murray 1991; Gulin 2000b; Hay et al. 1990), geochronological reconstruction of the recent deposition events and pollution history (Gulin 2000a; Gulin et al. 1997). The most recent findings were the tracing of secondary radioactive contamination of the Black Sea due to remobilization of the Chernobyl-derived radionuclides from the seabed sediments (Gulin et al. 2013b), and the long-distance penetration of the Danube water into the abyssal Black Sea waters via the submarine Viteaz (Danube) Canyon (Gulin et al. 2015).

Environmental Capacity, Self Purification and Half-Lives

One of the first quantitative definitions of the limits for accumulation of hazardous substances in aquatic environments was fulfilled by Agre and Korogodin (1960) and their followers (Korogodina et al. 1996; Kutlakhmedov et al. 2006), who have developed the concept of radiocapacity of water bodies subjected to radioactive contamination. This capacity was defined as maximum amount of radionuclides in the aquatic ecosystem which does not yet violate its trophic functions, i.e. abundance, diversity, biomass and productivity of biota.

An approach for computation of the environmental (assimilation) capacity was described by Izrael and Tsiban (1983), who applied the terms of a threshold concentration of pollutant and the averaged rate of its removal from the water column.

Polikarpov and Egorov (1986) have developed a concept of 'ecological capacity' regarding the pollution, which is a dynamic parameter showing the balance between pollutant inflow to marine basins and their capability for self-purification.

The main tool for assessment of the self-purification rate of aquatic environments is the value of environmental half-life of radioactive and non-radioactive pollutants. Such values combine the effective and ecological half-lives. The first denotes the halving of concentration of nuclear and non-nuclear pollutant in a particular biotic or abiotic compartment of marine ecosystem by interaction of radioactive decay (or degradation of the chemical toxicant) and the processes of self-purification of the water column, while the ecological half-life implies the decrease in the pollutant concentration only due to the self-purification (e.g., Florou et al. 2002).

An excellent opportunity for direct estimation of the half-lives is provided by data on time-series measurement of concentration of the man-made radionuclides in the upper marine waters, which were subjected to radioactive contamination as a result of global fallout and of the local nuclear accident, such as the Chernobyl disaster. These fallouts were relatively short-term, acting as a pulse input, and the only factor of radionuclide degradation was the radioactive decay that is independent of any environmental factors. An additional feature of radioactive contamination is a possibility to identify the main sources of artificial radionuclides using their activity or atomic ratios, which are well documented for both releases from the accidental reactors and nuclear weapons testing. All of these provide exceptional conditions for reliably interpreted tracking of the spatial and temporal trends of the seawater pollution, not only radioactive, but also chemical (Osvath et al. 2007).

One of such estimations was carried out during the long-term radioecological studies and monitoring of concentration of the Chernobyl-derived radionuclides, both well-soluble ^{137}Cs and ^{90}Sr, and sorption-reactive plutonium isotopes, in the Black Sea (Egorov et al. 2010; Gulin et al. 2012, 2013a) . The data have showed that post-Chernobyl evolution of the man-made radioactivity in the Black Sea was controlled, except radioactive decay, by the vertical water mixing, which dilutes the surface radionuclide concentrations with the less contaminated deep-sea waters; the loss through the Bosporus Strait connecting the Black Sea with the Mediterranean; and the removal of particle-bound radionuclides on sinking fossils of plankton and suspended material including those come from the land due to erosion of soils within the watersheds. Because of the effects of these factors, as well as due to the radioactive decay, the initial inventory of the Chernobyl-derived radionuclides has decreased abruptly, reaching the pre-accident level, except estuarine and some near-shore zones, particularly in front of the mouths of the Danube and Dnepr Rivers, which account for 75 % of the total river runoff to the Black Sea.

The summarized data on the temporal evolution of concentration of the Chernobyl-derived radionuclides in the surface Black Sea waters is given in Fig. 1 showing that initial concentration of ^{137}Cs (~2700 TBq) decreased rapidly during one year period after the Chernobyl fallout due to fast mixing of the water masses within the upper layer, and then more gradually to around 500–600 TBq in

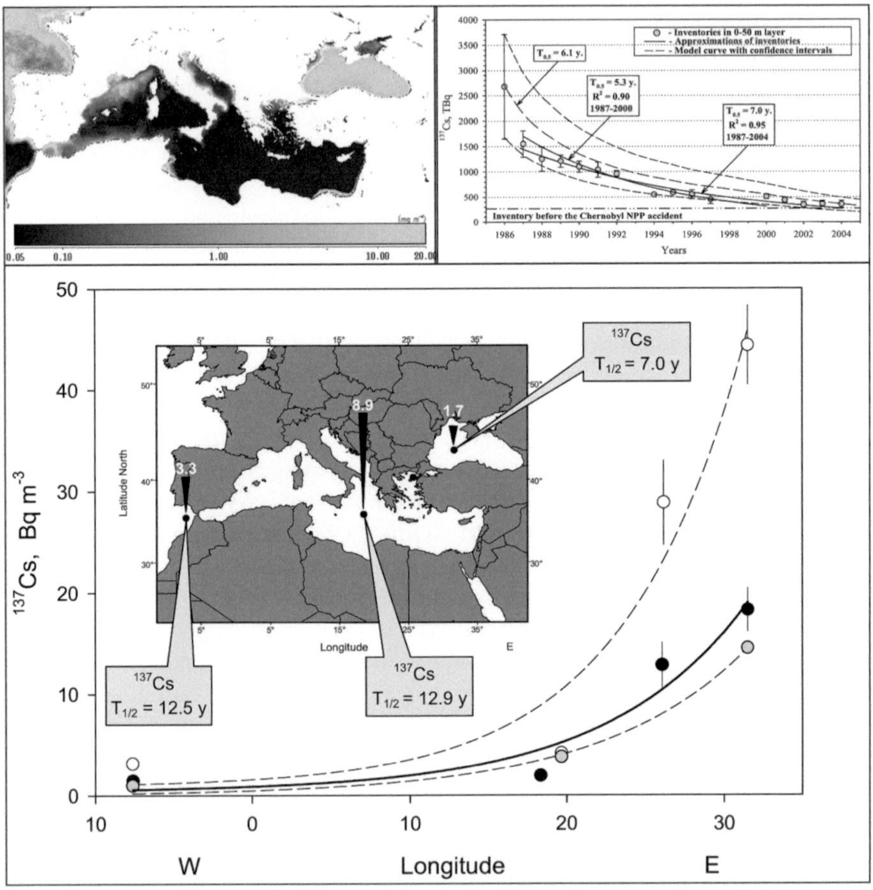

Fig. 1 Longitudinal distribution of ^{137}Cs and $^{239+240}$Pu (*black inverted triangular bars* in the map) in surface waters of the western Black Sea, Aegean Sea (near Dardanelles), Ionian Sea and in the Atlantic area adjacent to the Gibraltar Strait measured in 2002–2003 (*filled circle*) in comparison with the post-(*open circle*) and pre-Chernobyl (*filled circle*) data obtained in 1988–89 and in the early 1980s, respectively. *Numbers* above the *black triangular bars* relate to the values of $^{239+240}$Pu concentration, mBq/m^3. The *left upper panel* shows satellite map of annual mean surface chlorophyll concentration for 2002 (the *grey-scale* coding shows pigment concentration in mg•m^{-3}). The *right upper panel* represents temporal evolution of ^{137}Cs inventory in 0–50 m layer of the Black Sea after the Chernobyl accident

1998–2000 and 350 ± 60 TBq in 2001–2004 (Egorov et al. 2010; Gulin et al. 2012, 2013b, 2015). The exponential function approximating the time-series measurements of ^{137}Cs concentration in the upper 50-m Black Sea water resulted in the effective half-lives for decrease of ^{137}Cs inventory between 5.3 years in 1987–2000 (Egorov et al. 2010), 7.0 years in 1987–2004 (Egorov et al. 2010) and 8.6 years over the period of 1987–2011 (Gulin et al. 2013b, 2015) or 6.6 years on the average. The increase of these half-lives with time may be caused by growing influence of remobilization of ^{137}Cs from the seabed sediments and due to a slow migration from watersheds of the adsorbed ^{137}Cs with the eroded soil particles (Gulin et al. 2002, 2003, 2013b). The given value of effective half-life for ^{137}Cs in the Black Sea surface water is considerably shorter than those found earlier in the central part of Mediterranean (mid-Ionian Sea) and in the NE Atlantic area adjacent to the Gibraltar Strait (Fig. 1), where half-lives for ^{137}Cs in the surface water were 12.9 ± 1.0 and 12.5 ± 2.1 years, respectively (Gulin et al. 2012, 2013a). The shorter half-life, which was found in the Black Sea, is most likely caused by a faster dilution of ^{137}Cs in the surface seawater with riverine waters, as well as by a loss of ^{137}Cs via the Bosphorus Strait. Also, the higher self-purification rate of the Black Sea waters, relative to Mediterranean and NE Atlantic, has been revealed for particle-reactive $^{239+240}Pu$, whose surface water concentration measured in 2002 was considerably lower in the Black Sea than in Ionian Sea (Fig. 1) despite the fact that the latter is located much far away from the accident site. These data, in comparison with results of the previous measurements of plutonium concentration in the Black Sea waters (Egorov et al. 2010), suggest the effective half-life of $^{239+240}Pu$ of ~6.5 years on the average. In this case, the rapid decrease of $^{239+240}Pu$ surface water concentration is most likely caused by a higher rate of plutonium scavenging on sinking particles in the mesotrophic Black Sea basin, in contrast to oligotrophyc Mediterranean waters (Fig. 1).

Radiochemoecological Model for Environmental Regulation

Traditionally, the assessment of environmental state of marine ecosystems is based upon the use of maximum allowable concentrations (MAC) of toxic substances in seawater and biota or threshold limit values (TLV) of pollutants as the measures of anthropogenic pressure (e.g., GESAMP 1990 and references therein). The MAC is a ceiling concentration whereas the TLV is a time-weighted average of the concentration of the hazardous substance in the environment (Tsuchiya and Harashima 1965; Polikarpov 1966). But, the MAC and TLV values are static and they cannot be applied directly for estimation of permissible level of the pollutant input to water basins because they do not take into account the removal of toxic substances, their biogeochemical transformation, decay or degradation, immobilization and deposition in seabed sediments. The influence of self-purification processes on the pollutant concentration may be described with the following simple model (Gulin and Egorov 2013):

$$\frac{dC_W}{dt} = pC_W - \lambda C_W - s_{Ecol}C_W + rC_W = C_W(p - \lambda - s_{Ecol} + r), \quad (1)$$

where p defines the specific rate of the pollutant inflow; λ—constant of radioactive decay; s_{Ecol}—specific rate of the decrease in radionuclide concentration with time due to the processes of self-purification; r—specific rate of the radionuclide remobilization from seabed sediments and resulting from degradation of suspended matter or decomposition of dead organisms.

One may readily find the solution of Eq. (1) to be:

$$C_W(t) = C_{W_o}e^{k \cdot t}, \quad (2)$$

where $k = p - \lambda - s_{Ecol} + r$; C_{Wo}—is the initial concentration of radionuclide in seawater after the fallout or discharge of the pollutant.

Since the rates of self-purification of seawater against the suspended and dissolved forms of the pollutants may differ, the dynamics of its total concentration is to be expressed as:

$$\frac{dC_W}{dt} = \frac{d(C_p + C_d)}{dt} = C_p k_p + C_d k_d = aC_W k_p + (1 - a)C_W k_d \quad (3)$$

$$C_W(t) = C_{p_0}e^{k_p \cdot t} + C_{d_0}e^{k_d \cdot t} = aC_{W_0}e^{k_p \cdot t} + (1 + a)C_{W_0}e^{k_d \cdot t}, \quad (4)$$

where C_p and C_d—concentrations in seawater of particulate and dissolved fractions of the pollutant; k_p and k_d—constants of the rate of their changes with time t; a—percentage of the pollutant in particulate form.

If to set that pollutant concentration cannot exceed its maximum allowable concentration: $C_W(t) \leq MAC$, then specific rate of the maximum allowable input of the pollutant over one year period (t = 1) may be determined as:

$$p = \ln\left(\frac{MAC}{C_{W_o}}\right) + \lambda + s_{Ecol} - r \quad (5)$$

If to apply, as the measure of self-purification rate, the above-mentioned values of the effective half-lives of the decrease in concentration of well-soluble and particle-reactive fallout radionuclides (6.5–6.6 years on the average in both cases, see above), then we can express the integrated value of specific rate of the self-purification by combining it with specific rates of decay and remobilization: $S_{Eff} = S_{Ecol} + \lambda - r$. This allows simplification of the Eq. (5) as follows:

$$p = \ln\left(\frac{MAC}{C_{W_o}}\right) + s_{Eff} \quad (6)$$

The given value shows that specific rate of the input of pollutant depends on the ratio between its maximum allowable concentration and initial level of the pollution in water, and that the upper limit of the rate of the pollutant input may increase with the respective increasing of the self-purification rate. Thus, this parameter combines characteristics of toxicity of pollutant and an integrated rate of the self-purification of seawater. It, therefore, could be applied for purposes of environmental regulation of the maximum allowable anthropogenic pressure with regards to the nuclear and non-nuclear pollutants, applying the value of specific rate of the maximum allowable annual input of the pollutant, at which its concentration will not exceed the MAC level over one year period. But after that period, according to Eqs. (2) or (4), concentration of pollutant will continue to increase progressively. In order to avoid this, an upper limit of the pollutant concentration, to be equal to the MAC, should be included in Eq. (1), e.g. by transforming it to the logistic sigmoid (Verhulst) function:

$$\frac{dC_W}{dt} = kC_W \cdot \left(1 - \frac{C_W}{MAC}\right), \quad k = p - s_{Eff}, \tag{7}$$

assuming that specific rates of self-purification of seawater against the suspended and dissolved pollutant are comparable (i.e. $k_p \approx k_d$ as it was in the above-mentioned case with the decreasing of concentration of the well-soluble ^{137}Cs and particle-reactive $^{239+240}Pu$).

Equation (7) implies that if concentration of pollutant is much lower than MAC, then its dynamics may be described as the exponential growth similarly with Eq. (1), and, therefore, the value p, given in Eq. (6), can be applied also in Eq. (7). But the closer this concentration is approaching MAC, the slower it should be risen and has to stop completely ($dC/dt = 0$) when C has reached the MAC for infinite time (or come closer to reaching in finite time). It is important to stress that this model corresponds to a case when self-purification of the water body is not sufficient itself to maintain the pollutant concentration at the level not exceeding the MAC, and therefore an additional restriction of the discharge of the polluting substance is required as a mean to avoid the excess contamination. The solution of the logistic model (7) is:

$$C_W(t) = \frac{C_{W_o} \cdot MAC}{C_{W_o} + (MAC - C_{W_o}) \cdot e^{-k \cdot t}} \tag{8}$$

As can be seen from an example of the logistic curve plotted in Fig. 2, the beginning of this function shows an exponential rise that is slowing to the linear growth and then approaches the MAC value with an exponentially decaying curve. One of the most important consequences of the Eqs. (7, 8) is that the rate of pollutant discharge into the water body (dC_W/dt) should be decreased when the contamination level has reached exactly 1/2 MAC (Fig. 2). An example of such calculations is illustrated regarding the pollution of the Black Sea waters with mercury using our data obtained in 2011. The highest Hg concentration in the upper seawater layer was of 83 ng L^{-1}, while the basin-wide average value was much lower, 18 ng L^{-1}. The maximum allowable concentration of mercury, which is set for natural surface waters, is varied in different Black Sea countries around of 100 ng L^{-1} (e.g., UNEP 2002). Applying these values, as well as the above-mentioned value of half-life $s_{Eff} = 6.4$ years derived from radionuclide data, we obtain from Eq. (6) the value of specific rate of the maximum allowable annual input of mercury (p) of 4.5 $year^{-1}$. According to the bell-shaped curve of the function dC_W/dt in Fig. 2 and taking into account the volume of the upper 50-m layer of the Black Sea water $\approx 2.16 \times 10^{13}$ m^3 (e.g., Bogdanova 1959), the rate of the mercury input may be increased up to the maximal level of $\sim 1.4 \times 10^3$ t $year^{-1}$ corresponding to the time $t_{0.50}$ when the mercury concentration reaches a half of the MAC (Fig. 2). After this period, the rate of the annual input of mercury to the surface Black Sea waters must be steadily reduced down to the full stop when its concentration comes closer to reaching the MAC level in $t_{0.99}$ years after the fallout or discharge of the pollutant (Fig. 2). It should be noted, however, that the abovementioned maximum rate of the mercury input (1.4×10^3 t $year^{-1}$) is the highest admissible value, which will lead to the rising of the mercury concentration to the level of its maximum allowable concentration (100 ng L^{-1}) within the whole 50-m thickness of the upper Black Sea waters. But, if we aim to keep a steady state level of the contamination ($dC_W/dt = 0$), then the maximum input, at which its concentration will maintain at a current level $C(t) = C_0 = \text{const} = 18$ ng L^{-1}, may be estimated as a value inverse of the rate of self-purification using the formula derived from Eqs. (1) and (3). Applying the above-mentioned volume of the upper 50-m layer of the Black Sea water yields the basin-wide 'steady-state-providing' assimilation capacity of this water mass to be

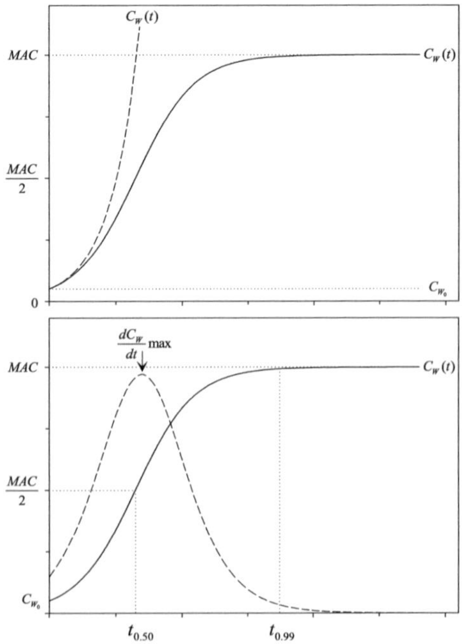

Fig. 2 Exponential (*top panel, dashed line*) and logistic (*both panels, solid line*) growth of the pollutant concentration in seawater and the dynamics of its change dC_W/dt (*bottom panel, dashed bell-shaped line*)

nearly 40 t year^{-1}. This value fits well to the range of previous indirect assessments of such capacity of the Black Sea waters regarding its contamination with mercury, which varied between 20 and 600 t year^{-1} (Polikarpov and Egorov 1986, and references therein). It should be emphasized, however, that the given value of relatively 'safe' mercury input (40 t year^{-1}) has already become comparable with the annual inflow of this high-toxic heavy metal from the Danube River alone, which is estimated at 60 t year^{-1} (e.g., EU Parliamentary Assembly 2008), pointing the necessity of an immediate reduction of amount of the mercury discharge into the Black Sea basin.

Thus, the post-Chernobyl data on temporal evolution of the man-made radioactivity in the Black Sea water provides an opportunity to obtain integrated assessments of the rate of its self-purification against soluble and particle-reactive pollutants, to evaluate directly the levels of maximum allowable discharge of pollutants into the basin, and to assess their capacity regarding the contamination as a quantitative measure for environmental regulation of human impacts allowing to maintain it at a steady state or at the level not exceeding the MAC.

Conclusion

One of the promising ways to solve the problems that have to be addressed in the context of environmental regulation of pollution of marine environments is the radiochemoecological evaluation of the capability of marine environments for self-purification against radioactive and non-radioactive pollutants. Data on temporal evolution of the man-made radioactivity in seawater provides opportunity to evaluate directly the levels of maximum allowable discharge into marine environments of nuclear and non-nuclear pollutants, and to assess capacity of aquatic ecosystems regarding the assimilation of contamination to be used as a quantitative measure for environmental regulation of human impacts allowing to maintain it at a steady state or at the level not exceeding the maximum allowable concentration. The developed approach to evaluate the maximum allowable annual input of the toxic contaminants may be a useful tool for environmental regulation of pollution of seawater.

References

Agre AL, Korogodin VI (1960) On the distribution of radioactive contamination in the non-flow-through water body. Med Radiol 1:67–73 (Russian)

Anderson RF, Fleisher MQ, LeHuray AP (1989) Concentration, oxidation state, and particulate flux of uranium in the Black Sea. Geochim Cosmochim Acta 53:2215–2224

Bogdanova AK (1959) Water exchange through the Bosporus Strait and its role in the mixing of the Black Sea waters. Proc Sevastopol Biol Stat 12:401–420 (Russian)

Buesseler KO, Benitez CR (1994) Determination of mass accumulation rates and sediment radionuclide inventories in the deep Black Sea. Deep-Sea Res I 41(11/12):1605–1615

Buesseler KO, Livingston HD, Casso SA (1991) Mixing between oxic and anoxic waters of the Black Sea as traced by Chernobyl cesium isotopes. Deep-Sea Res 38(Suppl 2):S72–S745

Calvert SE, Karlin RE, Toolin LJ et al (1991) Low organic carbon accumulation rates in Black Sea —sediments. Nature 350:692–695

Crusius J, Anderson RF (1991) Immobility of ^{210}Pb in Black Sea sediments. Geochim Cosmochim Acta 55:327–333

Egorov VN, Povinec PP, Polikarpov GG et al (1999) ^{90}Sr and ^{137}Cs in the Black Sea after the Chernobyl NPP accident: inventories, balance and tracer applications. J Environ Radioact 43:137–155

Egorov VN, Gulin SB, Polikarpov GG et al (2010) Black Sea. In: Atwood DA (ed) Radionuclides in the environment. Wiley, Chichester, pp 430–452

EU Parliamentary Assembly (2008) The fight against harm to the environment in the Black Sea. Report of Committee on the Environment, Agriculture and Local and Regional Affairs, 23 June 2008 (20th Sitting), Doc. 11632, Recommendation 1837, F-67075 Strasbourg

Florou H, Kritidis P, Vosniakos F et al (2002) Dispersion of ^{137}Cs in the Eastern Mediterranean and the Black Sea: the time evolution in relation to the sources and pathways. J Environ Protect Ecol 3(1):30–36

GESAMP (1990) The state of the marine environments. In: Reports and studies. GESAMP, 39, UNEP, New York

Gulin SB (2000a) Recent changes of biogenic carbonate deposition in anoxic sediments of the Black Sea: sedimentary record and climatic implication. Mar Environ Res 49(4):319–328

Gulin SB (2000b) Seasonal changes of ^{234}Th scavenging in surface water across the western Black Sea: an implication of the cyclonic circulation patterns. J Environ Radioact 51(3):335–347

Gulin SB, Egorov VN (2013) Self-purification of seawater: a measure for environmental regulation. In: White MR (ed) Seawater: geochemistry, composition and environmental impacts. Nova Science Publishers, New York, pp 93–126

Gulin SB, Aarkrog A, Polikarpov GG et al (1997) Chronological study of ^{137}Cs input to the Black sea deep and shelf sediments. Radioprotect 32(C2):257–262

Gulin SB, Polikarpov GG, Egorov VN et al (2002) Radioactive contamination of the north-western Black Sea sediments. Estuar Coast Shelf Sci 54(3):541–549

Gulin SB, Polikarpov GG, Martin J-M (2003) Geochronological reconstruction of ^{137}Cs transport from the Coruh river to the SE Black Sea: comparative assessment of radionuclide retention in the mountainous catchment area. Cont Shelf Res 23:1811–1819

Gulin SB, Egorov VN, Polikarpov GG et al (2012) General trends in radioactive contamination of the marine environment from the Black Sea to Antarctic Ocean. In: Burlakova EB, Naydich VI (eds) The lessons of Chernobyl: 25 years later. Nova Science Publishers, New York, pp 281–299

Gulin SB, Egorov VN, Polikarpov GG et al (2013a) Isotopes in hydrology, marine ecosystems and climate change studies. In: Proceedings of international symposium, vol 2, Monaco 27.03.-2.04.2011; IAEA Vienna, pp 535

Gulin SB, Mirzoyeva NYu, Egorov VN et al (2013b) Secondary radioactive contamination of the Black Sea after Chernobyl accident: recent levels, pathways and trends. J Environ Radioact 124:50–56

Gulin SB, Egorov VN, Duka MS et al (2015) Deep-water profiling of ^{137}Cs and ^{90}Sr in the Black Sea: a further insight into dynamics of the post-Chernobyl radioactive contamination. J Radioanal Nucl Chem 304(2):779–783

Hay BJ, Honjo S, Kempe S et al (1990) Interannual variability in particle flux in the southwestern Black Sea. Deep-Sea Res 37(6):911–928

Izrael YuA, Tsiban AV (1983) About assimilation capacity of the World ocean. Proc Acad Sci USSR 272(3):702–705 (Russian)

Korogodina VL, Korogodin VI, Kutlakhmedov YA (1996) Radiocapacity: prognosis of pollution after nuclear accidents. In: Proceedinga of IX international congress on radiation protection (IRPA9), IAEA Vienna

Kutlakhmedov Y, Korogodin V, Rodina V et al (2006) Radiocapacity: characteristic of stability and reliability of biota in ecosystems. In: Radiation risk estimates in normal and emergency situations, NATO Security through Science Series, pp 175–185

Nielsen SP, Lüning M, Ilus E et al (2010) Baltic Sea. In: Atwood DA (ed) Radionuclides in the environment. Wiley, Chichester, pp 415–436

Osvath I, Egorov V, Gulin S et al (2007) What can radiotracers tell us about the fate of Black Sea contaminants? Rapp du 38-e de la SIESM 38: 20

Polikarpov GG (1966) Radioecology of aquatic organisms. Reinhold, New York

Polikarpov GG, Egorov VN (1986) Marine dynamic radiochemoecology. Energoatomizdat, Moscow (Russian)

Sanchez A, Gastaud J, Noshkin V et al (1991) Plutonium oxidation states in the southwestern Black Sea: evidence regarding the origin of the cold intermediate layer. Deep-Sea Res 38:845–853

Stokozov NA, Buesseler KO (1999) Mixing model for the NW Black Sea using Sr-90 and salinity as tracers. J Environ Radioact 43(2):173–186

Timofeeff-Ressovsky NV (1957) Application of rays and emitters for experimental biocenology. Botanic J (USSR) 42(2):161–194 (Russian)

Tsuchiya K, Harashima S (1965) Lead exposure and the derivation of maximum allowable concentrations and threshold limit values. Br J Ind Med 22(3):181–186

UNEP (2002) Global mercury assessment, overview of existing and future national actions, including legislation, relevant to mercury. UNEP, New York

Wei CL, Murray JW (1991) ^{234}Th/^{238}U disequilibria in the Black Sea. Deep-Sea Res 38(Suppl. 2):855–873

Yücel M, Moore WS, Butler IB et al (2012) Recent sedimentation in the Black Sea: new insights from radionuclides and sulfur isotopes. Deep-Sea Res I 66:103–113

Zaitsev YP, Mamaev V (1997) Marine biological diversity in the Black Sea. A study of change and decline. United Nations Publications, New York

Some Aspects of Radioecology in the Areas Adjacent to Armenian NPP

Garnik E. Khachatryan, Valeriy B. Arakelyan, Nvard V. Simonyan, Nina I. Mkrtchyan, Tsovak M. Avakyan and Konstantin I. Pyuskyulyan

Introduction

It is known that the contribution of nuclear power plants (NPP) in the world production of electricity is significant and is growing year by year. For example, the share of Armenian Nuclear Power Plant (ANPP) in the energy supply of the country is about 40 %. There is a reasonable question regarding the possible impact of the NPP on the biological objects in the environment polluted by ^{137}Cs—the main component of emissions. Besides NPP, there are other sources of ^{137}Cs emissions into the environment. However, the contribution of NPP to the radioactive contamination of the environment, in particular radioactive contamination of the environment adjacent to the plant, is significantly smaller than that of the global fallout (Moiseev 1990). This complicates the problem of identifying the direct effects of the NPP on the environment on the background of global fallout. Therefore, to solve the task, we have developed an approach to estimate the contribution of radioactive fallout from the ANPP on the background of global fallout, the essence of which is as follows.

Since the radioactive contamination from global fallout in the nature takes place randomly, during the assessment of the radiation situation on the territory adjacent to the plant, with the same weather conditions and close characteristics of the soils, a uniform distribution of the radionuclides must be observed on the area of study.

G.E. Khachatryan (✉) · N.V. Simonyan · N.I. Mkrtchyan
A.I.Alikhanyan National Science Laboratory (Yerevan Physics Institute) Foundation, Yerevan, Armenia
e-mail: garnik@mail.yerphi.am

V.B. Arakelyan · T.M. Avakyan
Yerevan State University, Yerevan, Armenia

K.I. Pyuskyulyan
CJSC Armenian NPP, Metsamor, Armenia

V.L. Korogodina et al. (eds.), *Genetics, Evolution and Radiation*,
DOI 10.1007/978-3-319-48838-7_26

The presence of any fluctuation of any parameter, say along the wind rose, would indicate the presence of nuclear power plant emission.

It is known that the possible impact of the NPP on the environment can extend as far as 10 km (Il'in and Pavlovskiy 1983). Based on this fact, we have chosen an area around the ANPP with a radius of 17 km, have determined for it the wind rose (Report ANPP 2002), and have selected the points of sampling. Thus, in this area, the only factor that can cause irregular distribution of radionuclides released by the ANPP will be the wind rose. Consequently, the task of assessing the contribution of the NPP in the radioactive contamination of the environment against the background of global fallout will be limited to two subtasks: (i) identification of a correlation in the dynamics of change of ^{137}Cs concentrations in the soil depending on the distance to the NPP along the predominant wind direction; and (ii) identification of a correlation for the corresponding biological object.

Primarily, the aim of the second subtask was to choose a suitable bio-object of the environment and the detection of its reaction on the increased concentration of the pollutant factors, which was caused by the emission of radioactive aerosols. The biological object must be sufficiently sensitive to the low doses of radiation. Such an object could represent a community of soil microorganisms, since it is known that small doses of radiation can have a stimulating effect on biological objects (Kuzin 2002, 1989; Luckey 1991; Kudryashov 2004; Guidelines 1980); and the microorganisms are good model objects.

It should be noted that during the study of the reaction of microbiota on the change of the content of radionuclides in the soil, the evaluation of the number of individual species composing the microbial population of the soil is also informative. Equally informative are also the radiobiological parameters that can be revealed from the curves of survival of the cells of certain types of microflora. It is known that under natural conditions, different populations of living organisms, under the influence of enhanced natural background ionizing radiation and anthropogenic component, undergo some adaptive changes that mobilize their reparative capacity, which makes their populations more resistant to radiation or their radiobiological characteristics change. What is more, the probability of changing of the radiobiological parameters of radiosensitive organisms is higher than that of the radioresistant organisms (Kiselev et al. 1961; Shevchenko 1983).

Among all the species of the soil bacterial microflora in the area of monitoring, we have selected the genera of *Pseudomonas* and *Bacillus* as objects for the radiobiological research. This selection has been made based on the published data: *Pseudomonas*—as the most radiosensitive part of the bacterial population in the soil, and *Bacillus*—as the most radioresistant one (Kuzin and Kaushanskiy 1981; Hall 2000). They should be present at all the points of observation in order to make comparative assessment of the amount and radiosensitivity of the individual species that form the microbial population of the soil. However, the identification of cultures by the methods of routine microbiology is a labor consuming as well as a time consuming process. A set of fast procedures involving a limited number of relatively simple operations have been developed that allow identification of the sought-for cultures in the analyzed samples.

Approaches and Methodological Solutions

Estimation of the concentration of ^{137}***Cs in the soil***. The concentration of ^{137}Cs in the soil was measured using a low-background gamma spectrometer with a NaI detector in Marinelli geometry with software "PROGRESS.300". In addition, a pure germanium detector unit with GENIE software support (CANBERRA) was used. The samples were taken in accordance with the recommendations in (Guidelines 1980). The weight of each soil sample was from 1200 to 1400 g, depending on the soil type and its humidity. The minimum activity the installation can detect is equal to 2 Bq/kg. The specific content of ^{137}Cs was determined by a series of three measurements for each soil sample. Gamma spectrometric analysis of the samples showed that the soil contains natural radionuclides ^{40}K, ^{226}Ra, ^{232}Th, and ^{137}Cs. It should be noted that ^{137}Cs is the main dose-forming factor in the emissions of NPP, as its half-life is equal to 30.17 years and it is well accumulated in the environment (Moiseev 1990).

The environment may be contaminated by ^{137}Cs due to two factors: anthropogenic—global fallout and NPPs emissions, and natural—the result of the fission of U-238 and U-235, the contribution of which, however, is by an order of 3–4 smaller than that from the anthropogenic factors.

The observation points located along the predominant winds blowing in opposite directions and at different distances from the ANPP, from 800 m to 17 km, have been chosen as soil sampling sites. In addition to these points, we selected baseline control points far from the plant. The soil samples were collected from nine points located approximately on the same line passing along the Wind Rose, through the site with the ANPP in the middle of the line (Fig. 1).

Soil sampling for physical measurements was made using a special 5 cm deep steel frame with 10 × 15 cm sides. Each sample consisted of five frames of the soil

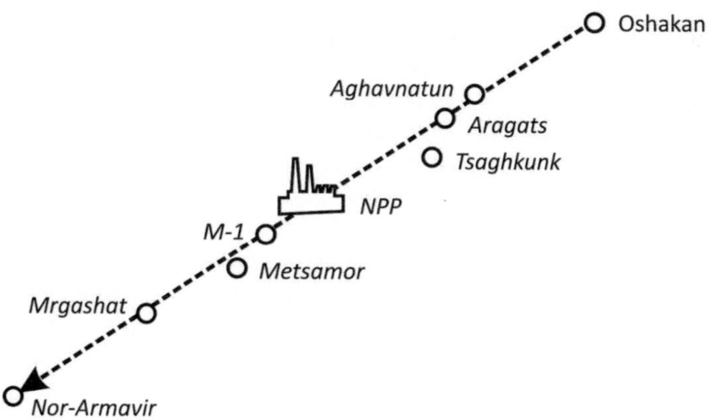

Fig. 1 Soil monitoring points. The *arrow* indicates the prevailing (south-west) direction of winds at the site of ANPP

sampled by the method of envelope. When sampling, the soil without vegetation and, if possible, of the same type were selected. They were mostly brown semi-desert soils. Before measurement, the sample of soil was dehydrated, ground up, and then weighed in accordance with the conventional techniques. The prepared soil sample, weighing from 700 to 750 g, was placed in the measuring chamber; the measurement time was 3–5 h.

Estimation of total bacterial population in the soil at the monitored points. Samples were taken from the same areas for both physical measurements and microbiological studies. In that case, the samples were taken (also by the method of envelope) using a special steel tube with a diameter of 10 cm and a length of 5 cm. The soil samples taken at each point weighed about 300 g of soil (a total of 1500 g from each site of monitoring). This amount meets the microbiological requirements (Zvyagintsev 1991).

Soil samples from each site of monitoring were thoroughly ground up, mixed up and then sieved through a sieve of 0.6 mm. The moisture of the sample was determined via heating a certain soil mass in a drying chamber for several hours at 105 °C until a constant weight was achieved. A weighed portion of soil to be analyzed, equivalent to 1 g of dry soil, was previously moistened by sterile tap water then ground up for 5 min in a sterile porcelain cup to a homogeneous pasty mass. The soil paste was resuspended by gradual addition of water up to a uniform suspension of final volume of 100 ml and thoroughly shaken for five minutes. Then the resulting suspension was allowed to stand for 30 s for settling of large particles, and 0.1 ml aliquot of the suspension supernatant was taken. From the resulting suspension—soil extract, which, in fact, presented itself the initial cell suspension, different cell suspensions were obtained via standard dilutions, thus allowing, after inoculation on the Petri dishes, to choose dilutions, suitable for visual analysis—counting the number of colonies formed (CFU).

It is known that for many reasons (Egorov 1983; Khachikyan 1998) it is not possible to determine by this way the absolute number of viable microbial cells per unit weight or area of the soil. However, since the necessary conclusions actually came to the comparison of the relative data values, this way was found to be satisfactory.

Estimation of the number of bacteria of the certain genera/species in the bacterial population of the soil at the points of monitoring. Since the balance of microorganisms in the microbial community of cultivated soils is always disturbed, the soil samples for the study were taken just from the virgin lands. To ensure the statistical validity of the data the samples from each district were taken three times.

To make the obtained results more reliable, from each initial tube with selected dilutions of cell suspension, parallel inoculations were made on Petri dishes. The experiments on each soil sample were repeated at least three times.

The methods for isolation and identification of microorganisms. The microorganisms were identified by traditional methods and techniques of microbiology, using classical selective and elective media and approaches (Anikeev and Lukomskaya 1977; Seliber 1962; Borozdina 2010; Gerhardt 1983a, vol. 1; Gerhardt 1983b, vol. 3; Mishustin and Emtsev 1970; Egorov 1983; Shaposhnikov 1947; Holbrook and Anderson 1980; Holt 1980; King et al. 1954; Mossel et al. 1967;

Williams 1936). Below are the steps developed to achieve this goal and a list of basic media used for their growth during the procedures of analysis of bacterial cultures. Others are mentioned as appropriate in the section Results and Discussion.

Pasteurization. As a first step towards the isolation of the sought-for bacteria, heat treatment of the original soil suspension was applied, allowing us to separate the spore-forming bacteria (*bacilli*) from the nonspore-forming ones. To do this, an aliquot of the appropriate dilutions of the bacterial suspension was heated in an ultrathermostate for 20 min at a temperature of 80 °C then inoculated on an appropriate medium (Gerhardt 1983a, vol. 1; Egorov 1983).

Growth on a medium with a triphenylmethane dye. Along with the heat treatment, the aliquot was inoculated on a selective medium containing triphenylmethane dye; on such media, only Gram-negative bacteria are able to grow. The medium was prepared on the base of Meat-Peptone Agar (MPA), and brilliant green was used as a dye in a ratio of 1 mg of dye per 200 ml of the medium (Gerhardt 1983a, vol. 1).

Pseudomonas agar medium F. Such inoculation of the medium makes it sure that the selected biotypes of the bacteria belonged to the species *P. putida* and *P. fluorescense*, dying their colonies as well as the medium on which they grew, to acquire a characteristic yellow-green color of different intensity (King et al. 1954).

Selective medium by Mossel. It contains 100 ml of egg emulsion and then polymyxin B added under sterile conditions to 900 ml of the original medium. This medium allows identifying *B. cereus* and *B. subtilis.* Having grown, their colonies get a specific color. The colonies of *B. cereus* are red and those of *B. subtilis* are yellow (Mossel et al. 1967).

Thus, the media as well as the techniques described above made it possible to develop a rapid method by which we could for a short time to determine the required culture in the test samples of soil.

Determination/estimation of radiosensitivity. The *B. subtilis* and *Pseudomonas* (biotypes of *putida* and *fluorescense*) cells isolated as objects of investigation were used in the experiments on determination of radiosensitivity. The cells were grown on slant MPA for 18 h at 28–30 °C.

The cell suspensions were exposed to X-rays at room temperature using RUM-17 at tube voltage of 165 kV, current 15 mA and a dose rate of 24 Gy/min. After making appropriate dilutions, the control and irradiated samples were inoculated on Petri dishes with MPA and incubated at 28–30 °C. The macro-colonies were counted in two days after inoculation. There were 4–5 runs of each experiment. The standard error of the mean cell survival values determined by averaging the results of different experiments was 5 % as a rule.

Results and Discussion

It is known that when radioactive aerosols get into the NPP's vent pipe and disperse, their minimal concentration is detected on the ground level at the foot of the pipe. The concentration increases with the distance, reaching its maximum at a certain distance, depending on weather conditions and the state of gas-aerosol mixture (point M-1 marked with an arrow in Fig. 2), and then decreases again (Netkov and Zlatanova 1984). At the notional point M-1, the maximum concentration of ^{137}Cs fallout and hence its maximum content in the soil is expected.

The distance from the vent pipe to the point of maximum deposition of radionuclides (point M-1, marked by the arrow in Fig. 1) is determined with account of the real methodological conditions, the terrain in the area where the ANPP is located and the characteristics of the ejected gas-aerosol mixture. The calculations were carried out according to the procedure specified in the works of Netkov and Zlatanova (1984), and Teverovskiy et al. (1985).

Figure 2 shows the concentrations of ^{137}Cs in soil samples at the points of monitoring (curve 1). The data from the village of Oshakan (−17.5 km, the first point on the line along the arrow in Fig. 1) with an abnormally high content of ^{137}Cs in the soil (compare with Table 1) is omitted on the abscissa axis. Other points correspond to those in Fig. 1.

As can be seen in Fig. 2, the concentration of ^{137}Cs at M-1 is maximal, and decreases with moving away from the station, decreasing to the values of the concentrations on the windward side. The curve is demonstrating the influence of the wind rose—a rather sharp rise and a gradual decline. At a distance of 10 km from the ANPP, on the windward side (the village of Aghavnatun in Fig. 1), the specific contents of ^{137}Cs in the soil almost matches its contents directly on the windward side of the industrial site, which is approximately 1.2 times less than at the point M-1. The content of ^{137}Cs in the soil in this area is only due to the global fallout. It means that the difference between the values of the concentrations of ^{137}Cs in the points on the windward side and in the area of M-1 is due to the influence of the emissions from the ANPP. The concentration of ^{137}Cs in the soil

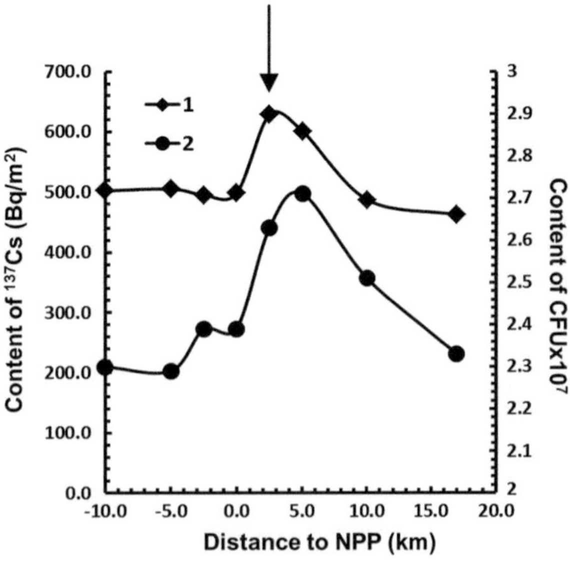

Fig. 2 The content of ^{137}Cs (1) and CFU (2) in the upper five-centimeter layer of soil in the area of the ANPP (point 0.0 is ANPP; negative coordinates correspond to points located on the windward side). The *arrow* shows the point M-1

Table 1 The quantitative content of ^{137}Cs, and the content of radiosensitive bacteria of the *Pseudomonas* genus and radioresistant—*Bacillus* in the upper five-centimeter layer of soil along the prevailing winds

Name of sampling area (village/city)	Content of ^{137}Cs in upper 5-cm layer of soil (Bq/m^2)	The number of colony-forming cells	
		Bacillus mesentericus in 1 g of dry soil ($\times 10^5$)	Biotypes of ***Pseudomonas*** in 1 g of dry soil ($\times 10^6$)
Oshakan	1630	2.0	3.0
Aghavnatun	502	2.5	5.0
Aragats	505	1.9	5.6
Tsaghkunk	495	1.6	3.5
ANPP	498	2.0	3.8
M-1	629	2.2	1.6
Metsamor	601	1.5	2.0
Mrgashat	487	1.8	4.1
Nor Armavir	462	2.2	5.0

decreases with the distance from the windward side of the ANPP, reaching the natural background values. It means that in the area with a radius of 10 km around the point M-1, in the direction of the maximum frequency of winds, the contribution of ANPP in the specific content of ^{137}Cs is approximately 124–155 Bq/m^2, which is about 20–25 % of the background due to the global fallout.

Thus, we can assume that the first sub-task is fulfilled and that the adopted approach has paid for itself. The next step involved a similar study of the correlation by bioindication.

The objective of this subtask was to count the total number of microorganisms obtained from the observation points.

The total quantity of bacteria in the soil samples taken from the points of monitoring located on the line from the northeast (windward side) to the south-west (leeward side) with the center at the site of the ANPP has been determined. Studies have shown that no significant variations in the integrated characteristics of the number of bacteria in the samples taken in the areas on the windward side of the ANPP had been registered. An exception is the territory adjacent to the village of Oshakan, as was mentioned before; the discussion on this issue follows.

Curve 2 (Fig. 2) shows a statistically significant increase in the total number of microorganisms in the soil samples taken from sites located 2.5 and 5.0 km away from the leeward side of the ANPP. Then with moving away from the ANPP, the number of bacteria decreases and at a distance of 17 km that number is close to that obtained for the immediate vicinity of the ANPP on the windward side and for the control area. A comparison of the results shown in Fig. 2 (curves 1 and 2) shows that near M-1 the increase in the maximum number of microorganisms in the soil is in full agreement with the highest concentrations of ^{137}Cs (points 6 and 7). This can be taken as confirmation of the assumption that small doses of radiation may stimulate the growth of microbial population in soils.

According to the data on the analyses of the control soil samples with similar soil characteristics taken at points of monitoring located at sufficient distances from the ANPP, away from the path of predominant winds (i.e. at the points where the impact of the station should not be felt in principle), the background concentration of ^{137}Cs was about 496 Bq/m^2.

For complete visualization of the detected pattern, the total number of bacteria in the areas chosen as control sites was determined. The soils with characteristics similar to those in the thirty-kilometer zone around the plant, namely—the type and aridity of the soil, topography, vegetation characteristics, etc., were being chosen.

The control samples were taken from two sites located at a distance of 6 and 7 km from the town of Vedi in the direction along the Vedi River between the villages Dashtakar and Urtsadzor. These points are located at a distance of approximately 57 km from the plant and away from the direction of prevailing winds, which excluded any impact from the ANPP. The total number of microorganisms in these points was 2.37×10^7 CFU.

As was mentioned above, in the study of certain types the focus on the genera *Bacillus* and *Pseudomonas* was made. With account of the fact that these genera are major taxonomic groups the members of which may be distributed unevenly in the soil at the points of monitoring, a matter of first priority was to choose such species which, according to the literature data, are fairly widespread in the surveyed terrain and which can be identified with the highest reliability. Initially, the range of our interests among the *bacilli* was turned to *B. cereus*, *B. subtilis* and *B. mesentericus* which, according to the literature (Khachikyan 1998), were common for these soils.

To identify the above-mentioned cultures, we have adopted some techniques and selective media described in the section Approaches and Methodological Solutions. We succeeded in isolating of the desired types of bacteria of the *Pseudomonas* and *Bacillus* genera by using different types of media (MPA, brilliant green, MPA-wort agar, medium of Mossel, sterile potatoes with chalk, gelatin medium, selective media for *Pseudomonas*: Agar F and Agar P, and a specific environment with three kinds of sugars) and by studying the cultural-morphological and physiological-biochemical characteristics of the bacteria isolated from the soil samples. The short description of the used methods and approaches is presented in what follows.

For isolation of Gram-negative bacteria, including *Pseudomonas* genus, it is suitable to use media that contain triphenylmethane dyes (i.e. Brilliant Green medium) the presence of which inhibits the growth of Gram-positive cultures. The combination of such media with the selective media of F and P makes it possible to identify the biotypes of *P. fluorescens* and *P. putida*. With the growth the medium around the colonies of these cultures acquires a specific yellowish-green color, making it easy to determine the biotype of interest. As a nutrient medium to estimate the number of bacteria of the Pseudomonas genus MPA and for the *bacilli* —MPA/wort agar in a ratio of 1:1 were used.

Being spore-forming cultures, the *Bacillus* genus were easily separated from the others by pasteurization (or selective inhibition by temperature), when the cultures are incubated in a water bath for 20 min at a temperature of 80 °C. These conditions killed practically all other bacterial cultures as well as vegetative cells of

Bacilli, not having time to form endospores. After incubation, the inoculation was performed on MPA/wort-agar medium and the number of colonies was counted. For the control of the results, inoculation from the initial soil suspension to a number of selective media was carried out in parallel. For example, on Mossel medium (Mossel et al. 1967) with polymyxin sulfate «B» added to it, the colonies of *B. subtilis* acquired a yellow and those of B. cereus a red color. The Williams medium (1936) allowed similar control of the sought-for *bacilli*; the colonies of *B. cereus* and *B. subtilis* having grown on it had acquired a yellow color on a violet background of the medium. For higher reliability, the isolated culture was compared with the control samples obtained by the classical methods of extraction using hay in the case with *B. subtilis*, and potatoes with chalk in the case with *B. mesentericus* (Anikeev and Lukomskaya 1977).

During the study the representatives of *Pseudomonas* genus of the specified biotype strains had been identified in the soil samples from all the points of monitoring. Concurrent inoculation of cultures on the media F and B (Gerhardt 1983b, vol. 3) allowed us to easily identify the biotype of *Pseudomonas.* The isolated strains of *Pseudomonas* were compared with the museum strains from the Culture Collection Museum (Pushchino, Russia).

Among the cultures of *Bacillus* genus, ultimately the choice had been stopped on the cultures of *B. mesentericus* and *B. subtilis*.

The *B. mesentericus* had been isolated in all the nine points of monitoring, while the *B. subtilis* species was isolated only in six points, which confirmed the existing literature data (Khachikyan 1998).

Table 1 shows the results of a quantitative study of the above-mentioned microorganisms in the sampling points.

In Table 1 one can see that the contents of the radiosensitive bacteria of the *Pseudomonas* genus in the soil decreases downwind ANPP to minimal in the point corresponding to the maximum content of ^{137}Cs in the soil. This indicates that the concentration of ^{137}Cs in the soil is sufficient to inhibit the growth of radiosensitive microorganisms. As for the content of radioresistant cells of *Bacillus mesentericus*, no certain regularity in the change of their contents at different points of observation had been recorded (Table 1).

Figure 3 shows the results of experiments on the determination of the survival of bacterial cells of *P. fluorescens*, isolated from the monitoring sites localized along the lines passing to the northeast (A—windward side) and southwest (B—leeward side) from the ANPP, depending on the dose of X-ray irradiation.

Table 2 shows the values of D_0 (dose that causes inactivation of 63 % of the exposed cells) on the survival curves of the cells of *P. fluorescens* bacteria shown in Fig. 3.

As it follows from the results of the experiments presented in Fig. 3 and Table 2, the dependence of the radiosensitivity of the cells of *Pseudomonas* on the dose of irradiation by X-rays is in all cases an exponential one. There is almost no difference between the D_0 of the cells of *Pseudomonas* bacteria isolated from the monitoring points at different distances from the ANPP and placed windward as well as alee of the prevailing winds. Exceptions are the cells of bacteria isolated

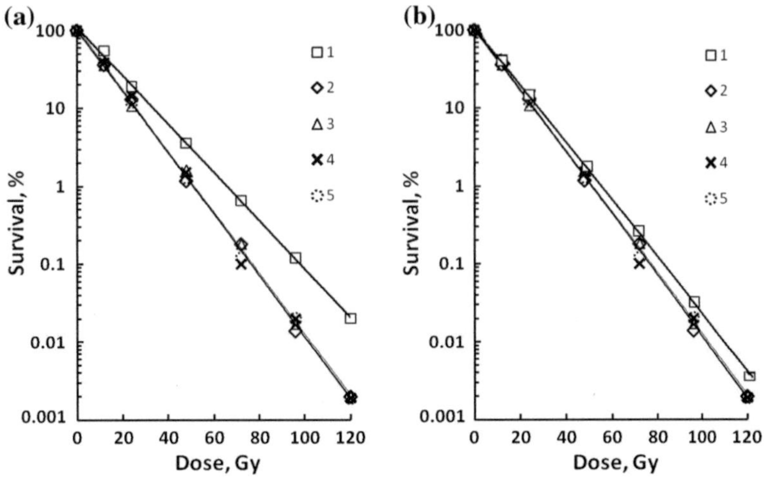

Fig. 3 Dose-response *curves* of the bacterial cells of *P. fluorescens*, isolated from the windward (**a**) and leeward (**b**) sites of monitoring. **a**: *1* Oshakan. *2* Aghavnatun, *3* Aragats, *4* Tsaghkunk, *5* ANPP; **b**: *1* M-1, *2* Metsamor, *3* Mrgashat, *4* Nor Armavir

Table 2 The value of D_0, or the value of radioresistance of bacterial cell of *P. fluorescens*, isolated from all monitoring points located on the windward (1–5) and leeward (6–9) side of the plant

Sample ##	Region of sampling	Distance from ANPP (km)	D_0, Gy
1	Oshakan	17	14.0
2	Aghavnatun	10	11.0
3	Aragats	5	11.0
4	Tsaghkunk	2.5	11.0
5	ANPP	0	11.0
6	M-1	2.5	12.5
7	Metsamor	5	11.0
8	Mrgashat	10	11.0
9	Nor Armavir	17	11.0

from the soil samples taken near the village of Oshakan and at the point M-1. The cells of bacteria isolated from these observation points are a little less radiosensitive.

We would like to discuss in more details the situation in the village of Oshakan with abnormally high concentration of ^{137}Cs where not only a high content of ^{137}Cs (1630 Bq/m^2) but also a significant reduction in the total amount of microorganisms versus the control, namely 1.34×10^7 against 2.37×10^7 CFU, had been detected. The results presented in Table 1 cannot be taken as casual since the analyses of the radioresistance of both biotypes of *Pseudomonas* isolated from the soil taken near

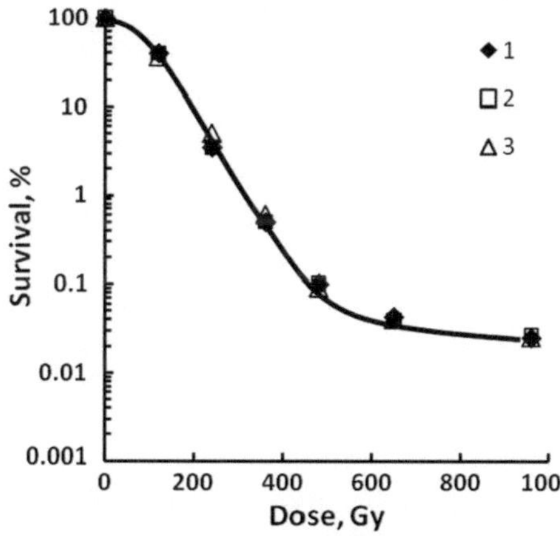

Fig. 4 Dose-response *curves* of cells of bacteria *Bacillus subtilis*, isolated from the following sites: *1* point M-1; *2* v.Aragats; *3* v.Oshakan

Oshakan clearly indicate their increased radioresistance versus the cultures isolated from the soils with a low content of ^{137}Cs (Fig. 3, curve 1).

A different picture is observed in the case of cells of *Bacillus*. Figure 4 shows the dependence of cell survival of bacteria *B. subtilis*, isolated from the soil samples collected at M-1, v.Oshakan, and v.Aragats on the radiation dose.

As can be seen from Fig. 4, the shape of the survival curves of cells of bacteria *B. subtilis* found in the points of monitoring with different content of ^{137}Cs in the soil is the same; the initial shoulder, then a straight line in the dose range from 120 to 480 Gy and, at higher doses, the kinetics of the inactivation process slows down.

As it is known, the shape of the curves that show the radiosensitivity of cells versus the dose of radiation is usually exponential or sigmoidal. The our obtained survival curves of *B. subtilis* cells have the form of a sigmoidal curve, with a so-called "tail". Such shape of the curve is an indicative of the irradiated cultures that are inhomogeneous by radiosensitivity. Our studies have been carried out on cultures in the exponential growth phase, which is dominated by vegetative cells, but there are also endospores. Thus, the shape of the our obtained curve can most likely be explained by the fact that the initial part of the curve shows the death of vegetative cells, and the end part shows the death of the spores that are much more radioresistant. The data presented in Fig. 3 show that the quantitative characteristics of the dose-response curves for *B. subtilis* cells isolated from different monitoring points are the same. Values of D_0 for all the curves are about 120 Gy.

Conclusion

There has been conducted monitoring of the ^{137}Cs content in the soil in the area of the Armenian nuclear power plant along the predominant winds. It has been shown that on a distribution curve of ^{137}Cs there is a maximum at a distance of 2.5 km downwind from the plant. The contribution of ANPP in the specific content of ^{137}Cs in the soil is equal to 124–155 Bq/m^2, which is about 20–25 % of the background due to global fallout.

The total number of bacteria in the soil samples taken from the same points of monitoring has been determined. Studies have shown that no significant variations of the integral characteristics of the bacterial population had been observed from the windward side of the ANPP. A statistically significant increase in the total number of microorganisms is found on the leeward side of the ANPP, the maximum being registered at the sampling point M-1 at a distance of 2.5 km from ANPP, which corresponds to the maximum concentration of ^{137}Cs. This can serve as a confirmation of the assumption that small doses of radiation may stimulate the growth of the microbial populations.

It is shown that the number of radiosensitive bacteria of the *Pseudomonas* genus along the predominant winds decreases downwind from ANPP with the minimum being registered at the point that corresponds to the maximum concentration of ^{137}Cs in the soil. As to *B. mesentericus*, the radioresistance of which is about 100 times higher than that of the bacteria of *Pseudomonas* genus, there has not been revealed any law of changing of their number that could correlate with the curve of distribution of ^{137}Cs in the soils around the NPP. No change in the radiosensitivity was observed in the comparative studies of the radiosensitivity of the cells of bacteria *B. subtilis* isolated from the soil samples taken at points of monitoring with different contents of ^{137}Cs in the soil.

The picture is different in the case with the radiosensitive *Pseudomonas*. On the dose versus response curves is given a somewhat higher value of the dose that inactivated 63 % of the irradiated cells (D_0), which had been isolated from the sampling points corresponding to the maximum on the curve of ^{137}Cs distribution in the soils around the ANPP (soils with relatively high concentration of ^{137}Cs in them).

Thus, the results indicate that the most vulnerable targets of a constant high radiation background, even within the permissible doses, are the radiosensitive organisms. To adapt to the new environment, in the cells there take place processes, which activate their reparation systems and make them more radioresistant.

Finally, an area was found with an abnormally high content of ^{137}Cs, where there has been fixed a low content (as compared with the control) of the total number of microorganisms. All the aforesaid about the radioresistant and radiosensitive bacterial cultures is true for the cultures isolated from the soil samples collected in the village of Oshakan. The results obtained from this region show that it is necessary to carry out special comprehensive radioecological studies.

Acknowledgments The work has been done under ISTC Project A-773. The authors express their gratitude to the directorate of the Centre.

References

Anikeev VV, Lukomskaya KA (1977) Guide to practical training in microbiology. Prosveschenie Publication, Moscow, pp 22–83 (Russian)

Borozdina IB (2010) Comparative characteristics of Pseudomonas bacteria when cultured on artificial media. Vestnik VSU: Chem Biol Pharm 2:67–71 (Russian)

Egorov NS (ed) (1983) Guide to practical training in microbiology. MSU, Moscow (Russian)

Gerhardt P (ed-in-chief) (1983a) Manual of methods for general bacteriology, vol 1. Mir Publishers, Moscow, pp 281–299 (Russian)

Gerhardt P (ed-in-chief) (1983b) Manual of methods for general bacteriology, vol 3. Mir Publishers, Moscow, pp 21, 23–24, 43, 82 (Russian)

Guidelines on sanitary inspection of the content of radioactive substances in the environment (1980) HM of USSR, Moscow, 336 p (Russian)

Hall EJ (2000) Radiobiology for the radiologist, vol 19. Williams & Wilkins, Baltimore, MD

Holbrook R, Anderson JM (1980) An improved selective and diagnostic medium for the isolation and enumeration of *Bacillus cereus* in foods. Canadian J Microbiol 26(7):753–759

Holt JG ed (1980) The shorter Bergey's manual of determinative bacteriology. Mir Publisher, Moscow, pp 112–118, 286–294 (Russian)

Il'in LA, Pavlovskiy OA (1983) Expected values of radiation exposure on the population of the Soviet Union in the course of implementation of plans for the development of nuclear power in the USSR, vol 1. Radiation safety in the operation of nuclear power plants. Energoatomizdat, Moscow, pp 69–79 (Russian)

Khachikyan LA (1998) The microflora of the basic types of soils in Armenia, vol 147. Davit Publisher, Yerevan (Russian)

King EO, Ward MK, Raney DE (1954) Two simple media for the demonstration of pyocyanin and fluorescein. J Lab Clin Med 44:401–407

Kiselev PN, Kashkin RP, YaB Boltaks, Vitivskaya GA (1961) Acquisition of radioresistance by microbial cells when inhabitation in an environment with high level of natural radiation. Microbiology 30:20 (Russian)

Kudryashov YuB (2004) Radiation biophysics. MSU Publishers, Moscow, p 33 (Russian)

Kuzin AM (1989) Features of the mechanism of action of ionizing radiation on the biota in small doses favorable for it. Preprint of NCBI of NAS USSR, Puschino, pp 19–23 (Russian)

Kuzin AM (2002) The role of natural background radiation and secondary biogenic emission in the phenomenon of the life. Nauka Publisher, Moscow, p 80 (Russian)

Kuzin AM, Kaushanskiy DA (1981) Applied radiobiology. Energoatomizdat, Moscow, p 48 (Russian)

Luckey TD (1991) Radiation hormesis. CRC Publisher, Boca Raton, pp 22–41

Mishustin EN, Emtsev VT (1970) Microbiology. Kolos Publisher, Moscow, p 218 (Russian)

Moiseev AA (1990) Cs-37, environment, human. Energoatomizdat, Moscow, p 144 (Russian)

Mossel DAA, Koopmann MJ, Jongerius E (1967) Enumeration of Bacillus cereus in foods. Appl Microbiol 15:650–653

Netkov TN, Zlatanova RI (1984) Radiation characteristics of the environment around the Kozloduy Nuclear Power Plant due to cesium-137 and strontium-90. Radiation safety in the operation of nuclear power plants, vol 5. Energoatomizdat, Moscow, pp 90–96 (Russian)

Report of Armenian NPP (2002) Characteristics of the area of the Armenian Nuclear Power Plant. Metsamor, p 45 (Russian)

Shaposhnikov VN (1947) Technical microbiology. Sovetskaya Nauka Publisher, Moscow, pp 22, 68, 126 (Russian)

Shevchenko VA (1983) The evolution of populations in the conditions of high radiation In: Molecular mechanisms of the genetic processes. Nauka Publisher, Moscow (Russian)

Seliber GL ed (1962) Large practicum for microbiology. Vysshaya Shkola Publisher, Moscow, pp 283–287 (Russian)
Teverovskiy EN, Artemova NE, Bondarev AA et al (1985) Allowable emissions of radioactive and chemical substances into the atmosphere. Energoatomizdat, Moscow, p 240 (Russian)
Zvyagintsev DG ed (1991) Methods of soil microbiology and biochemistry. MSU Publishers, Moscow, pp 7–176 (Russian)
Williams OB (1936) Tryptone medium for the detection of flat sour spores. Food Research 1 (3):217–221

Prediction of ^{137}Cs and ^{90}Sr Contamination in the Food Chain Following a Nuclear Accident

Arrigo A. Cigna

Introduction

When the radioactive fallout due to experimental nuclear weapons explosions some models were developed in order to foresee the impact in the human food chain. One of these models was developed and successfully tested with different sets of data covering several years of fallout contamination in Europe collected at the Euratom Joint Nuclear Research Centre of Ispra, Italy, (Van der Stricht et al. 1970, 1971). During such a time period the fallout deposition in that area was widely investigated and a large amount of data was available (De Bortoli et al. 1968).

Such a model took into account the local characteristics of the environment with reference to the history of the previous years and the most recent contribution. In the first days after the Chernobyl accident such a model was applied in order to have a reliable evaluation of the human food chain to be expected in the immediate future. At that time predictions without any scientific basis were widely reported (Cigna 1987, 2008) and a more realistic evaluation was needed.

The results of the measurements obtained on pooled samples of green vegetables and milk in North Italy, were at least one order of magnitude lower than the values predicted by the model, in the eight years after the accident when the contamination became lower than the minimum detectable concentration. The results were very useful to avoid groundless fears and, therefore, the application of countermeasures still stricter than those enforced. But in the following years the decrease of the contaminations was faster than the values predicted by the model. A sudden release due to an accident, in general, is also much more relevant with respect to the contribution due to any contamination accumulated in the past.

A.A. Cigna (✉)
Str. Bottino 2, Frazione Tuffo, 14023 Cocconato, Italy
e-mail: arrigocigna@tiscali.it

V.L. Korogodina et al. (eds.), *Genetics, Evolution and Radiation*,
DOI 10.1007/978-3-319-48838-7_27

The Original Model

Taking into account the fallout accumulated on the soil during the preceding years, the recently deposited fallout and the contamination found in milk and grass the original model was developed. In particular it describes the evolution of the contamination in milk and grass without the discrepancies found in previous models. Such an improvement was obtained by taking into account only the fraction of the deposition on soil, which is available for the transfer to the grass.

This model was of the type

$$C = p_r F_r + p_d F_d$$

where:

C	is the mean contamination of milk (Bq/L) for a given year, from May 1st to April 30th
F_r	is the recently deposited fallout (MBq/km^2), for a given year, from May 1st to October 31st
F_d	is the fallout accumulated on the soil during the preceding years (MBq/km^2)
p_r and p_d	are proportionality coefficients

The details for the evaluation of the proportionality factors and the time periods to be taken into account may be found in the original paper. What is relevant here is that this model described very accurately the evolution of the contamination of milk and grass during the period of important fallout (the '60s) around Ispra, Italy (Van der Stricht et al. 1970). It must be stressed that this model was developed aiming to study both the contribution of recent and delayed deposition to herbage and milk. For this reason the spring deposition was taken into account since it was relevant for the transfer to herbage and successively to milk.

^{137}Cs	p_r	p_d
Milk	0.00885	0.002107
Vegetables	0.01054	0.01625
^{90}Sr	p_r	p_d
Milk	0.00139	0.000495
Fr	^{137}Cs	14,800 MBq/km^2
Fr	^{90}Sr	260 MBq/km^2

In the very next days after the Chernobyl accident this model was applied to the region around Saluggia, Italy, where all the factors were obtained from the regular surveys carried on around the nuclear centre. The comparison between the model and the value measured in pooled samples in the years after the accident shows in general a concentration in the samples around one order of magnitude lower. This result is due to the fact that the contamination deposited on the ground in the previous years was negligible in comparison with the deposition due to the accident.

The New Model

First of all it must be emphasized that the difference with the original one is limited to a new set of coefficients, which have been evaluated in order to obtain a good agreement between the values given by the model and the samples of vegetables and milk.

The ^{137}Cs values in both kind of samples were obviously more abundant since the gamma ray spectrometry did not required any time consuming preparation. On the contrary a lesser number of ^{90}Sr values were available. In addition while the milk measurements are already the result of a pooled samples, the vegetables values, in addition to be scanty, are scarcely representative from a statistical point of view. In fact the concentration of ^{90}Sr in vegetables did not show any coherent trend. For this reason it was not possible to calculate a reliable model for such concentrations.

The role played by the release from an accident has a fast decrease in the successive years because the radionuclides deposited into the soil become less available with time since some are removed by herbage or move deeper while some, as Cs, are strongly captured by clay.

In the equation reported above the average deposition due to the Chernobyl accident, F_r was 14,800 (MBq/km^2) for ^{137}Cs and 260 (MBq/km^2) for ^{90}Sr.

The proportionality coefficients of the original model have been modified in order to have a better agreement between the models themselves and the observed measurements as reported above. In this paper the coefficients adopted are the following:

^{137}Cs	p_r	p_d
Milk	0.01	0.085
Vegetables	0.02	0.0015
^{90}Sr	p_r	p_d
Milk	0.0037	0.00085

where:

C is the mean contamination of milk (Bq/L) for a given year
F_r is the fallout (MBq/km^2) deposited by the accident
F_d is the fallout accumulated on the soil during the preceding years (MBq/km^2)
p_r and p_d are proportionality coefficients reported above

In Figs. 1, 2 and 3 the comparison of the model for milk and vegetables with the concentration values for ^{137}Cs in milk and vegetables and ^{90}Sr in milk.

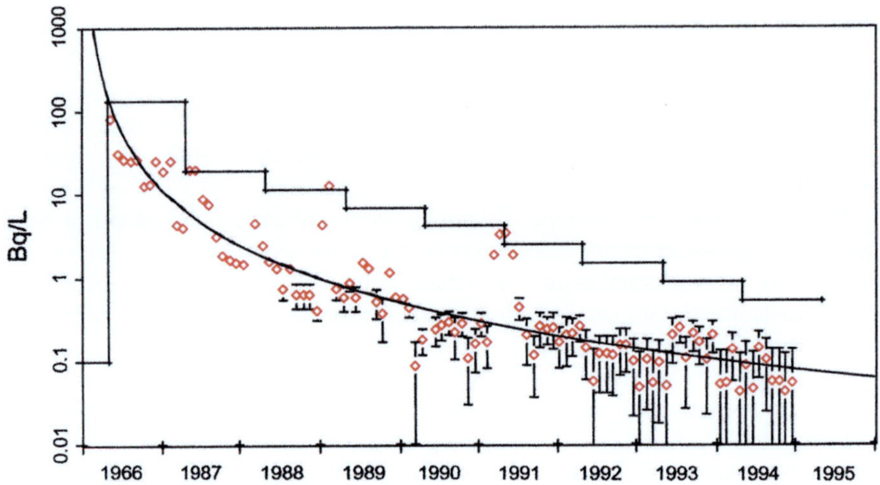

Fig. 1 Comparison between the model prediction and the ^{137}Cs concentration in milk. A best fit is also reported

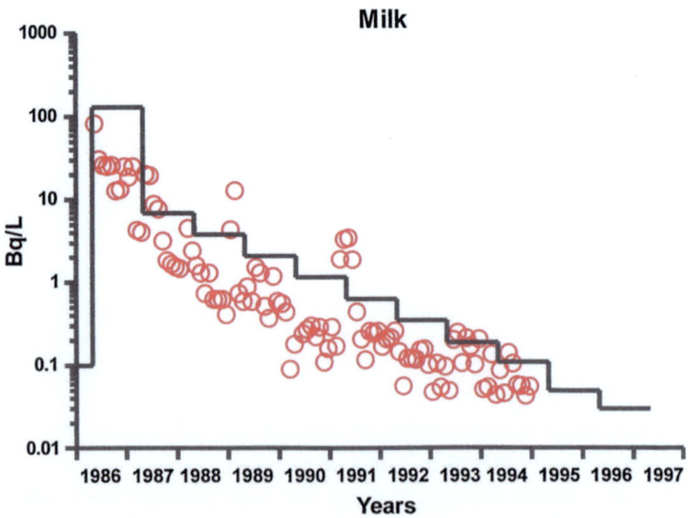

Fig. 2 New model and measured values of ^{137}Cs concentration in milk

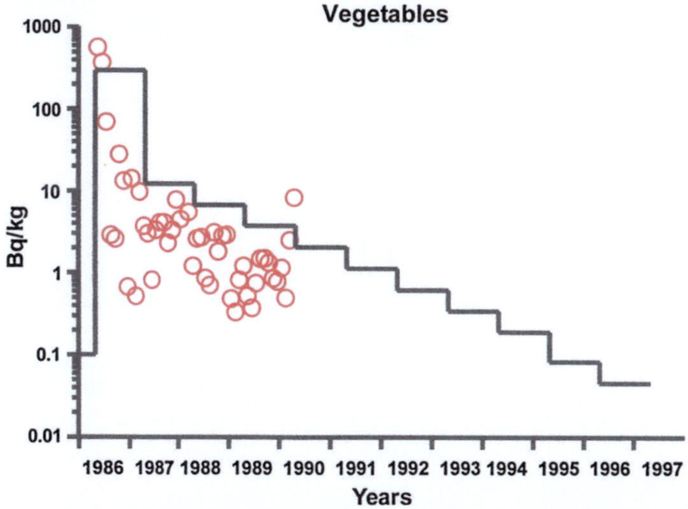

Fig. 3 New model and measured values of ^{137}Cs concentration in vegetables

Results

In case of a sudden release, as for instance due to an accident, it is instrumental to foresee the behaviour of the radioactive contamination in the food chain. Unless the area of interest has already been investigated by a radioecological survey, many environmental parameters are not known and their evaluation would require time and efforts frequently not available. Therefore a simple model derived from an experimental investigation could be very convenient to apply for approximate but reliable information.

As it can be observed in Figs. 2, 3 and 4, the predictions supplied by the new model here described give an upper value of the concentrations leaving outside very few values with a prediction of the contamination values to be expected in more than 90 % of cases. This model for milk and green vegetables may solve the problem to supply an immediate tool for the management of the consequences of the accident. Since ^{137}Cs is easily measured by gamma ray spectrometry without any delay the methodology here described allows an immediate view of the levels to face in the following years.

The release of ^{90}Sr during the Chernobyl accident was about fifty times less that the ^{137}Cs release and therefore the few measurements of the concentration in vegetables did not allow the evaluation of a reliable model for ^{90}Sr also for vegetables. Anyway, if necessary, a rough evaluation may be obtained by a comparison with the results reported for milk.

It must be stressed the danger of over-protection, because a relatively small reduction of the doses delivered to the population is obtained at an unjustified cost.

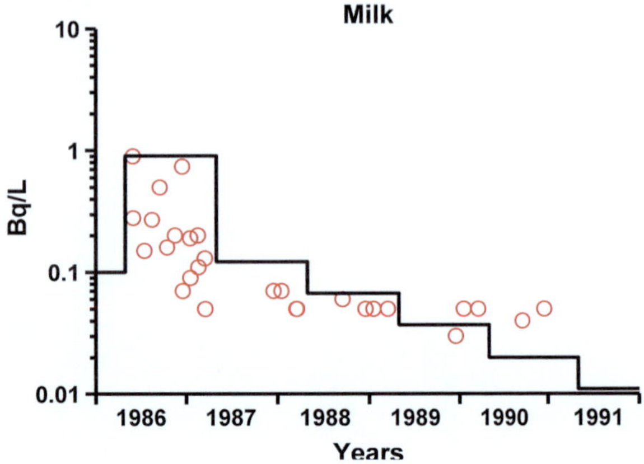

Fig. 4 Model and measured values of ^{90}Sr concentration in milk

Table 1 Induced abortions or decrease in births in some European countries after the Chernobyl accident (details in: Cigna 2009)

Country	Increase in abortions		
	Measured	Estimated	Decrease in births
Denmark	230	ND	ND
Finland	0	ND	350
Greece	ND	2500	ND
Hungary	0	?	About 8000
Italy	ND ND 2600–8000	About 2000 ND ND	About 3500 5800 ND
Norway	0		About 100
Sweden	500–600	ND	ND

The real problem to face is the successful communication with the authorities involved. After the Chernobyl accident, notwithstanding the existence of reliable information and experts, the countermeasures established in Italy by the Minister of Health caused high costs and the loss of thousand of human lives due to induced abortions (Cigna 2009). Unfortunately such a tragedy was also common in other European countries as reported in Table 1.

These numbers are the proof of the horrible effects due to unjustified countermeasures resulting in consequences worst than the effects of the accident itself. The final conclusion is the necessity of every effort to educate authorities to rely on competent organizations with a good specific experience and avoiding the trend of media based on emotion only.

Conclusion

As it has been pointed out previously, in case of an emergency it is necessary to have the best tools for the implementation of the countermeasures. In particular it is relevant to avoid those overprotection that may produce effects sometimes worst than the accident itself, as it was reported before. Therefore a strong effort must aim to have the competent authorities avoid taking emotional decisions instead of scientifically founded ones.

Acknowledgments I am very grateful to Dr. Sandro Ridone for the help in the preparation of diagrams and Dr. Daniela Pani for the revision of the English text. This paper is devoted to the memory of Piero Gaglione, unforgettable colleague and friend, who greatly contributed to the development of the original model.

References

Cigna AA (1987) A simple model for the prediction of ^{90}Sr, ^{134}Cs and ^{137}Cs concentrations in the food chain. In: Proceedings of international seminar on foodstuffs intervention levels following a nuclear accident, Luxembourg, CCE Report EUR 11232, pp 239–249

Cigna AA (2008) Comments on "the role played by stakeholders in the public debate that brought Italy out of the club of nuclear energy producers". Health Phys 94(5):484–485

Cigna AA (2009) Impact of the Chernobyl accident in Western Europe. Environ Protect Ecol 10 (2):476–487

De Bortoli M, Gaglione P, Malvicini A, Van der Stricht E (1968) Five years experience of ^{90}Sr and ^{137}Cs herbage to milk transfer under field conditions. In: Proceedings of 1st international congress IRPA, Rome 2, pp 1053–1063

Van der Stricht E, Gaglione P, De Bortoli M (1970) Valutazione della contaminazione dell'erba e del latte dovuta al ^{137}Cesio nelle ricadute radioattive. G fis sanit prot raziaz 14(2):53–60

Van der Stricht E, Gaglione P, De Bortoli M (1971) Predictions of ^{90}Sr levels in milk on the basis of deposition values. Health Phys 21(2):217–226

Principles and Methods of Radiocapacity Assessment of Ecology Systems

Yury Kutlakhmedov, Gennady Polikarpov and Vladimir Korogodin

In the Ural laboratory of N.W. Timofeeff-Ressovsky the extensive radioecology (radiation biogeocenology) investigations were provided (Timofeeff-Ressovsky 1957) These investigations resulted in the concept of distribution, redistribution and migration of radionuclides in natural conditions (Fig. 1).

Agre and Korogodin (1960) offered the concept of radiocapacity of ecosystems to assess the role of bottom sediments in the removal of radionuclides from the water. Later the concept of radiocapacity was developed (Polikarpov and Egorov 1986) . The idea of ecological capacity as a measure of ecosystem reliability (Polikarpov 1985) is based on the assumption the dual function of ecosystem. To be reliable any ecosystem has to reproduce its biota and to serve as a self-conditioning. Conditioning capacity is determined by the flow of pollutants, which does not disrupt the ecosystem. The reproduction function supports a balanced level of biomass species, their abundance, biodiversity, etc. (Polikarpov and Egorov 1986). It is necessary to do two assesses of reliability of ecosystem: the probabilities of its reproduction (P_1) and conditioning (P_2).

For the first time the article was published in the book "Heuristic Radiobiology", Kiev, Naukova Dumka, 1988, 109-115. It was presented by the M.G. Kholodny Institute of Botany, AS USSR, Kiev the A.O. Kovalevskiy Institute of Biology of the Southern Seas, AS USSR, Sevastopol, Ukraine Joint Institute for Nuclear Research.

Y. Kutlakhmedov (✉)
Institute of Cell Biology and Genetic Engineering, NAS of Ukraine
(former "the M.G. Kholodny Botany Institute"), Kiev 03022, Ukraine
e-mail: ecoetic@yandex.ru

G. Polikarpov
The A.O. Kovalevskiy Institute of Biology of the Southern Seas,
AS Ukraine SSR, Sevastopol, Ukraine

V. Korogodin
Joint Institute for Nuclear Research, Dubna, Russia

V.L. Korogodina et al. (eds.), *Genetics, Evolution and Radiation*,
DOI 10.1007/978-3-319-48838-7_28

Fig. 1 Arrival of N.W. Timofeeff-Ressovsky on the Biological Department of the MSU (1960). From *left* to *right*: M. Aslanian, G.G. Polikarpov, A.N. Tyuryukanov, N.W. Timofeeff-Ressovsky, V.I. Korogodin, V.M. Glazer (from the archive of V.L. Korogodina)

The assessment of reproduction reliability of ecosystem should take into account species diversity; overall indexes of biomass, growth rate, species density, and lifespan of the individuals of different species. The overall biomass in ecosystem is determined by the formula

$$\bar{B} = s \sum_{i=1}^{n} (B_i / B_{\max}) N, \tag{1}$$

where N is maximum number of species; B_{max} is maximum biomass among species; B_i is biomass of ith species; σ is a rank relevance, which is evaluated by expert assessments and equal to 0.567. The other indexes can be similarly determined according to their rank value. The reproduction reliability of ecosystem can be estimated as the ration of sum of their indices to sum of their ranks.

Let us evaluate a probability of conditioning (P_2) of ecosystem experienced effects of k pollutants. We suppose that the capacity assessment for each pollutant is known (F_i.). Understanding of "the weakest link" allows us to estimate the overall reliability (F) by the formula

$$F = \prod_{i=1}^{k} (1 - F_i), \tag{2}$$

where k is number of pollutants; F_i is a factor of capacity for each pollutant in ecosystem. Synergism and antagonism of pollutants will increase or decrease their effect.

In radioecology the main assessment of ecological capacity is its radiocapacity. It is determined by the amount of radionuclides which can be kept for a long time in the ecosystem. Radiocapacity can be quantitatively counted as a radionuclides' share held in the ecosystem in comparison with the total incoming activity. This percentage may vary from 0 to 1. It depends on the characteristics of the ecosystem and, primarily, accumulation coefficients of its living and nonliving components.

Consider two extreme values of the radiocapacity. If radiocapacity is equal to 1, it means that the nonliving and living components are fully capable to hold all incoming radioactivity in this ecosystem. The radionuclides somehow remain in the ecosystem, i.e. in soil, in sediments of water bodies, in plants, animals, etc. This case relates to the ideal ecosystem. Another extreme case when radiocapacity equal to 0, i.e., ecosystem is not capable of holding radioactive substances. What radionuclide enters the ecosystem, and so it exits.

This "transit" type of ecosystems is possible in the case of radionuclides of noble gases, or extremely rapid relief a large number of radionuclides, when all the basic mechanisms of conditioning did not manage to largely operate, or when radiocapacity ecosystem has already been exhausted and coming anew radioactive substances in it is almost not held.

Let's consider the principles and methods for assessing radiocapacity of simple ecosystem of descending series (cascade) of freshwater sedimentation reservoirs. A.L. Agre and V.I. Korogodin (Agre and Korogodin 1960) offered to calculate the radiocapacity of a freshwater reservoir by formula

$$F = Kh/(H + Kh), \quad (3)$$

where H—the average depth of the reservoir; h—the average thickness of sorbent layer of soil; K—average weighted sorption coefficient in this layer (accumulation factor). In the steady state radiocapacity factor F describes a part of the radioactivity that is bound up with its sediments, and the $(1 - F)$ value presents radioactivity in the water which goes through the upper discharge flowing into the pond next stage. It can be expected equilibrium steady state in the reservoirs for the system of slowly flowing ponds. It is possible to estimate total radiocapacity of this cascade, knowing the parameters of reservoirs and their own radiocapacity. Sequential equilibrium system of ponds can be considered as independent entities and, therefore, it lets count their radiocapacity independently and sequentially.

Table 1 presents the parameters of the system of slowly flowing reservoirs and the calculated value of the radiocapacity factor F, which varies from 0.6 to 0.8.

In the first pond, the sediments and benthic communities sorb and accumulate 0.7 part of the total radioactivity (A), and the rest ($0.3A$) remains in the water, plankton and neuston organisms and goes to the second reservoir; 0.6 part of the radionuclides which were came in the second reservoir ($0.6 \cdot 0.3A = 0.18A$), goes into the its sediments, and the rest $(0.3 - 0.18 = 0.12)A$ goes into in the third

Table 1 Parameters: area (S), volume (V), thickness of sorbent layer (h), and characteristics: coefficient of radionuclide accumulation (K), radiocapacity factor (F), share of the total radioactivity in the reservoir' water (A) of the cascade of settling-ponds

Reserv. number	Area (*S*), m^2	Volume (*V*), km^3	Average depth (*H*), m	Thickness sorb. layer (*h*), cm	Coef. accumul. (*K*)	Radio-capacity factor (*F*)	Radioact. in water, share (*A*)
1	920	3.7	4	10	100	0.7	0.3
2	680	2.6	4	10	50	0.6	0.12
3	2250	13.5	6	10	800	0.8	0.024
4	570	2.4	4	10	100	0.7	0.007
5	410	3.3	8	10	230	0.7	0.002
6	2150	18.2	8	10	280	0.7	0.0006

$F_{cascade} = 0.9994$

reservoir cascade, and so on. The total radioactivity of the water of the sixth, the latter stage of the reservoir will be equal to 0.0006*A*, and the overall radiocapacity cascade of six reservoirs is a significant value of 0.9994.

This high capacity of the cascade of reservoirs reduces the radioactivity of the water by 3–4 orders. It suggests a high conditioning capacity of the cascade of reservoirs in relation to the discharge of radioactive substances in it. The proportion of radionuclides from the initial their amount that passes through the cascade is defined as the product of probabilities of going through every single body of water $(1 - 0.7) \cdot (1 - 0.6) \cdot (1 - 0.8) \cdot (1 - 0.7) \cdot (1 - 0.7) \cdot (1 - 0.7)$. Radiocapacity of each reservoir is an independent value, which leads to effective filtration of radionuclides and their decrease by 3–4 orders. Of course, it is necessary to take into account the radionuclide releases which depend on season changing.

The following are estimates of reliability function of self-reproduction of the first model reservoir under the simplified scheme of calculation. Data on its phytoplankton in three generalized parameters were obtained in the course of long-term observations. Species diversity is estimated by the formula

$$\bar{N} = \frac{N}{50}a, \tag{4}$$

where N is number of species in an ecosystem; α is significance coefficient of the criterion (0.77). According to the preliminary estimates, 50 is the number of species in an ecosystem, which is sufficient for its normal functioning. If $N \geq 50$, than the value of $\overline{N}$ is assumed to be 1. Generalized biomass is calculated by the formula (1).

Similarly, a generalized numbers is calculated by the formula

$$\bar{K} = g\sum_{i=1}^{N}(R_i/R_{\max})N, \tag{5}$$

where N—numbers of the specific species (per unit volume); R_{max}—maximum numbers among species in this ecosystem; γ—significance coefficient of the criterion (0.64).

The total capacity of self-reproduction of the ecosystem is determined by the formula

$$P_i = (\bar{N} + \bar{B} + \bar{K})/(\sigma + \alpha + \gamma) = \\ = 0.39N/50 + 0.29\sum_{i=1}^{N}(B_i/B_{\max})N + 0.32\sum_{i=1}^{N}(R_i/R_{\max})N \quad (6)$$

The P_1 values were obtained by years of observations of ecosystem and may serve as a simplified assessment of the ecology capacity, ie the reliability of a given ecosystem. In this assessment the radiocapacity factor F can be accounted to evaluate radiation reliability ($P = P_1F$).

Tables 2 and 3 present the criteria and capacity of phytoplankton ecosystem for different observation time and points in space of the reservoir. Attention is drawn to the stability of the value of capacity on productivity (P_1) and its independence from the seasons of observation. Table 3 shows that ecosystem reliability changes insignificantly (≈0.5) in the middle site (1965–1967). In the period from 1967 to 1969, a decline of ecological capacity to the value of 0.3 is observed, which corresponds to the period of strong anthropogenic impacts on the water body, and then there is the restoration of the values of P_1 to 0.5. The findings of the simplified calculation of the ecological capacity of the reservoir model showed the possibility

Table 2 The criteria values of phytoplankton ecosystem of the model reservoir at different points in a separate area: number of species (*N*), numbers of species (*K*), biomass (*B*), ecology capacity (P_1)

Date, 1981	Number of species, N	Numbers of species, K	Biomass, $\overline{B}$	Ecology capacity, P_1
26. $\overline{\text{V}}$	74	0.038	0.039	0.47
29. $\overline{\text{V}}$	46	0.071	0.10	0.53
1. $\overline{\text{VI}}$	60	0.034	0.06	0.48
4. $\overline{\text{VI}}$	68	0.030	0.05	0.47
7. $\overline{\text{VI}}$	77	0.044	0.08	0.51
15. $\overline{\text{VII}}$	79	0.038	0.036	0.47
18. $\overline{\text{VII}}$	60	0.033	0.034	0.45
21. $\overline{\text{VII}}$	28	0.096	0.13	0.45
24. $\overline{\text{VII}}$	34	0.058	0.046	0.38
9. $\overline{\text{X}}$	35	0.041	0.064	0.38
14. $\overline{\text{X}}$	24	0.13	0.21	0.48

Average value $\overline{P}_1$ = 0.045

Table 3 Ecology capacity of phytoplankton self-reproduction of the model reservoir from 1965 to 1981

Site	1965	1966	1967	1968	1969	1970	1977	1978	1981
Lower	0.33	0.4	0.38	0.44	0.5	0.5	0.5	0.45	0.45
Middle	0.5	0.4	0.5	0.3	0.4	0.4	0.45	0.5	0.48
Upper	0.6	0.55	0.44	0.4	0.5	0.5	0.51	0.41	0.58
1-й tributary	0.53	0.51	0.54	0.6	0.53	0.5	0.5	0.61	0.55
2-й tributary	0.46	0.63	0.53	0.4	0 4	0.6	0.5	0.55	0.57
Average	0.48	0.5	0.48	0.43	0.47	0.5	0.49	0.5	0.53

of using this approach to assessing the reliability of the ecosystem on its self-reproduction.

The reliability of the first model reservoir in conditions of radioactive contamination can be determined by the product $P = P_1 \cdot P_2 = 0.5 \cdot 0.7 = 0.35$. As a whole, this value reflects the high ecological capacity, and, respectively, the reliability of the first model of the reservoir.

Cascade ponds will work well as a decontamination system for a number of years, until exhausted radiocapacity sediment. The degree of depletion of the radiocapacity of the reservoir is determined primarily by the rate of dumping of radioactivity in it. If the rate of dumping does not exceed the rate of sorption and bioaccumulation, the accumulation of radionuclides in sediments and benthos may occur almost linearly over 3–20 years without significant deterioration of water quality in the reservoir (Timofeeva-Ressovskaya and Timofeeff-Ressovsky 1960).

Further studies on specific ecosystems allow us to determine the usefulness of the proposed heuristic approach to the assessment of their radiocapacity and reliability.

References

Agre AL, Korogodin VI (1960) The distribution of radioactive contamination in landlocked body of water. Med Radiol 1:63–73 (Russian)

Egorov VN, Gulin SB, Polikarpov GG et al (2010) Black Sea. In: Atwood DA (ed) Radionuclides in the environment. Wiley, Chichester (UK), pp 430–452

Gulin SB, Egorov VN (2013) Self-purification of seawater: a measure for environmental regulation. In: White MR (ed) Seawater: geochemistry, composition and environmental impacts. Nova Science Publishers, New York, pp 93–126

Kutlakhmedov YA (2015) Way to the theoretical radioecology. Phytosociocentre Publication, Kiev, 360 p (Russian)

Polikarpov GG (1985) Environmental introduction to the problem of the reliability of self-cleaning function of the aquatic ecosystem. In: Reliability of the biological systems. Naukova Dumka, Kiev, pp 172–179 (Russian)

Polikarpov GG, Egorov VN (1986) Marine dynamic radiochemoecology. EnergoatomIzdat, Moscow, 476 p (Russian)

Timofeeff-Ressovsky NW (1957) The use of radiation and radiation emitters in the experimental biogeocenology. Botanical J 42(3):161–194 (Russian)

Timofeeva-Ressovskaya EA, Timofeeff-Ressovsky NW (1960) Distribution of trace elements in the components of water bodies. Proc Inst Biol UD AS USSR 12:189–238 (Russian)

Part V
Radiation and Man

Fundamental Mechanisms Underlying the Ill Health and Chronic Fatigue Syndrome Suffered by Atomic and Gulf War Veterans: A Unifying Hypothesis

Carmel Mothersill and Colin Seymour

Introduction

These are two key hypotheses we wish to propose which will be discussed below.

1. CFIDS is caused by low dose chronic or acute exposure to stressors including radiation. The stress signalling leads to epigenetically sustained genomic instability in stem cell microenvironments producing the various immune, neurological and gastrointestinal symptoms.
2. Alleviation of CFIDS will involve understanding the initial signalling process and deactivating it or resetting it using inhibitors or epigenetic modifiers.

Background

CFIDS is a debilitating illness of unknown aetiology. It often manifests suddenly after exposure to stress or a flu-like illness (Katz and Jason 2013; Glaser and Kiecolt-Glaser 1998; Morris and Maes 2013; Bansal et al. 2012). However the symptoms which include multiple manifestations of both neurological, gastrointestinal and immune dysfunction (Maes et al. 2009; Maes and Twisk 2009; Moss-Morris et al. 2013), strongly suggest an underlying malfunction in the ability of the cells and tissues to manage cellular stress (Nicolson 2007; Hanley and Van de Kar 2003). We suggest that stem cells are a particular target and that regarding

C. Mothersill (✉) · C. Seymour
Department of Medical Physics and Applied Radiation Sciences, McMaster University, Hamilton, ON L8S 4K1, Canada
e-mail: mothers@mcmaster.ca

C. Seymour
e-mail: seymouc@mcmaster.ca

V.L. Korogodina et al. (eds.), *Genetics, Evolution and Radiation*,
DOI 10.1007/978-3-319-48838-7_29

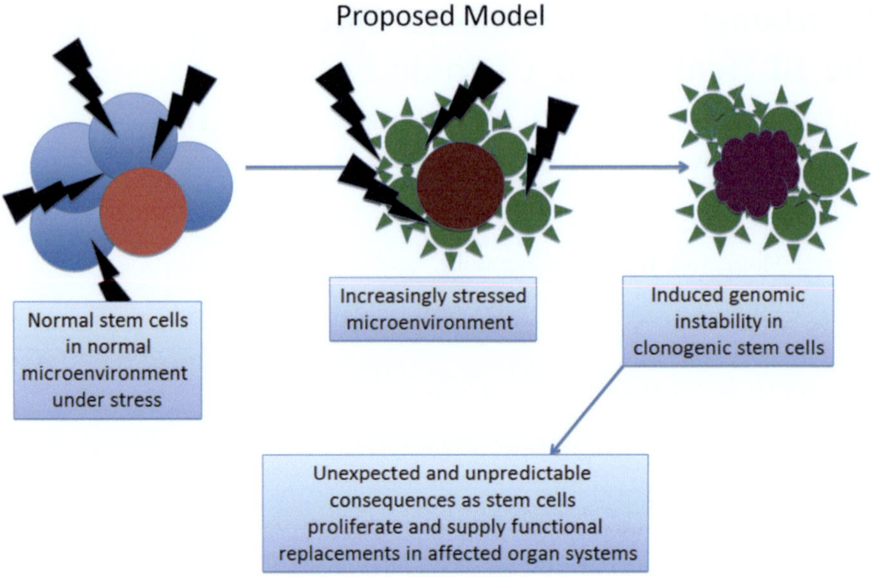

Fig. 1 Proposed model

CFIDS as a malaise of the stem cells might provide new insights into potential treatments (see Fig. 1 for the model).

Similar syndromes to CFIDS include the complex set of symptoms seen in Atomic Veterans and Gulf War survivors which are often dismissed as psychological (Ferguson and Cassaday 2001–2002; Israeli 2012; Garcia 1994; Murphy et al. 1990) because the small numbers and varied symptoms are not amenable to epidemiological analysis (McCauley et al. 2002; Coughlin 2013). However the variability in symptoms would imply that the abnormality may be unexpressed until a major stress event challenges the tissues to regenerate or otherwise become activated. The stress event could be biological e.g. viral infection, vaccine challenge, radiation or toxin exposure, or psychological e.g. war, catastrophic accidental event occurs which triggers the emergence of the full-blown syndrome (Maloney et al. 2013; Glaser et al. 1998; Witthöft et al. 2013). There is convincing but largely anecdotal evidence that chronic exposure to low doses of ionizing radiation (LDIR) is associated with neurological effects such as memory impairment, chronic fatigue syndrome and cognitive decline in humans (Marazziti et al. 2012; Ballesteros-Zebadúa et al. 2012; Huang 2012). The Cold War veterans represent the best example of this (Hansen and Schriner 2005) but because the prevailing theory in radiation biology and radiation protection did not recognise these low dose effects, the cohorts have not been well studied and are now too old to be systematically studied.

Major advances have now been made in our understanding of low dose effects in the laboratory and it is clear that these studies could significantly advance our understanding of chronic ill health syndromes in general and CFIDS in particular. Recent evidence mainly funded by NASA, who are concerned about long term exposure to space radiation on cognitive function, fatigue and other non-cancer effects in astronauts (Cucinotta et al. 2013; Ushakov et al. 2011) implicates the neural stem cells in the process since these are extremely sensitive to LDIR (Kim et al. 2013; Etienne et al. 2012). The immune dysfunction symptoms could result from compromise of the bone marrow stem cells for which there is ample evidence linking radiation exposure to genomic instability in the stem cells (Mothersill and Seymour 2012; Szumiel 2014). Similarly gastrointestinal stem cells are also known to be extremely sensitive to low doses of radiation (Mothersill and Seymour 2012; Szumiel 2014). The mechanisms are unclear and obviously in order to treat or prevent the impacts of LDIR, mechanistic understanding of the process is crucial. This proposal aims to test the overall hypothesis that chronic fatigue and immune dysfunction syndrome (CFIDS) can become established in latent form as genomic instability maintained epigenetically though signalling (bystander effect) either due to genetic/epigenetic predisposition or as a result of chronic exposure to multiple stressors including LDIR. A trigger stress can then lead to full-blown expression of the disease.

Radiation biology has undergone a paradigm shift in recent years (for reviews see Mothersill and Seymour 2012; Szumiel 2014). The old idea was that DNA damage caused by radiation developed in a dose dependent manner and was usually efficiently repaired meaning that low doses caused very low or no effects (Hall 2011). Cancer was considered to be the only low dose effect and doses below 100 mGy were not considered to cause incidence above background (Morgan and Sowa 2013). Chronic doses and doses due to "internal emitters" i.e. ingested or inhaled radioisotopes were largely ignored (Mothersill and Seymour 2013). This all changed in the late 1980s and early 1990s when there were major breakthroughs which showed that the effects of ionizing radiation were very different after low doses and that even after high doses, repair was not an all-or-nothing affair. The phenomena of Lethal Mutations (Seymour et al. 1986), Genomic instability (Kadhim et al. 1992) and the Bystander Effect (Nagasawa and Little 1992) all suggested that effects of LDIR were very complex and very long lasting. Later research implicated the immune system (Hilgers and Frank 1994; Ojo-Amaize et al. 1994; Yancey and Thomas 2012; Lorusso et al. 2009) and linked cellular oxidative stress and inflammatory response to the perpetuation of a chronic dysfunction in affected cells, which could be inherited via epigenetic mechanisms (Landmark-Høyvik et al. 2010; Smith et al. 2008). Now it is accepted that these mechanisms are fundamentally important to the understanding of cellular and organismal responses to LDIR and many other stressors such as metals, organic toxins and pathological agents. Many mechanisms, particularly those involved in the initiation of the signalling pathways are still largely unknown although recent evidence from our laboratory implicates LDIR induction of UVA photons in exposed biological material and cells (Ahmad et al 2013; Le et al. in press). The UVA is thought to generate reactive oxygen species (ROS), which have long been known to compromise mitochondrial function and disrupt the sensitive control of inter and intra cellular signalling (Festjens et al. 2006; Formigari et al. 2013; Frank et al. 2012; Morris and Maes 2014; Meeus et al. 2013; Voloboueva and Giffard 2011).

The theme, which comes through much of the literature about CFIDS is that it appears to have no one cause or clear set of symptoms. Its appearance can appear random and may be distant in time from an alleged triggering stress event. This is why its existence is sometimes denied by Western Medicine. However, this is exactly what would be predicted to happen if LDIR or other low dose toxic exposures were involved. It is particularly true if there is a two-stage etiological mechanism involving chronic low dose multiple stressor exposure followed by (or possibly preceded by) an acute mental or physical stress. One idea we will pursue throughout this research program is that CFIDS would have no clear cause or set of symptoms if it were a system level disease consequential upon systems not communicating or interacting as they should. Symptoms and causes would then be system specific. Individual predisposition to sensitivity would then be a most important factor (hence our interest in the impacts of chronic stressors during developmental). This individual, underlying sensitivity would be exacerbated by acute stressor exposure acting on the same or an interacting system. The mode of action of the stressors would also be important e.g. whether they act locally or systemically and whether exposure is local or systemic. A major focus of our research and possibly the major key to understanding CFIDS mechanisms is to understand the role of signalling processes in converting initially localised responses into systemic responses via processes such as "bystander signalling". Currently our research is suggesting that physical signals including photons in the UV range and phonons (sound vibrations) are key early initiating events when cells or organisms are exposed to LDIR (Mothersill et al. 2012). These signals may not initially be recognised as related to the LDIR exposure but we have shown they result in oxidative stress leading to inflammatory responses, which are of course systemic and deleterious and may underlie CFIDS pathology.

The section which follows is built around three quite specific research questions which we think should be explored to identify possible metabolic pathways which could be manipulated to reset the systemic disorder which underlies CFIDS.

What Are the Basic Mechanisms Involved in the Communication of Stress Between Cells in Tissues?

We need to explore how the physical electromagnetic energy (EM) associated with low dose exposures to electromagnetic radiation, is transduced to lead to biological responses in affected human patients. We know that UVA is involved and intend to explore what molecules absorb the UV energy and what results from this absorption. We suspect that the perturbation of delicately balanced voltage gated ion channels is involved and there is direct evidence for both sodium (Na v 1.5) and calcium (5HT-3A) ion gated channels being involved in the mechanism. We know that the energy perturbation results in activation of H-Ras and MAPK stress response pathways but that instead of apoptosis occurring (which would remove damaged cells), a process known as genomic instability results. This is a permanent increase (reset) in the tolerance of the stem cell population for mutations. It is driven by bystander signals and is an example of an epigenetic control mechanism which results in chronically compromised stem cells in an abnormal microenvironment.

It will be important to investigate how mitochondria in stem cells coordinate function and decide response after toxin challenge. Of particular interest here is the putative photoreceptor in the mitochondrial membrane thought to be part of complex IV. This links in with our laboratory's recent work on UVA emission by cells when exposed to ionizing radiation which was discussed earlier. We do not know if other stressors such as metals can cause UVA emissions. However the common mechanisms seen in metal, bacterial and radiation challenged tissues and organisms strongly suggest a common underlying stress response which could involve UVA leading to ROS generation and consequent mitochondrial activity.

This area of research could be progressed by using co-culture systems where acutely stressed cells can signal to chronically stressed or control cells. Endpoints to be measured include bystander signal production, calcium and sodium fluxes across membranes using patch clamping, mitochondrial numbers, location and function, expression of key stress response proteins and UVA emission. Now that we have the antibody to the sodium channel Na v 1.5, associated with the ciguatera toxin which also is affected by chronic LDIR exposure, we will have a direct link between bystander signalling and channel activation in clonogenic stem cells.

Treatment of CFIDS Will Involve Elimination of Stress Signal Production and Resetting of the Supportive Microenvironment Which Allows the Aberrant Cells to Thrive

Several lines of evidence implicate LDIR exposure to internal emitters such as polonium, strontium, caesium, uranium or radium in the production of chronic ill-health and chronic fatigue syndromes. Classic examples are the ill-health of atomic veterans who were exposed during the Cold War, Gulf War personnel and inhabitants of Gulf War or nuclear bomb test countries e.g. the Marshall Islanders. While Governments mostly deny a causal link between ultralow dose radiation exposure and chronic ill health, because of weak epidemiology, a perfectly logical explanation is that the exposures lead to genomic instability driven by oxidative stress in the extremely sensitive stem cell compartments in the body which causes chronic inflammatory processes to occur and injures the microenvironment leading to a "self-sustaining" disease prone state. These processes are known to underlie many of the chronic conditions including CFIDS from which these cohorts suffer.

An approach which might prove useful would be to culture stem cells from key tissues such as brain (hippocampus), gastrointestinal crypts and bone marrow in compromised micro-environments of support cells, which has been conditioned by LDIR exposure to signal stress to neighbouring cells. The behaviour of the stem cells in these conditions could be monitored using the assays suggested in Section "What Are the Basic Mechanisms Involved in the Communication of Stress Between Cells in Tissues?". It would also be useful to test the effectiveness of interfering with UVA production, with ROS, with MAPK pathways and with p53 pathways to see if this will prevent damage to the stem cells. Finally, there are established treatments for CFIDS promoted by practitioners of alternative medicine with proven effectiveness in eliminating symptoms such as the terpene D-limonene or curcumin. It would be very important to see if these substances are effective at preventing genomic instability from occurring or being expressed in the stem cells. This would provide useful confirmation of the correctness of our hypotheses.

Probing the Neurodevelopmental Processes as Affected by LDIR or Trace Metals and Microtoxins Thought to Be Involved in the Aetiology of CFIDS

As discussed earlier we now have substantial evidence that neurotransmitters and their inhibitors and activators can modulate response to toxin or radiation challenge even in non-neural cells. This is logical given that embryonic stem cells are pluripotent. It stands to reason that all cells have the potential to develop or express receptors if induced by the correct stimuli. In the case of stem cells in differentiated tissues, this potential is very great if the epigenetic information forthcoming. These mechanisms, if fully understood could lead to development of new drugs for modulating metal or radiation effects in cells, thus providing hope for CFIDS patients.

Membranes in the body function as barriers maintaining differential concentrations of ions on either side of the membrane by creating and maintaining a potential difference or electrical gradient. Channels in the membranes regulate the passage of sodium, potassium, calcium and chloride ions. These are known as ion channels and the passage of ions through them is tightly regulated.

The specific case of voltage-gated ion channel (VGIC) manipulation in the nervous system is of particular interest to those interested in LDIR because both radioactive metal ions similar to sodium, potassium or calcium (e.g. strontium, caesium, radium) and non-particulate ionizing radiation (e.g. x or gamma rays) would, by exciting or ionizing the membrane, upset delicate VGIC balances. This could lead to complex or chaotic responses. Channelopathies are linked to at least 20 diseases now including CFIDS (Kass 2005; Waxman and Ptacek 2000; Chaudhuri et al. 2000). A particular focus of our research will be the 5HT type 3A channel with we already know to be involved in the calcium response to LDIR (Fazzari et al. 2012; Mothersill et al. 2010; Saroya et al. 2009; Poon et al. 2007; Mothersill et al. 1976) and the sodium channel Na v 1.5 which is involved in internalising ionizing radiation (radioisotopes) and is known to be altered in CFIDS patients (Fulle et al. 2003; Beyder et al. 2010; Abriel 2010). We hypothesise that very low doses of ionizing radiation can cause excitation of cell membranes leading to aberrant ion fluxes and initiation of inappropriate signalling—a systemic consequence. We postulate that the cognitive defects in CFIDS patients may result from this leading to a hypersensitive state in chronically stressed individuals.

Because we think that CFIDS is related to early exposure to chronic stress during development followed by an acute triggering exposure during later life, we suggest that human embryonic stem cells should be used for these experiments. These can be induced to differentiate along neural, gastrointestinal or bone marrow stem cell pathways and development can be tracked using a number of specific markers for tissue specific development and differentiation (Curtis et al. 2014; Wu and Wang 2012). Agonists and specific inhibitors should interfere with the differentiation pathway and allow us to pinpoint metabolic pathways implicated in the development of CFIDS.

Conclusion

This paper presents a working hypothesis which seeks to provide an explanation for a distressing collection of symptoms associated with low dose radiation exposure which result in Atomic Veterans syndrome, Gulf War syndrome, Chronic Fatigue and Immune Deficiency Syndrome and Myalgic Encephalomyelitis. Also presented is a possible research approach to improving treatment for sufferers of these complex diseases.

Acknowledgments We thank the Canadian National Science and Engineering Research Council and the Canada Research Chairs Programme for financial support. We are especially grateful to Alan Cocchetto of the National CFIDS Foundation Inc. for financial support and inspiration.

References

Abriel H (2010) Cardiac sodium channel Na v 1.5 and interacting proteins: physiology and pathophysiology. J Mol Cell Cardiol 48(1):2–11

Ahmad SB, McNeill FE, Byun SH, Prestwich WV, Mothersill C, Seymour C, Armstrong A, Fernandez C (2013) Ultra-violet light emission from HPV-G cells irradiated with low let radiation from (90)Y; consequences for radiation induced bystander effects. Dose Response 11:498–516

Ballesteros-Zebadúa P, Chavarria A, Celis MA, Paz C, Franco-Pérez J (2012) Radiation-induced neuroinflammation and radiation somnolence syndrome. CNS Neurol Disord: Drug Targets 11 (7):937–949

Bansal AS, Bradley AS, Bishop KN, Kiani-Alikhan S, Ford B (2012) Chronic fatigue syndrome, the immune system and viral infection. Brain Behav Immun 26(1):24–31

Beyder A, Rae JL, Bernard C, Strege PR, Sachs F, Farrugia G (2010) Mechanosensitivity of Nav1.5, a voltage-sensitive sodium channel. J Physiol 588(Pt 24):4969–4985

Chaudhuri A, Watson WS, Pearn J, Behan PO (2000) The symptoms of chronic fatigue syndrome are related to abnormal ion channel function. Med Hypotheses 54(1):59–63

Coughlin SS, McNeil RB, Provenzale DT, Dursa EK, Thomas CM (2013) Method issues in epidemiological studies of medically unexplained symptom-based conditions in veterans. J Mil Veterans Health 21(2):4–10

Cucinotta FA, Kim MH, Chappell LJ, Huff JL (2013) How safe is safe enough? Radiation risk for a human mission to Mars. PLoS One 8(10):e74988

Curtis DJ, Sood A, Phillips TJ, Leinster VH, Nishiguchi A, Coyle C, Lacharme-Lora L, Beaumont O, Kemp H, Goodall R, Cornes L, Giugliano M, Barone RA, Matsusaki M, Akashi M, Tanaka HY, Kano M, McGarvey J, Halemani ND, Simon K, Keehan R, Ind W, Masters T, Grant S, Athwal S, Collett G, Tannetta D, Sargent IL, Scull-Brown E, Liu X, Aquilina K, Cohen N, Lane JD, Thoresen M, Hanley J, Randall A, Case CP (2014) Secretions from placenta, after hypoxia/reoxygenation, can damage developing neurones of brain under experimental conditions. Exp Neurol pii: S0014–4886(14)00137-X. doi:10.1016/j.expneurol.2014.05.003 (Epub ahead of print)

Etienne O, Roque T, Haton C, Boussin FD (2012) Variation of radiation-sensitivity of neural stem and progenitor cell populations within the developing mouse brain. Int J Radiat Biol 88 (10):694–702

Fazzari J, Mersov A, Smith R, Seymour C, Mothersill C (2012) Effect of 5-hydroxytryptamine (serotonin) receptor inhibitors on the radiation-induced bystander effect. Int J Radiat Biol 88 (10):786–790

Ferguson E, Cassaday HJ (2001–2002) Theoretical accounts of gulf war syndrome: from environmental toxins to psychoneuroimmunology and neurodegeneration. Behav Neurol 13(3–4):133–147

Festjens N, Vanden Berghe T, Vandenabeele P (2006) Necrosis, a well-orchestrated form of cell demise: signalling cascades, important mediators and concomitant immune response. Biochim Biophys Acta 1757(9–10):1371–1387

Formigari A, Gregianin E, Irato P (2013) The effect of zinc and the role of p53 in copper-induced cellular stress responses. J Appl Toxicol 33(7):527–536

Frank M, Duvezin-Caubet S, Koob S, Occhipinti A, Jagasia R, Petcherski A, Ruonala MO, Priault M, Salin B, Reichert AS (2012) Mitophagy is triggered by mild oxidative stress in a mitochondrial fission dependent manner. Biochimica et Biophys Acta (BBA)—Mol Cell Res 1823(12):2297–2310

Fulle S, Belia S, Vecchiet J, Morabito C, Vecchiet L, Fanò G (2003) Modification of the functional capacity of sarcoplasmic reticulum membranes in patients suffering from chronic fatigue syndrome. Neuromuscul Disord 13(6):479–484

Garcia B (1994) Social-psychological dilemmas and coping of atomic veterans. Am J Orthopsychiatry 64(4):651–655

Glaser R, Kiecolt-Glaser JK (1998) Stress-associated immune modulation: relevance to viral infections and chronic fatigue syndrome. Am J Med 105(3A):35S–42S

Hall EJ (2011) Radiobiology for the radiologist, 7th edn.

Hanley NR, Van de Kar LD (2003) Serotonin and the neuroendocrine regulation of the hypothalamic–pituitary-adrenal axis in health and disease. Vitam Horm 66:189–255

Hansen D, Schriner C (2005) Unanswered questions: the legacy of atomic veterans. Health Phys 89(2):155–163

Hilgers A, Frank J (1994) Chronic fatigue syndrome: immune dysfunction, role of pathogens and toxic agents and neurological and cardial changes. Wien Med Wochenschr 144(16):399–406

Huang TT (2012) Redox balance- and radiation-mediated alteration in hippocampal neurogenesis. Free Radic Res 46(8):951–958

Israeli E (2012) Gulf war syndrome as a part of the autoimmune (autoinflammatory) syndrome induced by adjuvant (ASIA). Lupus 21(2):190–194

Kadhim MA, Macdonald DA, Goodhead DT, Lorimore SA, Marsden SJ, Wright EG (1992) Transmission of chromosomal instability after plutonium alpha-particle irradiation. Nature 355 (6362):738–740

Kass RS (2005) The channelopathies: novel insights into molecular and genetic mechanisms of human disease. J Clin Invest 115(8):1986–1989

Katz BZ, Jason LA (2013) Chronic fatigue syndrome following infections in adolescents. Curr Opin Pediatr 25(1):95–102

Kim JS, Yang M, Kim SH, Shin T, Moon C (2013) Neurobiological toxicity of radiation in hippocampal cells. Histol Histopathol 28(3):301–310

Landmark-Høyvik H, Reinertsen KV, Loge JH, Kristensen VN, Dumeaux V, Fosså SD, Børresen-Dale AL, Edvardsen H (2010) The genetics and epigenetics of fatigue. PM R 2 (5):456–465

Le M, McNeill F, Seymour C, Rainbow A, Mothersill C An observed effect of ultraviolet-A radiation emitted from β-irradiated HaCaT cells upon non-β-irradiated bystander cells. Rad Res (in press)

Lorusso L, Mikhaylova S, Capelli E, Ferrari D, Ngonga GK, Ricevuti G (2009) Immunological aspects of chronic fatigue syndrome. Autoimmun Rev 8(4):287–291

Maes M, Twisk FN (2009) Why myalgic encephalomyelitis/chronic fatigue syndrome (ME/CFS) may kill you: disorders in the inflammatory and oxidative and nitrosative stress (IO&NS) pathways may explain cardiovascular disorders in ME/CFS. Neuro Endocrinol Lett 30(6): 677–693

Maes M, Mihaylova I, Kubera M, Uytterhoeven M, Vrydags N, Bosmans E (2009) Coenzyme Q10 deficiency in myalgic encephalomyelitis/chronic fatigue syndrome (ME/CFS) is related to fatigue, autonomic and neurocognitive symptoms and is another risk factor explaining the early mortality in ME/CFS due to cardiovascular disorder. Neuro Endocrinol Lett 30(4):470–476

Maloney CD, Jensen S, Gil-Rivas V, Goolkasian P (2013) Latent viral immune inflammatory response model for chronic multisymptom illness. Med Hypotheses 80(3):220–229

Marazziti D, Baroni S, Catena-Dell'Osso M, Schiavi E, Ceresoli D, Conversano C, Dell'Osso L, Picano E (2012) Cognitive, psychological and psychiatric effects of ionizing radiation exposure. Curr Med Chem 19(12):1864–1869

McCauley LA, Joos SK, Barkhuizen A, Shuell T, Tyree WA, Bourdette DN (2002) Chronic fatigue in a population-based study of gulf war veterans. Arch Environ Health 57(4):340–348

Meeus M, Nijs J, Hermans L, Goubert D, Calders P (2013) The role of mitochondrial dysfunctions due to oxidative and nitrosative stress in the chronic pain or chronic fatigue syndromes and fibromyalgia patients: peripheral and central mechanisms as therapeutic targets? Expert Opin Ther Targets 17(9):1081–1089

Morgan WF, Sowa MB (2013) Non-targeted effects induced by ionizing radiation: mechanisms and potential impact on radiation induced health effects. Cancer Lett pii: S0304-3835(13) 00662-9. doi: 10.1016/j.canlet.2013.09.009

Morris G, Maes M (2013) A neuro-immune model of myalgic encephalomyelitis/chronic fatigue syndrome. Metab Brain Dis 28(4):523–540

Morris G, Maes M (2014) Mitochondrial dysfunctions in myalgic encephalomyelitis/chronic fatigue syndrome explained by activated immuno-inflammatory, oxidative and nitrosative stress. Metabolic brain disease. Springer, Berlin

Moss-Morris R, Deary V, Castell B (2013) Chronic fatigue syndrome. Handb Clin Neurol 110:303–314

Mothersill C, Seymour C (2012) Changing paradigms in radiobiology. Mutat Res 750(2):85–95

Mothersill C, Seymour C (2013) Uncomfortable issues in radiation protection posed by low-dose radiobiology. Radiat Environ Biophys 52(3):293–298

Mothersill C, Heffron JJ, McLoughlin JV (1976) Inhibition of actomyosin ATPase by high concentrations of 5-hydroxytryptamine. Possible basis of lesion in 5HT-induced experimental myopathy. Enzyme 21(6):481–487

Mothersill C, Saroya R, Smith RW, Singh H, Seymour CB (2010) Serum serotonin levels determine the magnitude and type of bystander effects in medium transfer experiments. Radiat Res 174(1):119–123

Mothersill C, Smith RW, Fazzari J, McNeill F, Prestwich W, Seymour CB (2012) Evidence for a physical component to the radiation-induced bystander effect? Int J Radiat Biol 88(8):583–591

Murphy BC, Ellis P, Greenberg S (1990) Atomic veterans and their families: responses to radiation exposure. Am J Orthopsychiatry 60(3):418–427

Nagasawa H, Little JB (1992) Induction of sister chromatid exchanges by extremely low doses of alpha-particles. Cancer Res 52(22):6394–6396

Nicolson GL (2007) Metabolic syndrome and mitochondrial function: molecular replacement and antioxidant supplements to prevent membrane peroxidation and restore mitochondrial function. J Cell Biochem 100(6):1352–1369

Ojo-Amaize EA, Conley EJ, Peter JB (1994) Decreased natural killer cell activity is associated with severity of chronic fatigue immune dysfunction syndrome. Clin Infect Dis 18(Suppl 1): S157–S159

Poon RC, Agnihotri N, Seymour C, Mothersill C (2007) Bystander effects of ionizing radiation can be modulated by signaling amines. Environ Res 105(2):200–211

Saroya R, Smith R, Seymour C, Mothersill C (2009) Injection of resperpine into zebrafish, prevents fish to fish communication of radiation-induced bystander signals: confirmation in vivo of a role for serotonin in the mechanism. Dose Response 8(3):317–330

Seymour CB, Mothersill C, Alper T (1986) High yields of lethal mutations in somatic mammalian cells that survive ionizing radiation. Int J Radiat Biol Relat Stud Phys Chem Med 50(1): 167–179

Smith AK, Dimulescu I, Falkenberg VR, Narasimhan S, Heim C, Vernon SD, Rajeevan MS (2008) Genetic evaluation of the serotonergic system in chronic fatigue syndrome. Psychoneuroendocrinology 33(2):188–197

Szumiel I (2014) Ionising radiation-induced oxidative stress, epigenetic changes and genomic instability: the pivotal role of mitochondria. Int J Radiat Biol 17:1–55

Ushakov IB, Petrov VM, Shafirkin AV, Shtemberg AS (2011) Problems of ensuring human radiation safety during interplanetary flights. Radiats Biol Radioecol 51(5):595–610

Voloboueva LA, Giffard RG (2011) Inflammation, mitochondria, and the inhibition of adult neurogenesis. J Neurosci Res 89(12):1989–1995

Waxman SG, Ptacek LJ (2000) Chronic fatigue syndrome and channelopathies. Med Hypotheses 55(5):457

Witthöft M, Hiller W, Loch N, Jasper F (2013) The latent structure of medically unexplained symptoms and its relation to functional somatic syndromes. Int J Behav Med 20(2):172–183

Wu Q, Wang X (2012) Neuronal stem cells in the central nervous system and in human diseases. Protein Cell 3(4):262–270

Yancey JR, Thomas SM (2012) Chronic fatigue syndrome: diagnosis and treatment. Am Fam Physician 86(8):741–746

Relevance of the Chernobyl Research for the Evaluation of Genetic Radiation Risks in Humans

Inge Schmitz-Feuerhake and Sebastian Pflugbeil

Introduction

The most serious effects of ionizing radiation—hereditary defects in the descendants of exposed parents—had been already detected in the 1920s by Herman Joseph Muller. He exposed flies—*Drosophila*—to X-rays and found malformations and other disorders in the following generations. He concluded from his investigations that low dose exposure, and therefore even natural background radiation, is mutagenic. Already in the thirties, the idea arose that cancer is initiated by a single cell transformation, a "somatic" mutation. Likewise, Muller concluded that there is no harmless dose range for cancer induction either (Muller 1936). His work was honoured by the Nobel Prize for medicine in 1946.

Muller—as a famous expert for radiation—was designated for a speaker at the Atomic Conference in Geneva in 1955 where the large-scale, so-called peaceful, use of nuclear energy was announced by U.S. President Eisenhower. But then they became aware that he had warned against the deterioration of the human genetic pool by the production of huge amounts of artificial radioactivity, and the invitation was cancelled (his manuscript is still available).

Nevertheless, previously, genetic effects in descendants were thought to be the most significant injuries caused by radiation. From this results the protection of the gonads, if possible, during x-raying for diagnostic purposes.

The International Committee on Radiological Protection ICRP, however, substantially decreased their risk estimate in 2007 (Table 1).

I. Schmitz-Feuerhake (✉)
University of Bremen, Bremen, Germany
e-mail: ingesf@uni-bremen.de

S. Pflugbeil
German Society for Radiation Protection, Berlin, Germany
e-mail: pflugbeil.KvT@t-online.de

V.L. Korogodina et al. (eds.), *Genetics, Evolution and Radiation*,
DOI 10.1007/978-3-319-48838-7_30

Table 1 ICRP recommendations 2007

Detriment adjusted nominal risk coefficients for radiation effects in an exposed population		
	Present (% per Sv)	ICRP 1990 (% per Sv)
Heritable effects	0.2	1.3
Cancer deaths	5.5	6.0

The value of 0.2 % per Sv means that, if a population is exposed to a gonadal dose of 1 Sv, 0.2 % of the persons will have a genetic disorder caused by radiation.

The ICRP claims that there is no direct evidence that children of exposed parents will suffer from heritable diseases. They refer to their human reference group, the Japanese survivors of the atomic bomb explosions in Hiroshima and Nagasaki in 1945. The American-Japanese Institute in Hiroshima did not find mutations in the descendants of the survivors. Because of the evidence of such effects in animal studies, they derive their current risk figure from experiments in mice. They consider only dominant mutations in the first generation. Their result corresponds to a doubling dose of about 2 Sv which is similar to the evaluation by the UNSCEAR committee (2001).

There exists a variety of arguments against the Japanese survivors as a suitable reference for common populations; the most relevant in the case of hereditary effects may be the high dose rate by the bomb explosion in comparison to situations of chronic exposure where all stages of sperm development are continuously affected.

Effects in populations exposed by Chernobyl fallout are excluded by the official committees, because they claim that the doses are much too low in order to generate statistically observable increase. This, however, is certainly wrong, because we know from many studies of chromosome aberrations, that the exposures calculated by UNSCEAR are much too low.

Hereditary disorders are classified in 4 groups (Uma Devi et al. 2000):

1. **Mendelian disorders** due to defective single genes which follow Mendel's laws of inheritance
2. **Chromosomal disorders** as e.g. Down's syndrome
3. **Polygenic disorders** to be detected in clusters in families
4. **Non-chromosomal inheritance** which has nothing to do with genes.

Examples for diseases are shown in the following list where those which have been found after low level exposures were printed by us in italics:

1. **Mendelian**

   Autosomal dominant; examples:
   Huntington's chorea, polycystic kidney, multiple polyposis, cerebellar ataxia, myotonic dystrophy

Congenital abnormalities as ***syndactyly, brachydactyly, polydactyly***, taste for the chemical PTC (taste is dominant to non-taste), acondroplasia, bilateral aniridia, osteogenesis imperfecta
Autosomal recessive; examples:
Cystic fibrosis, phenylketonuria, lactose intolerance, adrenal hyperplasia
Sex-linked; examples:
X-linked dominant/Duchenne muscular dystrophy, haemophilia A, some forms of colour blindness, fragile-X associated mental retardation, X-linked retinitis pigmentosa
X-linked recessive/birth ***deficit of females***.

2. **Chromosomal**

   Aneuploidy (numerical chromosomal anomaly); examples:
   Down syndrome (trisomy 21), Turner syndrome (X0), Klinefelter syndrome (XXY)
   Structural anomalies; examples:
   Cri du chat syndrome (deletion in chromosome 5), ***preimplantation loss, embryonal death, foetal abortions***.

3. **Polygenic**

   Cluster in families; examples:
   Congenital abnormalities as ***neural tube defects, heart defects, pyloric stenosis, cleft lip with or without cleft palate, undescended testes***
   Common disorders of adult life of varying severity. Among the serious conditions are ***schizophrenia, multiple sclerosis, epilepsy, acute myocardial infarction***, systemic lupus erythematosus. Moderately serious conditions include ***psychoses***, Graves'disease, ***diabetes mellitus***, gout, glaucoma, ***essential hypertension, asthma***, peptic ulcer, rheumatoid arthritis. The least severe diseases include varicose veins of the lower extremities and allergic rhinitis.
   Cancer.

4. **Non-chromosomal inheritance**

   Cytoplasmic inheritance, mosaicism, imprinting etc.

Findings in Children Born After the Chernobyl Accident and in Kazakhstan

We have formerly made a compilation of findings about foetal deaths, perinatal mortality, and congenital malformations after Chernobyl (Busby et al. 2009). Table 2 shows the results about congenital malformations. They appeared not only in the area of the exploded reactor but also in Turkey, Bulgaria, Croatia, and Germany.

Table 2 Increase of congenital malformations after exposure by the Chernobyl accident

Country	Effects	Reference
Belarus National Genetic Monitoring Registry	Anencephaly, spina bifida, cleft lip and/or palate, polydactyly, limb reduction defects, esophageal atresia, anorectal atresia, multiple malformations	Lazjuk et al. (1997), Feshchenko et al. (2002)
Highly exposed region of Gomel	Congenital malformations	Bogdanovich (1997), Savchenko (1995), Petrova et al. (1997)
Chechersky district (Gomel region)	Congenital malformations	Kulakov et al. (1993)
Mogilev region	Congenital malformations	Petrova et al. (1997)
Brest region	Congenital malformations	Shidlovskii (1992)
Ukraine Polessky district (Kiev region)	Congenital malformations	Kulakov et al. (1993)
Lugyny region	Congenital malformations	Godlevsky and Nasvit (1998)
Turkey	Anencephaly, spina bifida	Akar et al. (1988, 1989), Caglayan et al. (1990), Güvenc et al. (1993), Mocan et al. (1990)
Bulgaria, region of Pleven	Malformations of heart and central nervous system, multiple malformations	Moumdjiev et al. (1994)
Croatia	Malformations by autopsy of stillborns and cases of early death	Kruslin et al. (1998)
Germany FRG, Central registry	Cleft lip and/or palate	Zieglowski and Hemprich (1993)
malformations	Cleft lip and/or palate	Scherb and Weigelt (2004)
Bavaria	congenital malformations	Korblein (2004)
Ann. health report W. Berlin 1987 City of Jena (Registry GDR)	Malformations in stillborns Isolated malformations	Government of West Berlin Lotz et al. (1996)

These findings were mainly interpreted as effects induced in utero. Because men and women were both exposed continuously to radioactive fallout, the genetic effects are not always clearly distinguishable from such, which can be generated by exposure of embryos and foetuses in utero.

The authors in Belarus, however, came to another interpretation. A central registry for congenital anomalies had been established there by the Ministry of Health in 1979 for continuous follow-up. The rates of anomalies before and after the Chernobyl accident could be compared. Results in the 17 most contaminated regions are shown in Table 3.

Unfortunately, the data are not continuously published, at least until 2004 there was no decrease (Yablokov et al. 2009). The authors think these effects are

Table 3 Percentage increase in congenital malformations in 17 most contaminated regions of Belarus in the period 1987–1994 in percent (from Lazjuk et al. 1997)

Kind of malformation	Increase (%)
Anencephaly	39
Spina bifida	29
Cleft lip/palate	60
Polydaktyly	910*
Limb reduction	240*
Esophageal atresia	13
Rectal atresia	80*
Multiple malformations	128*

*Significance ($p < 0.05$)

genetically induced because it is not plausible that doses in pregnant women rose in the period of decreasing environmental contamination and decreasing food contamination after the accident. The genetic origin is confirmed in those anomalies which are combined with a recognized mutation that is not present in either of the parents (Lazjuk et al. 1999).

The national registry of Belarus was also evaluated in 1995 by a Belarussian-Israeli group of scientists (Lomat et al. 1997). They found the following high rates of diseases—supposedly originated by polygenetic effects—in children of Chernobyl-exposed parents:

- Hematological diseases (6-fold)
- Endocrine diseases (2-fold)
- Diseases of digestive organs (1.7-fold).

Wertelecki (2010), Wertelecki et al. (2014) found increased rates of congenital malformations in the years 2000–2009—more than 14 years after the accident—in the Ukrainian province (oblast) Rivne, about 200 km west of Chernobyl. The authors interpret the effect as induced in utero. Predominantly in the highly contaminated northern part Polissia, there are significant increases in comparison to the southern part:

Congenital malformations **in Rivne Oblast, Ukraine**
Study of 145,437 live births between 2000–2009
1.6 % congenital anomalies
Neural tube defects

Rivne Non-Polissia	1.6 per 1000 live births (Europe mean 0.94)
Polissia district	2.6 per 1000 live births

Microcephaly, microphthalmos increased

A region where the population has also been exposed to large amounts of radioactivity is near the former Soviet nuclear test site near the town Semipalatinsk (now in Kazakhstan). The tests above ground occurred between 1949 and 1963. Sviatova et al. (2001) studied congenital malformations in three generations of inhabitants, investigating births between 1969 and 1997. They found significantly increased rates of malformations as a whole, including Down’s syndrome, microcephaly and also multiple malformations in the same individual.

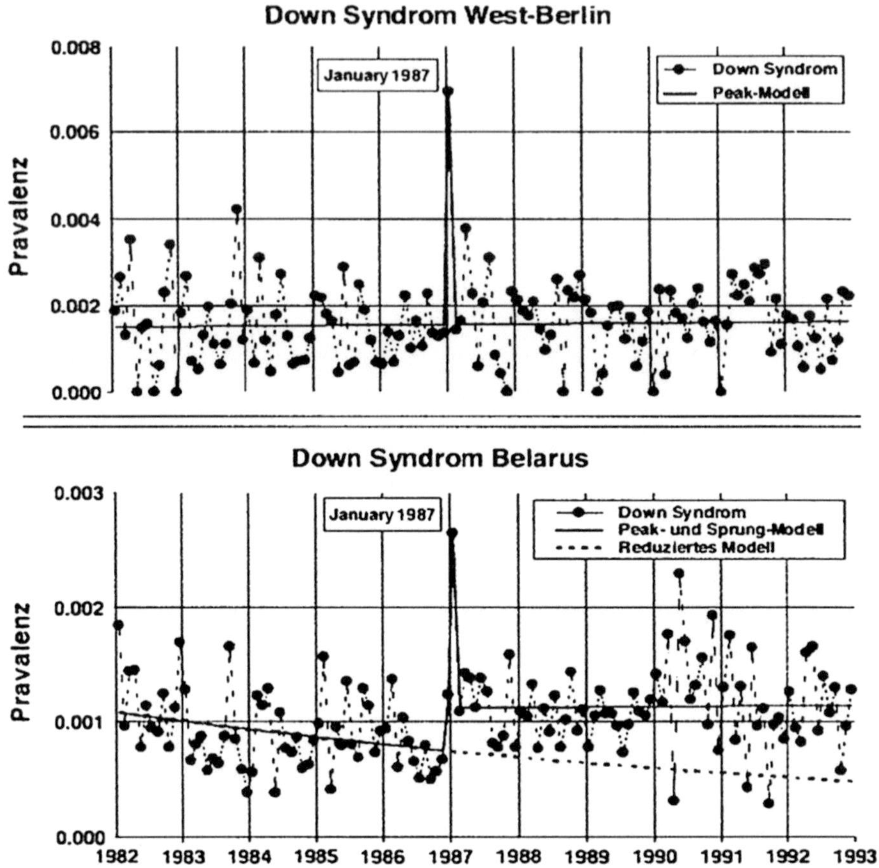

Fig. 1 Down's syndrome before and after the Chernobyl accident (from Scherb and Sperling 2011)

Down's syndrome as a certain genetic effect increased in several contaminated European countries (Busby et al. 2009; Sperling et al. 2012). Examples are shown in Fig. 1. In Berlin West, which was a kind of closed island at that time, the geneticist Sperling registered a sharp and significant increase in cases exactly 9 months after the accident. A very similar situation was observed in Belarus (Zatsepin et al. 2004).

Hereditary Effects in Children of Exposed Mothers

If a population is exposed, genetic effects will occur in the gonads of fathers as well as of mothers. In Germany, an investigation was done in women who were occupationally exposed to radiation which showed a 3.2-fold significant increase in congenital abnormalities, including malformations, in their offspring (Wiesel et al. 2011). The authors interpret this effect as generated in utero but do not prove such a connection because it appears improbable given the short sensitive phase in pregnancy and the ban on pregnant women working in high risk environments.

Although the study was funded by the Federal Ministry of Environment, Protection of Nature and Nuclear Safety, these alarming results have not resulted in any action. The findings confirm early results in the Department of Medical Genetics of the Montreal Children's Hospital where the genetic effects of diagnostic X-rays were investigated (Cox 1964). The author observed the offspring of mothers who had been treated in childhood for congenital hip dysplasia since 1925 and were X-rayed for several times in the pelvis region. The ovarian dose was estimated to lie between 60–200 mSv. In 201 living births of these women 15 individuals showed severe malformations and other congenital distortions or Down's syndrome, who required hospitalization, and 11 cases occurred with other abnormalities (all congenital abnormalities 12.9 %), while the control group showed less than half of this rate. The latter was chosen from a large group of descendants whose parents were unexposed siblings of the study group.

Findings in the Descendants of Occupationally Exposed Men

Congenital anomalies: Studies in children of exposed men where the mothers were not exposed will show definite hereditary effects. There were only very few of them in occupationally exposed cohorts before 1986, when the accident of Chernobyl occurred. Exposures below the official dose limits were thought to be too low to produce statistically recognizable effects. A compilation of results for congenital malformations is given in Table 4.

The findings in Hanford and Sellafield (Nos. 2 and 3 in Table 4) lead to heavy discussions in the scientific community. The registered doses of the workers were very low. The alarming findings did not lead to further studies about hereditary sequels in the American or European populations concerned.

The anomalies seen in the descendants of liquidators also indicate unexpectedly high radiation sensitivity. These studies are most important and should be continued.

Cancer: In 1984, an exceptionally high level of leukaemia cases in children and juveniles was reported in Sea scale, a village near the British Nuclear Fuels reprocessing plant in Sellafield in Cumbria, UK. These were then explained by Martin Gardner et al. (1990) as a hereditary effect, because the fathers of the patients had worked in the plant. This result has been discussed in the literature for years and was confirmed or denied in several subsequent studies. The effect, however, had been described in principle already in experimental studies (Nomura 1982, 2006), and has also been found after X-ray diagnostic exposures (Table 5).

Table 4 Congenital anomalies, especially malformations, in descendants (1st generation) of occupationally exposed men

No.	Cohort of fathers	Kind of defect	Dose	References
1	Radiologists U.S.A. 1951	Congenital malformations increase 20 %		Machtand Lawrence (1955)
2	Workers of the Hanford Nuclear facility, U.S.A.	Neural tube defects significantly increased by 100 %	In general <100 mSv	Sever et al. (1988)
3	Radiation workers at Sellafield nuclear reprocessing plant, U.K.	Stillbirths with neural-tube defects significantly increased by 69 % per 100 mSv	Mean 30 mSv	Parker et al. (1999)
4	Radiographers in Jordan	Congenital anomalies significantly increased 10-fold		Shakhatreh (2001)
5	Liquidators from Obninsk (Russia), 300 children	Congenital anomalies increased 1994–2002	Mainly 10–250 mSv	Tsyb et al. (2004)
6	Liquidators from Russia, Bryansk region	Congenital anomalies increased about 4-fold		Matveenko et al. (2005)
7	Liquidators from Russia 2379 newborns	Significant increase for: Anencephaly 310 % Spina bifida 316 % Cleft lip/palate 170 % Limb reduction 155 % Multiple malformations 19 % All malformations 120 %	5–250 mSv	Lyaginskaja et al. (2009)

McKinney et al. (1991) found a 3.2-fold increase in leukaemia and lymphomas in children of occupationally exposed men in three British regions in a case-control study. The research of Hicks et al. (1984) concerned exposed service men in the air force.

Statistical investigations in Belarus and the other highly contaminated neighbouring states of Chernobyl show increased cancer deaths in children who were born many years after the accident (Yablokov 2006; Yablokov et al. 2009). Higher rates of leukaemia and other cancers were also observed in children of liquidators (Tsyb et al. 2004).

Sex Ratio and X-Linked Lethal Factors

Normally, it is not possible to study how many inseminated oocytes (zygotes) will be aborted after irradiation of the gonadal cells, in humans. There is however, one way to prove such an effect. It is observed that men who were exposed before fathering will have fewer daughters than sons as normally, i.e., the male/female sex ratio increases with dose.

Table 5 Cancer in children after preconceptional low-dose exposure of parents

Exposed collective	Malign disease	Gonadal dose/mSv	Relative risk	Doubling dose/mSv
Seascale fathers (Gardner et al. 1990)				
all stages of spermatogenesis	Leukaemia + lymphoma	200	7	33
6 months before conception	Leukaemia + lymphoma	10	7	1.7
Sellafield workers (Dickinson and Parker 2002)			1.9	
Further occupational exposure of fathers				
Military jobs (Hicks et al. 1984)	Cancer		2.7	
Regions of U.K. (McKinney et al. 1991)	Leukaemia + lymphoma		3.2	
Preconceptional X-ray diagnostics in	Leukaemia			
Fathers (Graham et al. 1966)			1.3	
Fathers (Shu et al. 1988)			1.4–3.9	
Fathers (Shu et al. 1994)			3.8	
Mothers (Stewart et al. 1958)			1.7	
Mothers (Graham et al. 1966)			1.7	
Mothers (Natarajan and Bross 1973)			1.4	
Mothers (Shiono et al. 1980)			2.6	

Gene mutations may be responsible for the death of the zygote and will also occur in the sex chromosomes where they will predominantly affect the greater X-chromosome. The X-chromosome of the male can only be transmitted to a daughter. A dominant lethal factor will then lead to the death of the female zygote. Recessive lethal factors in the X-chromosome are much more frequent than dominant ones (Vogel et al. 1969). They also affect only female births.

Studies in large exposed populations can show this effect. A very impressive result was obtained in workers of the British nuclear fuel reprocessing plant at Sellafield in West Cumbria (Table 6).

A similar effect is detected in an investigation of cardiologists, who undertook interventional angiographic procedures in patients, which involve relatively high x-ray exposures at the workplace (Fig. 2). The portion of female descendants declines significantly with higher exposures of the father.[1]

[1]The authors wanted to study the reverse, however, if the male births will decline with dose.

Table 6 Sex ratio for births in Cumbria (Dickinson et al. 1996)

All Cumbrian children	All fathers employed[a] at Sellafield	Fathers employed at Sellafield >10 mSv[b]
1.055	1.094	1.396

[a]Employed before conception
[b]Dose 90 days preconceptional

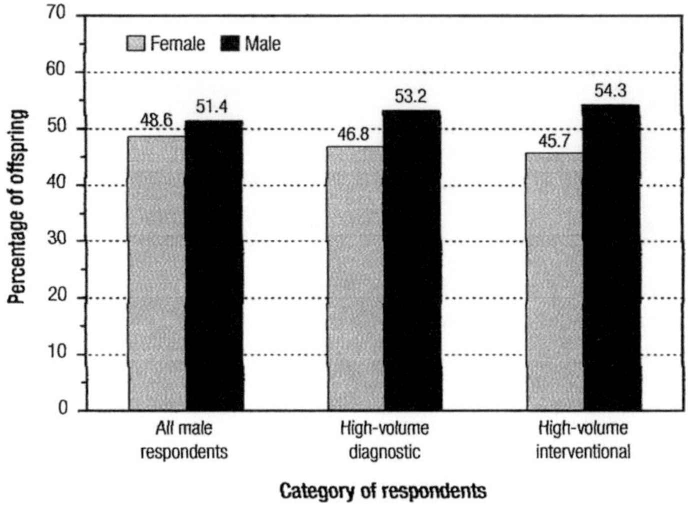

Fig. 2 Percentage of male and female offspring among cardiologists (Choi et al. 2007)

German scientists Hagen Scherb, Kristina Voigt (Helmholtz Centre Munich) and co-workers have shown that exposure of both parents in a population may also lead to a decline in female births. They studied different groups of inhabitants in a variety of countries after the Chernobyl accident for hereditary effects and found radiation-induced foetal deaths and early mortality, Down's syndrome and alterations of the sex ratio in newborn children.

The sex ratio was investigated by them as a consequence of:

- Nuclear tests above ground which affected U.S. inhabitants
- Chernobyl emissions in Europe
- Living near German and Swiss nuclear plants.

They found significant decreases in the female birth rate in all these conditions (Scherb and Voigt 2007, 2011; Scherb et al. 2012).

Sex ratio is a very relevant parameter. It shows that genetic alterations are induced in the germ cells of men by very low doses, and it proves to be a sensitive indicator for exposures of the population.

Summary and Discussion

Genetically induced malformations, cancers, and numerous other health effects in the children of populations who were exposed to low doses of ionizing radiation have been proved in scientific investigations.

The question arises as to why the ICRP and UNSCEAR are denying the findings. Their reference population, the A-bomb survivors of Hiroshima and Nagasaki, is not suitable for persons being chronically exposed.

The committees claim that the exposures of the population due to the Chernobyl accident were extremely low. The UNSCEAR committee (1988) has done calculations about it. Even in the most contaminated regions of the area with more than 37 kBq/m^2 ^{137}Cs surface activity the mean dose in people is assumed to be not higher than 10 mSv (effective life time dose). In Turkey and the more distant countries of central Europe the mean dose is estimated at below 1.2 mSv.

Their simple conclusion is then, that such low doses are not able to produce statistically recognizable radiation effects. Many studies, however, of chromosome aberrations in the populations, equivalent to "biological dosimetry" (Schmitz-Feuerhake 2009), show that the exposures are about 100-fold higher.

References

Akar N, Cavdar AO, Arcasoy A (1988) High incidence of Neural Tube defects in Bursa, Turkey. Paediatr Perinat Epidemiol 2:89–92

Akar N, Ata Y, Aytekin AF (1989) Neural Tube defects and Chernobyl? Paediatr Perinat Epidemiol 3:102–103

Bogdanovich IP (1997) Comparative analysis of the death rate of children, aged 0–5, in 1994 in radiocontaminated and conventionally clean areas of Belarus. In: Medicobiological effects and the ways of overcoming the Chernobyl accident consequence. Collected book of scientific papers dedicated to the 10th anniversary of the Chernobyl accident. Minsk-Vitebsk, p 4

Busby C, Lengfelder E, Pflugbeil S et al (2009) The evidence of radiation effects in embryos and fetuses exposed to Chernobyl fallout and the question of dose response. Med Confl Surviv 25:20–40

Caglayan S, Kayhan B, Mentesoglu S et al (1990) Changing incidence of neural tube defects in Aegean Turkey. Paediatr Perinat Epidemiol 4:264–268

Cox DW (1964) An investigation of possible genetic damage in the offspring of women receiving multiple diagnostic pelviv X-rays. Am J Hum Genet 16:214–230

Choi JW, Mehrotra P, MacDonald LA et al (2007) Sex proportion of offspring and exposure to radiation in male invasive cardiologists. Proc (Bayl Univ Med Cent) 20:231–234

Dickinson HO, Parker L (2002) Leukaemia and non-Hodgkin's lymphoma in children of male Sellafield radiation workers. Int J Cancer 99:437–444

Dickinson HO, Parker L, Binks K et al (1996) The sex ratio of children in relation to paternal preconceptional radiation dose: a study in Cumbria, northern England. J Epidemiol Community Health 50:645–652

Feshchenko SP, Schroeder HC, Mueller WEG et al (2002) Congenital malformations among newborns and developmental abnormalities among human embryos in Belarus after Chernobyl accident. Cell Mol Biol 48:423–426

Gardner MJ, Snee MP, Hall AJ et al (1990) Results of case-control study of leukaemia and lymphoma among young people near Sellafield nuclear plant in West Cumbria. Brit Med J 300:423–429

Godlevsky I, Nasvit O (1998) Dynamics of health status of residents in the Lugyny district after the accident of the ChNPS. In: Imanaka T (ed) Research activities about the radiological consequences of the Chernobyl NPS accident and social activities to assist the sufferers by the accident. Research Reactor Institute, Kyoto University, KURRI-KR-21, pp 149–156

Government of West Berlin, Section of Health and Social Affairs. Annual Health Report 1987

Graham S, Levin ML, Lilienfeld AM et al (1966) Preconception, intrauterine, and postnatal irradiation as related to leukemia. Natl Cancer Inst Monogr 19:347–371

Güvenc H, Uslu MA, Güvenc M et al (1993) Changing trend of neural tube defects in Eastern Turkey. J Epidemiol Community Health 47:40–41

Hicks N, Zack M, Caldwell GG et al (1984) Childhood cancer and occupational radiation exposure in parents. Cancer 53:1637–1643

ICRP, International Commission on Radiological Protection (2007) The 2007 Recommendations of the International Commission on Radiological Protection. ICRP-Publication 103, Ann ICRP 37 Nos. 2–4

Korblein A (2004) Fehlbildungen in Bayern nach Tschernobyl. Strahlentelex 416–417:4–6. www.strahlentelex.de

Kruslin B, Jukic S, Kos M et al (1998) Congenital anomalies of the central nervous system at autopsy in Croatia in the period before and after Chernobyl. Acta Med Croatica 52:103–107

Kulakov VI, Sokur TN, Volobuev AI et al (1993) Female reproduction function in areas affected by radiation after the Chernobyl power station accident. Environ Health Persp 101(Suppl. 2):117–123

Lazjuk GI, Nikolaev DL, Novikova IV (1997) Changes in registered congenital anomalies in the Republic of Belarus after the Chernobyl accident. Stem Cells 15(Suppl 2):255–260

Lazjuk G, Satow Y, Nikolaev D et al (1999) Genetic consequences of the Chernobyl accident for Belarus Republic. In: Imanaka T (ed) Recent Research Activities on the Chernobyl NPP Accident in Belarus, Ukraine and Russia, KURRI-KR-7 (Kyoto University, Kyoto), pp 174–177

Lomat L, Galburt G, Quastel MR et al (1997) Incidence of childhood disease in Belarus associated with the Chernobyl accident. Environ Health Persp 105(Suppl 6):1529–1532

Lotz B, Haerting J, Schulze E (1996) Veränderungen im fetalen und kindlichen Sektionsgut im Raum Jena nach dem Reaktorunfall von Tschernobyl. Oral presentation at the Int Conf of the Society for Medical Documentation, Statistics, and Epidemiology. Bonn, Germany (Available on request)

Lyaginskaja AM, Tukov AR, Osipov VA et al (2009) Congenital malformations at posterity of the liquidators of the consequences from Chernobyl accident. Radiat Biol Radioecol 49:694–702 (Russian)

Macht S, Lawrence P (1955) National Survey of congenital malformations resulting from exposure to Roentgen radiation. Am J Roentgenol 73:442–466

Matveenko EG, Borovykova MP, Davydow GA (2005) Physical characteristics and primary morbidity in liquidator's children. In: Proceedings of Conference on "Chernobyl 20 Years After" (cited from Yablokov et al. 2009: 148)

McKinney PA, Alexander FE, Cartwright RA et al (1991) Parental occupations of children with leukemia in west Cumbria, north Humberside, and Gateshead. Brit Med J 302:681–687

Mocan H, Bozkaya H, Mocan ZM et al (1990) Changing incidence of anencephaly in the eastern Black Sea region of Turkey and Chernobyl. Paediatric Perinatal Epidemiol 4:264–268

Moumdjiev N, Nedkova V, Christova V et al (1994) Influence of the Chernobyl reactor accident on the child health in the region of Pleven, Bulgaria. In: 20th Int Congr Pediatrics Sept 6–10, 1992 in Brasil: 57 (Cited by Akar N. Further notes on neural tube defects and Chernobyl (Letter). Paediatr Perinat Epidemiol 8:456–457)

Muller HJ (1936) Über die Wirkung der Röntgenstrahlung auf die Erbmasse. Strahlentherapie 55:207–224

Natarajan N, Bross IDJ (1973) Preconception radiation leukemia. J Med 4:276–281

Nomura T (1982) Parental exposure to X-rays and chemicals induces heritable tumours and anomalies in mice. Nature 296:575–577

Nomura T (2006) Transgenerational effects of radiation and chemicals in mice and humans. J Radiat Res 47(Suppl):B83–B97

Parker L, Pearce MS, Dickinson HO et al (1999) Stillbirths among offspring of male radiation workers at Sellafield nuclear reprocessing plant. Lancet 354:1407–1414

Petrova A, Gnedko T, Maistrova I et al (1997) Morbidity in a large cohort study of children born to mothers exposed to radiation from Chernobyl. Stem Cells 6(Suppl 2):141–150

Savchenko VK (1995) The ecology of the Chernobyl catastrophe. In: Scientific outlines of an international programme of collaborative research. Man and the Biosphere Series 17. UNESCO Paris: 83

Scherb H, Weigelt E (2004) Cleft lip and cleft palate birth rate in Bavaria before and after the Chernobyl nuclear power plant accident. Mund Kiefer Gesichtschir 8:106–110

Scherb H, Voigt K (2007) Trends in the human sex odds at birth in Europe and the Chernobyl Nuclear Power Plant accident. Reprod Toxicol 23:593–599

Scherb H, Voigt K (2011) The human sex odds at birth after the atmospheric atomic bomb tests, after Chernobyl, and in the vicinity of nuclear facilities. Environ Sci Pollut Res 18:697–707

Scherb H, Sperling K (2011) Heutige Lehren aus dem Reaktorunfall von Tschernobyl. Naturwiss Rundschau 64 Heft 5:229–239

Scherb H, Voigt K, Kusmierz R (2012) Modelling human radiation risks around nuclear facilities in Germany and Switzerland: a case study. In: Seppelt R et al (eds) Int Congr Environ Modelling and Software Managing Resources of a Limited Planet. Sixth Biennial Meeting, Leipzig, Germany http://www.iemss.org/society/index.php/iemss-2012-proceedings

Schmitz-Feuerhake I (2009) How reliable are the dose estimates of UNSCEAR for populations contaminated by Chernobyl fallout? A comparison of results by physical reconstruction and biological dosimetry. In: Proceedings of ECRR International Conference on "Criticisms and Developments in the Assessment of Radiation Risks" in Molyvos, Lesbos, Greece, 2011, pp 70–85

Sever LE, Gilbert ES, Hessol NA et al (1988) A case-control study of congenital malformations and occupational exposure to low-level ionizing radiation. Am J Epidemiol 127:226–242

Shakhatreh FM (2001) Reproductive health of male radiographers. Saudi Med J 22:150–152

Shevchenko VA, Platonov ES, Snigiryova GP (1997) Genetic consequences of disaster at Chernobyl nuclear plant on April 26, 1986. In: Nesterenko VB (ed) Reasons and consequences. International association for restoration of the environment and for safe living of people "Senmurv" (Minsk-Moscow-Kiev, Chernobyl Accident, Minsk, pp 213–245

Shidlovskii PR (1992) General morbidity of the population in districts of the Brest region. Zdravoohranenie Belorussii (Minsk) 1:8–11 (Russian)

Shiono PH, Chung CS, Myrianthopoulos NC (1980) Preconception radiation, intrauterine diagnostic radiation, and childhood neoplasia. J Natl Cancer 65:681–686

Shu XO, Gao YT, Brinton LA et al (1988) A population-based case-control study of childhood leukemia in Shanghai. Cancer 62:635–644

Shu XO, Reaman GH, Lampkin B et al (1994) Association of paternal diagnostic x-ray exposure with risk of infant leukemia. Cancer Epidemiol Biomarkers Prev 3:645–653

Sperling K, Neitzel H, Scherb H (2012) Evidence for an increase in trisomy 21 (Down syndrome) in Europe after the Chernobyl reactor accident. Genet Epidemiol 36:48–55

Stewart A, Webb J, Hewitt D (1958) A survey of childhood malignancies. Brit Med J 1:1495–1508

Sviatova GS, Abil'dinova GZ, Berezina GM (2001) Frequency, dynamics, and structure of congenital malformations in populations under long-term exposure to ionizing radiation. Russ J Genetics 37:1696–1704 (Russian)

Tsyb AF, Souchkevitch GN, Lyasko LI et al (2004) General characterization of health in first-generation offspring born to liquidators of the Chernobyl NPP accident consequences. Int J Radiat Med 6 (1–4):116–121. http://www.mns.gov.ua (Special Issue)

Uma Devi P, Nagarathnam A, Satish Rao BS (2000) Introduction to radiation biology. B.I. Churchill Livingstone PVT Ltd., New Delhi

UNSCEAR (1988) United Nations Scientific Committee on the Effects of Atomic Radiation: Sources, effects and risks of ionizing radiation. 1988 Report to the General Assembly, United Nations, New York

UNSCEAR (2001) United Nations Scientific Committee on the Effects of Atomic Radiation: Hereditary effects of radiation. 2001 Report to the General Assembly, United Nations, New York. www.unscear.org

Vogel F et al (1969) Strahlen Genetik der Säuger, 1. Fortschritte der allgemeinen und klinischen Humangenetik. In: Becker E et al (eds). Georg Thieme Verlag Stuttgart

Wertelecki W (2010) Malformations in a Chernobyl-impacted region. Pediatrics 125:836–843

Wertelecki W, Yevtushok L, Zymak-Zakutnia N et al (2014) Blastopathies and microcephaly in a Chornobyl impacted region of Ukraine. Congenital Anomalies 54:125–149

Wiesel A, Spix C, Mergenthaler A, Queißer-Luft A (2011) Maternal occupational exposure to ionizing radiation and birth defects. Radiat Environ Biophys 50:325–328

Yablokov AV (2006) The Chernobyl catastrophe—20 years later. In: Busby CC, Yablokov AV (eds) European Committee on Radiation Risk ECRR. Chernobyl: 20 years on. Green Audit www.euradcom.org

Yablokov AV, Nesterenko VB, Nesterenko AV (2009) Chernobyl—Consequences of the catastrophe for people and the environment. Ann NY Acad Sci 1181 (Boston)

Zatsepin IO, Verger P, Robert-Gnansia E et al (2004) Cluster of Down's syndrome cases registered in January 1987 in Republic of Belarus as a possible effect of the Chernobyl accident. Int J Radiat Med 6(1–4):57–71. http://www.mns.gov.ua (Special Issue)

Zieglowski V, Hemprich A (1993) Facial cleft birth rate in former East Germany before and after the reactor accident in Chernobyl. Mund Kiefer Gesichtschir 3:195–199 (German)

Fundamental Difficulties in Dose Calculation

Alexey V. Yablokov

Introduction

Radioactive contaminations are the most inconspicuous in appearance, but at the same time the most dangerous man-made changes in the biosphere. They started on a large scale as a result of the creation and testing of nuclear weapons in the 1950s, and for a decade, covering especially the Northern hemisphere increased so much that in 1963 it was necessary to conclude an agreement on the cessation of nuclear explosions in the atmosphere due to the catastrophic health effects.

Since the 1960s radiation exposure associated with the development of the medical use of ionizing radiation and nuclear energy began to increase. The level of man-made radioactive contamination has now once again become comparable to that of natural background radiation. It is repeatedly suggested that anthropogenic radioactive contamination of the biosphere—one of the main reasons for cancer epidemic in the world (taking millions of lives each year).

The modern system of radiation protection is based on the calculations of effective doses. These doses are not physically measured, but only calculated. Some basic assumptions underlying these calculations are effective only under strictly controlled conditions (almost exclusively for the personnel, but not to the public).

Summary of the "Dose" Concept

The amount of external and internal exposure determines the radiation dose. Internal exposure is dependent on the residence time of radionuclides in the body and the place of their concentration. External exposure is determined by the transfer of the energy of ionized air through the human body (absorption).

A.V. Yablokov (✉)
N.K. Koltzoff' Institute of Developmental Biology RAS, Moscow, Russia
e-mail: Alexey.ablokov@gmail.com

V.L. Korogodina et al. (eds.), *Genetics, Evolution and Radiation*,
DOI 10.1007/978-3-319-48838-7_31

The basic postulates of the concept of dose (Yablokov 2013):

- the effect of each radionuclide constant in time and space;
- the level of external radiation can be determined as residence time in the ionized environment;
- the level of internal exposure can be determined by amount of the absorbed radionuclides with water, air and food;
- biological effectiveness of X-rays and all γ- and beta β-emitters are identical, biological efficiency of slow neutrons tripled, and α-emitters and ultra-fast neutrons—20 times higher;
- relative radiosensitivity of organs and tissues: the gonads—0.2; red bone marrow, stomach, colon, lungs—0.12; breast, liver, esophagus, bladder, thyroid gland—0.05; skin, bone surface—0.01; all other organs—a total of 0.05;
- homogeneous phantom can simulate radiation exposure;
- individual effective dose is the sum of doses of internal and external radiation (in addition to the natural);
- the higher the dose, the greater biological effect;
- safe level of exposure—1 mSv/person/year.

Other formulations can be described of the dose concept, and the proposed form of the above postulates is selected only for ease of analysis of methodological correctness and procedural (practical) feasibility of the dose calculations.

Analysis of the Main Postulates of the "Dose" Concepts

Let us consistently consider the above assumptions underlying the radiation safety standards.

The Impact of Each Radionuclide Is not Uniform in Time and Space

Field radiation is always heterogeneous. Radionuclides determining exposure are not in the soil at a single location. There is their constant vertical and horizontal migration—as immersion and removal of wind, water, plants and animals. The rate of such annual immersion (millimeters–centimeters per year) at each location for different specific radionuclide depends, inter alia, on the chemical properties of the element, in particular its solubility in water of the compounds, the composition of the soil and climatic characteristics.

As a result of vertical migration of radionuclides, the level of radiation in the surface atmosphere is first reduced. Then, when radionuclides get in root soil layers (15–30 cm), they are partially trapped by the roots of plants and brought back to the surface. Moles, earthworms and other burrowing animals can also take radionuclides from deep soil layers to the surface, thereby affecting the level of radiation.

As a result, the horizontal migration of radionuclides can spread hundreds of kilometers (e.g., as a result of forest fires and high winds). In September 1992, as a result of wind transport from the 30-km Chernobyl zone the concentration of ^{137}Cs in the vicinity of Vilnius (Lithuania) in a few hours increased a hundredfold. In August 2010, the concentration of ^{137}Cs in the vicinity of Moscow increased 24 times as a result of the transfer of radionuclides from the burning forests of the Bryansk region (Yablokov et al. 2011).

As a result of the natural process of radioactive decay of short-lived radionuclides, the level of surface radiation after a release of radionuclides decreases rapidly at first and then more slowly during the year. Data on Chernobyl show that the level of ionizing radiation in radioactively-contaminated areas can vary by more than 10 000 times during the year (Kryshev and Ryazantsev 2000).

Physical characteristics of the soil which affect the migration of radionuclides and, therefore, radiation in the near-surface layer of the air do not remain constant. Some are regular daily and seasonal changes in humidity and density of the soil; others are irregular—such as those associated with precipitation and wind. Therefore, even in one fixed point in space an abundance of the value of exposure can vary at different times of day, different days, weeks and months.

Radionuclides, for many reasons, are not uniformly distributed in the soil; they are always spotty. Figure 1 shows the actual distribution of ^{137}Cs and ^{144}Ce in the forest soil near the Chernobyl NPP.

It is known that the concentration of radionuclides in the region may be different from a distance of tens of meters. We know about a hundredfold difference in the content of radionuclides in the surrounding habitats (between top of a hill and in the valley, marsh and meadow). Without exception, detailed studies have shown patchy distribution of radionuclides in all the studied areas. This spotting defines multiple changes in the intensity of the radiation in space, even on the scale of tens or hundreds of meters. The heterogeneity of spatial and temporal distribution of

Fig. 1 Spotting concentration (Ci/km^2) ^{137}Cs (*left*) and ^{144}Ce (*right*) on the forest floor in the 30-km zone of the Chernobyl NPP. Scale 1:600 (Tscheglov 1999)

radionuclides and radiation causes heterogeneity irradiation of individuals and groups of people. Under these conditions, data averaging can significantly distort the real value of the exposure.

The Level of External Exposure Cannot Be Accurately Determined Basing on the Timing of Human Habitation in the Conditions of Radiation Exposure

Any person at times can be shielded from radiation (the walls of the house, the car body, etc.), and therefore the level of exposure can be repeatedly changed, even for a short period of time.

As a person from time to time bends, rises or falls on the stairs, the intensity of external radiation generated in the soil by beta radionuclides, will somewhat vary (for γ-emitters such changes are insignificant).

The most detailed memories of the time spent away from home a few days or weeks are insufficient for the reconstruction of the level of external radiation. It is very difficult (if at all feasible) to quantify changes in exposure dose associated with the position of the human body in relation to the β-contamination substrate.

The inevitable space—time radiation heterogeneity and the significant variation of the individual exposure resulted in individual dose of any person for a year which can both increase and decrease repeatedly. In these circumstances, its one-time (or even over multiple years) measuring and averaging cannot adequately reflect the radiation load for a person or the population.

The Level of Internal Exposure Cannot Be Correctly Calculated by the Quantities of Radionuclides Absorbed with Food, Water and Air

The exact calculation of the amounts of radionuclides on the basis of a diet is hardly possible because of the considerable variability in the levels of radionuclides in each kind of the food. This amount depends on:

- radioactive contamination of a specific area in which the meat, dairy products, root crops and leafy vegetables, grains and legumes fruits and berries, etc. are obtained. Amount of radionuclides in the same products in different areas can differ at many times;
- the concentration coefficient of different radionuclides vary in different species, different varieties of the same species, in different years and seasons.

The exact calculation of the average intake of radionuclides with food is difficult:

- the individual food preference, which is different in men and women and changes with age;
- availability of seasonal and local food preferences (Fig. 2: an example).

It is difficult, if possible at all, to pinpoint any specific coefficients of transition of the radionuclides from the soil into the food chain (for each species and variety) plants, animals and fungi, for each type of soil, for different seasons and different years.

The polls cannot be restored with precision how much and what kind of dairy products, leafy vegetables, root crops, fruits, berries, how many and what kind of animal products (meat, fish, etc.) a person ate, how much and what quality he/she drank water a week and a month ago. The results obtained in these surveys data on the volume and the range of food consumed differ by one or two orders, which renders meaningless data averaging. In addition, such surveys of radiation contamination of a large territory are impossible to cover the entire population. For example, official estimates of levels of internal exposure of the population of Belarus after the Chernobyl disaster are based on interviews about the nature of individual consumption of milk and vegetables only among few thousands of people—about 0.1 % of the population (Yablokov et al. 2011). The calculation of the average intake of radionuclides with water and air is less wrong than for food, but also cannot be accurate because of the age and sex of individual variability in metabolism.

There are the great individual variations of the biological half-life of the radionuclides in a body and, therefore, their contribution to the internal dose. It depends on the physiological condition of the person, the age, sex and course of the diet. For example, the half-life of Cesium-137 in a small group (5 people) ranged from 36 to 124 days, and the ICRP of this period offers average to 70 days (Yablokov 2002). It turns out that for a large part (perhaps even—for the most) of the exposed group of the population calculations for such averaging will significantly underestimate the magnitude of internal exposure.

Using the average values in half-life of the radionuclide is incorrect because of different radionuclides derived from different organs and tissues. For example, the

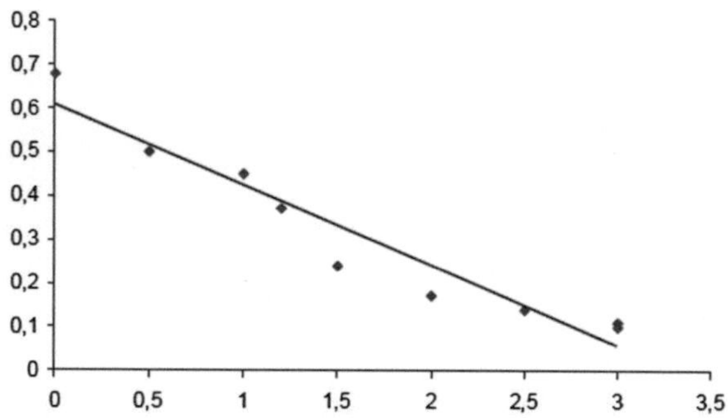

Fig. 2 The dependence of the mean internal dose of the distance of the village from the forest (Vetka distr, Gomel reg, Belarus) (Visenberg 2008). x-axis: distance to forest, km; y-axis: mean individual dose per year, mSv

average half-life of ^{90}Sr for all body considers about 40 days, but 10 % of the radionuclide, which fall into the bone tissue—the half-life of about 50 years.

The Ideas that (1) the Biological Effectiveness of X-Rays and All γ- and β-Emitters Are Identical, (2) Biological Efficacy of Slow Neutrons Tripled, and (3) α-Emitters and Ultra-Fast Neutrons Are 20 Times Higher—Are Too Simplified

These assumptions were made in the early period of radiobiology. It is clear now that the biological efficacy of each of γ-, β- and α-emitters are specific. It is determined not only by the number of emerging electrons (β-decay), γ-rays (with γ-decay and X-rays) or α particles (the α-decay), but microdistribution transmitted by these particles/quanta of energy as direct cellular structures and intracellular fluid, as well as the specific chain decay of each radionuclide. At the same time the energy of particles of different radionuclides forms a substantially continuous range from 2.5 keV to tens of MeV (thousandfold rather than 20-fold difference). Besides, some radionuclides have both alpha- and beta-decay: for instance, bismuth-212 produces thallium-208 as the result of alpha-decay and polonium-212 as the result of simultaneous beta-decay.

The possibility of an adequate application of the "weighting factors" for a variety of radiation sources is doubtful, and as the effect of α-emitters is due almost exclusively to the internal radiation, γ- and β-emitters can be sources of both internal and external irradiation, whereas X-rays can only be obtained from an external source.

Such physically significant different types of irradiation as X (γ-) and β-radiation cannot be uniform in their biological effects and consequently, it is impossible to impart uniform weighting coefficient of "1". There is no convincing argument for the claim that the biological effect of α-radiation is 20 times stronger than that of X-ray, γ- and β-radiation. Probably, in some miniscule number of pairwise comparisons of some emitters it is possible, but it is difficult to assume that the average biological effects of α-radiation are 20 times stronger than the effect of γ- and β-radiation.

It remains unclear how to use the "weighting factors" in case of the transmutation of some radionuclides.

Finally, the biological effectiveness of radiation can vary with the dose. Table 1 shows an example of such a change for the cosmic radiation: determined by the number appearing in the blood of astronauts chromosomal aberrations (dicentric and concentric rings) "biological dose" differs significantly from the physical dose: in the range of 2–6 mGy this difference is 20-fold, but in the range of 29–71 mGy this difference is much less.

Table 1 The relationship between the physical dose (mGy) and "biological" (frequency of chromosome aberrations) for the Russian cosmonauts after irradiation with different intensity (Snigireva 2009)

Physical dose, mGy	Biological dose, mSv	Biological/physical
2 (1–3)	53 (27–80)	26.5
6 (2–11)	107 (47–167)	17.8
29 (18–46)	113 (67–160)	3.9
71 (7–109)	227 (187–267)	3.2

Biological effectiveness of β-radiation at a dose of 30 mGy is almost half at a dose of 1 Gy (Snigireva 2009).

It has to be mentioned that even exactly the same quantitative level of ionization in different ways affects the cell depending on what stage of the cell cycle impact takes place.

All three assumptions underlying the provisions of the relative biological effectiveness of different types of radiation cannot be considered valid at the present time.

The Radiosensitivity of Organs and Tissue Does not Form a Continuous Series of Values

The assertion that the relative radiosensitivity of human organs and tissues form distinct deterministic series ("weighting factor" for the gonads—0.2, red bone marrow, stomach, colon, lungs—0.12; breast, liver, esophagus, urinary bladder, thyroid gland—at 0.05, skin, bone surface—0.01; the adrenal glands, the brain, the upper section of the large intestine (cecum, ascending and transverse portion of the colon), small intestine, kidney, muscle, pancreas, spleen, thymus and uterus—a total of 0.05), simplifies the situation to the loss of the biological meaning of this division.

This statement is based on the inevitable assumptions:

- biological effects of external and internal exposure to specific organs are the same;
- biological effects of different radionuclides on each organ are equal and consistent;
- the radiosensitivity of each organ and tissue is the same in different people;
- radiosensitive organs and tissues of animals determined for experiment under control in laboratory condition, adequately reflect the radiosensitivity of human organs and tissues;
- human organs and tissues are a conglomerate of independent structures;
- radiosensitivity of eyes and nose, mouth, upper respiratory tract, and a number of other structures that are not included in the fixed list, is negligible.

None of the six above-mentioned assumptions corresponds to the actual state of things, and this makes the idea of a "weighting factors" for various organs inadequate.

It Is Impossible to Correctly Calculate and Summarize the Dose from All Sources

To determine individual effective dose it is necessary to summarize internal and external exposure received from all the (additional to natural) sources of radiation, but to do so in many cases it is practically impossible because of the difficulty of taking into account the impact of all radionuclides.

In any nuclear reactor several hundred radionuclides are produced of which several dozen go into the environment (in small amounts during normal operation of nuclear power plants and the processing of spent nuclear fuel, and in large numbers - in case of accidents). Characteristics of the "Chernobyl" radionuclides are presented in the Tables 2, 3 and 4: the features of a huge number of radionuclides which were thrown out from the fourth Chernobyl reactor (Table 2); the dynamics of reducing the activity of the Chernobyl radionuclides for the first 12 years (Table 3); the Chernobyl radionuclides and their actual recorded concentrations in different countries (Table 4).

During the Fukushima disaster (2011), the environment has got tens different radionuclides, among them ^{31}I, ^{134}Cs, ^{137}Cs, ^{132}Te, ^{137m} Ba, ^{110m}Ag, ^{102}Rh, ^{212}Pb, ^{238}Pu, $^{239+240}$Pu, ^{131m} Xe, ^{234}U, ^{235}U, ^{90}Sr. The composition and activity (Bq) of the main radionuclide is presented in Table 5.

After Chernobyl and Fukushima, the focus in the early days and weeks was paid to the presence of ^{131}I, although sometimes it was not the main dose-related radionuclide. The same applies to the ^{137}Cs concentrations which are considered to be the main dose—a few months after these disasters. Thus it was overlooked that such radionuclides as ^{140}Ba, ^{136}Cs, ^{110m}Ag, ^{141}Ce, ^{103}Ru, ^{89}Sr, ^{95}Zr, ^{144}Ce, ^{106}Ru, ^{134}Cs and ^{90}Sr are no less, and had a total of more significant role than the ^{137}Cs in

Table 2 Chernobyl radionuclides and their half-life (hours/h, days/d, months/m, years/y) (Pshenichnikov 1996)

^{135}I	6.6 h	^{131}I	8.0 d	^{89}Sr	50.5 d	^{106}Ru	1.0 y	^{240}Pu	6537 y
^{133}I	20.8 h	^{133}Xe	10.5 d	^{124}Sb	60.2 d	^{134}Cs	2.1 y	^{99}Tc	213,000 y
^{131m}Te	30 h	^{147}Nd	11 d	^{95}Zr	64.0 d	^{125}Sb	2.8 y	^{239}Pu	24065 y
^{140}La	40.3 h	^{140}Ba	12.4 d	^{242}Cm	162.8 d	^{85}Kr	10.7 y	^{26}Cl	301,000 y
^{115}Cd	53.5 h	^{136}Cs	13.1 d	^{65}Zn	243.9 d	^{241}Pu	14.4 y		
^{239}Np	59 h	^{141}Ce	32.5 d	^{110m}Ag	249.9 d	^{90}Sr	29.1 y		
^{99}Mo	66 h	^{129m}Te	33.6 d	^{144}Ce	284.3 d	^{137}Cs	30.1 y		
^{132}Te	78.2 h	^{103}Ru	39.3 d	^{144}Pm	1.0 y	^{238}Pu	87.7 y		

Table 3 Dynamics of reduction of the activity of some β- and γ-radionuclides in Chernobyl plumes (Pshenichnikov 1996)

	5 days	10 days	30 days	2 month	6 month	1 year	3 years	5 years	12 years
^{143}Ce	0.43	0.04							
^{105}Rh	0.14	0.01							
^{149}Pm	0.18	0.04							
^{239}Np	5.76	1.32							
^{99}Mo	1.59	0.45							
^{132}Te	0.88	0.3							
^{132}I	0.88	0.3							
^{131}I	1.96	1.27	0.23	0.02					
^{147}Nd	1.59	1.16	0.33	0.05					
^{140}Ba	3.6	2.75	0.93	0.18					
^{140}La	3.6	2.75	0.93	0.18					
^{143}Pr	4.12	3.20	1.16	0.26					
^{141}Ce	4.71	4.23	2.76	1.46	0.11				
^{103}Ru	4.16	3.81	2.68	1.58	0.18				
^{89}Sr	1.73	1.62	1.24	0.83	0.16	0.01			
^{91}Y	2.31	2.17	1.71	1.20	0.28	0.03			
^{95}Zr	4.30	4.07	3.28	2.37	0.63	0.09			
^{95}Nb	4.30	4.07	3.28	2.37	0.63	0.09			
^{144}Ce	3.11	3.07	2.93	2.72	2.02	1.29	0.22	0.04	0.000
^{144}Pr	3.11	3.07	2.93	2.72	2.02	1.29	0.22	0.04	0.000
^{106}Ru	2.08	2.06	1.98	1.88	1.49	1.05	0.27	0.07	0.001
^{106}Rh	2.08	2.06	1.98	1.88	1.49	1.05	0.27	0.07	0.001
^{134}Cs	0.14	0.14	0.14	0.13	0.12	0.10	0.05	0.03	0.002
^{147}Pm	0.89	0.88	0.87	0.85	0.78	0.68	0.40	0.24	0.038
^{90}Sr	0.21	0.21	0.21	0.21	0.21	0.21	0.20	0.19	0.156
^{90}Y	0.21	0.21	0.21	0.21	0.21	0.21	0.20	0.19	0.156
^{137}Cs	0.28	0.28	0.28	0.28	0.28	0.27	0.26	0.25	0.214

the overall level of ionizing radiation in the first years after the disasters in some areas.

Of particular importance in determining the overall level of radiation at a given location are trans-uranic α-emitters—^{242}Cm, ^{241}Am, $^{238-240}$Pu, and ^{235}Uranium. These radionuclides have little effect on the ionization of the air, but they can be decisive in the formation of a total (sum of the external and internal radiation) dose. The short range in air α- (first centimeters) and β-particles (a few meters) impedes their detection tool, but the biological effect of such powerful factors is that without taking them into account in the human body (where they fall from the air and water, and through skin) cannot be an accurate representation of the true magnitude of ionizing radiation of any person.

The distribution of α- and β-radionuclides in space may differ significantly from the distribution of ^{137}Cs as a base in the preparation of the official maps of

Table 4 Actual recorded concentrations of different "Chernobyl" radionuclides

Nuclide	Activity	Nuclide	Activity
Finland (Sinkko et al 1987)			
^{131}I	223,000	^{103}Ru	2880
^{133}I	48,000	^{99}Mo	2440
^{132}Te	33,000	^{136}Cs	2740
^{137}Cs	11,900	^{239}Np	1900
^{134}Cs	7200	^{131m}Te	1700
^{140}Ba	7000	^{127}Sb	1650
^{129m}Te	4000	^{106}Ru	630
Vicinity of Krakow, May 1, 1986. Soil (0–5 cm) Bq/m^2 (Broda 1987)			
^{132}Te	29,300	^{140}Ba	2500
^{132}I	25,700	^{140}La	2400
^{131}I	23,600	^{99}Mo	1700
^{129m}Te	8,000	^{106}Ru	1300
^{103}Ru	6100	^{127}Sb	800
^{137}Cs	5200	^{136}Cs	700
^{134}Cs	2700	Total	Up to 360 000

Nuclide	*Aesculus hippocastanum*	*Tilla cordata*	*Betula verrucosa*	*Pinus silvestris*
Kiev, the end of July 1986. Plant leaves, dry weight Bq/kg (Grodzinsky 1999)				
^{144}Pm	58,800	146,150	10,800	–
^{141}Ce	18,000	–	6500	4100
^{144}Ce	63,300	–	21,800	18,800
^{140}La	1100	1930	390	660
^{137}Cs	4030	–	3400	4300
^{134}Cs	2000	–	1540	2100
^{103}Ru, ^{103}Rh	18,350	36,600	10,290	7180
^{106}Ru	14,600	41 800	400	5700
^{95}Zr	35,600	61 050	11,400	6500
^{95}Nb	53,650	94 350	18,500	9900
Total	312,000	399 600	101,400	70,300

US and Canada, radionuclides (Larsen et al. 1986; Roy et al. 1988)						
^{7}Be	^{60}Co	^{95}Zr	^{131}I	^{136}Cs	^{140}La	Pu
^{54}Mn	^{95}Nb	^{103}Ru	^{132}I	^{137}Cs	^{141}Ce	Am
^{59}Fe	^{95}Mo	^{106}Ru	^{134}Cs	^{140}Ba	^{144}Ce	

radioactive contamination and, accordingly, for the calculation of the dose. For example, in all the samples of breast milk in the Mogilev and Brest regions, Belarus (dose calculations were based only on ^{137}Cs) Strontium-90 was also found in large quantities (Zubovich et al. 1998).

Table 5 The composition and activity (Bq) of the main radionuclide have been released into the environment as a result of Fukushima catastrophe in 2011

Nuclide	Activity	Nuclide	Activity	Nuclide	Activity
^{133}Xe	1.1×10^{19}	^{89}Sr	2.0×10^{15}	^{147}Nd	1.6×10^{12}
^{131}I	1.6×10^{17}	^{127m}Te	1.1×10^{15}	^{241}Pu	1.2×10^{12}
^{132}Te	8.8×10^{16}	^{90}Sr	1.4×10^{14}	^{242}Cm	1.0×10^{11}
^{133}I	4.2×10^{16}	^{129}Sb	1.4×10^{14}	^{238}Pu	1.9×10^{10}
^{134}Cs	1.8×10^{16}	^{239}Np	7.6×10^{13}	^{103}Ru	7.5×10^{09}
^{137}Cs	1.5×10^{16}	^{141}Ce	1.8×10^{13}	^{99}Mo	6.7×10^{09}
^{127}Sb	6.4×10^{15}	^{95}Zr	1.7×10^{13}	^{239}Pu	3.2×10^{09}
^{131m}Te	5.0×10^{15}	^{132}I	1.3×10^{13}	^{240}Pu	3.2×10^{09}
^{129m}Te	3.3×10^{15}	^{144}Ce	1.1×10^{13}	^{106}Ru	2.1×10^{09}
^{140}Ba	3.2×10^{15}	^{143}Pr	4.1×10^{12}		
^{135}I	2.3×10^{15}	^{91}Y	3.4×10^{12}		

The total picture of radioactive contamination—and, accordingly, the level of exposure would be incomplete without taking into account the effect of "hot particles", when melted fuel from the damaged reactor emitted not only gases and aerosols, but also—tiny ceramic particles of enriched of some radionuclides. After Chernobyl and Fukushima catastrophes such hot particles include not only ^{95}Zr, ^{140}La, ^{144}Ce, but also such β-emitters as $^{103,\ 106}Ru$, ^{140}Ba, and α-emitters that flew thousands of kilometers. Most of the routine methods of radiation dose control do not detect these particles, but their contribution to total human exposure can be significant.

To correct calculation of the total effective dose there should be a detailed map of the distribution of tens of radionuclides, not only Iodine and Cesium. However, such mapping is practically impossible. It is not possible to determine the contribution to the total dose received by each person from α-emitters, and it is very difficult to determine the contribution of the β-emitters. As a result, doses calculated after any nuclear disasters are only an unknown part of the real average individual effective dose.

The Homogeneous Phantom Does not Adequately Simulate the Effects of External Irradiation

All doses calculations obtained up to 2010 were made on the basis of modeling the distribution of radionuclides in the human body using a homogeneous phantom "reference person". This "reference person" had the average characteristics of a healthy 20-year-old white male weighing 70 kg.

The model "reference person" is inadequate as it cannot be is applicable to the vast majority of the exposed population due to significant intraspecific (group and individual) variability of radiosensitivity (see review in: Yablokov 2002; Makhijani 2009).

Within any group of people, homogenous by race, nationality, sex, age, physiology, at any given time there are always differences of radiosensitivity of individuals—individual variability of radiosensitivity. For example, the radiation sensitivity of people with genotype Hp 2-2 haptoglobin is more than three times higher than radiosensitivity of people with genotype Hp 1-1 and Hp 2-1 (Tel'nov and Sotnik 2001). The rate of accumulation of ^{137}Cs in Rh-positive individuals was significantly higher (Bandazhevsky 2001).

Apparently, in any human population 14–20 % are significantly hypo-radiosensitive, 10–20 % -super-radiosensitive (Kovalev and Smirnova 1996). The difference in radiosensitivity of these groups can be deeper in the range of low doses (Plachinda 2001).

The fact that since 2010 the ICRP recommends wage calculations separately for men (phantom "Golem") and women (phantom "Laura") (ICRP 2009) may slightly change the position only in the future, until the radiation safety standards (1 mSv/year) are the same for men and women. All national regulations of radiation protection based on inadequate simulation external dose calculations, excluding population and individual variations of radiosensitivity.

The statement that the higher the dose, the greater biological effect, is incorrect. This assumption (the linear no-threshold hypothesis) is true only for relatively high exposure levels (above 100 mSv). In the area of low doses (at the level of cell structures, and at the level of the whole organism) is shown the effect of ultra-high impact of small doses—the effect Petkau—Burlakova (ECRR 2003 and many other).

Discussion and Conclusion

It is possible to expand the list of postulates underlying the basic concept and add to those outlined above, which are not clearly inconsistent with the facts. But this does not alter the conclusion that the main provisions of the concept of dose (and practical dose calculations) raise a lot of questions.

All complex calculations to determine the "individual effective dose" as a generalized population characteristic are justified only if the result ("dose") adequately reflects the actual level of exposure. However, numerous assumptions underlying the dose calculations (only briefly mentioned above) are a total of uncertainty that the whole calculation is meaningless in terms of the use of dose indicators for the organization of radiation protection. Those determined basing on the concept of dose levels of irradiation ("dose") are calculated, virtually, basing on extrapolation of average data, and refer to the unreal "average" person. Defined in this way the dose is "the average temperature of patients in the hospital." This averaging leads to the fact that a significant portion of the exposed population is non-protected from hazardous radiation exposure.

These are the fundamental difficulties of using the methodology of calculating "individual effective dose" to determine the individual level of exposure. Errors and uncertainties associated with the averaging "weighting factor" for organs and tissues, the average consumption of food and water, external dose estimates in this case, may be even greater than for population characters.

Dose calculation was originally developed to solve the problems of protection of troops in the conditions of use of nuclear weapons, and security of the nuclear weapons complex. The main criterion for radiation protection on the battlefield was to maintain the combat capability of the soldier for a few hours, maximum a few days. Personal radiation protection has been facilitated by the need to control the impact of only a few radionuclides in controlled conditions of exposure in the workplace.

Dose concept originated in the 1950s before the discovery of the functioning of intracellular structures (DNA replication, the functioning of intracellular membranes, macromolecules of enzymes and hormones, and others), until the discoveries of complex responses of cells, tissues, organs and the whole body at different exposure, before many discoveries in biophysics and physics of ionizing radiation (see reviews: Burlakova 2002; ECRR 2003). Historically the modern dose concept was created by numerous additions and resulted in growing complexity. Today, the dose concept resembles a colossus with feet of clay.

The inadequacy of the dose concept to protect the public became clear in the early 1970s, after the discovery of the negative effects for public health of the radioactive fallout from nuclear explosions. To-day the wrong calculations of the dose after nuclear catastrophes like Chernobyl and Fukushima impede the analysis of the medical consequences and hamper the choice of the most effective direction for treatments.

Some of the basic assumptions underlying simplify the calculations of radiation dose to the public are not allowed, or are contrary to the facts. For effective public protection against the consequences of the nuclear disasters the dose concept has to be seriously reshaped.

References

Bandazhevsky YI (2001) Radicesium and congenital malformations. In: Proceedings of International Conference on "Health effect of the Chernobyl accident: results of 15-years follow-up studies", Kyiv, Ukraine, 4–8 June

Broda R (1987) Gamma spectroscopy analysis of hot particles from the Chernobyl fallout. Acta Physica Polica B18:935–950

Burlakova EB (2002) Low doses of radiation: are they dangerous? Nova Science Publisher, Huntington, 329 p

ECRR (2003) Recommendations of the European Committee on radiation risk. Health effects of ionizing radiation exposure at law doses for radiation protection porpoises. Regulators' Edition. Brussels, Green Audit Book, Aberystwyth, 186 p

Grodzinsky DM (1999) General situation of the radiological consequences of the Chernobyl

ICRP (2009) Adult reference computational phantoms. ICRP Publication 110. Ann ICRP 39(2). http://www.icrp.org/publication.asp?id=ICRP%20Publication%20110

Kovalev EE, Smirnova OA (1996) Estimation of radiation risk based on the concept of individual variability of radiosensitivity. AFRRI Contact Rep 96—1. Bethesda, 1996. V + 202 p
Kryshev II, Ryazantsev EP (2000) Environmental safety of Russian nuclear power complex. "IzdAT", Moscow, 384 p (Russian)
Larsen RJ, Sanderson CJ, Rivera W, Zamicvhieli M (1986) The characterization of radionuclides in North America and Hawaiian surface air and deposition following the Chernobyl accident. Compend Environ Measur Lab Кузю EML-460, US Department of Energy, New York, 104 p
Makhijani A (2009) The use of reference man in radiation protection standards and guidance with recommendations for change. IEER Tech Rep. http://ieer.org/wp/wp-content/uploads/downloads/reports/referenceman.pdf
Plachinda YI (2001) The use of discriminant analysis in biological dosimetry. In: Abstract 3rd International Conference on Health effects of the Chernobyl disaster: the results of 15 years of research. 4–8 June 2001, Kiev, Ukraine, pp 271–272 (Russian)
Pshenichnikov BV (1996) Low doses of radiation and radiation sclerosis. Kiev, "Soborna Ukraina" Publ: 40 p (Russian)
Roy GC, Cote JE, Machfound A et al (1988) On the transport of Chernobyl radioactivity to Eastern Canada. J Environ Radioact 6:121–130
Sinkko A, Altonen H, Mustonen R et al (1987) Airborn radioactivity in Finland after Chernobyl accident in 1986. Report/STUK-A56.Helsinki: 56 p
Snigireva GP (2009) The effects of ionizing radiation: cytogenetic changes in human blood lymphocytes. Doctoral thesis, MSU, 264 p (Russian)
Tel'nov AI, Sotnik NV (2001) The relative nature of the genetic mechanisms of the human radiosensitivity. In: Abstract Conference no medical aspects of radiation and chemical safety, 11–12 Oct 2001, SPb: 257 (Russian)
Tscheglov AI (1999) Biogeochemistry of the anthropogenic radionuclides in the forest ecosystems. Nauka Publication, Moscow, 268 p (Russian)
Visenberg YV (2008) Villages: social, environmental and ecological factors affecting irradiation doses. Probl Ecol Health (Gomel) 17:30–38 (Russian)
Yablokov AV (2002) The myth of the safety of low-dose radiation. Moscow, Center Rus Environ Policy 178 p (Russian)
Yablokov AV, Nesterenko VB, Nesterenko AV, Preobrazhenskaya NM (2011) Chernobyl: consequences of the Catastrophe for people and Nature. "Universarium" Publication, Kiev, 589 p (Russian)
Yablokov AV (2013) A review and critical analysis of the "effective dose of radiation" concept. J Health Pollut 3(5):34–40
Zubovich VK, Petrov GA, Beresten SA et al (1998) Characters of human milk and infant health in contaminated regions of Belarus. Zdravookhranenie (Healthcare, Minsk) 5:28–30 (Russian)

Radiation-Induced Aging and Genetic Instability of Mesenchymal Stem Cells: An Issue for Late Health Effects?

Michael Rosemann

Introduction

Adult stem cells represent a relatively small fraction of the somatic cells of an organism. For a long time their existence could only be deduced from quantitative analysis of transplantation studies or from the in vitro growth pattern of single-cell derived clones. Although it became possible now to enrich rare adult stem cells by selecting for particular protein markers expressed on their cell surface, it is still a challenge to identify and study them in their normal tissue environment.

Whereas the stem cells of the developing embryo (forming the ectoderm, endoderm and mesoderm germ layer) are extensively studied and characterized in terms of their role in the formation of the entire organism, the potential of adult stem cells for the maintenance of health or their role in the development of diseases still requires much research.

The best studied adult stem cells are those of the hematopoietic system (HSC), which produces cells of the myeloid and lymphoid lineages, forming the peripheral blood and immune system (Barnes et al. 1966). The radiation response of HSCs and the lineage-committed precursor cells derived from them have been extensively studied to better understand the factors that influence long-term success of bone-marrow transplantation after radiation accidents or after high-dose chemotherapy (Gale et al. 2014).

Although the existence of stem cells in adult solid tissues was suggested already in 1867 by J Cohnheim, their role in the regeneration of solid tissues could be studied only when histological labelling techniques using cell-type specific antibodies or autoradiography after isotope incorporation became available. The best studied non-hematopoietic stem cells of the adult organism are those that form the

M. Rosemann (✉)
German Research Center for Health and Environment, Institute of Radiation Biology, Helmholtz Zentrum München, Neuherberg, Germany
e-mail: rosemann@helmholtz-muenchen.de

V.L. Korogodina et al. (eds.), *Genetics, Evolution and Radiation*,
DOI 10.1007/978-3-319-48838-7_32

crypts of the small intestine, and their involvement in the occurrence of post-irradiation gastro-intestinal syndrome is well established (Potten 1998).

With the rise of in vivo fluorescence labelling methods for single cells and their progeny, even the involvement of specific cellular signaling pathways (such as the Wnt-pathway) in the dynamics of small-intestine stem cell regulation could be discovered.

Stem cells that contribute to adult skin regeneration are also widely studied, in part because the hair follicles in the skin form a very prominent histological pattern. Similar to the situation in the small intestinal crypts, this highly reproducible tissue pattern allows the visual identification of single stem cells, the committed precursor cells derived from them and the fate of terminally differentiated cells. The role of skin stem cells in the induction of radiation induced skin tumors (basal cell carcinoma and squamous cell carcinoma) is well understood and the involvement of specific pathways (such as the Hedgehog pathway) established.

Two other types of adult stem cells belong to organs that are usually not considered to be as radiation-sensitive: the brain and the connective tissue. Neural stem cells (NSC) and mesenchymal stem cells (MSC) are much more difficult to identify in their normal tissue context, since they don't reside at morphologically defined sites within the tissue architecture. Only by transgenic expression of SC specific fluorescence-tagged proteins, or induced expression of fluorescence proteins, followed by the analysis of stem-cell derived daughter cell lineages can the existence of these rare cells with long term repopulating potential (LTRP) in these two tissue types be analyzed. Nevertheless, NSCs and MSCs are being extensively studied because of the expectation that a better understanding of their function will help to prevent or to cure neurodegenerative diseases, age-related disabilities, cardiovascular diseases or even cancer.

There has not been much interest in the response of NSCs or MSCs to radiation exposure. This might be due to the fact that for tumors of the CNS and of connective tissue (derived from mesenchymal cells) the radiation-associated excess relative risk (ERR) among the A-bomb survivors was much lower than for carcinoma (of skin, mammary gland, lung, thyroid, and colon) or for leukemia (Preston et al. 2003). This picture looks much different, however, when cancer patients are studied who were cured from a primary malignancy by (high dose) radiotherapy and who developed a secondary tumor later in life. Here it became clear that external beam radiotherapy confers a relatively higher risk for the induction of therapy-associated secondary sarcomas (derived from mesenchymal cells Berrington de Gonzalez et al. 2012) and medulloblastoma of the brain (in the case of Gorlin syndrome patients) as compared to carcinoma or leukemia.

In the following chapters I will try to explain why adult stem cells, in particular MSCs, should be considered an important target for the development of radiation-associated disease. This is also relevant in view of an increasing number of clinically approved MSC based therapies. These involve the collection of MSCs from various anatomical sites of a patient (including sites that might have been exposed to diagnostic, therapeutic or occupational ionizing radiation), followed by forced in vitro expansion of the cells prior to autologous re-engraftment.

Biology of Adult Stem Cell

Adult stem cells have the unique capacity of long-term repopulating tissues and organs. For this purpose they undergo self-renewal at every cell division. This implies that at least one of their daughter cells retains the long term repopulating potential (LTRP) of the parental stem cell. Depending on the tissue of origin, ASCs are multipotent in the sense that in addition to self-renewal they can produce daughter cells (committed precursor cells) that give rise to transiently amplifying cells and finally to mature cells that serve different biological functions.

The decision of stem cells to undergo symmetric cell division (only self-renewal and generation of two identical ASCs) or asymmetric cell division (self-renewal of the stem cell plus the generation of a committed precursor cell) depends on external triggers such as the presence of growth factors or their contact to neighboring cells (Fig. 1). The dynamics of stem cell proliferation and the fate of the daughter cells is a focus of current research that requires highly sophisticated methods of life single-cell analysis.

When ASCs divide asymmetrically, committed precursor cells or the terminally differentiated cells change their cellular program as compared to the parental stem cell.

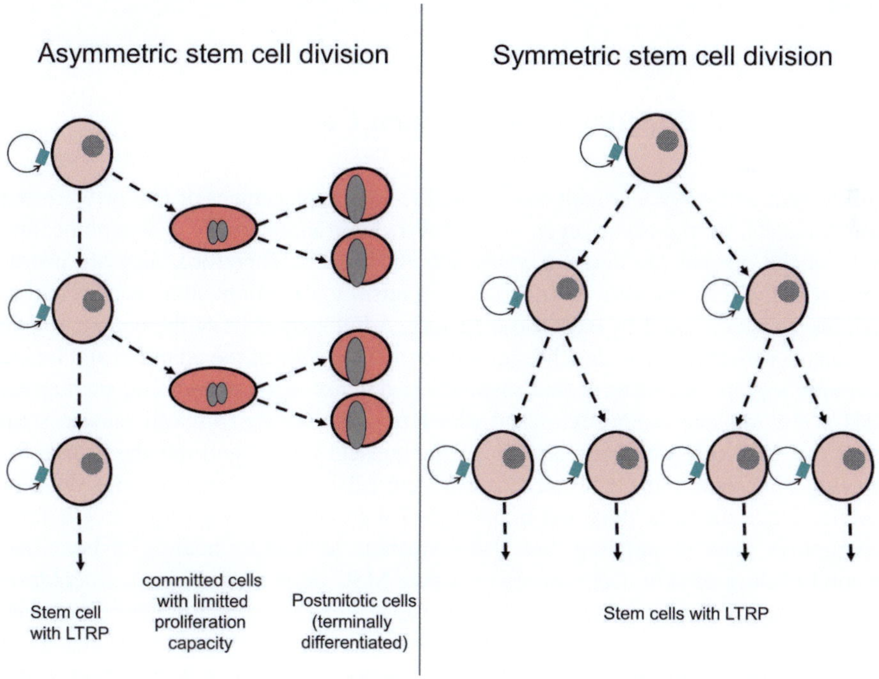

Fig. 1 Symmetric versus asymmetric division of adult stem cells determines the dynamics of self-renewal, generation of committed precursor or differentiated cells and the potential increase of the number of stem cells

This affects their metabolism (number of mitochondria, ATP production), expression of genes involved in cell signaling, DNA repair and telomere maintenance, their capability to express structural proteins and to undergo apoptosis. There is good evidence that stem cells that undergo lineage commitment or terminal differentiation do so by following epigenetic changes in large parts of their genome (affecting chromatin organization, methylation, and expression of non-coding RNA).

To maintain the capacity for long-term and high fidelity tissue regeneration, ASCs have to protect their genome from endogenous damage (by ROS or replication errors) and from cellular stress by exogenous noxae. An important role in protecting ASCs from genotoxic and cellular stress is played by supporting cells of the so-called "stem cell niche". The precise function of this niche is not completely understood yet, but it is known that cells that have a direct contact to stem cells are instrumental in directing symmetric or asymmetric stem cell division, most likely by providing a topological orientation and governing cell polarity during mitosis. When this well organized morphology of the stem cell niche is compromised, for instance by chronic inflammation or radiation-induced cell death, ASCs might respond with unscheduled symmetric cell division. This would give rise to an increased number of stem cells, and interfere with normal tissue homeostasis by so-called accelerated repopulation. One can only speculate about the potential impact of such an accelerated stem cell proliferation for their long term genetic stability.

Physiological Function of Adult Stem Cells

Adult stem cells play a crucial role in the life-long replacement of lost or worn-out somatic cells, for the tissue regeneration following trauma, injury or wound healing and for the normal plasticity of many somatic tissues. Therefore, they are instrumental in the response of a multicellular organism to environmental changes and to cellular stress imposed by exogenous factors. Adult stem cells of the hematopoietic system, for instance provide a life-long reservoir for cells of the adaptive and innate immune system, ensuring a fast responding defense against invading pathogens. HSCs also replace erythrocytes and platelets continuously. Smooth muscle stem cells (satellite cells) become activated after muscle injury, and are the source for healing of skeletal muscle. Mesenchymal stem cells also replenish various types of connective tissue in healthy and in pathological situations. They provide cells for a continuous bone remodeling, but also contribute to fracture healing of bone and wound healing of skin and connective tissue. MSC derived fibroblasts, osteoblasts and adipocytes also constitute the bone marrow niche for hematopoietic stem cells (Fig. 2). During wound healing MSCs also secrete paracrine factors that exert immune-modulating and anti-inflammatory response. This has been shown to be also an important mechanism to prevent graft-versus-host disease after allogenic bone-marrow transplantation.

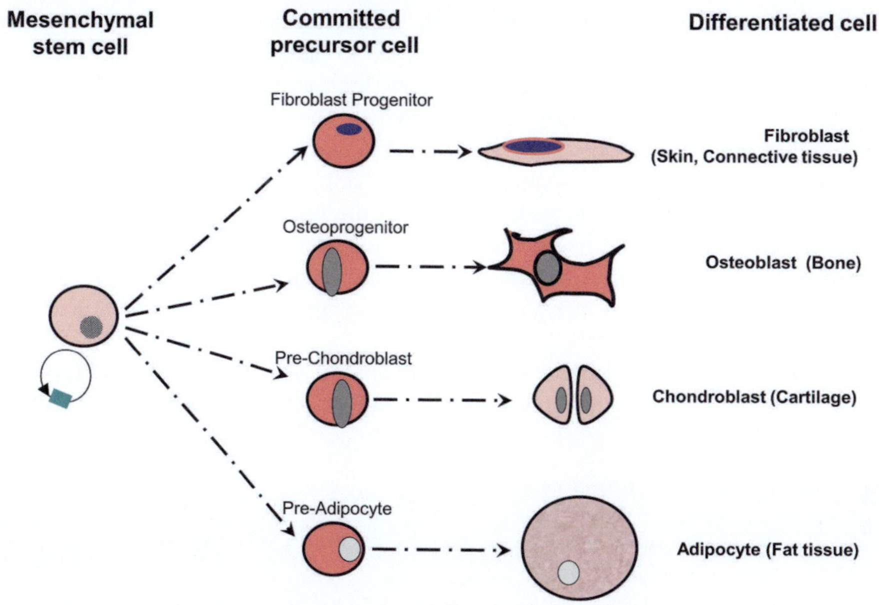

Fig. 2 Cells of the connective tissue compartment derived from MSC

Since MSCs have a multipotent differentiation capacity, they contribute to tissue plasticity under environmental changes. In response to changes in the caloric uptake, MSCs can up- or down-regulate the number of pre-adipocytes. An increase in mechanical load to bones will induce MSCs to generate more osteoblast precursor cells. Since these functions of ASCs are all associated with cell division, a stable genome in cells with a long term proliferation potential is essential.

Adult Stem Cells as Radiation Targets

As compared to the total number of somatic cells in an organism, tissue specific adult stem cells are very low in number. Depending on the organ or tissue under consideration, between 0.1 and 0.005 % of all cells are estimated to be adult stem cells with LTRP. In some organs, such as in the hematopoietic bone marrow, the numbers of proliferative stem cells seem to shrink with age. It has been shown that from a total of $\sim$1000 pluripotent HSCs in young subjects only a few (in some studies only one single HSC) were still present in patients older than 70 years (McKerrell and Vassillou 2015). Nevertheless, despite this potential loss of adult stem cells with increasing age, the cellular life time of an individual adult stem cell can in theory be as long as the entire life span of the organism itself. Here it is important to note that adult stem cells such as pluripotent HSC seem to be rather radio-resistant in terms of their clonogenic survival (Ploemacher et al 1992). In radiation accidents where subjects were exposed to radiation doses high enough to cause an acute complete bone marrow failure, long-term auto-transplantation and recovery of the entire blood forming hierarchy was observed sometimes after years, also suggesting that the most primitive stem cells are more radio-resistant than the committed precursors or transiently amplifying cells (Baranov et al. 1994), probably because of the long cell cycle turnover of the former.

In another well studied tissue system, the small intestinal crypt, at least a subpopulation of slowly proliferating clonogenic stem cells was found to be radio-resistant and able to repopulate the entire crypt, even after gamma doses as high as ~10 Gy (Potten 1998).

To adequately weight the factors that could qualify adult stem cells as target cells of a radiation-induced malignant transformation, one can do the following comparison:

Relative abundance of target cells:

$$\text{ASC/differentiated or precursor cells} : 10^{-3}\ldots10^{-4}$$

Time window available for the accumulation of radiation-induced mutations

ASC: 70 years

differentiated or precursor cells: 2 weeks

→ *Relative length of time window available for finishing carcinogenic progression*:

ASC/differentiated or precursor cells: 1860.

This means that ASCs in a 70 year old person had the chance to accumulate almost 2000 times more radiation-induced mutations in their genome than short-living committed or differentiated cells, assuming exposure to the same radiation dose-rate. In the case of the natural background radiation (~3 mSv/a in Germany), every ASC in a 70 year old person would have been exposed to a cumulative dose of ~200 mSv. In the context of the multistep process of environmental tumorigenesis (Trott and Rosemann 2000) this would also mean that long-term slowly proliferating ASCs have a much higher chance to acquire a rare, complex pattern of cancer-driving mutations in their genome as compared to the faster growing, but short living precursor cells (Fig. 3).

This calculation shows that the low abundance of ASCs as cancer-forming target cells can be easily out-weighted by their higher chance of acquiring tumorigenic mutations over a long time.

Using next-generation sequencing to identify somatic mutations in single cells it was recently estimated that haematopoietic precursor cells accumulate about 10 random mutations per year, and that those mutations which by chance confer a

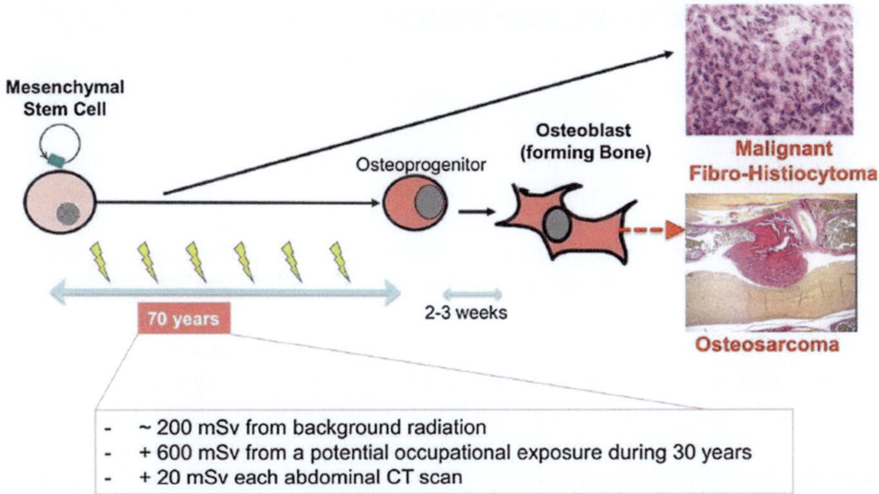

Fig. 3 Cumulative radiation dose in mesenchymal stem cells. After life-long exposure to chronic IR of various sources, MSCs in an adult person could accumulate several hundred mSv. This is in contrast to committed precursor cells, which have a proliferative life span of just a few weeks. Undifferentiated MSCs can give rise to malignant fibrous histiocytoma/pleomorphic undifferentiated sarcoma upon malignant transformation, whereas transformed osteoblasts form osteosarcoma

growth or survival advantage to the daughter cells will eventually give rise to predominant clones (Xie et al. 2014). In blood donors above the age of 70 years, a few such mutant clones were found to dominate the entire population of blood cells. The authors made the case that this is consistent with a model of long-living haematopoietic precursor or stem cells accumulating somatic mutations throughout the life span of a person. The age-associated increase of the mutational load to an organ or a tissue is also known to happen in solid tissues, such as skin (Goodell and Rando 2015), and the resulting growth of mutant clones with a concomitant loss of wildtype cells has been termed "clonal collapse". This shows clearly that throughout the entire life of an organism, stem cells accumulate more and more mutations that are passed to their progeny cells. Although most of these somatically mutated cell clones do not seem to progress to full malignancy within a life time, there is a clear association with an increased risk for myeloid-dysplastic syndrome, aplastic anaemia, leukaemia (Jaiswal et al. 2014) as well as with the frequency of other age related diseases (Bonnefond et al. 2013).

At the moment one can only speculate about the contribution of chronic background radiation (2–5 mSv/a) to this age related increase of stem cell mutations, or if repeated exposures to diagnostic X-rays ($\sim$2 mGy for a conventional X-radiography or $\sim$20 mGy for a CT) throughout life would further add to this mutation load.

Chronic Disease by Dysfunctional SC

As outlined earlier in this article, ASCs represent a life-long cell reservoir for tissue repair. Their loss or functional deprivation will therefore always cause a higher risk for chronic diseases, in particular for those characterized by tissue degeneration. They can affect neuronal tissues (Alzheimer disease, Parkinson disease, depression, stroke, dementia, hearing loss, macular degeneration or other retinopathies), the blood and immune system (aplastic anaemia, chronic inflammation, impaired immune response and increased infection risk), cardio-vascular organs (arteriosclerosis, hypertension, ischaemic infarct), the metabolic organs (metabolic syndrome, diabetes), or the musco-skeletal system (osteoporosis, osteoarthritis).

Since most of terminally differentiated cells rely on a high expression of specific proteins and/or on a sufficiently high ATP production in their mitochondria, cellular stress by endogenous radical oxygen species (ROS) and by the aggregation of misfolded proteins (or a combination of both) are likely to cause a time-limitation of their functionality. Therefore, similar to a building that requires regular maintenance and repair work in-order to ensure its long-time stability, most organs and tissues in a healthy organism can maintain their physiological function only if stem cells continuously provide fresh functional cells to replace old ones. If adult tissue stem cells are depleted or funtionally impaired, the homeostasis between newly generated cells and lost dysfunctional cells is disrupted, most likely leading to degenerative health complications.

It has been shown that chronic low dose irradiation (that does not induce acute cell death) has the capacity to trigger premature senescence in human epithelial cells (Yentrapalli et al. 2013). In murine MSCs we observed loss of stem cell capacity by increased premature adipogenesis also at doses that do not induced cell death (Höfig et al. 2016) (Fig. 4).

This shows that despite being refractory to radiation-induced cell death, stem cells can lose their long term repopulating potential by several other mechanisms as well.

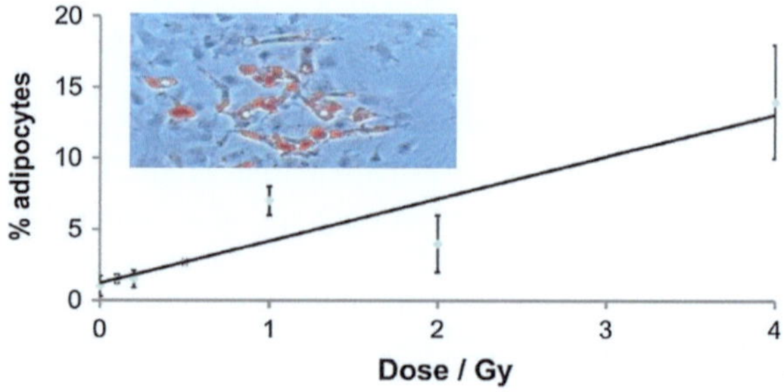

Fig. 4 Premature adipogenesis in murine MSCs after gamma irradiation. Adipogenic differentiation as detected by Oil-red staining in a culture of murine MSCs 3 weeks after gamma-irradiation

When Too Much of a Good Thing Becomes Bad: Malignant Transformation

At the opposite end of the spectrum of disturbances of the stem-cell proliferative capacity one can finds their malignant transformation. Here, a gain-of-function alteration in the cellular genome causes the stem cell to divide in an uncontrolled manner. This generates cellular progeny which escape the normal cellular program to differentiate or to undergo apoptotic cell death and which acquire proliferative immortality them self. The neoplasia that are the clinical manifestation of this transformation are an extremely heterogeneous class of disease, often require sophisticated forms of therapy depending on the genetic alterations and molecular biology the malignant cells. Despite tremendous progress in the field of personalized cancer medicine, with novel therapies using tumour-directed antibodies, tumour-targeting immune cells or therapeutic compounds directed against specific cellular pathways, on average 35 % of all malignancies will finally cause death of the patient.

In a recently published study by Tomasetti and Vogelstein (2015) an association was found between the abundance of adult stem cells in various organs and the frequency of tumours. It would be to early, however, to blame only transformed adult stem cells for each and every tumour. Hahn et al. (1999) demonstrated in various in vitro models that differentiated human cells can readily be transformed into fully malignant cells, just by the targeted genetic manipulation that affects two or three cellular signalling pathways. This clearly shows that immortality can be easily acquired by mutations in non-stem cells as well.

The Case of MSCs as Target Cells for Radiation Carcinogenesis

Post-irradiation sarcoma is a prominent late effect following therapeutic irradiation. In the case of external beam radiotherapy with healthy connective tissue in the radiation field, a large portion of sarcoma are undifferentiated malignant fibrous histiocytoma (MFH, recently termed pleomorphic undifferentiated sarcoma). When most of the radiation dose is absorbed by the bone surface (the area where pre-osteoblasts are located), as in the case of therapy with bone-seeking short-lived alpha-emitters such as Ra223, the predominant type of late arising secondary tumour is osteosarcoma. This suggests that after high-dose therapeutic irradiation the sarcoma type is determined by the type of target cell in the radiation field. On the other hand, whole body irradiated subjects that survived moderate doses (0.2–3 Gy) during the A-bomb blasts were recently found to have an increase of osteosarcoma, but not of MFH, implying that even when MSCs are in the radiation field they become more important as target cells for a malignant transformation only after higher doses. It is possible that MSCs and pre-osteoblasts (the transiently

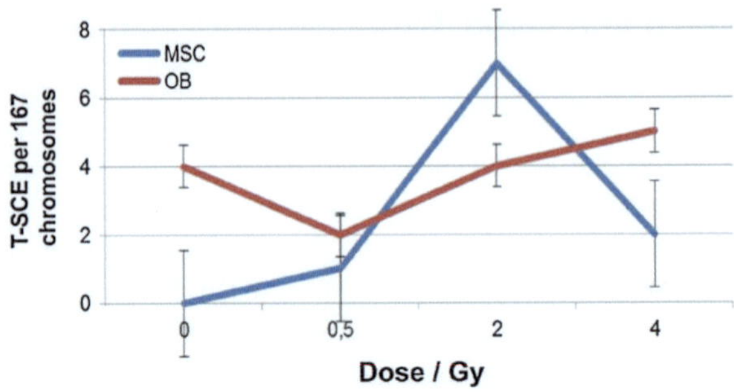

Fig. 5 Telomere sister-chromatid exchange in murine MSCs and osteoblasts. Metaphase chromosomes were differentially stained for 3′ and 5′ telomeres after gamma-irradiation and BUdR incorporation. For each experiment, 167 chromosomes were scored for the occurrence of 3′–5′ telomere exchange

amplifying cells that after malignant transformation give rise to osteosarcoma) have different mechanisms to respond to the genotoxic effect of ionizing radiation. Analyzing the impact of gamma radiation onto the degree of telomere sister chromatid exchange we found, that although non-irradiated MSCs had a much lower level of this cytogenetic abnormality than osteoblasts, the frequency went up sharply after 2 Gy gamma-irradiation (Kilinc et al., unpublished) (Fig. 5).

Telomere stability in osteoblasts has been found to be impaired by loss of Rb1 expression (Gonzalez-Vasconcellos et al. 2013), and this is assumed to contribute to the high risk of Rb1 mutation carriers for the development of radiation-induced osteosarcoma (Rosemann et al. 2014). In contrast to this, MSCs derived from the same Rb1 ±mice as those that yielded the osteoblasts did not show an impaired telomere stability. The telomere length during extended proliferation was always higher in MSCs than in osteoblasts, and it was independent on the Rb1 gene status (Fig. 6).

Fig. 6 Telomere length in murine MSCs and osteoblasts of different Rb1 gene status. DNA extracted from in vitro growing cells of Rb1 ±knockout mice was used measure average telomere length by qRT-PCR

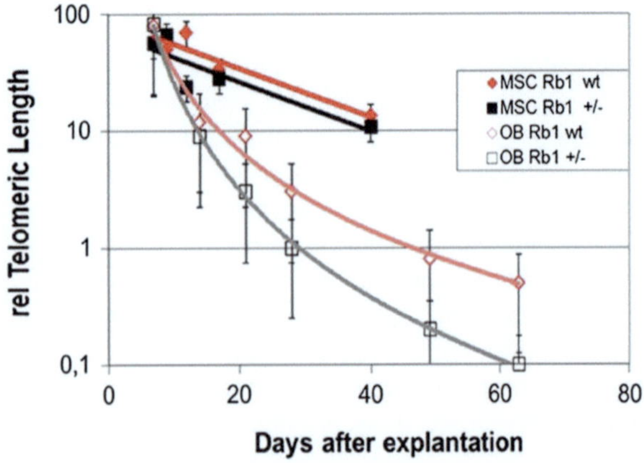

This suggests that the radiation-risk for these two histological types of bone sarcoma can differ significantly, depending on whether the target cell that undergoes transformation is a stem-cell or a committed precursor cell.

A similar picture was seen in the case of myeloid leukaemia, where acute myeloid leukemia (AML) (characterized by mutations in genes such as Aml1, Flt3 or Pou1) exhibit a high excess relative risk in IR exposed cohorts, whereas chronic myeloid leukemia (CML) incidence (characterized by the recurrent Bcr-Abl translocation) remains almost at the background level as seen in unexposed cohorts (Gale et al. 2014). This suggests that cells which belong to the same stem cell lineage, but differ in their differentiation stage, can vary dramatically in terms of their potential to undergo a radiogenic transformation.

Conclusion

Adult stem cells represent only a small portion of somatic cells. This, together with earlier findings showing that ASCs are relative radio-resistant, led to the assumption that they might not be important for the expression of radiation-induced late effects. With the availability of sophisticated single-cell assays to identify and analyse stem cells, however, it now becomes more and more obvious that rare ASCs can be important target cells for different forms of health impairment following acute or chronic radiation exposure. Since it has been shown that ionizing radiation to ASCs can impair their genetic stability and their stem-cell capacity, their very distinct regulation of cellular and molecular processes should be considered as a potential factor that modulate the radiation risk for late health effects in various organs.

Acknowledgement This study was supported by a grant for the joint EU project "RISK-IR" (EURATOM contract 323267).

References

Baranov AE, Selidovkin GD, Butturini A, Gale RP (1994) Hematopoietic recovery after 10-Gy acute total body radiation. Blood 83(2):596–599

Barnes DW, Bungay GT, Mole RH (1966) Delayed mortality after mid-lethal exposures to whole body irradiation and its modification by treatment with syngeneic lymph-node or bone-marrow cells. Int J Radiat Biol Relat Stud Phys Chem Med 11(5):409–427

Berrington de Gonzalez A, Kutsenko A, Rajaraman P (2012) Sarcoma risk after radiation exposure. Clin Sarcoma Res 2(1):18

Bonnefond A, Skrobek B, Lobbens S et al (2013) Association between large detectable clonal mosaicism and type 2 diabetes with vascular complications. Nat Genet 45(9):1040–1043

Gale RP, Hlatky L, Sachs RK, Radivoyevitch T (2014) Why is there so much therapy-related AML and MDS and so little therapy-related CML? Leuk Res 38(10):1162–1164

Gonzalez-Vasconcellos I, Anastasov N, Sanli-Bonazzi B, Klymenko O, Atkinson MJ, Rosemann M (2013) Rb1 haploinsufficiency promotes telomere attrition and radiation-induced genomic instability. Cancer Res 73(14):4247–4255

Goodell MA, Rando TA (2015) Stem cells and healthy aging. Science 350(6265):1199–1204

Hahn WC, Counter CM, Lundberg AS, Beijersbergen RL, Brooks MW, Weinberg RA (1999) Creation of human tumour cells with defined genetic elements. Nature 400(6743):464–468

Höfig I, Ingawale Y, Atkinson MJ, Hertlein H, Nelson PJ, Rosemann M (2016) p 53-dependent senescence in mesenchymal stem cells under chronic normoxia is potentiated by low-dose γ-irradiation. Stem Cells Int (in press)

Jaiswal S, Fontanillas P, Flannick J et al (2014) Age-related clonal hematopoiesis associated with adverse outcomes. N Engl J Med 371(26):2488–2498

McKerrell T, Vassiliou GS (2015) Aging as a driver of leukemogenesis. Sci Transl Med 7 (306):306fs38

Ploemacher RE, van Os R, van Beurden CA, Down JD (1992) Murine haemopoietic stem cells with long-term engraftment and marrow repopulating ability are more resistant to gamma-radiation than are spleen colony forming cells. Int J Radiat Biol 61(4):489–499

Potten CS (1998) Stem cells in gastrointestinal epithelium: numbers, characteristics and death. Philos Trans R Soc Lond B Biol Sci 353(1370):821–830

Preston DL, Shimizu Y, Pierce DA, Suyama A, Mabuchi K (2003) Studies of mortality of atomic bomb survivors. Report 13: solid cancer and non cancer disease mortality: 1950–1997. Radiat Res 160(4):381–407

Rosemann M, Gonzalez-Vasconcellos I, Domke T, Kuosaite V, Schneider R, Kremer M, Favor J, Nathrath M, Atkinson MJ (2014) A Rb1 promoter variant with reduced activity contributes to osteosarcoma susceptibility in irradiated mice. Mol Cancer 13:182

Tomasetti C, Vogelstein B (2015) Cancer etiology. Variation in cancer risk among tissues can be explained by the number of stem cell divisions. Science 347(6217):78–81

Trott KR, Rosemann M (2000) Molecular mechanisms of radiation carcinogenesis and the linear, non-threshold dose response model of radiation risk estimation. Radiat Environ Biophys 39 (2):79–87

Xie M, Lu C, Wang J, McLellan MD et al (2014) Age-related mutations associated with clonal hematopoietic expansion and malignancies. Nat Med 20(12):1472–1478

Yentrapalli R, Azimzadeh O, Sriharshan A et al (2013) The PI3K/Akt/mTOR pathway is implicated in the premature senescence of primary human endothelial cells exposed to chronic radiation. PLoS One 8(8)

Significance of Cytogenetic Study for Estimation of Biological Effects of Low-Dose Irradiation of People

Irina E. Vorobtsova and Alexey Semenov

Introduction

The application of ionizing radiation for military, industrial and medical purposes, various emergency situations has dramatically increased the amount of people irradiated at the low doses. This required forecasting of possible somatic and genetic effects for the human population. In the case of low dose exposure it seems to be difficult to estimate the risk of negative effects for people's health using direct epidemiological observations. The studies of disorders raised in cells as a response to such radiation, allow understanding of the initial stages of the processes inducing development of cancer and noncancerous forms of radiation pathology in exposed people and in their progeny.

Since 1986 we study the cytogenetic effects in persons who suffered from the Chernobyl accident, as well as in people called "veterans of special risk division", exposed in the past to radiation during nuclear tests, nuclear waste burial, various radiation accidents. People who had no professional or any random contact with ionizing radiation and their children were also studied, as well as children of Hodgkin's disease patients who underwent radio- and chemotherapy (the positive control) before conception.

The investigations were performed in the following main directions:

- The structural/mutational state of genome: frequencies of unstable and stable chromosomal aberrations, (ChA), micronuclei (MN), GPA and HPRT-mutations in blood cells.

I.E. Vorobtsova (✉) · A. Semenov
Federal State Institution Russian Research Center for Radiology and Surgical Technologies, St. Petersburg, Russia
e-mail: radgenetika@mail.ru

V.L. Korogodina et al. (eds.), *Genetics, Evolution and Radiation*,
DOI 10.1007/978-3-319-48838-7_33

- The functional state of genome: whole genome stability, chromosomal sensitivity of lymphocytes to challenge irradiation in vitro and the age dynamics of stable ChA.
- A correlation between the frequency of unstable ChA and the risk of somatic pathology development.
- Comparison of dose-responses for ChA in lymphocytes irradiated in vitro and in vivo.
- Radiation induced bystander effect (RIBE) on the novel model—mix culture of lymphocytes from opposite gender donors.

Materials and Methods

Several groups of people exposed in the past to low dose irradiation were studied: LQ—liquidators of 1986–1990 years (174 persons, 24–66 years old); LC—liquidators' children (60 persons, 0.5–7 years old); EA—adults evacuated from radionuclide contaminated regions (19 persons, 26–45 years old); EC—evacuated children (64 persons, 2–19 years old); VT—veterans of special risk divisions—people who survived irradiation due to nuclear bomb testing or accidental nuclear waste burial (22 persons, 48–76 years old); PC—children of Hodgkin's disease patients born after exposure to radio- and chemotherapy (15 persons, 3–13 years old); AC—adult control (93 persons, 20–85 years old); CC—children control (52 persons, 2–18 years old). For comparison of in vitro and in vivo dose-responses of ChA blood from 5 cancer patients exposed to the whole body fractionated irradiation at low dose during 1 week was used. In the study of RIBE mix and mono cultures of lymphocytes from 4 pairs of male and female donors of the age range 23–32 years were started. An individual form was filled in for every person under test who gave an informed consent to participate in this investigation.

Cultivation of lymphocytes, harvesting procedure and slide preparation were performed in accordance with standard protocols, using BrdU for cell cycle control. Unstable ChA were scored on slides stained by 6 % Gimsa dye, stable ChA—on slides prepared by FISH technique using fluorescent DNA probes for 3 pairs of chromosomes covered about 20 % of genome DNA as described (Vorobtsova et al. 2001a). MN were identified in binucliated lymphocytes using cytochalasin block. MN in erythrocytes was determined on blood smears. The methods of registration of GPA and HPRT-mutations were described in details elsewhere (Jones et al. 2002).

For the estimation of genome stability, the donor blood was exposed in vitro to γ-radiation from ^{137}Cs source at the dose 1.5 Gy.

To generate dose-response curve of ChA for in vivo irradiated lymphocytes the blood from cancer patients exposed every other day to whole body γ-irradiation from ^{60}Co source (at a single dose of 0.115 Gy up to total dose of 0.57 Gy) was sampled after each exposure. The technique of whole body irradiation was applied as a pretreatment antitumor therapy. To construct dose-response of ChA in lymphocytes irradiated in vitro blood of the same cancer patients sampled before this treatment, was irradiated at the dose range of 0.1–0.6 Gy of γ-radiation.

In the study of RIBE blood of male/female donors before mixing was irradiated at the dose of 1 Gy of X-rays when the spontaneous frequency of ChA was used as an endpoint. If an adaptive response (AR) was used for estimation of RIBE G_0 or early G_1 male/female lymphocytes were pre-irradiated at the dose of 0.05 Gy of γ-radiation from ^{60}Co source, the challenge irradiation was performed at the dose of 1 Gy of X-rays on late G_1 lymphocytes in mix culture.

The significance of differences was estimated using Student's t-test, Fisher's exact test and χ^2criterion. The regression analysis of the data was performed with computer software Poisson Iteratively Reweighed Least Squares and Statistics for Windows 6.0.

Results

Cytogenetic, Mutational and Functional State of Genome of People Exposed to Low Dose Radiation in the Past

Spontaneous level of various chromosomal damages was increased in all exposed groups (Fig. 1). The frequency of dicentrics—the specific radiation marker—was increased only in directly irradiated persons (LQ, EA, EC and VT).

The number of single fragments exceeds the control level not only in exposed persons but in the children of irradiated parents as well (LC, PC). Since this very type of ChA is typical for spontaneous chromosomal mutagenesis, one may suggest that it proceeds more intensively than in norm, both in directly irradiated persons and in their children due to genome instability induced by the radiation impact. The direct estimation of genome state was performed in groups LC, PC and EC using functional criteria. These groups were characterized by the increased chromosomal sensitivity of lymphocytes to challenge irradiation in vitro at the dose 1.5 Gy as compared to CC group (Fig. 2). The elevated frequency comparing to norm of HPRT-mutations (15.9 vs. 12.0 per 10^6 cells), but not of GPA ones was observed in group of LQ (Jones et al. 2002).

One more approach to the estimation of the genome functional state was the comparison of the age dynamics of stable ChA in exposed and control groups (Vorobtsova et al. 2001b). The individual yields of translocations fitted well the quadratic age-response function. The resulting equations were: $Y = (0.21 \pm 0.08) + (0.32 \pm 0.05)X^2$, for the control donors, and $Y = (0.42 \pm 0.11) + (0.51 \pm 0.06)X^2$, for the exposed people, where Y is the whole genome translocation number (per 100 cells); X is the age (years). The comparison of β-coefficients from these equations showed that exposed persons aged faster as estimated by the frequency of translocations than in norm i.e. their "biological age" appeared to be greater than the calendar one.

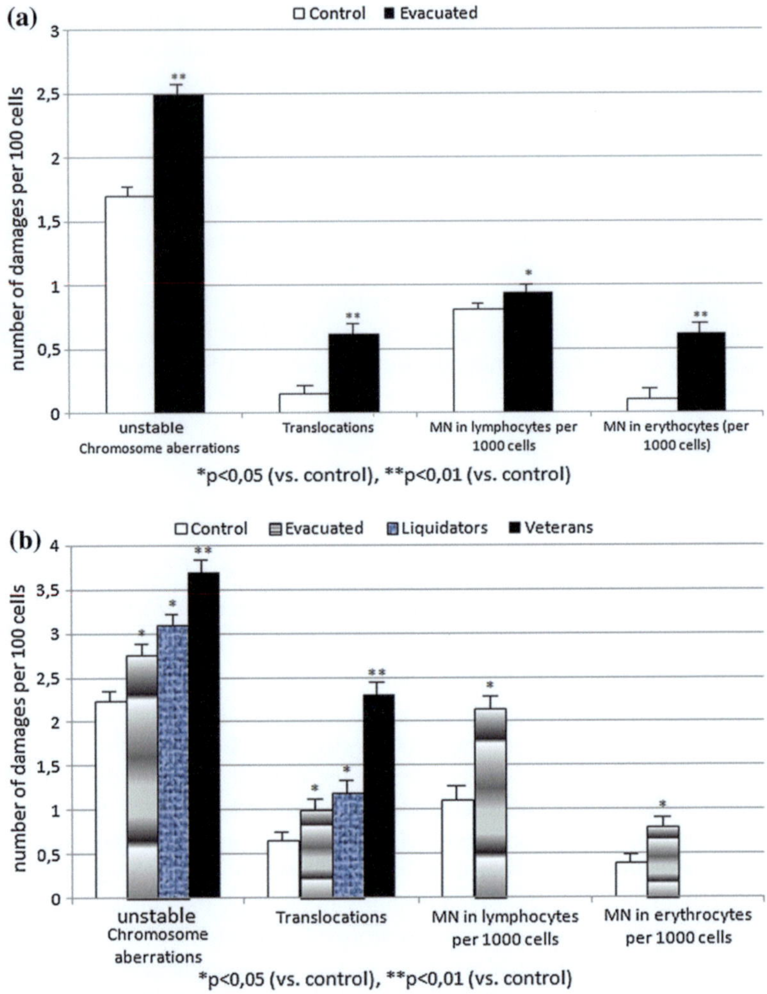

Fig. 1 Different chromosomal damages in exposed people of various groups. **a** Children. **b** Adults

Functional inferiority of damaged genome was shown also by the comparison of the morbidity parameters among persons with low and high levels of mutation events in cells. The estimation method was based on the division of the examined group to quartiles (in accordance with the individual ChA level). In the quartile with the lowest (L) and with the highest (H) level of ChA the frequencies of the most voluble systemic pathology: cardiovascular and gastrointestinal ones were registered. The frequencies of these diseases in L and in H group were 23 and 45 %, correspondingly (Vorobtsova and Semenov 2006). The positive correlation between ChA frequency and the risk rate of various diseases was also observed by others (Shevchenko and Snigireva 2006).

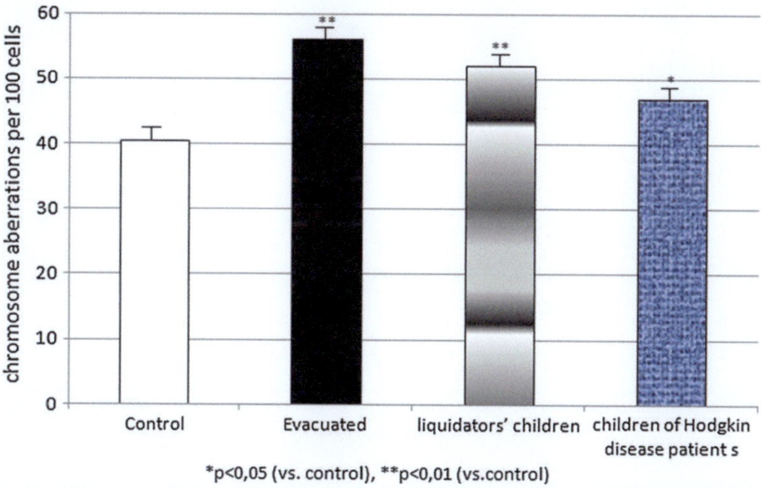

Fig. 2 Chromosomal aberrations in lymphocytes irradiated in vitro (1. 5 Gy) in children of various groups

Comparison of Dose-Responses for ChA in Lymphocytes Irradiated in Vivo and in Vitro

The in vivo and in vitro dose-responses for dicentrics fitted well the linear model in the dose-range 0.1–0.6 Gy. The corresponding equations were calculated as $Y = (0.35 \pm 0.09) + (3.13 \pm 0.22) \times D$ (in vivo) and $Y = (0.31 \pm 0.09) + (6.85 \pm 0.41) \times D$ (in vitro), correspondingly, were Y is the number of dicentrics (per 100 cells), D is the value of dose (Gy). The efficiency of in vivo irradiation of lymphocytes per unit of dose was lower than that of vitro exposure (Fig. 3). It means that reconstruction of individual dose performed by means of in vitro dose-response calibration curve for dicentrics underestimates the real absorbed dose. It was possible to confirm this proposal for one cancer patient who received in total 1.15 Gy of fractionated whole body irradiation. The dose estimated on the base of his number of dicentrics by means of in vivo equation was 1.2 Gy and by means of in vitro one −0.57 Gy. That is a more correct value of absorbed dose was obtained using in vivo dose-response curve. For translocations the in vivo and in vitro dose-responses were practically equal.

Radiation Induced Bystander Effect

The bystander effect was estimated by the ability of non-irradiated female/male lymphocytes to develop AR in mixed culture with pre-irradiated at the low dose lymphocytes of opposite gender donor, using two time schedules: with adaptive

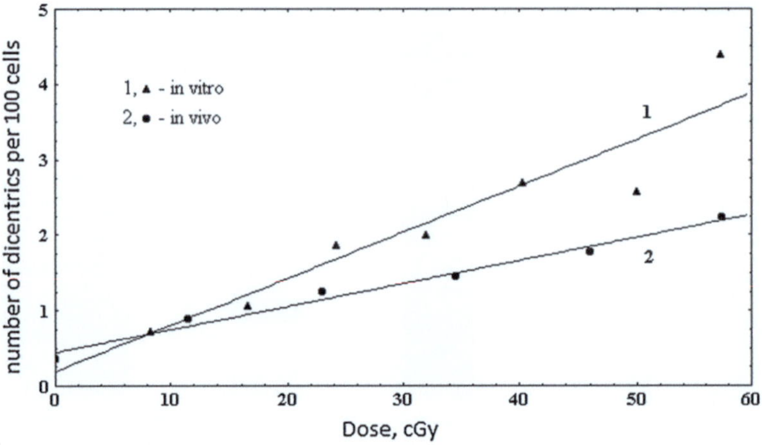

Fig. 3 The dose-response curves for dicentrics in lymphocytes irradiated in vitro and in vivo (Vorobtsova et al. 2001b)

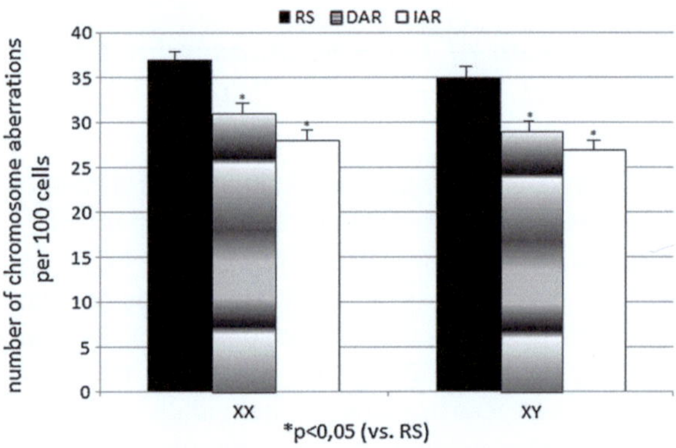

Fig. 4 Direct (DAR), indirect (IAR) adaptive response and own radiosensitivity (RS) of male (XY)/female (XX) lymphocytes from mix culture

irradiation of G_0 male/female lymphocytes and challenge exposure—of G_1 lymphocytes in mix culture; with adaptive irradiation of male/female early G_1 lymphocytes and challenge exposure of the late G_1 lymphocytes in mix culture (Vorobtsova and Kolesnikova 2007; Kolesnikova and Vorobtsova 2011).

In Fig. 4 an example of RIBE observed with the 2^d-time schedule is presented. Radiosensitivity (RS) of pre-irradiated male/female lymphocytes themselves estimated by the level of induced ChA was decreased as compared to their own RS i.e. direct adaptive response (DAR) was observed in these cells. At the same time lymphocytes neighboring in mix culture the pre-irradiated cells displayed the reduced RS as well, i.e. developed the indirect adaptive response (IAR).

When the spontaneous level of ChA was used as an endpoint of RIBE the number of ChA in nonirradiated male/female lymphocytes cultivated together with irradiated (1 Gy) ones was increased as compared to that in non-irradiated monoculture (4.66 ± 0.66 vs. 3.07 ± 0.53 correspondingly). On the other side the decreased number of ChA was observed in irradiated cells when they were cultivated together with non-exposed ones (Kolesnikova 2012).

Conclusion

The data presented here on the level of different mutational events in various exposed groups are rather a digest from those published earlier (Vorobtsova and Semenov 2006). Together with the results obtained by others they demonstrate that long after initial low dose irradiation human genome continues to be spoiled.

The damages observed in irradiated genome are accompanied by its instability. It is worth noting that we were the first to propose and demonstrate in the 1970s on animals and then on humans that radiation induced genome instability could be transmitted via germ line and is accompanied by an increased cancer risk in the progeny of irradiated parents (Vorobtsova 1989, 2006). Both these effects of parental irradiation manifest themselves most clearly when provoke (promote) factors are applied to prezigotically irradiated genome. Instability of irradiated genome confirmed later in many studies (Dubrova 2003; Pilinskaya et al. 2005) became nowadays universally adapted. The genome instability in people who underwent the direct low dose irradiation is likely, possibly, to explain the acceleration of ageing processes in them as it was shown by an increased rapidity of stable ChA accumulation with age. It means that age somatic pathology in exposed persons will appear earlier as compared to norm due to their higher biological age.

Some evidence of the functional role of genetic damages in cells gives the fact that people with the high level of ChA have an increased risk of cancer (Hagmar et al. 1994) and noncancerous diseases (Vorobtsova and Semenov 2006; Shevchenko and Snigireva 2006). Thus, the frequency of ChA could be regarded as a surrogate marker of risk for negative health effects induced by low dose irradiation.

For biological dosimetry based on the frequency of ChA, their in vitro dose response calibration curves are usually applied. As it follows from our data, in vitro dose response curve for dicentrics underestimates the absorbed dose in the case of protracted exposure.

The RIBE is regarded now as a possible reason for radiation induced non-stochastic adverse health effects and as a mechanism of second tumors origin after primary antitumor radiotherapy. The mix culture of male/female lymphocytes presents the simple, convenient and cheap model for observation of RIBE, using various endpoints. Thus cytogenetic studies are very important in different aspects of radiobiology and medicine.

Acknowledgments Partly the investigations were carried out in the framework of the international collaborative projects, as well as National programs and grants of RFBR. The authors are grateful to the scientists who participated in these studies: M. Vorob'yeva, A. Bogomazova, N. Timofeeva, A. Kanaeva and I. Kolesnikova.

References

Dubrova YE (2003) Radiation-induced transgenerational instability. Oncogene 22:7087–7093

Hagmar L, Brogger A, Hansteen I-L et al (1994) Cancer risk in humans predicted by increased levels of chromosomal aberrations in lymphocytes: nordic study group on the health risk of chromosome damage. Cancer Res 54:2919–2922

Jones IM, Galick H, Kato P et al (2002) Three somatic genetic biomarkers and covariates in radiation-exposed Russian cleanup workers on the chernobyl nuclear reactor 6–13 Years after exposure. Radiat Res 158:424–442

Kolesnikova IS (2012) The effect of mixed cultivation of lymphocytes irradiated at a dose of 1 Gy and non-irradiated lymphocytes on the chromosomal aberration frequency. Radiat Biol Radioecol 52(3):247–251 (Russian)

Kolesnikova IS, Vorobtsova IE (2011) Radiation-induced "bystander effect" revealed by means of adaptive response upon mixed lymphocyte culturing from humans of different gender. Radiat Biol Radioecol 51(5):542–548 (Russian)

Pilinskaya MA, Dibskiy SS, Dibskaya OV et al (2005) Elucidate chromosomal instability in offspring of fathers irradiated by the Chernobyl disaster. Tsitol Genet 39:32–40

Shevchenko VA, Snigireva GP (2006) The significance of cytogenetic examination for assessment of the Chernobyl catastrophe consequences. In: Burlakova EB (ed) 20 years after the Chernobyl accident. Nova Science Publishers Inc NY, pp 1–10

Vorobtsova IE (1989) Increased cancer risk as a genetic effect of ionizing radiation. In: Perinatal and multigeneration carcinogenesis, vol 96. IARC publication, Lyon, pp 389–401

Vorobtsova IE (2006) Transgenerational transmission of radiation induced chromosome instability. Radiat Biol Radioecol 46(4):441–446 (Russian)

Vorobtsova IE, Kolesnikova IS (2007) Investigation of radiation-induced "bystander effect" using model of adaptive response in mixed lymphocyte culture from humans of different gender. Radiat Biol Radioecol 47(6):645–649 (Russian)

Vorobtsova IE, Semenov AV (2006) Complex cytogenetic characteristics of people suffered from the chernobyl accident. In: Burlakova EB (ed) 20 years after the Chernobyl accident. Nova Science Publishers Inc NY, pp 55–74

Vorobtsova I, Darroudi F, Semyonov A et al (2001a) Analysis of chromosome aberrations by FISH and Giemsa assays in lymphocytes of cancer patients undergoing whole-body irradiation: comparison of in vivo and in vitro irradiation. Int J Radiat Biol 77(11):1123–1131

Vorobtsova I, Semenov A, Timofeeva N et al (2001b) An investigation of the age-dependency of chromosome abnormalities in human populations exposed to low-dose ionizing radiation. Mech Ageing Develop 122:1373–1382

Regularities and Mechanisms of Radiation Effects on Cancer Stem Cells In Vitro and In Vivo

Irina Zamulaeva, Olga Matchuk, Elena Selivanova, Sergey Makarenko, Vyacheslav Andreev and Andrey Kaprin

Introduction

The idea of a hierarchical structure of malignant tumors and the existence of special cells that are progenitors of heterogeneous clones, making up the tumor, is actively developing in recent years. These cells differ from the rest of the tumor cells by tumorigenic activity (ability to restore the growth of the original tumor when transplanted into immunodeficient mice) and similar to the stem cells of normal tissue in a number of molecular biological, structural and functional features (gene expression profile, a set of protein markers on the cell surface, intense removal of drugs and lipophilic dyes, etc.). These cells are called cancer stem cells (CSCs); in various literary sources they are also called stem-like or tumor-initiating cells. Currently, the existence of CSCs is demonstrated in malignancies of different localization and stable tumor cell lines of human and animals, despite the known difficulties of identification of these cells.

Study of radiobiological properties of CSCs covers a fairly wide range of solid tumors and cell lines of different origin, including glioblastoma, melanoma, cancers of the upper respiratory tract (URT) and gastrointestinal tract, cancer of the female reproductive system (ovary, breast prostate, uterus and uterine cervix), prostate and others (Diehn and Clarke 2006; Koch et al. 2010; Nguyen et al. 2011; Dubrovska 2014; Rycaj and Tang 2014).

The vast majority of publications indicates a much higher resistance of CSCs to low-LET radiation compared to the rest mass of cells. It is therefore believed that CSCs determine adverse results of radiation or combined treatment of cancer patients, including the occurrence of relapses and metastasis. At present, this

I. Zamulaeva (✉) · O. Matchuk · E. Selivanova · S. Makarenko · V. Andreev · A. Kaprin
A. Tsyb Medical Radiological Research Centre—Branch of the National Medical Research Radiological Centre of the Ministry of Health of RF, Korolev str. 4, Obninsk 249036, Russia
e-mail: zamulaeva@mail.ru

V.L. Korogodina et al. (eds.), *Genetics, Evolution and Radiation*,
DOI 10.1007/978-3-319-48838-7_34

assumption based on extensive experimental data is subjected to intensive research in the clinical setting.

In general, the accumulated amount of data on CSC properties justifies the need for further study of regularities and mechanisms of response of these cells to various anticancer exposures in order to improve treatment of cancer patients. In particular, the elucidation of the effects of high-LET radiation exposure to CSCs in various model systems in vitro and in vivo, as well as in the treatment of cancer patients is of great interest. In the literature, only a few publications are presented on this subject (Zhang et al. 2013; Cui et al. 2011).

The effect of photodynamic treatment on CSCs is almost unstudied. In addition, analysis of the literature has shown that the mechanisms of chemo- and, especially, radioresistance of CSCs have not been studied enough, in spite of the high importance of this problem, the solution of which can become a basis to develop new effective antitumor agents or methods and improve existing therapies.

The research carried out at A. Tsyb Medical Radiological Research Centre over the last few years have been directed to figure out these issues in model systems in vitro and in the course of cancer patient treatment (Matchuk et al. 2012, 2013; Zamulaeva et al. 2014). The objects for the study were stable cultures of melanoma line B16, breast cancer line MCF-7, and cervical cancer line HeLa, as well as biopsies of patients with URT cancer. We used two main methods of CSC identification known in the literature: immunophenotyping of surface markers and evaluation of fluorescent dye Hoechst 33342 exclusion from the cells. As it was shown in many studies, CSCs (unlike the other sells) have the ability to efflux specified dye due to high expression of various ATP-binding transporters on the plasma membrane. Due to this ability CSCs can be identified by flow cytometry as «side population» (SP) in some stable tumor cell lines, including the ones used in this study. The term non SP (NSP) applied in this study is often used to indicate the remaining cells. SP method was used in our study of cell cultures, and the method of immunophenotyping—in the analysis of biopsy material of malignant tumors.

Effect of Ionizing Radiation on SP Cells In Vitro

According to our data, the proportion of SP cells was on the average (±SE) 0.90 ± 0.04, 1.81 ± 0.04 and 5.40 ± 0.60 %—in intact cell culture lines B16, MCF-7 and HeLa, respectively. SP of all the three lines had a higher resistance to ^{60}Co γ—radiation exposure than NSP as it was shown by multiple criteria including clonogenic survival, the absolute and relative number of viable cells. For example, in B16 cell line, the average value of D0 for SP and NSP was 2.3 ± 0.3 and 1.4 ± 0.2 Gy, respectively ($p < 0.05$).

Survival of SP and NSP cells was not different after exposure of B16 and MCF-7 cell cultures to neutron radiation (14.8 MeV). For cell line B16 the relative biological effectiveness of neutron radiation in comparison with γ-radiation was 2.6 for SP and 2.1 for NSP at the level of D10.

The features of SP cells were revealed in relation with their higher resistance to low-LET radiation compared to NSP. This research was made taking into account the common factors of radioresistance of cells, whose role was demonstrated for many biological objects, but for SP (or CSCs detected by other techniques) was studied insufficiently or practically unexplored. In particular, a comparative study of SP and NSP cells included such indicators as proliferative activity, distribution over different cell cycle phases, the number of spontaneous and radiation-induced double-strand DNA breaks, as well as constitutive and radiation-induced levels of various heat shock proteins (HSP27 and HSP70). The results showed significant differences in the distribution of SP and NSP cells over the phases of the cell cycle, as well as quiescence of SP cells melanoma line B16 (compared to NSP cells). For example, the total proportion of cells in S, G2, M-phases was 4.2-fold less in SP, than NSP: 14.5 ± 3.7 % vs. 61.6 ± 5.5 %, respectively ($p < 0.01$). Similar results were obtained for SP and NSP cells of breast cancer line MCF-7. The proportion of cells in S + G2 + M phases was significantly lower in the SP, than in NSP: 16.4 ± 0.6 % vs. 48.3 ± 1.1 % ($p < 0.01$). Thus, SP cells of the studied cultures were characterized by the state of relative proliferative quiescence.

To compare the number of DNA double-strand damage in the studied populations, we applied the method of detecting γH2AX foci, formed in the places of these damages. The object of the study was pre-sorted SP and NSP cells of B16 line before exposure and 1 h after the γ-irradiation at a dose of 3 Gy. Radiation exposure significantly increased the number of γH2AX-foci in both groups, although to different degrees. In NSP amount of γH2AX- foci increased by 10 times compared to the control and after the irradiation averaged to 40.3 ± 3.4 per cell. However, the amount of γH2AX foci in SP after irradiation was only 3 times higher than the control value and was 24.4 ± 3.7 on the average per cell. By this criterion the difference between the compared populations was highly statistically significant ($p < 0.01$).

The expression of HSP27 and HSP70 in the pre-sorted SP and NSP cells of MCF-7 line was studied in the intact state and after exposure to γ-radiation at the dose of 5 Gy by means of the immunocytochemical method using the laser scanning microscopy.

The constitutive expression level of HSP27 was not different in SP and NSP cells ($p > 0.05$). However, after exposure the HSP27 content increased significantly higher in SP than NSP cells. Thus, the radiation-induced increase of this indicator in the SP and NSP cells was 63.8 ± 6.2 and 25.0 ± 5.5 rel. units, respectively ($p < 0.01$).

The constitutive level of HSP70 expression was higher in NSP than SP cells: 164.4 ± 14.4 vs. 100.0 ± 9.7 rel. units, respectively ($p < 0.01$). However, SP cells showed the increase in HSP70 expression after irradiation, whereas in NSP cells the content of the protein did not change. As a result the post irradiation expression of HSP70 was higher in the SP cells in comparison with that in the NSP ($p < 0.01$).

Importantly, the content of HSP27 increased to a greater extent in CSCs than in the rest of the tumor cells after γ-irradiation. The expression of HSP70 increased after irradiation only in CSCs. These findings are the first evidence of HSP participation in forming the increased resistance of CSCs to the low-LET radiation.

Overall, the results of our study have shown that a higher resistance of SP (compared to NSP) to low-LET radiation can be explained by a number of reasons, including low proliferative activity, low amount of radiation-induced DNA double-strand breaks, high levels of radiation-induced expression of HSP70 and especially HSP27.

Effect of Photodynamic Treatment on SP Cells In Vitro

The aim of this part of the study was the elucidation of the regularities of photodynamic treatment on CSCs in vitro (for melanoma line B16 and breast cancer line MCF-7). The time and concentration dependences of the intracellular accumulation of photosensitizer of chlorine series (Photolon) in SP and NSP cells were compared. SP cells accumulated photosensitizer by 1.5–1.8 times less than NSP in the both cell lines. But the level of destruction after the photodynamic treatment did not significantly differ in the populations of cancer stem and non-stem cells despite the differences in the content of the photosensitizer and known differences in the regulation of death of these cells under the influence of other damaging effects. Thus, the results for the first time have substantiated the high efficiency of photodynamic treatment in terms of the CSC elimination as a fraction resistant to conventional anti-tumor treatment (low-LET radiation and many chemotherapy drugs).

Quantitative Changes of CSCs In Vivo (in the Course of Combined Treatment of Patients with URT Cancer)

The radiobiological properties of CSCs detected in vitro have allowed us to assume that the number and/or the reaction of cells to radiotherapy may have a predictive value. Preliminary results of the verification of these assumptions are presented below. The study group consisted of 74 patients with primary URT cancer aged from 31 to 85 years (mean 57 ± 9 years). Laryngeal cancer was diagnosed in 45 patients (60 %), cancer of the hypopharynx—in 16 (21 %), cancer of the oropharynx—in 12 (16 %), cancer of the tonsils and throat—in 2 (3 %). Most often T3 stage was observed (34 patients). T1 stage was detected in 3 individuals, T2—in 24, T4—in 13 patients. Regional lymph nodes were involved in the neoplastic process in 32 persons (43 %).

All studied patients had biopsies before the treatment, and 27 of them—after γ-irradiation at the total focal dose (TFD) of 10 Gy with concurrent chemotherapy. The biopsies of patients were tested for the proportion (percentage) of $CD44^{+}CD24^{-}CD45^{-}$ and $CD44^{+}CD24^{low}CD45^{-}$ CSCs by flow cytometry.

The average value (±SE) proportion of $CD44^{+}CD24^{-}CD45^{-}$ cells before the treatment amounted to 3.4 ± 0.5 %, the proportion of $CD44^{+}CD24^{low}CD45^{-}$ cells—1.7 ± 0.2 %. Statistically significant differences by the proportion of these cells in the groups of patients with different stages of disease and histological types of tumor were not defined. The degree of tumor regression was known for all the patients after the first stage of the split-course radiotherapy (TFD 30–40 Gy) concurrent with chemotherapy. It was found that the proportion of $CD44^{+}CD24^{-/low}CD45^{-}$ cells in the biopsies before the treatment was not significantly different in patients with varying degrees of tumor regression (more or less than 50 %). Data on the presence/ absence of relapses and metastasis within 1 year after the treatment were available for 61 patients, and within 2–3 years—for 48 patients. Proportion of $CD44^{+}CD24^{-/low}CD45^{-}$ CSCs before the treatment did not differ in groups of patients with varying efficiency of treatment, evaluated by the disease-free survival for the stated periods of observation.

The data on changes in the proportion of CSCs in URT tumors after fractionated irradiation at TFD of 10 Gy are of a special interest. The results have indicated considerable individual variability of the observed changes in the CSC population. In most cases the increase in the proportion of CSCs took place after irradiation in vivo, as well as under experimental conditions of single irradiation in vitro. Thus, increasing the proportion of $CD44^{+}CD24^{-/low}CD45^{-}$ cells was observed in 16 patients, the reduction of this indicator—in 9 patients. In the remaining 2 patients, this figure almost did not change. The character of the observed changes in the CSC proportion did not correlate with the clinical and morphological parameters of the tumor process. It is important to note that the response to irradiation of CSC population differed in radiosensitive and radioresistant tumors, divided by criterion of the 50 % regression after the first stage of radiotherapy. Thus, in patients with radioresistant tumors (regression was less than 50 %) the statistically significant increase in the average proportion of CSCs was shown after irradiation at the dose of 10 Gy, compared with that before the treatment: 9.3 ± 4.0 % vs. 4.6 ± 2.0 %, respectively ($p < 0.05$).

In patients with radiosensitive tumors (regression was more than 50 %), this figure remained unchanged after radiation exposure, which appears to indicate the same sensitivity of CSCs and other tumor cells to irradiation in these cases. Thus, the obtained results have shown the relationship of quantitative changes of CSCs with the immediate short-term outcomes of URT cancer treatment. The results of the clinical and experimental parts of the work have confirmed the assumption of the predictive value of CSC quantitative changes after the first fractions of irradiation, but not the proportion of CSCs before the treatment. Taking into account the small number of patients studied in the dynamics, the results require increased sampling and further studies of groups of patients, homogeneous by primary tumor localization, disease stage, morphological characteristics of the tumor and treatment regimens. In addition, it is necessary to continue the clinical observation for the subsequent analysis of disease-free survival in patients with different CSC responses to the first fractions of irradiation, and to evaluate the predictive value of quantitative CSC changes for long-term outcomes.

Conclusion

In general, the literature analysis and the results of our own investigations of CSC radiobiological properties have shown prospects for further work in this direction to develop new antitumor methods of treatment (which are effective for CSC elimination), improving radiotherapy of cancer patients and elaborating methods of personalized medicine.

References

Cui X, Oonishi K, Tsujii H, Yasuda T, Matsumoto Y, Furusawa Y, Akashi M, Kamada T, Okayasu R (2011) Effects of carbon ion beam on putative colon cancer stem cells and its comparison with X-rays. Cancer Res 71:3676–3687

Diehn M, Clarke M (2006) Cancer stem cells and radiotherapy: new insights into tumor radioresistance. J Natl Cancer Inst 98:1755–1757

Dubrovska A (2014) Report on the international workshop 'Cancer stem cells: the mechanisms of radioresistance and biomarker discovery'. Int J Radiat Biol 90:607–614

Koch U, Krause M, Baumann M (2010) Cancer stem cells at the crossroads of current cancer therapy failures—radiation oncology perspective. Semin Cancer Biol 20:116–124

Matchuk ON, Zamulaeva IA, Selivanova EI, Lipunov NM, Pronjushkina KA, Uljanenko SE, Lychagin AA, Smirnova SG, Orlova NV, Saenko AS (2012) Sensitivity of melanoma B16 side population to low- and high-LET radiation. Radiats Biol Radioecol 52:261–267(Russian)

Matchuk ON, Zamulaeva IA, Kovalev OA, Saenko AS (2013) Radioresistance mechanisms of side population cells in mouse melanoma cell line B16. Cell Tissue Biol 7:556–562

Nguyen GH, Murph MM, Chang JY (2011) Cancer stem cell radioresistance and enrichment: where frontline radiation therapy may fail in lung and esophageal cancers. Cancers (Basel) 3:1232–1252

Rycaj K, Tang DG (2014) Cancer stem cells and radioresistance. Int J Radiat Biol 90:615–621

Zamulaeva IA, Matchuk ON, Selivanova EI, Andreev VG, Lipunov NM, Makarenko SA, Zhavoronkov LP, Saenko AS (2014) Increase in the number of cancer stem cells after exposure to low-LET radiation. Radiats Biol Radioecol 54:256–264 (Russian)

Zhang X, Lin SH, Fang B, Gillin M, Mohan R, Chang JY (2013) Therapy-resistant cancer stem cells have differing sensitivity to photon versus proton beam radiation. J Thorac Oncol 8:1484–1491

Part VI
Laws of Evolution

Evolution of the Genomic Universe

Eugene V. Koonin

Introduction

The determination of the DNA structure in 1953 started the great scientific revolution that in a series of momentous discoveries ushered in the modern era in biology (Watson and Crick 1953a, b). The Watson-Crick model revealed how the genetic material was organized and how it could be replicated and inherited. Subsequent studies on the molecular biology of replication have fully validated and elaborated this mechanism, vindicating the early, prescient ideas on the template nature of the gene that were first systematically presented by Timofeeff-Ressovsky et al. (1935). In parallel, the elucidation of the genetic code revealed the key features of the route from the genetic information to the phenotype (Crick 1958, 1968, 1970). The discovery of the fundamental fact that genetic information is encoded in the sequence of bases in the genomic DNA immediately implies that this sequence evolves, can be subject to selection and thus potentially could bear witness to the course of evolution (Crick 1958). However, from this fundamental insight alone, it is not obvious that sequence comparison can yield a meaningful reconstruction of the evolution of life. The feasibility of such reconstruction depends on the character and rate of genome evolution. The first indications that molecular evolution is indeed tractable were obtained by Zuckerkandl and Pauling in 1962, the year when the discovery of the DNA structure was awarded by the Nobel Prize and the experiments on the genetic code were proceeding at full steam (Zuckerkandl and Pauling 1965). However, it took another 30 years and the invention of the modern sequencing methods to show that sequence comparison was powerful enough to successfully explore and reconstruct the evolution of complete genomes (Koonin 2011b).

E.V. Koonin (✉)
National Center for Biotechnology Information, National Library of Medicine,
National Institutes of Health, Bethesda, MD, USA
e-mail: koonin@ncbi.nlm.nih.gov

V.L. Korogodina et al. (eds.), *Genetics, Evolution and Radiation*,
DOI 10.1007/978-3-319-48838-7_35

At the time of this writing (the New Year day, 2016), exactly two decades since the release of the first complete genome of a cellular organism (the bacterium *Haemophilus influenzae*) (Fleischmann et al. 1995), thousands of genomes from all walks of life are available, providing ample material for comparative analysis (Clark et al. 2016). This genomic collection is by no account complete and might not even be representative of the true diversity of life given that over 99 % of bacteria and archaea so far have not been grown in culture (Kaeberlein et al. 2002; Vartoukian et al. 2010). The genomes from this dominant "dark matter" of life are only starting to trickle in thanks to the latest advances of metagenomics and single cell genomics (Handelsman 2004; Kodzius and Gojobori 2016). Nevertheless, even the genome collection now available is extremely diverse with respect to the taxonomy, genome size and organismal biology, which may justify some generalizations on the principles of genome evolution (Koonin 2011a, b). In this chapter, I address such emerging general characteristics of the structure and dynamics of the genomic universe.

Orthologous Gene Sets as Atomic Units of Evolution

Gene Orthology, the Cornerstone of Comparative Genomics

Whether or not the genomic universe is amenable to informative dissection critically depends on the level of evolutionary conservation of gene sequences and existence of distinct gene families. The global evolutionary resilience of genes, manifested primarily in the conservation of protein and RNA sequences, became apparent in the very first comparisons of sequenced prokaryotic and eukaryotic genomes, the bacteria *Haemophilus influenzae* and *Mycoplasma genitalium*, the archaeon *Methanocaldococcus jannaschii*, and the eukaryote, yeast *Saccharomyces cerevisiae* (Tatusov et al. 1997). A central generalization of comparative genomics is that genes are not simply conserved through varying evolutionary spans but constitute distinct units of evolution, namely orthologous gene lineages (Koonin and Wolf 2008). Orthologs are the genes in different genomes that descend from a single ancestral gene in the last common ancestor of the respective organisms (Fitch 1970; Koonin 2005) (whenever we mention "the same" gene in different organisms, we actually mean orthologs). In the current collection of sequenced genomes (Galperin et al. 2015), orthologs in distant taxa are found for the substantial majority of protein-coding genes in each genome (Fig. 1). Thus, a major part of the genomic universe can be organized into clusters of orthologous genes (COGs) albeit at different phylogenetic depths. When the genes in a genome are classified by their apparent "age", i.e. the phylogenetic depth at which orthologs are detectable, the resulting breakdown turns out to be similar for distant organisms, with a clear excess of genes that are at least hundreds of million years old (Wolf et al. 2009). As discussed below, the organization of orthologous gene clusters is far from

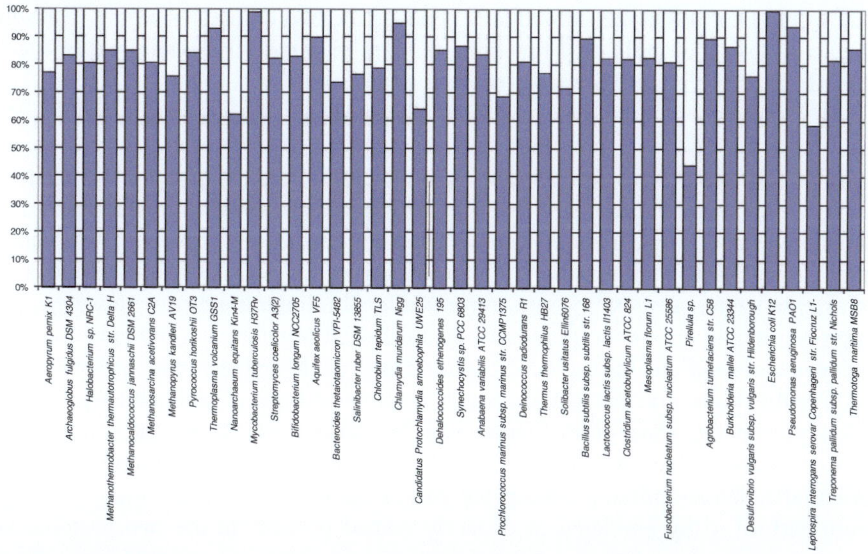

Fig. 1 Evolutionary conservation of genes: coverage of archaeal and bacterial genomes by clusters of orthologs (COGs). Modified with permission from (Koonin and Wolf 2008)

simple due to frequent gene loss and lineage-specific duplication which leads in the formation of families of paralogs. Nevertheless, the long-term conservation of the identity of most genes in the form of the stable orthologous clusters provides for the possibility of a highly informative quantitative analysis of the genomic universe.

Minimal Gene Sets, Non-orthologous Gene Displacement and the Elusive Essential Core of Life

Sequencing of the genomes of symbiotic and parasitic bacteria prompted the attractive idea that their gene repertoires could approximate the "minimal gene set", that is, the set of genes that is both necessary and sufficient to sustain a simple (i.e., prokaryotic) cell under the most favorable conditions that can be created outside other cells (Fraser et al. 1995; Mushegian and Koonin 1996). The latter qualification is critical because "minimal gene sets" are inevitably contingent on environmental conditions in which the respective organism exists (or would exist, in the case of computationally derived "conceptual" genomes). As soon as the first two bacterial genomes became available, the second one being the genome of *Mycoplasma genitalium* a wall-less parasitic bacterium with only ~570 genes, the obvious idea presented itself, that comparison of these two differentially specialized genomes of bacterial pathogens belonging to different bacterial phyla would naturally yield the "true" minimal set (Mushegian and Koonin 1996). More precisely, one would expect that the orthologous genes in the two organisms would represent the set of essential biological functions that are required for the survival of a cell regardless of the unique life style of each organism.

The comparison of the gene sets of *H. influenzae* and *M. genitalium* yielded 240 pairs of orthologous genes which encompassed most of the apparently essential cellular functions; however, several such functions were conspicuously missing from the conserved gene set. Defining the minimal set of essential biological functions is not trivial. It is tempting to go about this task by "reverse evolutionary engineering", that is, purely by comparative genomics, and define the minimal set of essential genes as those that are conserved in all cellular life forms. However, this approach would ignore the possibility that different organisms could have evolved independent solutions for the same essential task. So to delineate the minimal set of cellular functions, one has to additionally employ the logic of biochemistry and cell biology, and the state of the knowledge in these fields is indeed sufficient to produce a reasonable catalogue of essential activities. Under the assumption that the essential functions missing among the 240 *H. influenzae-M. genitalium* orthologs were performed by unrelated or distantly related proteins in the two bacteria, an attempt was made to infer candidates from the remaining genes of *M. genitalium*. This straightforward effort on the derivation of a minimal gene set by a combination of comparative genomics and biological reasoning seems to have been reasonably successful and might approximate the functional repertoire of the simplest bacterial cell capable of independent growth under the best possible conditions. Subsequent gene knockout experiments have confirmed that most of the genes included in the minimal set are essential for bacterial survival, and the genes from the minimal set were found to be conserved in many (although not necessarily all) newly sequenced bacterial genomes (Delaye and Moya 2010; Koonin 2003).

The minimal set of bacterial genes is heavily dominated by genes encoding proteins involved in information transmission in the cell, that is, replication, transcription and above all translation. Metabolic enzymes and transport systems are much more sparsely represented as one might expect for an organism growing on the richest possible media. In that respect, the minimal gene set is dramatically different from the full set of COGs but resembles the set of essential bacterial genes, i.e., the genes the inactivation of which kills the bacterium (Koonin 2003). This striking preferential evolutionary conservation of the information transmission systems is one of the core generalizations of comparative genomics, and below we shall repeatedly return to this subject.

Perhaps, the most consequential outcome of the minimal gene set exercise was the finding that several essential functions were missing from the list of readily detectable orthologs. This observation was dramatically reinforced by the comparison of the bacterial genomes with the first archaeal genome (*Methanocaldococcus jannaschii*) which revealed a number of additional glaring gaps in the set of conserved essential functions (Bult et al. 1996; Koonin et al. 1997). These findings have been conceptualized in the notion of non-orthologous gene displacement (NOGD), an evolutionary scenario under which unrelated or distantly related genes (not orthologs) become responsible for the same essential function in different organisms (Koonin 2003). The actual evolutionary scenario for NOGD is easy to imagine: an evolving lineage acquires an alternative, functionally redundant gene for a particular essential role and so goes through an intermediate state in which both implementations of the function in question are present (such redundancy is often observed in organisms with more complex genomes), followed by the loss of the original version (Koonin and Mushegian 1996). With the growth of the genome collection, it is becoming increasingly common to find organisms—typically, those with larger genomes—in which both incarnations of various functions are represented, thus adding weight to the above scenario of NOGD evolution.

As the genome database grows, it is becoming clear that NOGD reaches across most of the functional systems and pathways such that there are very few functions that are truly "monomorphic", i.e. represented by genes from the same orthologous lineage in all organisms that are endowed with these functions. Accordingly, the universal core of life has shrunk almost to the point of vanishing. All that remains ubiquitous are some 30 genes for proteins involved in translation and 3 large RNA polymerase subunits, along with an approximately equal number of structural RNA genes (rRNAs and tRNAs). Even when parasitic bacteria are disregarded, the list of universal genes does not expand much (Koonin 2003). Thus, with the notable exception of a miniscule core of genes involved in the key steps of information transmission, there is no universal genetic core of life, owing to the (near) ubiquity of NOGD. The concept of a relatively small, universal set of functions that are required to sustain a cell remains viable but, considering the combinatorics of NOGD, there seems to be a vast variety of gene ensembles that can fill the same minimal set of functional niches.

The Units of Evolution and the Fundamental Fractal Structure of the Genomic Universe

The results of comparative genomics lead to a key generalization that allows one to conduct productive evolutionary studies: the fundamental units of evolution can be clearly defined as evolving orthologous gene (domain) lineages (Koonin and Wolf 2009). All the complexity of the evolution of many genes notwithstanding, the COGs are natural elements of the genomic universe, and quantitative analysis of the patterns among these elements potentially can tell us a lot about the structure of that universe. Orthology is most easily traced between genes of prokaryotes, and here I discuss the prokaryotic domain of the genomic universe. The trends among eukaryotes are similar in principle but are more complicated because of widespread multidomain organization of proteins and extensive paralogy. In the genomic spacetime, the COGs show a unique "skewed U-shaped" distribution across the genomes that can be well approximated with three exponential functions which

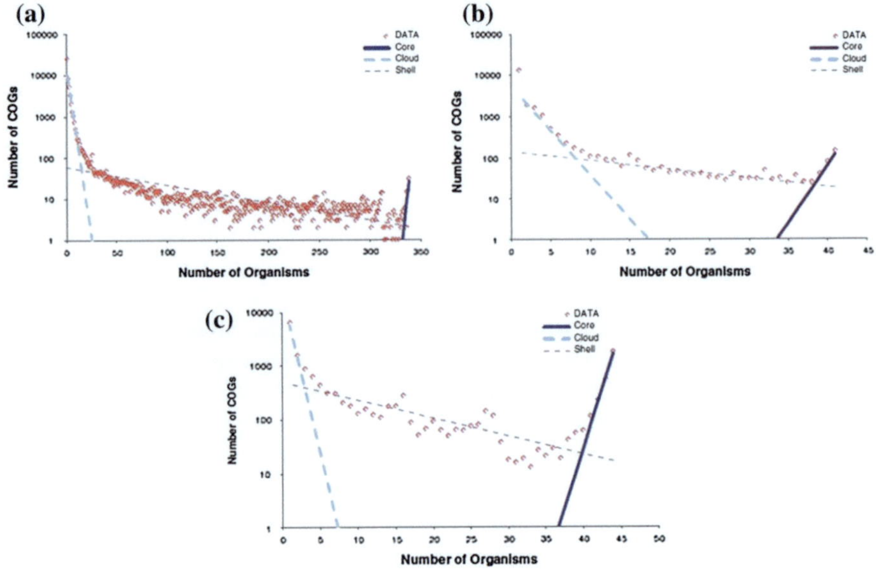

Fig. 2 The universal gene frequency distribution. **a** A set of 338 genomes representing all major phyla of archaea and bacteria. **b** A set of 41 genomes representing all major phyla of archaea. **c** A set of 44 genomes from the closely related bacterial genera *Escherichia, Shigella* and *Salmonella.* Each component of the distribution is approximated by an exponential function. The plot is rendered in semi-logarithmic coordinates

partition the gene population into 3 distinct classes (Koonin and Wolf 2008) (Fig. 2).

1. The (nearly) universal genes, those that are represented in all or most of the genomes of cellular life forms, make up but a tiny part of the genomic universe: altogether, this core of cellular life includes only about 100 genes (Fig. 2). In each particular genome, the fraction of these core genes is no greater than 10 %, even in the smallest of the genomes of cellular life forms (parasitic bacteria like *M. genitalium*) but typically, is closer to 1 % of the genes or less (Fig. 3).
2. The moderately conserved gene "shell", that is, COGs represented in a broad variety but not overwhelming majority of genomes; analysis of the available prokaryotic genomes puts the number of "shell" COGs at about 5000. The shell genes comprise the bulk of the gene complement in any genome (Figs. 2 and 3).
3. The poorly conserved, large "cloud" of COGs that are limited to narrow groups of organisms along with "ORFans" (Open Reading Frames without detectable homologs; this widely used acronym is a wordplay on ORFs and "orphANS"), i.e. genes so far identified in one genome only (but for which homologs are usually detected when additional related genome sequences become available). The size of the "cloud" is at least several times that of the "shell". The "cloud" genes usually account for 10–30 % of the genes in any particular genome (Fig. 3).

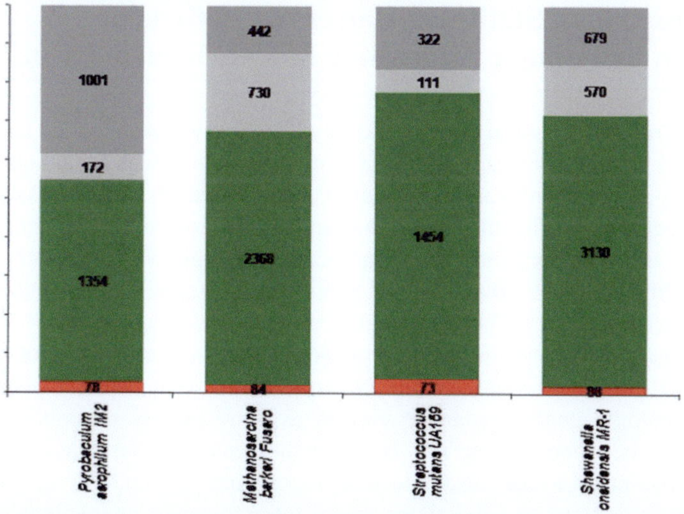

Fig. 3 Breakdown of the genes from individual genomes into the core, shell, cloud and ORFan components. The data are presented for two archaeal (*P. aerophilum* and *M. barkeri*) and two bacterial (*S. mutans* and *S. oneidensis*) genomes. The evolutionary classes of genes are shown from *bottom to the top* in the order of decreasing conservation (core genes at the *bottom*). Modified with permission from (Koonin and Wolf 2008)

Remarkably, this structure is self-similar, or fractal: the same three components, namely the tiny core, the larger shell and the comparatively huge cloud, appear at any level where the gene space-time is dissected, from the entire prokaryote world to narrow groups of bacteria (Fig. 2). An evolutionary model explaining the fundamental fractal pattern of gene frequency distributions and its biological implications are discussed below.

There is an appearance of a paradox in the distribution of the COGs in the gene space. Although in each individual genome, the majority of the genes belong to the shell that is shared with distantly related organisms, when the entire gene universe is considered, the core and shell genes (or more precisely, COGs) are but a small minority (compare Figs. 2 and 3). Obviously, this difference is due to the fact that the shell COGs are represented in many genomes whereas the cloud COGs and ORFans are rare or unique. Given this distinctive structure of the gene universe, evolutionary reconstructions inevitably yield a highly dynamic picture of genome evolution, with numerous genes lost and many others gained via horizontal gene transfer (HGT), at least, in prokaryotes (Doolittle 1999a, b, 2000; Hilario and Gogarten 1993).

Gene Ensembles, Statistical Universals of Genome Evolution and Simple Generative Models Behind Them

The advent of genomics and systems biology have forever changed the landscape of life sciences by producing a wealth of quantitative data to which mathematical and statistical approaches are readily applicable. Simple analysis of this massive data on gene and genome architecture, evolution and expression reveals several universal regularities that connect genomic and molecular phenomic variables and potentially might qualify as quantitative laws of genomics. If genes or their evolutionarily autonomous portions can be reasonably construed as "atomic units" (or "molecules") of genome evolution, then genomes are statistical ensembles of such units (Koonin 2011a, b). Extending this oversimplified but potentially productive physical analogy, genomes can be viewed as more closely resembling gases or perhaps liquids, where interactions between molecules are variable and important but weak compared to the intramolecular interactions that underlie the stability of molecules, in contrast to solid states where intermolecular interactions are strong and define the properties of this state of matter. It is textbook knowledge that the behavior of ensembles of weakly interacting particles (molecules) follows, to a good approximation, simple and universal statistical regularities, such as for instance the Boltzmann distribution of particle velocities (Frank 2009a; Frank and Smith 2011). The analogy between ensembles of genes that constitute genomes and ensembles of molecules that form gases and liquids prompts the search for universal statistical patterns in genome function and evolution. Moreover, this line of thinking makes one predict with some confidence that these statistical patterns should come in the form of mathematically simple, universal distributions of the values of certain variables that describe the process of evolution. As described below, such evolutionary universals indeed emerge from quantitative analysis of genomes and molecular phenomes.

Two groups of key questions immediately come up when it comes to such universals. First, is the analogy between genomes (and other "omes" such as transcriptomes and proteomes) and gases or liquids valid, or are these biological entities more like a solid state? In other words, can epistatic interactions between genes, proteins or other biological entities be disregarded in statistical descriptions of ensembles of such entities or are these interactions strong enough to make such descriptions pointless? Second, are these universal patterns trivial, biologically irrelevant consequences of aggregation of the outputs from a large number of weakly interacting entities ("particles")? Or else, are they manifestations of universal biological processes, i.e. laws of biology? Or is some kind of third way plausible, i.e. could these conserved patterns be stochastic, yet reflect pertinent biological reality? The typical approach used to address such questions involves attempts to develop the simplest possible model of the evolutionary process that would yield the observed universal behavior. The central problem with such models is, to what extent is a model constrained by the universal pattern it strives to explain, or put another way, can we discriminate, by applying statistics to the massive genomic and other "omic" data, between models that imply different biological mechanisms? Furthermore, given that the universals are undoubtedly shaped by evolutionary processes, a pressing question is, can these patterns be accounted for by neutral models or is selection an essential factor behind the universals?

The Universal Gene Frequency Distribution

It makes sense to begin our brief discussion of the universals of genomic evolution and function with the already familiar example of the U-shaped gene frequency distributions (Koonin and Wolf 2008) (Fig. 2). Undoubtedly, this broad distribution of gene frequencies is generated by processes of gene gain, primarily via HGT, and loss. The pertinent question is, can the observed distribution be explained by a neutral model with equal rates of gain and loss for all genes, or some distribution of these rates reflecting selection is required to account for the observations.

We addressed this question through explicit comparison of the fits to the empirically observed gene frequency distributions produced by simple evolutionary models with and without selection (Lobkovsky et al. 2013). Gene frequency distributions for almost 400 groups of 10 bacterial or archaeal species each, covering a broad range of evolutionary distances, were fit to steady-state, infinite allele models using a particular distribution of gene replacement rates and the phylogenetic tree of the species in each group. The fits of the theoretical frequency distributions to the empirical ones yielded optimal model parameters and estimates of the goodness of fit. Using rigorous statistical criteria, we found that the neutral model of genome evolution, with the same replacement rate for all genes, can be confidently rejected (Fig. 4).

We also compared three models that incorporate purifying selection in the form of a distribution of gene replacement that was derived using different approaches. Of these three models, the most complex and potentially most realistic one, in which the distribution of replacement rates was derived from a stochastic population model with additive per-gene fitness, yielded substantially better fits to the distributions obtained from all data sets than other models. The selection strength estimated from the fits declines with evolutionary divergence while staying well outside the neutral regime.

Thus, selection made a substantial contribution to the mechanisms that shaped the universal gene frequency distribution in prokaryotes. Certainly, it would be unreasonable to question the existence of selection affecting genes that are responsible for key biological functions. However, it was far from obvious whether the effect of selection and its strength could be detected and measured at the level of the overall frequency distribution, without turning to individual genes. The results of our modeling study show that although selection measurably affects only the

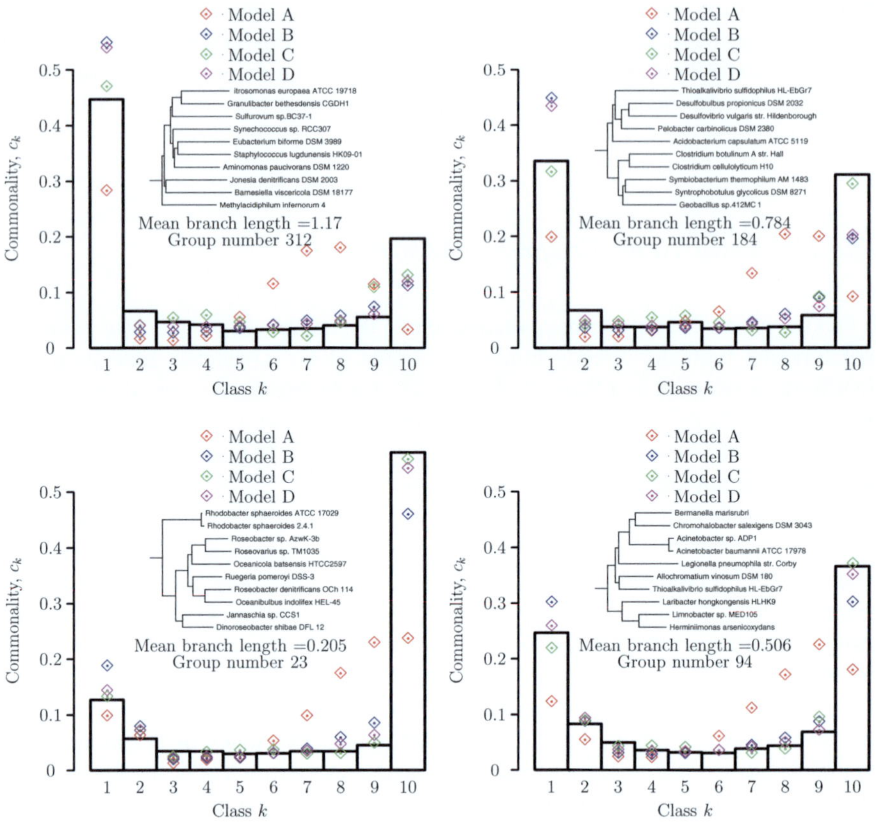

Fig. 4 The fit of different evolutionary models to the observed gene frequency distributions. The gene frequency distributions and model fits are shown for four groups of bacteria (10 species each). The evolutionary trees for the respective groups and the mean branch lengths are shown in the insets. Evolutionary models: **a** neutral model in which all genes have the same replacement rate (one parameter); **b** gamma-distributed replacement rate (two parameters); **c**, two class model in which a fraction **c** of the genome has one replacement rate and the remainder evolves with a different replacement rate (three parameters); **d**, replacement rate extracted from a stochastic population model (two parameters). Modified with permission from (Lobkovsky et al. 2013)

right end of the gene frequency distribution, which includes widespread genes, the effect is strong enough to confidently reject not only a neutral model but also over-simplified models with selection. The flip side of the coin, of course, is that the majority of the genes in the gnomic universe are subject only to weak selection and in many contexts can be viewed as circulating under a nearly neutral regime.

Distribution of Evolutionary Rates of Orthologous Genes and the Inverse Correlation Between Evolutionary Rate and Gene Expression

Protein-coding genes, at least, the non-synonymous positions that determine the encoded amino acid sequence, are among the most strongly constrained sequences in all genomes. However, already in the early days of molecular evolution studies, it has been realized that evolutionary rates of protein-coding genes differ within a broad range (Wilson et al. 1977). These broad distributions have been generally attributed to the wide spectrum of protein functions which differentially constrain the evolution of the respective genes. Indeed, it stands to reason that, for example, the function of a DNA polymerase, a sophisticated enzyme that catalyzes the template-dependent incorporation of nucleotides into the growing DNA chain, would constrain the gene sequence evolution to a much greater extent than, for instance, the function of a structural protein the only role of which is to maintain the integrity of the nuclear matrix. However, the prescient idea that evolution of protein-coding genes might not completely boil down to unique molecular details of the protein function emerged in these early days. Wilson and coworkers hypothesized in a seminal 1977 article that the evolution rate of a gene sequence depended both on the unique function of the encoded protein and on the importance of that protein for the survival of the organism (Wilson et al. 1977). However, at that time, there were no direct methods to study evolutionary constraints, so these ideas, however intriguing, belonged in the realm of speculation.

In the beginning of the third millennium, genomics and systems biology have completely transformed evolution research (Koonin 2009). With multiple genome sequences available, it has become possible to analyze and compare the distributions of evolutionary rates across complete sets of orthologous genes in different taxa, and also to examine the correlations between evolutionary rates of orthologs in different lineages. The distribution of the rates of evolution among non-synonymous sites in orthologous genes in any pair of compared genomes spans 3–4 orders of magnitude and is much broader than the distribution of the rates for synonymous sites (Koonin and Wolf 2010).

The shapes of the rate distributions for orthologous proteins are remarkably similar in all studied cellular life forms, from bacteria to archaea to mammals (Fig. 5) (Grishin et al. 2000; Wolf et al. 2009). All these distributions have the log-normal shape, i.e., the logarithm of the evolutionary rate is distributed approximately normally. In the theory of random processes, this shape of a distribution typically appears as a result of multiplication of many independent random variables. Given the dramatic differences in the functional organization and the actual number of genes between organisms, this universality of the evolutionary rate distribution appears unexpected and suggests the possibility of a fundamental, simple explanation which I discuss later in this chapter.

The progress of Systems Biology created the appealing opportunity to measure correlations between evolutionary rates and all kinds of “molecular phenomic”

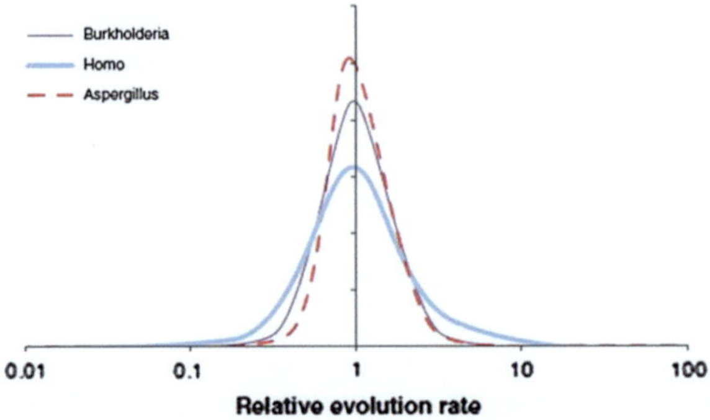

Fig. 5 The universal log-normal distribution of the evolutionary rates of orthologous genes in pairs of closely related genomes from diverse taxa. The evolutionary distances were calculated using the Jukes–Cantor correction and normalized so that the mean of each distribution was equal to 1. The probability density curves were obtained by Gaussian-kernel smoothing of the individual data points. Modified with permission from (Wolf et al. 2009)

variables, such as expression level, protein abundance, protein-protein interactions, the actual phenotypic effects of gene mutation, and more (Koonin 2011a, b; Koonin and Wolf 2006). Such "correlomics" studies have become almost a research area in its own although the ultimate goal, certainly, is not to simply describe the correlations or even their fine structure, but to develop physically sound and transparent models of genome and phenome evolution.

Among the correlations between evolutionary and molecular-phenomic variables, the strongest and universal one is the negative correlation between the rate of evolution of protein-coding genes and their expression levels: highly expressed genes typically evolve slowly, a dependence that was invariably observed in all model organisms for which expression data are available (Drummond et al. 2006; Drummond and Wilke 2008; Krylov et al. 2003; Pal et al. 2001). The discovery of the universal link between gene expression and evolution prompted a bold attempt on theoretical reinterpretation of protein evolution under which the primary causes of protein evolution have more to do with fundamental principles of protein structure and folding that are common across all life than with unique biological functions. It has been proposed, primarily, in the work of Drummond and Wilke, that the principal selective factor underlying the evolution of proteins is robustness to misfolding, owing to the deleterious effect of misfolded proteins that, in addition to the expenditure of energy, could be toxic to the cell (Drummond et al. 2005; Drummond and Wilke 2008). Without going into the details, this is an intuitively attractive model that quite naturally explains the anticorrelation between expression and sequence evolution: obviously, the deleterious effect of misfolding is expected to be stronger for abundant proteins than for those produced in small quantities, hence the genes for abundant proteins would be subject to strong constraints resulting in slow evolution. This model is compatible with the well-established preferential use of optimal codons (strong codon bias) in highly expressed and highly conserved protein-coding genes. Under the misfolding hypothesis, evolution of synonymous sites is constrained, at least in part, by the same factors as the evolution of proteins owing to the pressure for the preferential use of optimal codons in highly expressed proteins (hence fast translation) and in specific sites that are important for protein folding.

The universal distribution of the evolutionary rates of orthologous genes and the inverse correlation between the evolutionary rate and expression of a gene are two universal that potentially could be connected because both involve the same evolutionary rates. An attempt has been made to explain both these universal patterns within the framework of a single simple model (Lobkovsky et al. 2010). Indeed, an analysis of protein evolution that employed a simple, general model of protein folding yielded estimates of evolutionary rates under the assumption that misfolding was the only source of fitness cost. The results reproduced, with considerable accuracy, the universal distribution of protein evolutionary rates as well as the dependence between evolutionary rate and expression. These findings suggest that the universal rate distribution indeed is a consequence of the fundamental physics of protein folding.

Genome Evolution by Gene Duplication, Gene Birth and Death Models, and the Universal Distribution of Gene Family Sizes

Gene duplication is a major route of genome evolution in all walks of life, and conceivably, the main one in eukaryotes. Formally, evolution by gene duplication is a very simple process that can be readily encapsulated in straightforward mathematical models. The idea of duplication as a facile "method" of genome evolution is at the heart of our evolutionary thinking: deliberately trivializing the matter, it seems obvious that making new functional devices (proteins or RNAs) from pre-existing evolved entities by "tinkering" with them (Jacob 1977) is much easier than creating such devices de novo, from scratch. Genomics has put the concept of evolution by gene duplication on a firm quantitative basis by showing that the majority of the genes in any genome belong to families of paralogs (with the exception of the smallest genomes like those of *Mycoplasma* and other parasitic bacteria in which the fraction of "singletons" is greater) (Jordan et al. 2001). More detailed evolutionary reconstructions show that duplications occur, with varying intensity, at all stages of evolution, so any genome is a compendium of duplications of all ages. However we define a lineage—say, animals, chordates, mammals, primates etc.—we can find in a genome (e.g., our own) all classes of lineage-specific duplications: animal-specific, chordate-specific, mammalian-specific and so on (Lespinet et al. 2002). Prokaryotes and eukaryotes differ with respect to the major route of genome evolution. Phylogenomic analysis shows with confidence that eukaryotic genomes indeed evolve by duplication, whereas in prokaryotes, the main process that shapes gene families is HGT such that in many cases, genes that appear paralogous (i.e. evolved by duplication) actually are "pseudoparalogs" that have been acquired by a given lineage via independent HGT from different sources (Koonin and Wolf 2008; Makarova et al. 2005; Treangen and Rocha 2011)

The distribution of the size of gene families in any genome is another universal statistical pattern uncovered by comparative genomics (Fig. 6). The distributions for all genomes approximately follow a power law with a negative exponent: $y = ax^{-\gamma}$ (γ is a positive number, a is a coefficient) (Koonin et al. 2002; Luscombe et al. 2002). These distributions that conveniently become straight lines in double-logarithmic coordinates show that the majority of the families are small (including singletons) whereas a small fraction include numerous genes.

The emergence of the universal power law-like distribution of paralogous family size is described with remarkable accuracy by a simple mathematical model of the evolutionary process. The model comes from the mathematical theory of the birth and death processes and, in the case of evolution by gene duplication, is best described as a Birth, Death and Innovation Model (BDIM) (Karev et al. 2002, 2004; Koonin et al. 2002). Under BDIM, birth is gene duplication or acquisition of a family member via HGT yielding a new member of a gene family; death is gene loss; and innovation is the birth of a new family either via duplication followed by rapid divergence so that the "memory" of the old family is obliterated, or via HGT. The most striking result in the modeling of evolution by gene duplication was that a BDIM had to meet a set of precisely specified requirements to reproduce the observed distributions of gene family size. The rates of gene birth and death have to be (almost) equal but depend, in a specific manner, on the size of the family, that is, large families are more dynamic than smaller ones

Notably, the dynamics of gene family evolution is described by a purely stochastic model of the kind commonly used in statistical physics. However, for the model to be compatible with the data, there has to be a precise balance between domain birth, death and innovation rates, and this balance is likely to be maintained by natural selection.

Completing the discussion of the universal patterns uncovered by comparative genomics, it is appropriate to reiterate the question: do these universals reflect true "laws of biology" or are they simple manifestations of the maximum entropy principle whereby the values of any variable for a large ensemble of "particles" aggregate to the highest entropy distribution within the applicable constraints (Frank 2009a, b; Frank and Smith 2011)? It appears that the maximum entropy principle indeed applies but this in itself is not a trivial conclusion because it implies that the behavior of genes is reasonably approximated by the model of a "gas" of independent "particles". Beyond doubt, epistatic interactions are essential in biology but the observations discussed above imply that they do not dramatically affect the description of gene evolution and activity on the genome scale.

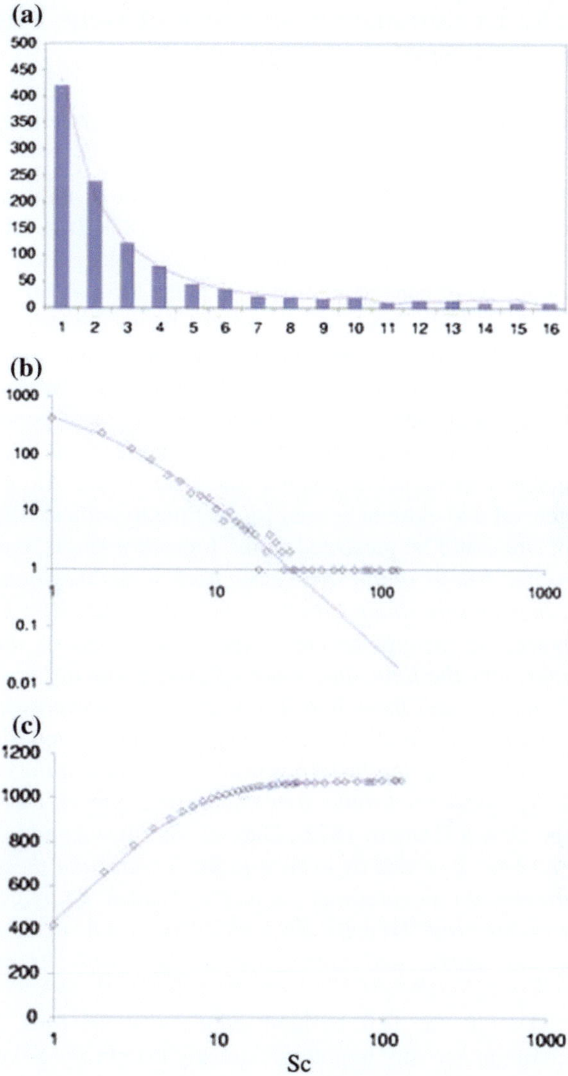

Fig. 6 The universal power law distribution of gene family size and the underlying gene birth-death-innovation model. Fit of empirical domain family size distributions to the second-order balanced linear BDIM: the yeast **Saccharomyces cerevisiae**. **a**. Distribution of the size of domain families grouped into bins **b**. Domain family size distribution in double logarithmic coordinates. **c** Cumulative distribution function of domain family size. The *line* shows the prediction of the second-order balanced linear BDIM. The *lines* show the fit to the model. From (Karev et al. 2002)

The maximum entropy principle is a "model free" description of reality, meaning that a variety of widely different evolutionary processes could lead to the same distribution of a given quantity. However, the observed distributions often subtly but significantly deviate from the maximum entropy distribution, and these deviations can be explained by biologically meaningful evolutionary models that, in particular, include effects of selection.

The Evolutionary Coherence of Gene Ensembles and the Statistical Tree of Life

The Tree of Life as a Concept and a Metaphor

The concept of the Tree of Life (TOL) in its modern meaning was introduced by Darwin in his notebooks as early as 1838 (the famous "I think" drawing) and 20 years later captured in the single illustration of the "*Origin of species*". Depiction of genetic relationships in the form of a tree certainly was not invented by Darwin. Trees have been used for many centuries to represent genealogies, i.e. actual histories of families, for example, royal ones, or longer, mythical family stories, such as those on the Bible. However, Darwin was the first to come up with the seminal idea that different species (not just generations within a family) were related by a tree in which the leaves correspond to extant species and the internal nodes correspond to extinct ancestral forms. Moreover, Darwin formulated the general and extremely bold for its time hypothesis that ultimately the entire history of life could be presented in the form of a single, enormous tree: "*The affinities of all the beings of the same class have sometimes been represented by a great tree. I believe this simile largely speaks the truth. The green and budding twigs may represent existing species; and those produced during each former year may represent the long succession of extinct species. .... The limbs divided into great branches, and these into lesser and lesser branches, were themselves once, when the tree was small, budding twigs; and this connexion of the former and present buds by ramifying branches may well represent the classification of all extinct and living species in groups subordinate to groups.*" (Darwin 1859) In the 6th edition of the *Origin* (Darwin 1872), Darwin went further and explicitly introduced the TOL: "*As buds give rise by growth to fresh buds, and these, if vigorous, branch out and overtop on all sides many a feebler branch, so by generation I believe it has been with the great* ***Tree of Life****, which fills with its dead and broken branches the crust of the earth, and covers the surface with its ever-branching and beautiful ramifications.*"

Despite the lack of hard evidence at the time, Darwin's sweeping universal common ancestry hypothesis quickly caught up. Several years after the publication of the *Origin*, Ernst Haeckel populated Darwin's conceptual TOL with real life forms that included almost exclusively animals, with "MAN" on top and some amoebae and "Monera" (the 19th century name for bacteria), at the roots (Haeckel 1905). Since then, the TOL became the centerpiece of evolutionary biology and, in a sense, of biology in general.

For nearly 140 years after Darwin and Haeckel, phylogenetic trees, that were initially constructed using phenotypic characters but, following the seminal work of Zuckerkandl and Pauling in the early 1960s (Zuckerkandl and Pauling 1965), increasingly relied on molecular sequence comparison, have been construed as a (more or less accurate) depiction of the evolution of the respective organisms. In other words, a tree built for a specific character or a gene was by default equated with a "species tree". The adoption of the 16S rRNA, a molecule that is universal in cellular life forms, as the golden standard for phylogenetic reconstruction yielded the three-domain TOL of Woese and colleagues, a striking culmination of the classic period of phylogenetics (Pace 1997, 2006; Woese 1987; Woese et al. 1990). The 16S tree included parts with excellent resolution of the branches, and although many other parts remained poorly resolved, especially deep within the tree, further improvement of phylogenetic methods along with the analysis of several additional universal genes, was expected to reveal the detailed, definitive topology of the TOL in a not so remote future (Pace 1997).

The views of evolutionary biologists on the status of the TOL in the face of the pervasive HGT spanned the entire range from: (i) continued denial of a significant role of HGT in the evolution of life, to (ii) a "moderate" revision of the TOL concept, to (iii) genuine uprooting whereby the TOL is declared meaningless as a representation of the evolution of organisms or genomes (O'Malley and Boucher 2005; O'Malley and Koonin 2011). With the accumulation of comparative genomic data indicative of massive HGT, the adherence to the traditional TOL concept is being quickly marginalized (O'Malley and Koonin 2011). The real debate seems to be between the "revisionist" and "radical uprooting" views (ii and iii). The moderate approach maintains that, all the differences between individual gene trees notwithstanding, the TOL remains relevant as a central trend that, at least, in principle, can be revealed through a comprehensive comparison of gene tree topologies (Wolf et al. 2002). The radical view counters that massive HGT obliterates the very distinction between the vertical and horizontal routes of genetic information transmission, so the TOL concept should be abandoned in favor of a network representation of evolution (Doolittle and Bapteste 2007; Gogarten et al. 2002).

The views of evolutionary biologists on the status of the TOL in the face of the pervasive HGT spanned the entire range from: (i) continued denial of a significant role of HGT in the evolution of life, to (ii) a "moderate" revision of the TOL concept, to (iii) genuine uprooting whereby the TOL is declared meaningless as a representation of the evolution of organisms or genomes (O'Malley and Boucher 2005; O'Malley and Koonin 2011). With the accumulation of comparative genomic data indicative of massive HGT, the adherence to the traditional TOL concept is being quickly marginalized (O'Malley and Koonin 2011). The real debate seems to be between the "revisionist" and "radical uprooting" views (ii and iii). The moderate approach maintains that, all the differences between individual gene trees notwithstanding, the TOL remains relevant as a central trend that, at least, in principle, can be revealed through a comprehensive comparison of gene tree topologies (Wolf et al. 2002). The radical view counters that massive HGT obliterates the very distinction between the vertical and horizontal routes of genetic information transmission, so the TOL concept should be abandoned in favor of a network representation of evolution (Doolittle and Bapteste 2007; Gogarten et al. 2002).

Below I discuss a comprehensive dissection of the evolution of prokaryotes into the tree-like and web-like components, in order to objectively determine the role and place of trees in our understanding of evolution, which should eventually settle the TOL controversy. Before turning to this quantitative analysis, I examine the roots of tree thinking at a conceptual level.

The Fundamental Units of Evolution and Its Intrinsic Tree-like Nature

Replication of the genetic material, a process that is intrinsically error-prone, is both the condition and the direct cause of evolution. The status of trees in biology is defined by the simple fact that replication and the evolution that it necessarily entails are inherently tree-like processes (Koonin and Wolf 2009). Indeed, a replicating molecule gives rise to two (in the case of semi-conservative replication of dsDNA that occurs in all cellular organisms and many viruses) or multiple (in the case of the conservative replication of viruses with ssDNA or ssRNA genomes) copies with errors, resulting in a tree-like process of divergence. In graph-theoretical terms, such a process can be isomorphously represented by a specific form of a directed acyclic graph known as arborescence, i.e., a generalized tree in which multifurcations are allowed and all edges are directed away from the root (hereinafter, I simply call such graph trees).

A potential major complication to the tree-like character of evolution is recombination that, if common, would turn the tree-like representation of the history of a replicating lineage into a network (or worse, a morass). Is it possible to determine a fundamental, "atomic" level of genetic organization at which recombination is negligible? This does not seem to be feasible in the case of homologous recombination that is extensive during co-replication of closely related sequences, in particular, in eukaryotes that engage in regular sex, and in "quasi-sexual" prokaryotes. Essentially, the unit of homologous recombination is a single base pair. However, homologous recombination cannot occur between distantly related sequences, so HGT between diverse prokaryotes involves only non-homologous (illegitimate) recombination complemented by more specific routes such as dissemination via bacteriophages and plasmids that have the potential to integrate into chromosomes. In contrast to homologous recombination, a strong preference for evolutionary fixation of non-homologous recombination events outside of genes or between parts of genes encoding distinct domains of multidomain proteins should be expected because preservation of gene integrity after non-homologous recombination within genes is extremely unlikely. The prevalence of intragenic recombination in the course of HGT between distantly related prokaryotes has not been studied in sufficient detail. Nevertheless, at least one study shows that regions encoding relatively small protein domains are significantly avoided by recombination (Chan et al. 2009). Hence a conclusion that appears important and credible even if not yet sufficiently supported by data: the evolutionary history of a gene or domain is reticulate on the micro scale due to homologous recombination but is largely tree-like on the macro scale.

Thus, trees cannot be banished from evolutionary biology for a fundamental reason: they are intrinsic to the evolutionary process. This being the case, the main pertinent question becomes: what are the fundamental genetic units of tree-like evolution? In the practice of evolutionary biology, trees are most often built for individual genes or for sets of genes that are believed to evolve coherently. However, it is typically either implied or stated explicitly that the ultimate goal is a species (organismal) tree. The lack of clarity about the basic unit to which tree analysis applies seems to be an important, if not the main, source of the entire TOL controversy.

Conceptually, the answer to the above question seems clear: the fundamental unit of evolution can be most adequately defined as the smallest portion of genetic material with a distinct evolutionary history, i.e., one that evolves independently of other such units through a substantial duration of evolution. Given the dominance of HGT in the evolution of prokaryotes, the unit that generally meets this requirement is a genomic locus that encodes an RNA or protein molecule (or a distinct evolutionary domain), i.e. a gene (except for the case of genes encoding multidomain proteins where a single gene may comprise two or more evolutionary units). The realization that individual genes, as opposed to genomes, are the "atomic" units of evolution undermines the TOL concept. However, as shown above, trees are inalienable from any description of evolution for the simple reason that replication of the genetic material is an intrinsically tree-like process. Together, these two fundamental observations lead to a straightforward conclusion as to what should replace the TOL: the "Forest of Life" (FOL), i.e., the collection of phylogenetic trees for all genes (with the obvious exception of genes without detectable homologs) (Puigbo et al. 2013). The reconstruction of the history of life requires charting the FOL in search of "groves" of similar trees that might reflect long term trends of coherent (vertical) evolution of gene ensembles and "vines" of HGT. I describe the results of such comparative analysis of gene trees in the next section.

The Forest of Life and the Nearly Universal Trees

In principle, the FOL includes trees for "all" genes. In practice, however, working with the thousands sequenced genomes is technically difficult because of error accumulation in sequence alignment as well as computational complexity of the nest methods for phylogenetic analysis. However, analysis of all available genomes is not essential. The dominance of HGT in the evolution of prokaryotes notwithstanding, core and shell genes in closely related organisms (defined, for example, by the high sequence similarity of their rRNAs or other core genes) most of the time evolve congruently (and only core and shell genes are widespread enough to yield meaningful trees). Thus, a carefully selected representative set of organisms should be sufficient to reveal major trends in the FOL (Puigbo et al. 2009). For the analysis discussed here, we constructed a representative set of 100 prokaryotic genomes, 41 archaeal and 59 bacterial. Trees were built for all sets of orthologs with more than 4 members (the minimum number of sequences required to make an unrooted tree), so altogether we obtained almost 7000 trees. Predictably, given the core-shell-cloud structure of the prokaryotic gene space described above, most of these trees are small: only 2040 trees included more than 20 species, and only a small set of 102 Nearly Universal Trees (NUTs) included >90 % of the analyzed prokaryotes.

Usually, phylogeneticists attempt to identify HGT by comparing trees of individual genes to a predefined "species tree". However, as we have seen in the preceding section, the very concept of a "species tree" is invalidated by the pervasive HGT because of which the fundamental units of tree-like evolution are individual genes. Thus, we attempted to chart the FOL without using any standard tree against which to compare the rest of the trees in the FOL. To this end, we analyzed the complete, all-against-all matrix of the topological distances between the trees which included almost 24 million pair-wise tree comparisons although many cells in the matrix are empty because the respective trees consist of non-overlapping sets of species. A convenient way to represent and analyze this matrix is a network graph in which the nodes are individual gene trees which are connected by edges if the overlap of the respective tree sets is sufficient to measure the distance, with the weights on the edges inversely proportional to the distance [the distances are calculated from the topological differences between the trees; I skip here the details of these calculations, see (Puigbo et al. 2012)]. In this network, the group of NUTs occupies a special position: about 40 % of the trees are highly similar to at least one NUT. As a control, under the same similarity cut-off, 102 randomized NUTs were connected only to ~0.5 % of the trees in the FOL. Thus, there is a high and non-random topological similarity between the NUTs and a large part of the FOL.

Knowing all the distances between the trees in the FOL, one can apply statistical methods for data clustering, i.e., determine whether the FOL is simply a cloud of randomly scattered points (trees in the topology space) or contains distinct clusters of trees with similar topologies. The applied statistical procedure partitioned the FOL into 7 clusters of trees; notably, all the NUTs formed a compact group within one of the clusters. The 7 clusters showed considerable differences in the distribution of the trees by the number of species, the distribution of archaea and bacteria, and the functional classification of the respective genes. So the clustering results indicate that the FOL can be partitioned into large, distinct groups of topologically similar trees; however, at this stage, it remains unclear how much of this clustering is due to "vertical" and how much to "horizontal" evolutionary processes. The key observation is that all the NUTs occupy a compact and contiguous region of the tree space, are not partitioned into distinct clusters, in contrast to the rest of the FOL, and are separated by approximately the same distance from all clusters of trees.

These findings lead to an important conclusion: the topologies of the NUTs are highly similar to each other and seem to represent a central evolutionary trend in the FOL. This statement might seem vague but actually reflects two simple, straightforward observations: (i) the topologies of the NUTs are similar to topologies of many other trees in the FOL, and (ii) the NUTs are approximately equidistant from the clusters of other trees, i.e. can be considered to occupy a central position in the FOL.

The NUTs correspond to the core in the tripartite organization of the genomic universe described above (Koonin and Wolf 2008). These genes encode ribosomal proteins and other highly conserved proteins involved in translation along with a several core subunits of the DNA-dependent RNA polymerase. These are the genes that are expected to be least of all prone to HGT according to the so called complexity hypothesis which postulates that genes encoding subunits of macromolecular complexes are mutually adjusted by evolution so that exchange of any subset of these genes is likely to be deleterious (Jain et al. 1999). Somewhat paradoxically, this set of nearly universal genes also encompasses some of the most notable examples of HGT, in particular, among the aminoacyl-tRNA synthetases, some of which are responsible for antibiotic resistance, but also among quite a few ribosomal proteins (Makarova et al. 2001; Wolf et al. 1999). Nevertheless, the analysis of the FOL unequivocally shows that the group of NUTs is internally topologically consistent and moreover is linked by topological similarity to numerous other trees in the FOL.

In light of the near ubiquitous HGT, there is no chance to revive the TOL as a single tree that would describe organismal evolution (species tree). However, if one seeks the best meaningful approximation of a TOL, the consensus topology of the NUTs appears to be the strongest candidate (O'Malley and Koonin 2011). Before settling on this conclusion, I describe a deeper quantitative analysis of the FOL that allowed us to assess the extent of vertical (tree-type) and horizontal (web-type) evolutionary processes.

The signal of tree-type evolution, defined as the consensus topology of the NUTs, seems to reflect a central trend of evolution in the FOL and is traceable throughout the entire range of phylogenetic depths despite the substantial rate of HGT. In contrast, the sum total of all evolutionary patterns that appear incompatible with the consensus NUTs topology, whether caused by HGT or by other processes such as parallel gene losses that are also common among prokaryotes, can be denoted the web-type signal. We developed a quantitative measure to directly estimate (on a 0–1 scale) the tree-type and web-type contributions to the evolutionary distances between species (Puigbo et al. 2010). The lower the score (that is, the closer to the distance expected by chance, under the assumption that genes are freely mixed), the more the relationship between the given pair of species is determined by web-type evolutionary processes. The Tree-Net map of the NUTs was dominated by the tree-like signal: the mean score for the NUTs was 0.63, so the evolution of the nearly universal genes of prokaryotes appears to be nearly "two-third tree-type". The exceptions are the radioresistant bacterium *Deinococcus radiodurans* that showed, primarily, web-like relationships with most of the archaea and several bacterial taxa (*Actinobacteria*, *Aquificae*, *Chloroflexi*, Cyanobacteria, *Firmicutes*, *Fusobacteriae*, *Thermotogae*) each of which formed a strongly connected network with other bacteria. In a stark contrast, the rest of the FOL is dominated by the web-type evolutionary processes, with the mean score of 0.39 (about "60 % web-like"). The major web-like areas observed among the NUTs recurred in the FOL, and additional ones became apparent including Crenarchaeota that showed a pronounced signal of a non-tree-like relationship with diverse bacteria as well as some Euryarchaeota. A more detailed dissection of the FOL shows that the web-type signal dominates the evolution of genes that are present in a small number of prokaryotes whereas the evolution of more wide spread genes is more tree-like and more closely resembles the pattern seen among the NUTs. This trend is compatible with the HGT optimization hypothesis according to which genes that are frequently lost during evolution should be also frequently transferred if the extinction of these genes and the overall mutational meltdown of microbial populations are to be avoided.

Different functional classes of genes showed major differences with respect to the tree-type (vertical) and web-type (horizontal) trends in their evolution, from the dominance of the vertical signal among genes for translation machinery components and proteins involved in intracellular trafficking to the almost fully horizontal evolution of genes for ion transport, signal transduction and especially defense system components. This pattern is generally compatible with the complexity hypothesis but also reveals a more nuanced picture, with substantial differences, for instance, between enzymes of nucleotide metabolism that evolve mostly vertically and proteins involved in amino acid or carbohydrate metabolism and transport, for which the web-type signal was much more prominent.

Thus, the quantitative analysis of the tree-type and web-type signals reveals an apparent paradox of prokaryote evolution: although the tree-type evolution is by far the strongest single trend in the FOL, quantitatively, evolution of prokaryotes is dominated by the combination of the web-type processes, such as HGT and lineage-specific gene loss. The tree-like pattern accounts for most of the evolution among the NUTs; however, because the FOL consists mostly of small trees among which the tree signal is barely detectable, the diverse, in the first approximation, random horizontal processes that govern the evolution of relatively small gene families are quantitatively dominant.

When Darwin introduced the TOL metaphor, his argument came from observations on the evolution of animals. However, he generalized the tree pattern of evolution to life in general with considerable confidence. In the narrow sense, Darwin was correct: no one denies that evolution of animals is tree-like. However, this is not a TOL but only a description of the evolution of a single, relatively small, tight group of eukaryotes. The straightforward generalization to the entirety of cellular life on earth fails because of the complex net of extensive HGT that is most common among prokaryotes but also prominently contributed to the evolution of eukaryotes, especially, via endosymbiosis (Keeling and Palmer 2008).

However, notwithstanding the newly discovered complex nature of evolution, with a major horizontal component, Darwin's metaphor reflects a deeper truth: trees remain the natural representation of the histories of individual genes given the fundamentally bifurcating character of gene replication and the substantially low frequency of intragenic recombination compared to intergenic recombination at long evolutionary distances. Thus, although no single tree can fully represent the evolution of complete genomes and the respective life forms, the realistic picture of evolution should necessarily combine trees and networks. These components can be revealed through the analysis of the FOL, the comprehensive collection of gene trees (Puigbo et al. 2010).

The quantitative dissection of the FOL reveals a complex landscape of tree-like and web-like evolution. The signals from these two types of evolution are distributed in a highly non-random fashion among different groups of prokaryotes, and among functional classes of genes. Overall, the web-type signal is quantitatively dominant, a finding that (almost literally) vindicates the concepts of "lateral genomics" or "net of life". These results are decidedly incompatible with the representation of prokaryote evolution as a TOL adorned with thin, random "cobwebs" of HGT (Ge et al. 2005; Kunin et al. 2005). However, the tree-type, vertical signal compatible with the consensus topology of the NUTs is also unmistakably detectable and seems to account for up to 40 % of the evolutionary processes in the prokaryote world. The crucial, even if somewhat paradoxical, feature of the evolution of prokaryotes is that, although web-type processes are quantitatively dominant, the single strongest trend is the vertical evolution reflected in the consensus tree topology of the NUTs that also largely recapitulates the rRNA tree. In principle, one could legitimately speak of this trend as a "statistical" TOL.

From the Structure to the Dynamics of the Genomic Universe

Genomes of bacteria and archaea (collectively, prokaryotes) appear to exist in incessant flux, expanding via horizontal gene transfer (HGT) and gene duplication, and contracting via gene loss. However, the actual rates of genome dynamics and relative contributions of different types of events across the diversity of prokaryotes remain largely unknown. Recently, we performed a comprehensive analysis of the genome dynamics in 35 groups of closely related microbial genomes by mapping the gene gain and loss events onto the evolutionary trees of each group and using a phylogenetic birth-and-death maximum likelihood model to reconstruct the evolutionary scenario and quantify the gain and loss rates(Puigbo et al. 2014) . The results of this analysis show that loss of gene families dominates the evolution of prokaryotes, occurring at approximately three times the rate of gain. The rates of gene family expansion and reduction are typically 7 and 20 times less than the gain and loss rates, respectively. Thus, the prevailing mode of evolution in bacteria and archaea is genome contraction that is partially compensated by the gain of new gene families via horizontal gene transfer. The rates of gene flux estimated from these reconstructions are extremely high such that several events of gene loss and gain typically occur over the time that it takes for a single nucleotide substitution to be fixed in an average gene in an evolving genome.

The rates of gene family gain, loss, expansion and reduction are also highly variable, with the most stable genomes showing rates about 25 times lower than the most dynamic genomes. There is little correlation between the rate of gene flux and the life style of microbes or the strength of selection that affects the protein sequences they encode. In contrast, there is a clear pattern in the distribution of gene loss and gain across the functional categories of genes: genes involved in anti-parasite defense as well as signal transduction are lost and gained much more often than genes coding for components of information transfer systems where metabolic enzymes and transporters show intermediate rates. Taken together, these findings clearly show that rapid gene flux involving extensive loss of genes and gene families, partially balanced by gain of new gene families via HGT, is the principal mode of microbial evolution. These conclusions are compatible with experimental results demonstrating bacterial genome contraction in real time (Nilsson et al. 2005).

Given the finding, compatible with previous observations made on larger evolutionary scales (Kunin and Ouzounis 2003; Snel et al. 2002), that gene family loss substantially prevails over gain in the evolution of microbial genomes, the question emerges, why genomes do not shrink out of existence? The answer is likely to be twofold. Some bacteria actually might be headed towards extinction as previously observed for the tiny genomes of intracellular parasites (McCutcheon and Moran 2012; Merhej et al. 2013; Merhej and Raoult 2011). However, the more common scenario would involve evolving prokaryotic lineages going through long phases of genome contraction, in which our analysis caught most of them, punctuated by shorter bursts of extensive gene gain, which are also detectable albeit, due to their short duration, only in a few groups, and compensate for the gradual gene loss.

Conclusions: The Dynamic Yet Structured Genomic Universe

The selected threads from recent evolutionary genomic studies discussed in this chapter seem to converge on a coarse grain but somewhat optimistic message: we are now in a position to decipher major trends not only in the evolution of particular groups of organisms but also in the entire genomic universe. Our ability to probe the structure and evolution of the genomic spacetime stems from a truly fundamental and non-trivial generalization yielded by comparative genomic, namely that genes are relatively stable atomic units of that spacetime, with their distinct evolutionary histories. Moreover, the discovery that key variables that characterize genome and molecular phenome evolution are distributed according to simple mathematical laws that follow the maximum entropy principle within the applicable constraints implies that treatment of genes and their products as weakly interacting particles is adequate for at least a crude description of the structure and evolution of the genomic universe. Deviations from the distributions dictated by the maximum entropy principle in some case allow one to choose a realistic mathematical model of the evolutionary process that may include a particular distribution of selection coefficients across a gene ensemble. In the new concept of evolution rooted in comparative genomics, the tree of life comes across a statistical trend of coherent gene evolution, one of the major patterns detectable in the genomic spacetime. Thus, the genomic universe appears to be extremely dynamic, yet amenable to quantitative analysis due to the existence of readily detectable, universal patterns. Refined analysis of these patterns and the underlying evolutionary models should allow us to better understand the evolution and ultimately the origin of the genomic universe.

Acknowledgments The author's research is supported by intramural funds of the US Department of Health and Human Services (to the National Library of Medicine).

References

Bult CJ et al (1996) Complete genome sequence of the methanogenic archaeon, Methanococcus jannaschii [see comments]. Science 273:1058–1073
Chan CX, Darling AE, Beiko RG, Ragan MA (2009) Are protein domains modules of lateral genetic transfer? PLoS ONE 4:e4524
Clark K, Karsch-Mizrachi I, Lipman DJ, Ostell J, Sayers EW (2016) GenBank. Nucleic Acids Res 44:D67–D72
Crick FH (1958) On protein synthesis. Symp Soc Exp Biol 12:138–163
Crick FH (1968) The origin of the genetic code. J Mol Biol 38:367–379
Crick F (1970) Central dogma of molecular biology. Nature 227:561–563
Darwin C (1859) On the origin of species. Murray, London
Darwin C (1872) Origin of species. The Modern Library, New York
Delaye L, Moya A (2010) Evolution of reduced prokaryotic genomes and the minimal cell concept: variations on a theme. BioEssays 32:281–287
Doolittle WF (1999a) Lateral genomics. Trends Cell Biol 9:M5–M8
Doolittle WF (1999b) Phylogenetic classification and the universal tree. Science 284:2124–2129
Doolittle WF (2000) Uprooting the tree of life. Sci Am 282:90–95
Doolittle WF, Bapteste E (2007) Pattern pluralism and the tree of life hypothesis. Proc Natl Acad Sci USA 104:2043–2049
Drummond DA, Wilke CO (2008) Mistranslation-induced protein misfolding as a dominant constraint on coding-sequence evolution. Cell 134:341–352
Drummond DA, Bloom JD, Adami C, Wilke CO, Arnold FH (2005) Why highly expressed proteins evolve slowly. Proc Natl Acad Sci USA 102:14338–14343
Drummond DA, Raval A, Wilke CO (2006) A single determinant dominates the rate of yeast protein evolution. Mol Biol Evol 23:327–337
Fitch WM (1970) Distinguishing homologous from analogous proteins. Syst Zool 19:99–113
Fleischmann RD et al (1995) Whole-genome random sequencing and assembly of Haemophilus influenzae Rd [see comments]. Science 269:496–512
Frank SA (2009a) The common patterns of nature. J Evol Biol 22:1563–1585
Frank SA (2009b) Natural selection maximizes Fisher information. J Evol Biol 22:231–244
Frank SA, Smith E (2011) A simple derivation and classification of common probability distributions based on information symmetry and measurement scale. J Evol Biol 24:469–484

Fraser CM et al (1995) The minimal gene complement of Mycoplasma genitalium. Science 270:397–403
Galperin MY, Makarova KS, Wolf YI, Koonin EV (2015) Expanded microbial genome coverage and improved protein family annotation in the COG database. Nucleic Acids Res 43:D261–D269
Ge F, Wang LS, Kim J (2005) The cobweb of life revealed by genome-scale estimates of horizontal gene transfer. PLoS Biol 3:e316
Gogarten JP, Doolittle WF, Lawrence JG (2002) Prokaryotic evolution in light of gene transfer. Mol Biol Evol 19:2226–2238
Grishin NV, Wolf YI, Koonin EV (2000) From complete genomes to measures of substitution rate variability within and between proteins. Genome Res 10:991–1000
Haeckel E (1905) The wonders of life: a popular study of biological philosophy. Harper & Brother, New York & London
Handelsman J (2004) Metagenomics: application of genomics to uncultured microorganisms. Microbiol Mol Biol Rev 68:669–685
Hilario E, Gogarten JP (1993) Horizontal transfer of ATPase genes–the tree of life becomes a net of life. Biosystems 31:111–119
Jacob F (1977) Evolution and tinkering. Science 196:1161–1166
Jain R, Rivera MC, Lake JA (1999) Horizontal gene transfer among genomes: the complexity hypothesis. Proc Natl Acad Sci USA 96:3801–3806
Jordan IK, Makarova KS, Spouge JL, Wolf YI, Koonin EV (2001) Lineage-specific gene expansions in bacterial and archaeal genomes. Genome Res 11:555–565
Kaeberlein T, Lewis K, Epstein SS (2002) Isolating "uncultivable" microorganisms in pure culture in a simulated natural environment. Science 296:1127–1129
Karev GP, Wolf YI, Rzhetsky AY, Berezovskaya FS, Koonin EV (2002) Birth and death of protein domains: a simple model of evolution explains power law behavior. BMC Evol Biol 2:18
Karev GP, Wolf YI, Berezovskaya FS, Koonin EV (2004) Gene family evolution: an in-depth theoretical and simulation analysis of non-linear birth-death-innovation models. BMC Evol Biol 9:32
Keeling PJ, Palmer JD (2008) Horizontal gene transfer in eukaryotic evolution. Nat Rev Genet 9:605–618
Kodzius R, Gojobori T (2016) Single-cell technologies in environmental omics. Gene 576:701–707
Koonin EV (2003) Comparative genomics, minimal gene-sets and the last universal common ancestor. Nature Rev Microbiol 1:127–136
Koonin EV (2005) Orthologs, paralogs and evolutionary genomics. Annu Rev Genet 39:309–338
Koonin EV (2009) Darwinian evolution in the light of genomics. Nucleic Acids Res 37:1011–1034
Koonin EV (2011a) Are there laws of genome evolution? PLoS Comput Biol 7:e1002173
Koonin EV (2011b) The logic of chance: the nature and origin of biological evolution. FT press, Upper Saddle River
Koonin EV, Mushegian AR (1996) Complete genome sequences of cellular life forms: glimpses of theoretical evolutionary genomics. Curr Opin Genet Dev 6:757–762
Koonin EV, Wolf YI (2006) Evolutionary systems biology: links between gene evolution and function. Curr Opin Biotechnol 17:481–487
Koonin EV, Wolf YI (2008) Genomics of bacteria and archaea: the emerging dynamic view of the prokaryotic world. Nucleic Acids Res 36:6688–6719
Koonin EV, Wolf YI (2009) The fundamental units, processes and patterns of evolution, and the tree of life conundrum. Biol Direct 4:33
Koonin EV, Wolf YI (2010) Constraints and plasticity in genome and molecular-phenome evolution. Nat Rev Genet 11:487–498
Koonin EV, Mushegian AR, Galperin MY, Walker DR (1997) Comparison of archaeal and bacterial genomes: computer analysis of protein sequences predicts novel functions and suggests a chimeric origin for the archaea. Mol Microbiol 25:619–637

Koonin EV, Wolf YI, Karev GP (2002) The structure of the protein universe and genome evolution. Nature 420:218–223
Krylov DM, Wolf YI, Rogozin IB, Koonin EV (2003) Gene loss, protein sequence divergence, gene dispensability, expression level, and interactivity are correlated in eukaryotic evolution. Genome Res 13:2229–2235
Kunin V, Ouzounis CA (2003) The balance of driving forces during genome evolution in prokaryotes. Genome Res 13:1589–1594
Kunin V, Goldovsky L, Darzentas N, Ouzounis CA (2005) The net of life: reconstructing the microbial phylogenetic network. Genome Res 15:954–959
Lespinet O, Wolf YI, Koonin EV, Aravind L (2002) The role of lineage-specific gene family expansion in the evolution of eukaryotes. Genome Res 12:1048–1059
Lobkovsky AE, Wolf YI, Koonin EV (2010) Universal distribution of protein evolution rates as a consequence of protein folding physics. Proc Natl Acad Sci USA 107:2983–2988
Lobkovsky AE, Wolf YI, Koonin EV (2013) Gene frequency distributions reject a neutral model of genome evolution. Genome Biol Evol 5:233–242
Luscombe NM, Qian J, Zhang Z, Johnson T, Gerstein M (2002) The dominance of the population by a selected few: power-law behaviour applies to a wide variety of genomic properties. Genome Biol 3: RESEARCH0040
Makarova KS, Ponomarev VA, Koonin EV (2001) Two C or not two C: recurrent disruption of Zn-ribbons, gene duplication, lineage-specific gene loss, and horizontal gene transfer in evolution of bacterial ribosomal proteins. Genome Biol 2: RESEARCH 0033
Makarova KS, Wolf YI, Mekhedov SL, Mirkin BG, Koonin EV (2005) Ancestral paralogs and pseudoparalogs and their role in the emergence of the eukaryotic cell. Nucleic Acids Res 33:4626–4638
McCutcheon JP, Moran NA (2012) Extreme genome reduction in symbiotic bacteria. Nat Rev Microbiol 10:13–26
Merhej V, Raoult D (2011) Rickettsial evolution in the light of comparative genomics. Biol Rev Camb Philos Soc 86:379–405
Merhej V, Georgiades K, Raoult D (2013) Postgenomic analysis of bacterial pathogens repertoire reveals genome reduction rather than virulence factors. Brief Funct Genomics 12:291–304
Mushegian AR, Koonin EV (1996) A minimal gene set for cellular life derived by comparison of complete bacterial genomes (see comments). Proc Natl Acad Sci USA 93:10268–10273
Nilsson AI, Koskiniemi S, Eriksson S, Kugelberg E, Hinton JC, Andersson DI (2005) Bacterial genome size reduction by experimental evolution. Proc Natl Acad Sci USA 102:12112–12116
O'Malley MA, Boucher Y (2005) Paradigm change in evolutionary microbiology. Stud Hist Philos Biol Biomed Sci 36:183–208
O'Malley MA, Koonin EV (2011) How stands the Tree of Life a century and a half after The Origin? Biol Direct 6:32
Pace NR (1997) A molecular view of microbial diversity and the biosphere. Science 276:734–740
Pace NR (2006) Time for a change. Nature 441:289
Pal C, Papp B, Hurst LD (2001) Highly expressed genes in yeast evolve slowly. Genetics 158:927–931
Puigbo P, Wolf YI, Koonin EV (2009) Search for a 'Tree of Life' in the thicket of the phylogenetic forest. J Biol 8:59
Puigbo P, Wolf YI, Koonin EV (2010) The tree and net components of prokaryote evolution. Genome Biol Evol 2:745–756
Puigbo P, Wolf YI, Koonin EV (2012) Genome-wide comparative analysis of phylogenetic trees: the prokaryotic forest of life. Methods Mol Biol 856:53–79
Puigbo P, Wolf YI, Koonin EV (2013) Seeing the Tree of Life behind the phylogenetic forest. BMC Biol 11:46
Puigbo P, Lobkovsky AE, Kristensen DM, Wolf YI, Koonin EV (2014) Genomes in turmoil: quantification of genome dynamics in prokaryote supergenomes. BMC Biol 12:66
Snel B, Bork P, Huynen MA (2002) Genomes in flux: the evolution of archaeal and proteobacterial gene content. Genome Res 12:17–25

Tatusov RL, Koonin EV, Lipman DJ (1997) A genomic perspective on protein families. Science 278:631–637

Timofeeff-Ressovsky NW, Zimmer KG, Delbruck M (1935) Uber die Natur der Genmutation. Nachrichten von der Gesellschaft von Wissenschaften zu Gottingen 1:190–245

Treangen TJ, Rocha EP (2011) Horizontal transfer, not duplication, drives the expansion of protein families in prokaryotes. PLoS Genet 7:e1001284

Vartoukian SR, Palmer RM, Wade WG (2010) Strategies for culture of 'unculturable' bacteria. FEMS Microbiol Lett 309:1–7

Watson JD, Crick FH (1953a) Genetical implications of the structure of deoxyribonucleic acid. Nature 171:964–967

Watson JD, Crick FH (1953b) Molecular structure of nucleic acids; a structure for deoxyribose nucleic acid. Nature 171:737–738

Wilson AC, Carlson SS, White TJ (1977) Biochemical evolution. Annu Rev Biochem 46:573–639

Woese CR (1987) Bacterial evolution. Microbiol Rev 51:221–271

Woese CR, Kandler O, Wheelis ML (1990) Towards a natural system of organisms: proposal for the domains Archaea, Bacteria, and Eucarya. Proc Natl Acad Sci USA 87:4576–4579

Wolf YI, Aravind L, Grishin NV, Koonin EV (1999) Evolution of aminoacyl-tRNA synthetases–analysis of unique domain architectures and phylogenetic trees reveals a complex history of horizontal gene transfer events. Genome Res 9:689–710

Wolf YI, Rogozin IB, Grishin NV, Koonin EV (2002) Genome trees and the tree of life. Trends Genet 18:472–479

Wolf YI, Novichkov PS, Karev GP, Koonin EV, Lipman DJ (2009) The universal distribution of evolutionary rates of genes and distinct characteristics of eukaryotic genes of different apparent ages. Proc Natl Acad Sci USA 106:7273–7280

Zuckerkandl E, Pauling L (1965) Evolutionary divergence and convergence of proteins. In: Bryson V, Vogel HJ (eds) Evolving gene and proteins, evolving gene and proteins. Academic Press, New York, pp 97–166

Microevolutionary Processes in Plant-Microbe Symbiosis

Igor A. Tikhonovich, Evgeny E. Andronov and Nikolai A. Provorov

Introduction

Considering the scientific heritage of Nikolai Vladimirovich Timofeev-Ressovsky, we usually distinguish two main issues: the impact of radiation on heredity and evolutionary patterns. He gave the following definition of microevolution: "… microevolution can be defined as a process that affects relatively small spaces … short periods, lower taxa; the intraspecific variability is mainly studied" (Timofeev-Ressovsky 2009). Among the definitions of micro- and macroevolution we would highlight the reversibility as the clearest distinction thereof. Once established macroevolution is irreversible and cannot return to its original state. Microevolution, at least some of its manifestations, such as changes in gene frequencies in populations, can return to almost initial conditions. Of the two aforementioned parts of the Timofeev-Ressovsky heritage, intensive experimental research into the radiation resulted in the fundamental discoveries. However, the second part of the heritage, namely the evolution, was much more difficult to study. Under the laboratory conditions, changes in allele frequencies of genes can be observed in populations, but in nature it is possible to find only evidences of the accomplished process, such as increasing the frequency of dark-colored moths around Manchester in response to the industrial pollution of the territory, leading to the darkening of the tree trunks.

In this paper we shall try to show the systems in which the microevolutionary processes can be directly observed and these processes are very important, since the productivity of the modern agricultural landscapes has greatly depended on them. These are the plant-microbe symbioses (MPS), whose significance for the modern agricultural production can be explained by the fact that genes of greatest interest

I.A. Tikhonovich (✉) · E.E. Andronov · N.A. Provorov
All-Russia Research Institute for Agricultural Microbiology,
Podbelsky Shausse, 3, St. Petersburg, Pushkin 196608, Russia
e-mail: ARRIAM2008@yandex.ru

V.L. Korogodina et al. (eds.), *Genetics, Evolution and Radiation*,
DOI 10.1007/978-3-319-48838-7_36

are those that rather control the resistance to environmental stresses than the productivity of cultivated organisms. This idea is quite obvious to experts in modern agriculture. According to the FAO CFS-HLPE report "Food security and climate change" (FAO CFS-HLPE Report 2012) "Efficient adaptation will require access (both physical and legal through appropriate intellectual property rules) to genetic resources, both of existing crops, livestock and their wild relatives, as well as varieties that may be used in the future. Crop genes encoding for drought and flood tolerance should be identified and shared. Yield stability traits of species under variable conditions are particularly important areas where more understanding and research is needed". To accept these challenges it is necessary to assess availability of these genes for selection and to find out what is the general adaptive potential of the eukaryotes.

In this respect we certainly admire the genius foreknowledge of Timofeev-Ressovsky. Decades before the first genome sequencing projects he had predicted the number of genes of higher organisms "… the number of genes cannot be equal to hundreds and millions… thousands, tens of thousands—that is what you need. … It seems that …the evolutionary progress relates not to the specialization and to the specialized adaptation to the specified condition, as it seems to many zoology and plant scientists, but just the opposite, morphophysiological differentiation relates to the maximum of omnipotence, i.e. with the sufficient degree of the absence of specific adaptations…" (Timofeev-Ressovsky 2009). Nowadays we know almost exactly the extent of genetic information in possession of the various kingdoms on our planet. Indeed, 20–30 thousands of genes comprise the entire genetic information maintained in chromosomes of the majority of eukaryotes. As for its diversity, it appears that genetic synteny does not have systematic limits and the choice of the new options is even less. These limitations of higher eukaryotes became especially apparent in the light of the recent achievements of metagenomics, basically dealing with prokaryotic diversity which operates several orders of magnitude more genetic information than in eukaryotes.

Genetic Design of Plant-Microbe Symbiosises

One of the largest reservoirs of prokaryotic diversity (taxonomic and functional) is the soil with its unimaginable genetic content. One gram of soil may harbor up to several billions of living prokaryotes representing several dozen thousands species with total DNA content reaching 10^{15}–10^{16} bp. Functional binning of soil metagenomes demonstrates that the most of coding sequences represent "unknown" proteins and two genomes belonging to the same species can differ in 10–40 % of unique genetic material. Even a simple presentation of the metagenomic data is an intractable problem demanding for fundamentally new solutions. One such attempt presenting 13-dimensional sequence space for 16S rRNA gene of Bacteria is shown on Fig. 1. (Pershina et al. 2014).

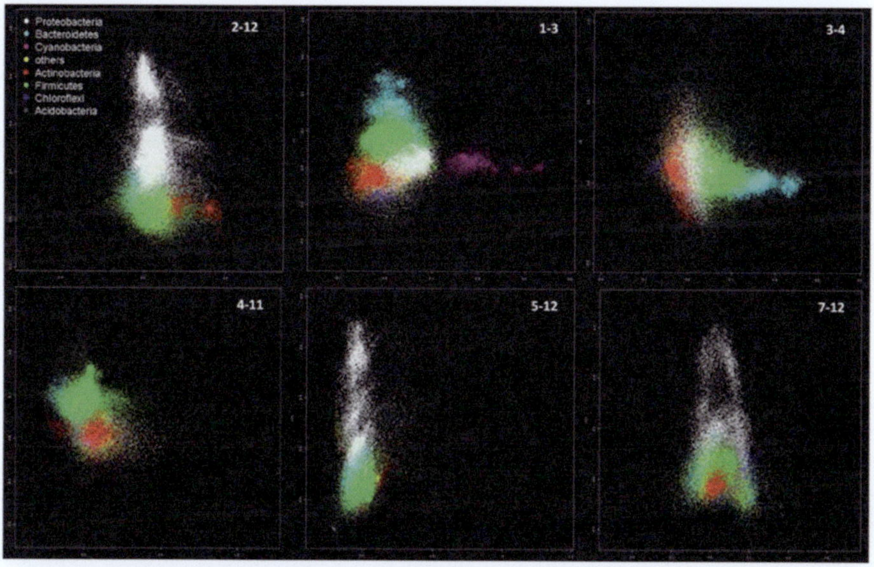

Fig. 1 The evolutionary space of the 16S rRNA gene of Bacteria. Results of the multidimensional scaling of a 16S rRNA database (more than 200,000 sequences) and 6 orthogonal sections through the 13-dimensional space. *Different colors* represent basic bacterial phyla

This figure demonstrates not only existing diversity but the trends of its expansion including "black holes" of used areas of the space, cyanobacterial ejection and other cosmology-like phenomena. Beyond these taxonomic frames there is a tremendous functional diversity capable to provide an adaptation with much greater extent than in the case of eukaryotic genomes only.

Therefore, an obvious question should be raised: how this prokaryotic diversity could became an element of the adaptive capacity of plants? Previously we formulated the principle of complementarity in the adaptation development in plants (Tikhonovich and Provorov 2009) which states that stable and limited by gene content eukaryotic genome may be complemented with microbial metagenome which is in its turn variable and unlimited with respect to the amount of genetic information. In this way the PMS system can increase the ability for the adaptation to environmental conditions. That is a field of "symbiogenetics" which is devoted to the analysis of integrated genomes, i.e. two or more organisms united in symbiotic associations. It is important to emphasize that it is not just a fact that plants are living together with microorganisms, but there is a new entity "symbiogenome", where genes from different genomes concertedly control a feature missing from both partners. Of course, these ideas have been stated much earlier in context of the origin of the eukaryotic cell, where most of the organelles are apparently of symbiotic origin (Margulis and Bermudes 1985). An interesting question arises if there still an area in nature for completely new symbioses or all these opportunities have already been realized? Probably not. We believe that there is a plenty of factors in the soil metagenome to be integrated in the plant genome (in symbiotic manner) providing highly essential environmental adaptations. Certainly, Timofeev-Ressovsky understood the importance of interspecific interaction in ecosystems. However he pointed out that their study would require completely new approaches: "... in populations and especially in biogeocoenosis we are dealing with, we face a damn bottomless pit of different factors... But even experimenting with relatively simple biogeocenosis we inevitably face a multi-factor system, which is hard to control. Frankly speaking, we almost never succeed in calculating variants, a factor or a component... And mathematicians can do it using their calculation models..." (Timofeev-Ressovsky 2009).

According to Timofeev-Ressovsky's opinion, the key factor capable of integration must be determinative for maintaining system as a whole. The most convincing example of such a factor is the symbiotic nitrogen fixation, which determines the possibility of life and the existence of proteins. As a result of nodulation in plant cells a temporary organelles "bacteroides" emerge (Fig. 2).

The ability to fix N_2 gives enormous advantages to organisms, but only few bacterial taxa have it. From the plant side only the Fabaceae (legume) family (with some exception) developed symbiotic ability to use atmospheric nitrogen as almost the only source of this element. Providing plants with ability to nitrogen fixation has been one of the main goals of genetic engineering, which still has not been implemented. Now there are many well-known useful features that microorganisms could pass on to their owners, namely optimization of phosphorus nutrition, hormone production, biocontrol effects, stress adaptation and others. We believe that an ability to use the microflora potential, eliminates the contradiction between the limited and unchangeable genetic information eukaryotes and the need to adapt to different environmental conditions. Here we can quote Timofeev-Ressovsky again:

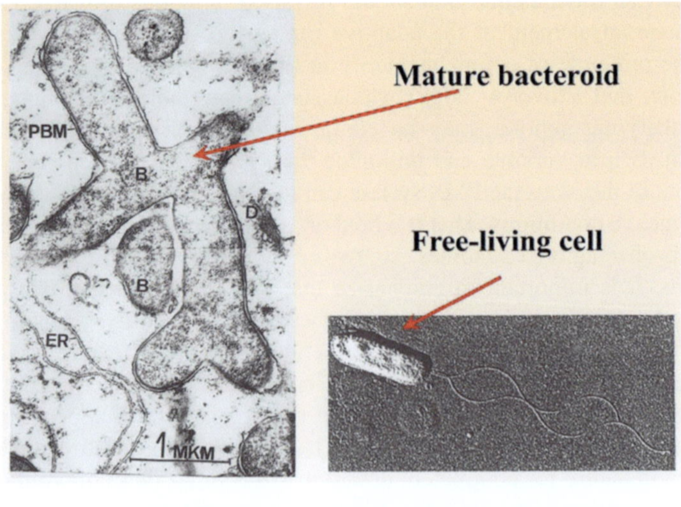

Fig. 2 Free living rhizobia and bateroides

".apparently evolutionary progress is associated … not so much with a major and specialized adaptation to certain conditions, as it seems to many zoologists and botanists, but on the contrary, it appears linked to morphophysiological differentiation associated with high evolutionary omnipotence i.e. with a sufficient degree of absence of special adaptations…" (Timofeev-Ressovsky 2009).

Studies of the genetic control of symbiosis show that plants follow this particular way. They use only about 50 hosting genes, which allow considerable expanding of their adaptive capacity, overcoming the limitations of their hosts' own genetic (chromosomal) information. In order to accomplish this task, microsymbionts should have the required functions to be able to penetrate and to colonize plant roots and other organs, survive in the soil during the time when the host is missing or does not need a microbial partner. Therefore, the question arises whether these microbes adapt to living in such different ecological niches as the soil and plant, and do it fast enough to have time to enter into a symbiosis promptly and develop effective symbiosis in a few days? Microevolutionary processes appear responsible for the adaptive diversification of rhizobia could underlie this phenomenon. In All-Russia Research Institute for Agricultural Microbiology (ARRIAM) we have obtained enough evidence that the genetic characteristics of soil nodule bacteria are quite different from those who live in the nodules (Fig. 3). These findings clearly indicate that the process of fine adaptation of soil rhizobia to the particular plant niche is driven by microevolutionary processes.

According to Timofeev-Ressovsky, there are four elementary evolutionary factors: mutation process, population waves, isolation and natural selection (Timofeev-Ressovsky 2009; Timofeev-Ressovsky et al. 1977). Symbiogenetic process is rather tightly controlled by the plant that at the first stage "defines" a need for interaction and its volumes.

Mutations in the plant genes controlling hosting result in a symbiosis failure, inefficient symbiosis or uncontrolled symbiosis when the plant produced ten times more nodules than it needs (Fig. 4).

It has been shown that the "long distance" signalling ("decision" on the formation of nodules made inside the plant leaf) is very similar to the regulatory meristem in the CLAVATA system (Osipova et al. 2012). To harmonize the nodulation process with the plant needs, signal peptides (CLE peptides) must interact with the CLAVATA complex producing a return signal, allowing nodule formation. The key to symbiosis development is a molecular signaling between partners. As a result of the binding plant flavonoids with rhizobial protein NodD the *nod*-operons of rhizobia are activated and so-called Nod factor is released to the environment. This factor is perceived by the receptor systems of plants, the first of which is a heterodimer of two receptor kinases, which is transmitting this signal downstream to another hosting genes. We succeeded in isolating a few genes that control the process of recognition in peas (Zhukov et al. 2008). The process of fine-tuning plant and rhizobial signals is reciprocal. A good example of coordinated adaptation of partners is the Afghan peas' case found in one of the gene center of Legumes by L.I. Govorov (1928) and Razumovskaya (1937). Analysis of the symbiotic specificity in this case (Afghan peas do not produce nodules with

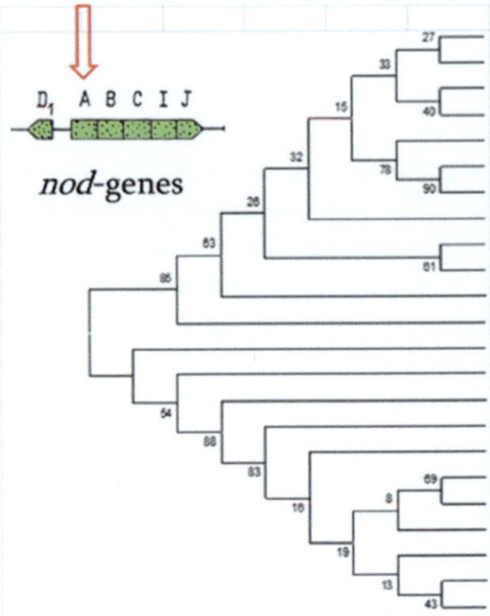

	Vic_nod	Vic_soil	Lath_nod	Lath_soil	Trif_nod	Trif_soil
OTU1	0	0	0	0	5	1
OTU3	3	0	0	0	5	10
OTU10	1	0	0	1	2	27
OTU11	0	0	0	1	10	10
OTU17	1	3	1	7	0	1
OTU12	0	0	0	1	0	2
OTU4	0	1	0	1	0	0
OTU23	77	589	9	675	3707	2812
OTU20	4	38	4	6	674	108
OTU7	0	2	0	1	4	5
OTU21	1	26	3	4	53	61
OTU19	0	2	0	0	7	1
OTU14	0	1	0	0	0	0
OTU5	0	2	1	0	0	1
OTU9	0	0	0	4	0	0
OTU2	2	1	1	0	0	0
OTU16	53	40	23	3	0	0
OTU13	0	0	2	0	0	0
OTU18	219	43	37	43	0	0
OTU0	16	0	6	2	0	0
OTU22	3402	2956	3540	3361	5	674
OTU15	280	45	23	1	0	0
OTU8	14	0	0	0	0	0

Fig. 3 nodA soil and nodule libraries demonstrating difference in distribution of OTU between soil and nodule population of rhizobia (Andronov et al. 2014) *Vicia* (Vic), *Lathyrus* (Lath) and *Trifolium* (Trif)

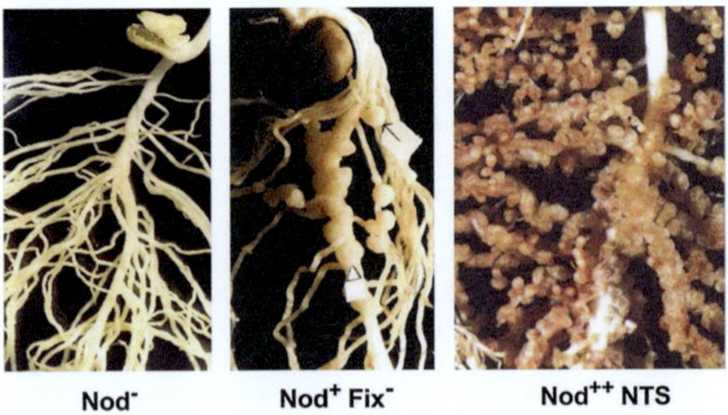

Fig. 4 Plant symbiotic mutants

rhizobial isolates specific for European peas) showed that this symbiotic phenotype determined by genetic factors of both macro- and microsymbionts. Nodule bacteria (*Rhizobium leguminosarum* bv. *viciae*) capable to inoculate Afghan peas possess an additional gene *nod*X, modifying the decor of the Nod factor by acetylation. The nature of the corresponding changes in the host receptor is not completely deciphered. However, V.A. Zhukov and his co-workers in ARRIAM proposed multi-stage recognition, including, apparently, the Afghan peas' previously unknown gene *Lyk*X, narrowing specificity of the Afghan peas and conferring on them an ability to distinguish Afghan rhizobia from European microsymbionts (Fig. 5).

When legumes moved to Europe, the need for such narrow specificity was lost as not needed in the European soils with lower diversity of the nodule bacteria. Thus, the corresponding receptor allele complex was eliminated from peas populations. Mutational analysis has allowed describing almost all hosting genes of legumes (Fig. 6). A total of only about 50 symbiotic regulatory genes were discovered in the genome of legumes.

This is consistent with the Timofeev-Ressovsky opinion of progressive evolution (Timofeev-Ressovsky 2009). It is important to note that the same genes are involved in the control of symbiosis with all three microbial partners: mycorrhizal fungi, rhizosphere-associated and nodule bacteria.

The plant seems to keep limited information potential, maintaining only a narrow range of the host genes. Thus, we propose to consider the symbiosis as a model

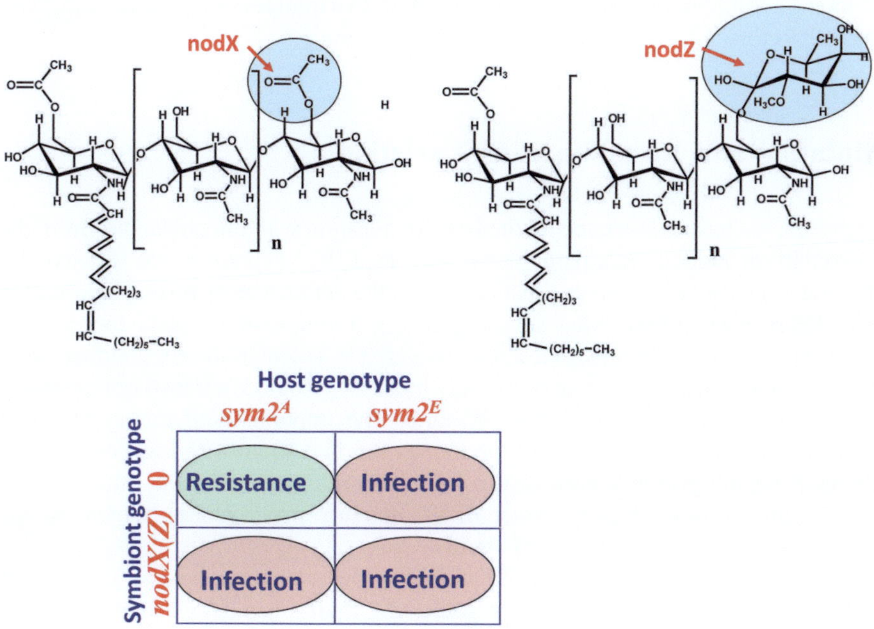

Fig. 5 Gene-for-gene interaction in Rhizobial-legume symbiosis. Structure of NodRlv factors required for nodulation of sym2A-carrying peas Modified scheme of Flor (1956)

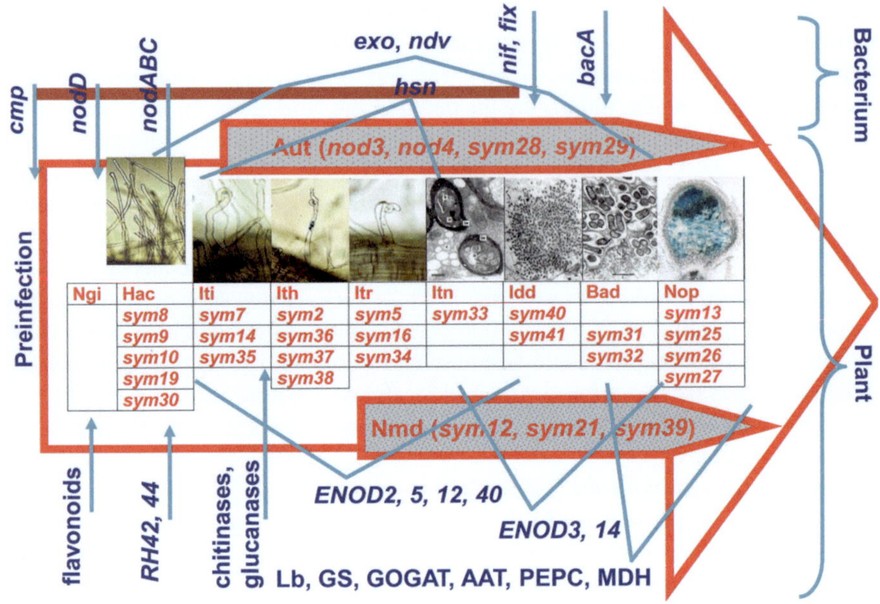

Fig. 6 Genetic dissection of root nodule morphogenesis. Plant symbiotic mutants

of micro-evolutionary processes that, according to Timofeev-Ressovsky, comprises several factors. Let's start with the genetic variation.

Mutation and Recombination Variation

For years, we have cooperated on the issue in question with our colleagues from the University of Leiden, and, particularly, with Prof. B. Lugtenberg. He showed that the mutation variability plays a major role in the settlement of rhizosphere niches (De Weert et al. 2004). After several passages through sterile plant root several bacterial mutants with enhanced ability to actively colonize the root surface were isolated. It has been shown that this ability arose as a result of a transposon insertion in reparation gene system of these bacteria. The importance of this system was confirmed by artificial insertion of the transposon in a strain with low colonization potential resulting in enhanced root-colonization ability. In this and other characteristics, in particular the emergence of rifampicin resistant clones, mutant strains were also superior to non-mutated strain.

The above effects are also clearly visible in the adaptation process to different soil conditions. For example, one of ARRIAM studies has demonstrated that rhizobia inhabiting salty marshes demonstrated very high frequency of IS-element insertion in the *bet*-region of rhizobial genome involved in salt tolerance. It is interesting that this insertion located exactly in the promoter region and resulted in enhanced expression of adjacent salt tolerance genes. Moreover, genomic microarray comparative hybridization showed that there was sufficient difference in the gene content between strains isolated in saline and normal soils (about 200 genes). We believe that such variability is also responsible for adaptive capacity of the whole PMS sufficient for the survival of nodule bacteria in different soil niches. However, other features crucial for symbiosis, have a very little impact on the environmental adaptations, namely, signaling, nodule development, nitrogen fixation, etc. Nodule bacteria usually do not fix nitrogen out of the plant and, therefore, the question arises as to how to ensure the creation of an effective system supporting all the necessary genes?

Let's consider the features of mutation process in the rhizosphere of the soil bacteria. As reported by several research groups, the nodule bacteria are distinguished by a high mutation rate, which is several orders of magnitude greater than, for example, in the case of *E. coli*. Earlier, a Russian scientist Nikolai Krasil'nikov showed that the properties of nodule bacteria could change under the influence of the plant (Krasil'nikov and Melkumova 1963). Although at that time this idea seemed controversial and even suspicious, we now see that the mutation process can be actually accelerated by an evolutionary driving force of plants in the framework of the symbiosis resulting in an increase in genetic diversity among the colonizers. Increasing diversity is achieved not only due to mutation. It is necessary to take into account the recombination processes. Timofeev-Ressovsky did not consider them important because recombinations in eukaryotes are not fixed in further generations, whereas recombination in prokaryotes is stably inherited and may be considered as a main contributor to increased diversity of nodule bacteria populations. Rhizobial genome is distinguished by high plasticity, which is mostly due to a large plasmid pool and different types of mobile elements including very large ones, the so-called "genome island", which very often can carry dozens and hundreds of genes and gene systems involved in environmental adaptations. At ARRIAM, Prof. B.V. Simarov and his colleagues showed that frequency of such islands in rhizobial populations inhabiting different stress-affected niches, such as saline soils, may indicate that acquisition of genome islands by recombination may confer adaptive benefit to its hosts (Muntyan et al. 2016). Thus, even in the soil without contact with plant, microevolutionary processes are actively employed by a nodule bacteria population. Adaptation to different soil conditions appear useful to all PMS, since stored in soil populations they ensure effective and sustainable symbiosis under a broad range of environmental conditions.

Evolutionary Factors

Another issue is to address the evolutionary factors responsible for fixation of novel genetic variants in populations which are also quite evident in symbiogenesis. Intensive exchange of signals, which takes place at early stages of the plant-microbe interaction is obviously a manifestation of a rhizobial partner selected by the plant. In response to flavones and flavonoids specific to the type of legumes, bacteria start to produce Nod factors in very low concentrations, which activate the genetic program in host plant, resulting in nodule formation giving a plant niche for rhizobia. The plant selects putative symbionts with precision, since no more than ten cells are requisite for sufficient inoculation. The whole amount and specificity of the pool of signals between bacteria and plants are hard to imagine. However, it is clear that signaling enables selection of not so much of bacteria with high fitness potential for survival in the soil, but rather of the genotypes that ensure symbiosis development. It is not a typical environmental adaptation, since the bacterial cells in question should provide for increased opportunities of a microbial-based plant adaptation for vegetation. Thus, there is an example of "kin selection" where the group is a whole PMS increasing survival and forming symbiosis. Surprisingly, to date there is no clear answer to the question how symbioses used for symbiotic bacteria living in the soil do not get into the plant? For a few rhizobial cells selected by the plant the transition to symbiotic state means a kind of altruism as bacterial cells inside of plant nodule are irreversibly converted into bacteroides (nitrogen-fixing temporary organelles) and inevitably die at the final stage of symbiosis. Even cells that enter the intercellular plant spaces and escape transformation into bacteroides, cannot be viewed as candidates to enrich the soil population due to the fact that passing through the plant appears unlikely for enhancing the adaptation to the life in the soil. Nevertheless amount of soil rhizobia is always higher after growing legume, which is difficult to explain. The hard specific selection raises the question of genetic drift. Availability of signaling genes does not guarantee that this strain contains the most effective alleles of other genes. For instance, this is true for genes, providing the symbiotic nitrogen fixation. Mutations that lead to a decrease in efficiency or strains unable to fix nitrogen are well known in rhizobia. Nodule bacteria are unable to fix nitrogen in the free-living state, so nitrogen fixation genes are not exposed to selection in the absence of plants. As a result, rhizobia effective in nodulation but ineffective in nitrogen fixation can freely enter the plant. Here, the selection operates separately in the nodules where two strains differ greatly in their ability to fix nitrogen together. In addition, the plant is able to limit a flow of energy sources to the ineffective nodules. Despite the selection, spontaneously inoculated leguminous plants in wild populations differ significantly in the effectiveness of the symbiotic relationship. Such PMS features are of particular practical interest.

Ensuring a transition to biologically fixed nitrogen without adaptation and productivity losses is possible when the PMS will function effectively, providing the plant with a sufficient amount of nitrogen with minimal energy consumption. In practice this is achieved by the use of preparations of nodule bacteria, which contain strains selected for maximum productivity of PMS. However, these strains cannot always successfully compete with soil aboriginal strains capable of actively colonize the rhizosphere, but inferior to the commercial strains on productivity. Since there is no hope that inoculant strains can occupy a dominant position in the soil, inoculation must be done each year, with hope that eventually the soil rhizobia populations will be enriched by alleles ensuring more competitive and effective symbiosis. Thus, dissection of the mechanisms of microevolution of bacteria is a way to increase legume inoculation efficiency.

Forming symbiosis with rare variants of rhizobia indicates that some rare alleles can get advantage caused by the pressure of selection, for example, signaling genes. Efficiency factors, including ineffective alleles are subjected to genetic drift leading to the increase of their frequency in the population. This situation is very similar to the "population waves" reviewed by Timofeev-Ressovsky, who wrote that "population waves as an elementary factor of evolution are evolutionary material suppliers… in the sense of increasing their concentrations to the sufficient limits … at least for the selection process" (Timofeev-Ressovsky 2009). When considering a symbiotic system we are unable to fully reconstruct all aspects of selection pressure inside the plant, but it is clear that it plays a defining role. Products of the "good" microevolutionary events are fixing in the populations. One of the most interesting models is the plant genes involved in signaling. Such genes can be protein kinase genes *Nfr*5 and *Nfr*1, which determine the production of the receptor heterodimer percepting rhizobial Nod factor. The analysis of populations of several types of legumes showed that there are many different allelic variants of these genes distinct in synonymous and nonsynonymous substitutions in the genetic code. Molecular modeling of interaction between plant receptors and Nod factor showed that sometimes substitution of one amino acid results in a clear-cut effect on this interaction (Fig. 7; Muntyan et al. 2012; Porozov et al. 2012).

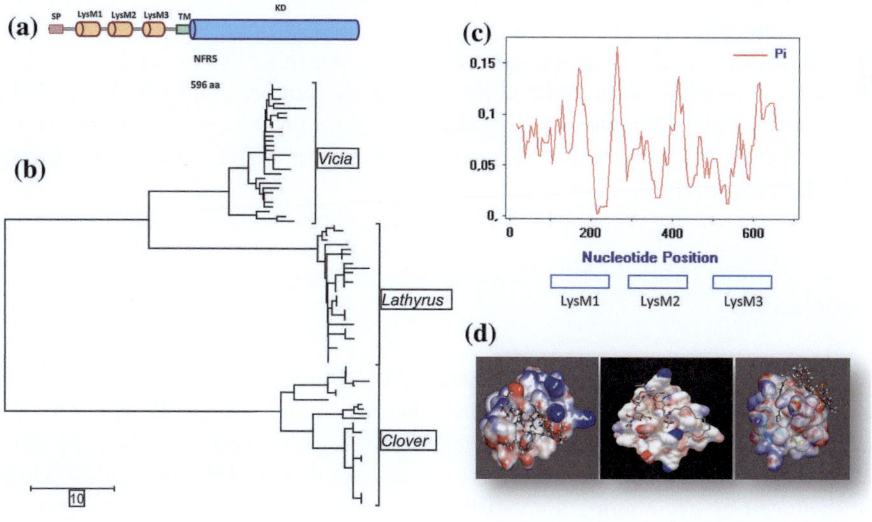

Fig. 7 Nucleotide diversity in plant receptor gene Nfr5 and their effect on the nod-factor docking. Nfr5 plant receptor domain structure (**a**), nucleotide phylogeny (**b**), nucleotide polymorphism (**c**), molecular docking with rhizobial nod-factor

Thus, microevolution of bacteria is probably accompanied by a slower process of enriching legume populations with receptor gene alleles, ensuring more precise and fine-tuned symbiotic interaction. As a result the identified strains of nodule bacteria entering the plant, increasing in number and formation of a specific structure of symbiotic bacterial population more adapted to the particular plant population.

Finally, isolation as the third Timofeev-Ressovsky condition is met. The isolation inside the plant provides independence from the soil bacteria population, whereas inside the nodule strains have much more opportunities for recombination, and the frequency of the recombination is much higher than that in free-living relatives. However, these recombination processes affecting a very limited material, as only a few rhizobia from the soil population can get into the nodule.

The process of PMS formation has all signs of microevolution both from the micro- and macro-symbionts providing rapid adaptation of the partners to each other and thereby secures the exclusion of nonspecific symbiotic and pathogenic strains. The greatest interest in this regard is the launch of the mutation process at the presymbiotic stage. Probably, there is a triggering mechanism to increase the genetic diversity of microsymbionts in course of plant-microbe interaction. There is a dangerous modern trend to link these processes with the "scientific" heritage of an evil genius of Soviet science T. Lysenko, who thought that plant deliberately changes (become "educated") under the impacts of bacteria inhabiting its rhizosphere to ensure the best fitness (Timofeev-Ressovsky 2009; Timofeev-Ressovsky et al. 1977).

Unfortunately, the data had been used to deny the existence of genes, even though actually the analysis of the genetic regulators allowed getting a real picture of the symbiogenesis microevolution. It may only seem that this is the scientific foresight of the notorious prosecutor of genetics, but in reality the purpose was to discredit the problem, after which serious scientists could risk of being involved in studies of this area. Timofeev-Ressovsky had been a prominent fighter against the pseudoscience; his work does not leave a chance to supporters of heredity change through "education".

Conclusion

In this paper we argue that development of PMS, which involves penetration of bacteria into the plant is accompanied by the switching on a genetic network that determines the microevolution of bacterial populations within the plant. Modern research gives a general outline of the genes operation in this process, which is substantiated by deciphering the structure and function of major symbiotic factors. The nature of symbiogenesis regulation is mostly negative: the number of infection threads greatly exceeds primordia forming nodules, and of these, only a few percent are converted into functioning nodules. The quantitative limits of symbiotic apparatus are controlled in the systemic manner, which optimizes the process of nitrogen fixation in accordance with the needs in this element and yield. The plant also imposes severe restrictions on the behavior of microsymbionts inside the root. Bacteria multiply the amount of DNA by amplification and stop the cell dividing. Up to 40 % of a bacteroid cell can consist of nitrogenase protein, cell greatly increases in the size and acquires a characteristic shape, specific for the bacteroid. We succeeded in isolating plant genes controlling the transformation of bacterial cells into bacteroids. Surprisingly, there are probably only two, one of which—*Sym31* the most likely candidate of this type (Zhukov et al. 2008). Control of microsymbionts may be regulated by the most important cellular processes such as division. Little is known about specific mechanisms of this control, but they certainly would be very interesting for understanding the fate of the bacterial cells inside the hosts. Bacteroides' persistence is also strictly controlled by plant and mutations with effect the «early senescence» are among most frequent alleles of regulatory genes of symbiogenesis. In the course of symbiosis development, a number of facts that further extend the Timofeev-Ressovsky theory of microevolution can be found. By the way, he clearly understood the importance of studying the evolutionary processes by using models of the interaction between living organisms. Speaking of biocenoses, he emphasized that understanding how to maintain consistency in the ecosystem is vital knowledge (Timofeev-Ressovsky 2009). At the same time, he realized that the biocenotic problems are multifactorial in their nature, and that research into isolated factors is hardly possible. The current knowledge of the nitrogen fixing symbiotic relationship allows us to infer a reduction of these relationships to one factor only, i.e. the nitrogen fixation, which is evident in the process of adaptation to symbiosis partners integrating their genetic resources in a network and creating an integrated symbiogenome with a clear biocenotic function.

Now it is clear that symbiogenesis is tightly linked with the processes of macroevolution and speciation, but this relationship is much more complex than it seemed previously. Also, we understand that in the evolution of symbiosis recombination plays more important role than mutations as a source of functional innovations (Provorov and Vorobyov 2010). Many parameters, details and features of micro- and macro-evolution, action of different types of selection can be simulated in computer experiments (Provorov et al. 2012). However, the importance of scientific heritage of Timofeev-Ressovsky is invaluable. The evolutionary concepts developed by him significantly contribute to modern understanding of causes and mechanisms of natural history.

Acknowledgements The work was supported by Russian Science Foundation (Grant No 14-24-00135, 2014–2016).

References

Andronov EE, Onishchuk OP, Kurchak ON, Provorov NA (2014) Population structure of the clover rhizobia *Rhizobium leguminosarum* bv. *trifolii* upon transition from soil into the nodular niche. Microbiology 83(4):500–510 (Russian)

De Weert S, Dekkers LC, Kuiper I et al (2004) Generation of enhanced competitive root-tip-colonizing Pseudomonas bacteria through accelerated evolution. J Bacteriol 186 (10):3153–3159

FAO CFS-HLPE Report #3 (2012) "Food security and climate change". A report by the high level panel of experts on food security and nutrition, Rome, Italy, p 102

Flor HH (1956) The complementary genic systems in flax and flax rust. Advan Genet 8:29–54

Govorov LI (1928) The peas of Afghanistan. Bull Appl Bot Genet Plant-Breed 19(2):497–522 (Russian)

Krasil'nikov NA, Melkumova TA (1963) Nodule bacteria variability inside the nodules of legume plants. Proc Ac Sci USSR 5:693–706 (Russian)

Margulis L, Bermudes D (1985) Symbiosis as a mechanism of evolution: status of cell symbiosis theory. Symbiosis 1:101–124

Muntyan AN, Andronov EE, Belova VS et al (2012) Linked symbiotic populations part I: analysis of the genetic diversity of the rhizobial component. Russ J Genet: Appl Res 10(1):3–11 (Russian)

Muntyan VS, Cherkasova ME, Andronov EE et al (2016) Prevalence of "islands" in genomes of *Sinorhizobium meliloti* native isolates. Russ J Genet in press (Russian)

Osipova MA, Mortier V, Demchenko KN et al (2012) Wuschel-related homeobox5 gene expression and interaction of CLE peptides with components of the systemic control add two pieces to the puzzle of autoregulation of nodulation. Plant Physiol 158(3):1329–1341

Pershina EV, Dolnik AS, Tamazyan GS et al (2014) The Evolutionary space model to be used for the metagenomic analysis of molecular and adaptive evolution in the bacterial communities. (Book Chapter) In: Evolutionary biology: genome evolution, speciation, coevolution and origin of life. Springer International Publishing, Switzerland, pp 339–355

Porozov YuB, Muntyan AN, Chizhevskaja EP et al (2012) Conjugate symbiotic populations part II: analysis of nfr5 receptor gene polymorphisms using molecular docking. Russ J Genet Appl Res 10(1):12–18 (Russian)

Provorov NA, Vorobyov NI (2010) Evolutionary genetics of plant-microbe symbioses Tikhonovichs IA (ed). NOVA Science Publishers, NY, p 290

Provorov NA, Tsyganova AV, Brewin NJ et al (2012) Evolution of symbiotic bacteria within the extra- and intra-cellular plant compartments: experimental evidence and mathematical simulation (mini-review). Symbiosis 58(1–3):39–50

Razumovskaya ZG (1937) Nodule formation in various pea cultivars. Russ J Microbiol 6(3):321–328 (Russian)

Tikhonovich IA, Provorov NA (2009) From plant-microbe interactions to symbiogenetics: a universal paradigm for the inter-species genetic integration. Ann Appl Biol 154(3):341–350

Timofeev-Ressovsky NV (2009) Genetics, evolution and methodology of natural sciences. Lectures delivered in Sverdlovsk in 1964. Tokmas-Press, Ekaterinburg, p 240 (Russian)

Timofeev-Ressovsky NV, Vorontsov NN, Yablokov AV (1977) Brief draft on the evolutionary theory. Nauka Publ Moscow, p 302 (Russian)

Zhukov V, Radutoiu S, Madsen LH et al (2008) The pea Sym37 receptor kinase gene controls infection-thread initiation and nodule development. Mol Plant Microbe Interact 21(12):1600–1608

The Animal Domestication Experiment as a Model of the Evolutionary Process: A New Insight into Evolution Under Selection Targeting Regulatory Systems

Ludmila N. Trut, Yury E. Herbek, Oleg V. Trapezov, Sergey A. Lashin, Yury G. Matushkin, Arcady L. Markel and Nikolay A. Kolchanov

How do biological systems evolve in time? A century ago there lived Crocodile No. 17, now there lives Crocodile No. 187, who is quite a different crocodile, so what have we got? We've got some "crocodilery", but what is that? .... a type of original control systems... this is fundamental to the modern concept that is absolutely required for a correct, efficient and fast-paced development of biology in the near future.

N.W. Timofeeff-Ressovsky ("*Genetics, evolution, significance of methodology in natural sciences*").

Introduction

Plant and animal domestication started out about 11–15 thousand years ago and in fact marked a transition from hunting and gathering to a new period of human evolution, the modern civilization. The development of the modern society proceeded side by side with the emergence and inclusion of domesticated animals and plants in human work activity and civil life, and the importance of these processes for human welfare is still as great. At the same time, the genetic mechanisms of domestication remain to be poorly studied. Pondering the matter of animal and plant domestication, Charles Darwin assumed that domestication is a result of selection, which can be seen as a model of evolution (Darwin 1883). It is therefore no wonder that the reconstruction and study of the history of domestic animals and plants may represent quite an efficient tool for studying the main factors and mechanisms of the evolutionary process. These considerations made Dmitry Konstantinovich Belyaev (Fig. 1) conceive and initiate in the 1950s a daunting experiment on the domestication of farmed silver foxes (*Vulpes vulpes*) (Belyaev

L.N. Trut · Y.E. Herbek · O.V. Trapezov · S.A. Lashin
Y.G. Matushkin · A.L. Markel (✉) · N.A. Kolchanov
Institute of Cytology and Genetics, SB RAS, Novosibirsk, Russia
e-mail: markel@bionet.nsc.ru

V.L. Korogodina et al. (eds.), *Genetics, Evolution and Radiation*,
DOI 10.1007/978-3-319-48838-7_37

1972)—“one of the century’s most intriguing systematic investigation of the nature of evolutionary processes”, as American biologists Coppinger and Feinstein (1991) put it. Similar studies were later begun with other objects, the American mink (*Mustela vison*) and the gray rat (*Rattus norvegicus*) (Belyaev and Borodin 1982; Plyusnina and Oskina 1997). Selection was performed in the opposite direction, for the preservation of the “wild” behavior. That resulted in the populations of foxes, minks and gray rats with markedly aggressive behavior. These works are still ongoing (Trut et al. 2009).

Performed on different animal species, selective breeding experiments allowed Belyaev and his co-workers to identify and state some of the common patterns of the evolutionary process that runs under strong selection acting to induce and fix the so-called domesticated type of behavior. The important result was the creation and development by Belyaev of the concept of a destabilizing function of selection for domestic behavior, the feature that marks the function of the main neuroendocrine regulatory systems, and about the role of stress responses in the evolutionary process (Markel and Borodin 1981; Belyaev and Borodin 1982).

Fig. 1 Director of the Institute of Cytology and Genetics of the Siberian Branch of the USSR Academy of Sciences (1959–1985), Full Member of the USSR Academy of Sciences Dmitry Belyaev in the company of domesticated foxes. A photo from the ICG archives

The Fox (Vulpes Vulpes)

Quite the unique result of Belyaev's experiment performed on the silver fox was that the domestication process had been put into as few as dozens of years and the wild foxes bred on commercial farms turned to pets, very similar to dogs in behavior and attitudes towards humans. The fact that foxes and dogs are in the same family (Canidae) made it possible to draw parallels between the dog, whose domestication began about 15 thousand years ago, and the fox, who it took only 50 years to turn into a true pet. These parallels are impossible to overlook, for they appear as well-seen morphological traits such as depigmentation areas, floppy ears, the curly tail, facial bone abnormalities, skeletal limb abnormalities, to mention a few (Trut 2001; Trut et al. 2004) (Fig. 2). Yet by far the most impressive changes are those in the behavior of the domesticated foxes.

Their behavior becomes truly dog-like (Belyaev et al. 1985; Kukekova et al. 2008) (Fig. 3). The foxes seek to be human-friendly, they develop many more social (including vocal) cues that promote that kind of communication (Gogoleva et al. 2010).

It can therefore be stated that there is an emerging new social structure that put together foxes and humans (Kukekova et al. 2006). In other words, it is crystal clear that major transformations have over an evolutionarily short period of time occurred and been genetically fixed in the population of domesticated foxes, these transformations applying to all the characteristics of these foxes' morphofunctional and behavioral organization (Figs. 4 and 5).

The rates at which these evolutionary changes occur are very difficult to explain by the general notions of the mechanisms underlying biological evolution. That made Belyaev to put forward a new concept, according to which selection becomes destabilizing, when it acts to change the functions of the main systemic regulators. That kind of selection was acting on the foxes under domestication. As is known, the foxes were under selection for the behavior (Trut 1988). However, that was not a primitive behavioral act like a simple reflex. That was a set of behavioral reactions that characterize the wild animals' attitude towards humans. This attitude has been in the making for millions of years of evolution and finally appears as a fixed set of cowardly and aggressive responses (Trut 1999). The aim of selection for domestication is to transform this solid emotional and behavioral set of responses in such a manner that humans stop being seen by the animal as enemies and turn into friends and even form part of a new zoosocial structure, which would otherwise have not been so comfortable for the animal (nor would it for man, afterwards). This transformation leads to changes in the functions of practically all systems in the organism. Having to withstand these harsh demands for animal domesticated generations it should be stressful. According to Belyaev, stress acts as one of the main mechanisms that implement the destabilizing function of selection for domesticated behavior (Belyaev and Borodin 1982; Markel and Borodin 1980; Markel and Trut 2011). This function induces high population variability. As has been pointed out, this variability applies to the most diverse traits: behavioral,

Fig. 2 Phenotypic parallelism between dogs and domesticated foxes

morphological, physiological (including those that should have been securely stabilized by natural selection, such as seasonal breeding), hormonal (including the function of the hypothalamic–pituitary–adrenal axis, which is the central regulator of stress responses) and the ontogenetic processes of maturation of functional systems (Belyaev et al. 1985). Rearrangements in the central neuromediating mechanisms of the brain, this integral regulatory center, appear to have a special role (Popova and Kulikov 1997; Trut et al. 2000). Belyaev wrote: "Explanations for the roots, rates and type of the variability that we observe should most probably be

Fig. 3 The behavior of foxes selected from commercially farmed populations as the founders for a selective breeding experiment: **a** a non-aggressive response towards humans; **b** an enhanced human-friendly response in first-generation foxes under selection for domestication

Fig. 4 Various emotionally positive responses towards humans become apparent as early as in the fourth-generation foxes under selection. **a** Pups are happy to see a human at the their youngest age; **b** "tame" foxes will remain human-friendly throughout all of their lives; **c** the behavior of "domestication elite" foxes resembles that of pet dogs

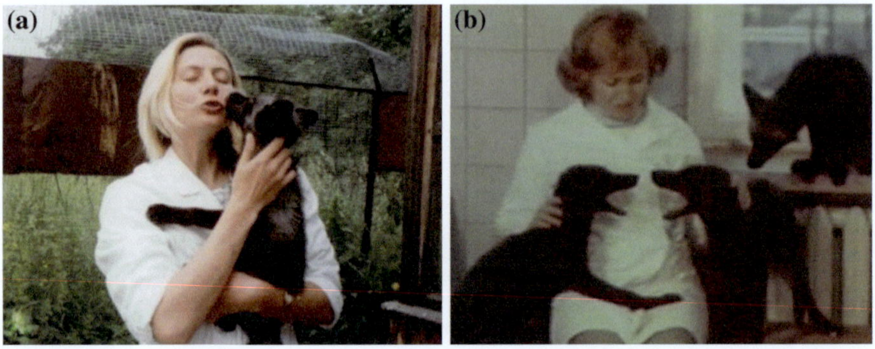

Fig. 5 **a** A domesticated fox in an experimenter's hands displays extreme friendliness, trying to lick the human's face. **b** Tame foxes rivaling for human attention, trying to roar each other away

looked for in the specific features of the functional system that is directly acted on by selection and the status of which serves as the main criterion for animal selection at the earliest stages of their domestication" (Belyaev 1979). Here, the primary target of selection is the neuroendocrine system, which is involved in the regulation of both behavior and many other functions and systems of the organism.

Belyaev (1979) hypothesized that it was nothing else but selection acting, in stressful conditions, on the main neurohormonal developmental pathways that induced as much variability as is observed under domestication. It is possible that stress and the associated destabilization of homeostasis represent an important mechanism of mobilization of hidden genetic variability (Belyaev et al. 1981). Additionally, mutation frequency and recombination probability can increase under stress (Belyaev and Borodin 1982). Evidence for a special role of stress in the development of an epigenotype can be found in literature (Nestler 2012). It can therefore be hypothesized that the intensification of the processes underlying epigenetic regulation of the functions of any and all genes under stress induced by domestication could be the main mechanism of rapid phenotypic transformations in animals under domestication. This could explain the extremely high rates at which the observed changes occurred and which cannot be accounted for by any mutational process, because such processes are slow.

In fact, the function of destabilizing selection is to mobilize genetic variability, which destabilizes ontogenesis and destroys previously correlated associations. As a result of intensive selection, a new harmony in the form of the two-component system "human—domesticated animal" emerges from the mess (Fig. 6).

Belyaev's idea about destabilizing selection and the role of stress in it is, beyond doubt, novel and important for understanding the evolutionary process, when it runs in unusual, harsh conditions. Needless to say, his idea was not something that came out of the blue, for certain prerequisites can be found in the works of evolutionist biologists, molecular biologists and endocrinologists.

It was many decades ago that the capability of organisms to become, when in unusual conditions, especially plastic came to Darwin's notice. He wrote: "There

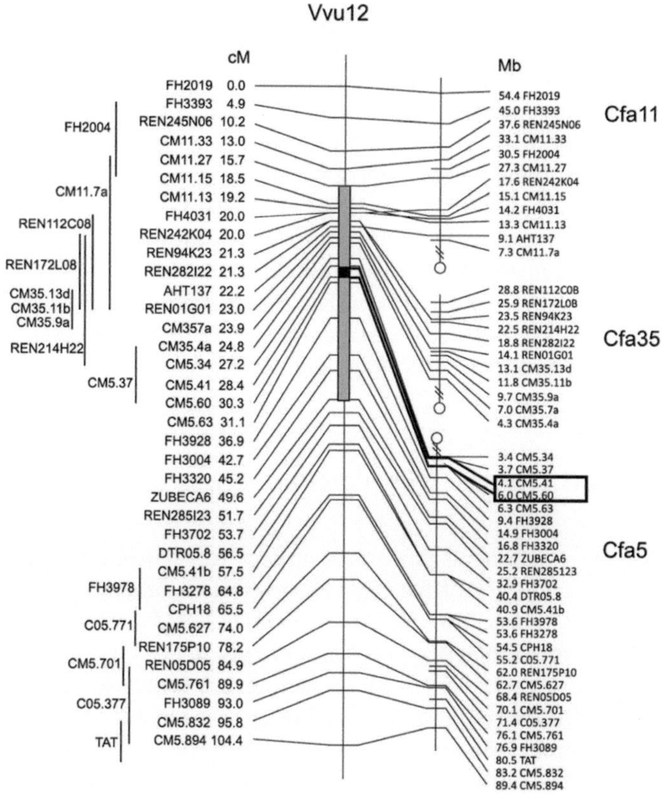

Fig. 6 The meiotic linkage map of fox chromosome 12 (Vvu12 on the *left hand side*) aligned against the nucleotide sequences of homologous dog chromosomes (Cfa11, Cfa35 and Cfa5 on the *right hand side*)

can, however, be no doubt that changed conditions induce an almost indefinite amount of fluctuating variability, by which the whole organization is rendered in some degree plastic" (Darwin 1927). The same mobilization of indefinite variability in unusual (that is, stressful) conditions was pointed out by Schmalhausen (1987). On the other hand, it is now well known that one of the main ways in which the effects of stress and hormones mobilized under stress operate is through regulation of genetic activity.

One of the main conclusions that can be made at this point is that stress—the commonest molecular-genetic and neuroendocrine phenomenon for all organisms faced with unusual environmental factors—can be an important factor of evolution. It appears as though any major evolutionary innovation comes to the scene when stress and destabilization are factors, for an all-new organization cannot be expected until the previous forms and functions have undergone considerable disintegration.

Belyaev exemplified the evolution of neuroendocrine mechanisms with that stage in primate evolution when human ancestors began to appear (Belyaev 1982).

It is the involvement of the neuroendocrine system in selection that may have strongly destabilized the most important correlation, which resulted in many weird and, most probably, dead-end evolutionary paths (Calow 1983). Admittedly, one of the experiments set up by Nature itself ended up leading the way to *Homo sapiens*. Furthermore, hominid brain growth may well have been one of the main reasons for the intensification of hominid evolution (Wilson 1985). This is consistent with the modern notions of self-domestication (Hare et al. 2012), which see migration to a new ecological and social niche with sponsorship from a well-developed nervous system as the most important factor of anthropogenesis.

The map and data are borrowed from Kukekova et al. (2011). At the center: the gray segment delimits fox QTLs; the black segment delimits the region homologous to the CFA 5 locus and relevant to the evolutionary transition from wolf to dog.

The Brown Rat (Rattus Norvegicus)

Selective breeding experiments performed on brown rats captured in the wild, too, exemplified the possibilities of the rapid domestication of quite a wicked kind of animal under intensive selection for non-aggressive behavior towards humans.

The brown rat (*Rattus norvegicus*) has a notorious reputation for being a dangerous, filthy and aggressive thing, pestering us humans wherever we are (Robinson 1965). On the other hand, brown rats are the founders of a large number of colonies and varieties of the so-called laboratory or tame (domesticated) rats, who have become as popular objects of the most diverse scientific research as laboratory mice and are even known to be kept as pets (Hedrich 2000). This is indicative of an extreme plasticity of the species *Rattus norvegicus*, of its capability to change significantly under natural and artificial selection and to be rapidly domesticated (Zeder 2012). The brown rat is seen by some authors as one of the most suitable species for studying the genetic and physiological mechanisms of domestication (Boice 1972). In the 1980s, this artificial reproduction of the spontaneous domestication of the wild brown rat, which dates back to 100 years ago, was initiated by Belyaev's suggestion at the Institute of Cytology and Genetics of the Siberian Branch of the former Academy of Sciences (Novosibirsk, Russia), first by Pavel Borodin (Belyaev and Borodin 1985) and later on by Irina Plyusnina and her associates (Plyusnina and Oskina 1997), with the aim to study the genetic mechanisms of domestication. Importantly, the rats captured in the wild were under selection in two opposite directions: some for the preservation and even enhancement of aggression towards humans and the others for the elimination of fear and aggressive responses towards humans. As a result, now there are two brown rat populations in the Animal Facility of the Institute of Cytology and Genetics: one aggressive and one tame (domesticated). Following selective pressure placed on the animals in two opposite directions for dozens of generations, differences in behavior towards humans between the aggressive and the tame rats have become genetically fixed and phenotypically clear: the aggressive rats cannot be handled

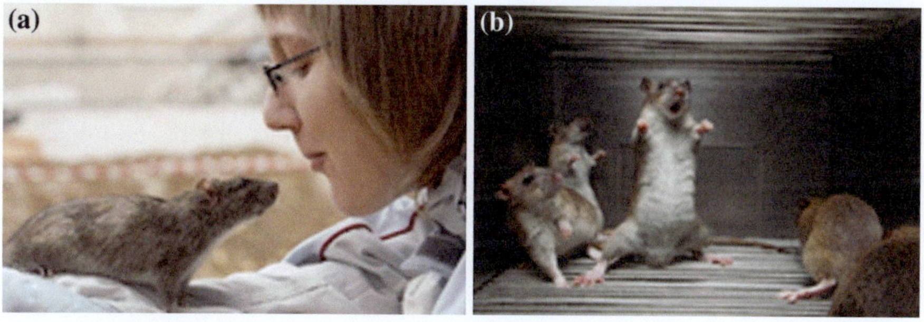

Fig. 7 Domesticated and aggressive brown rats (*Rattus norvegicus*)

safely unless your hands are protected by heavy duty gloves, while the tame rats are as easy to handle as laboratory Wistar rats. These behavioral nuances have been described in many publications and these two lines of behaviorally contrasting rat have been brought to the focus of many works on the study of the genetic and physiological mechanisms of aggression and domestication (see Konoshenko (2012) for plenty of literature) (Fig. 7).

The most important conclusion made is one about a similarity in the changes that occurred under domestication between the wild brown rats and the domesticated foxes. These are, first of all, multiple changes in coat color (the occurrence of depigmentation areas and changes in hair pigments) (Gulevich et al. 2010). The authors point out that depigmentation areas occur more frequently in brown rats with high scores of domesticated behavior. As in the foxes under domestication, multiple changes in the functions of the neuroendocrine systems (the pituitary-adrenal one in the first place) are observed to occur in the brown rats under domestication (Shikhevich et al. 2003), of which the reactivity is reduced not only in response to emotiogenic stimuli, but also in response to stimuli that alert the immune system. The authors associate the changes in the peripheral neuroendocrine systems with rearrangements in the central regulatory mechanisms, in particular, changes in the function of the brain serotonin system. In the domesticated rats, the density of some subtypes of serotonin receptors in the brain's parts is somewhat increased and the expression of the serotonin transporter gene is changed (Naumenko et al. 2009). Interestingly, in response to stress, the brain expression of the immediate early gene *c-Fos* in reduced only in the hypothalamus, which is immediately involved in the regulation of many neuroendocrine responses (Konoshenko 2012). Thus, the involvement of the stress response system in exposure of large amounts of variability at once is as well documented in the brown rats under domestication as in the domesticated foxes. To summarize this Section, cross-fostering experiments demonstrated that little do the changes observed in the progeny depend on ways in their aggressive or tame mothers' behavior; rather, these changes are largely due to selection, which acts to induce domesticated behavior (Plyusnina et al. 2009).

The American Mink (*Mustela Vison*)

Because Belyaev had a great interest in the effects that selection for domesticated behavior might produce on animals in different taxonomic groups, he initiated a selective breeding experiment on the American mink (*Mustela vison*) bread on the Experimental Farm of the Institute of Cytology and Genetics, with the aim to induce and fix non-aggressive and aggressive behavior towards humans (Trapezov and Belyaev 1986; Trapezov et al. 2009).

The results of selection for behavior for 18 generations are shown in Fig. 8. These results demonstrate the efficiency of that selection first of all. As can be seen from Fig. 8, the course of selection in two opposite directions is different. The main effect of selection for aggression becomes observable over the first two generations, with little variation in the mean afterwards. Selection for domesticated behavior is not that easy: the curve is saw-toothed in shape.

Additionally, in contrast to the animals under selection for aggression, those under domestication had throughout 15 generations under selection sex-dimorphic tame behavior [males were tamer than females on average ($p < 0.001$)]. What is even more important, the value of the phenotypic variance of the trait under selection, behavior, was 10 times as high in the population under domestication as in the population under selection for aggression.

The effort resulted in two mink populations with different defense responses to humans. A rapid divergence of the populations under selection suggests that the presence of aggressive responses in the original mink population was likely to be controlled by a small number of loci with a strong additive effect.

A complex pattern of change of the mean and phenotypic variance of the trait was demonstrated. The phenotypic variance is higher under selection for domesticated behavior than under selection for aggression.

One of the first correlated responses in the American minks under selection for domesticated behavior was, as in many other animals domesticated previously, increase in depigmentation (white spots) (Trapezov 2002). By contrast, the correlated response under selection for aggression was epidermal and hair hyperpigmentation (Fig. 9). As was demonstrated by molecular-genetic analysis, the

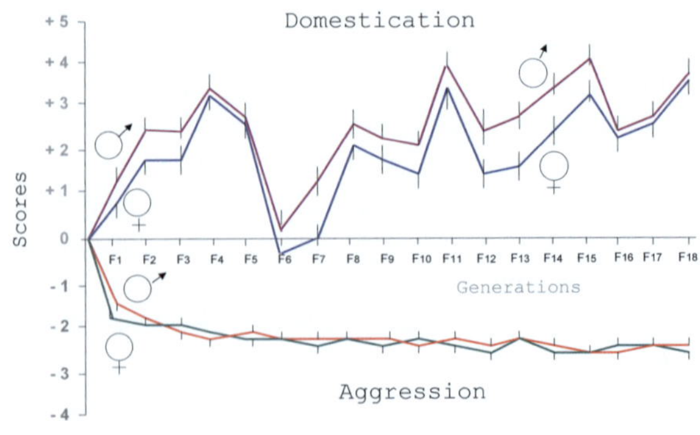

Fig. 8 Changes in mean behavioral score in the populations of minks under selection for domesticated and aggressive behavior for 18 generations

Fig. 9 Changes in pigmentation during the selection—driven transformation of behavior in minks: **a** a control animal, which was under no special selection for behavior; **b** *black*-and-*white* minks following selection for tame behavior have well-observed domestication markers; **c** epidermal and hair hyper pigmentation in the minks under selection for aggressive behavior

hyperpigmented animals have large duplications of genomic regions with various genes in them. One of them produces the neuropeptide neurotensin, which has relevance to aggressive behavior. It can be hypothesized that aggressive behavior can be enhanced in animals that possess additional copies of this gene. Additionally, this duplication contains the *KITLG* gene implicated in pigmentation, which may account for hyperpigmentation in aggressive animals.

Individuals with color phases never seen in the original population were observed to occur in the group under domestication at a frequency of 10^{-3}. A hybridological analysis of these phenotypic innovations demonstrated that they are semidominant (Figs. 10 and 11).

Selection for behavior was accompanied not only by the emergence of new color phases, but also by morphological aberrations. Minks with helically curled tails (in full analogy with foxes) were observed to occur (Fig. 12). The genetic control of that trait is complex and has yet to be studied.

It can be concluded that color and morphological innovations emerge due to selection for domestication. Not a single case of the color aberrations as described above has been recorded in the control population, which was under no special selection for behavior.

The experimental material lets us know that the genes acted upon by selection for domesticated behavior possess a function that influences developmental rates. The experiment on the selection-driven transformation of behavior in American minks in two opposite directions—towards aggression and towards domestication—demonstrated that eye-opening was an early event in the former case and belated in the latter. The difference in eye-opening between young aggressive and tame minks is 4.1 days ($p < 0.001$). The prolonged maintenance of this juvenile feature during early postnatal ontogenesis in the minks under domestication should be

Fig. 10 The depigmentation pattern in a new coat color form, *Silvery* (S^X/S^X), is a monogenic trait

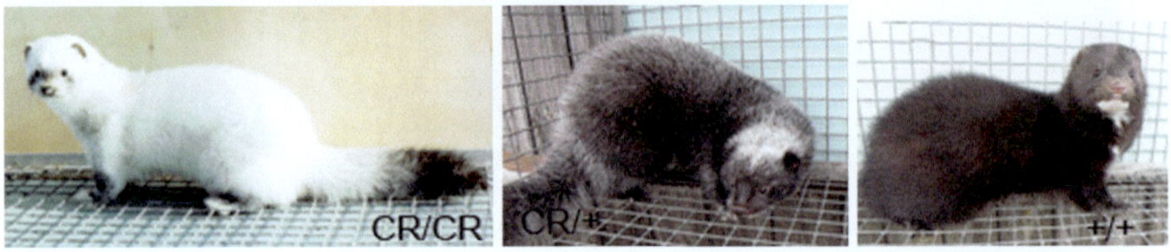

Fig. 11 A new coat color form, *BlackCrystal*(C_R/+), is a monogenic autosomal trait

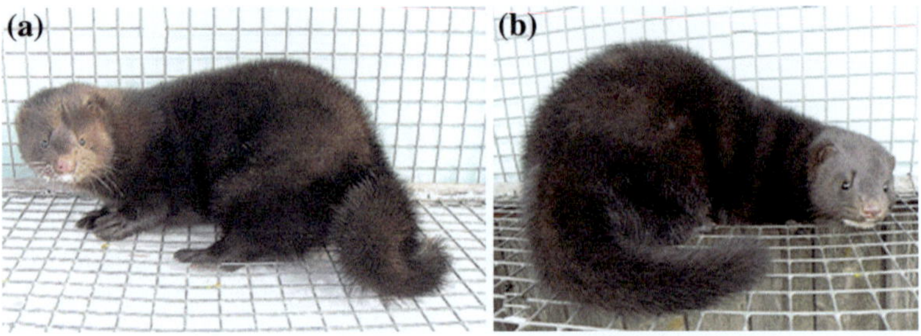

Fig. 12 **a** Helically curled tail; **b** normal tail

regarded as a case of neoteny. A similar prolongation of the period of primary socialization was observed in the foxes under domestication.

Effects of Negative Feedback Loops on Evolution Under Stabilizing and Destabilizing Selection

As was shown above, destabilizing selection was experimentally demonstrated by Belyaev's followers during domestication in foxes and mustelids (Trapezov 2008). Once the results were obtained, the question arose: What makes the phenotype under destabilizing selection (domestication) change so rapidly? Either new mutations that occur and are fixed due to a well-targeted selection or mutations that already exist in the genotypes and do not affect fitness, that is, are neutral.

We checked both assumptions using the theoretical and computational models of the function and evolution of gene regulatory networks. These models initially considered in (Kolchanov and Shindyalov 1991) were later reproduced on the modern simulation framework Diploid Evolutionary Constructor (DEC, Lashin and Matushkin 2013).

Models of regulatory circuits with negative feedback loops in gene regulatory networks at different hierarchical levels of their organization were considered. Our assumptions were that, first, any mutation would remain neutral for as long as it does not affect regulation in the gene regulatory network and, secondly, once the regulatory circuit is lost, all the available mutations would be tested as batches of randomly selected ensembles.

The stronger the negative feedback (NF) is, the narrower is the norm of reaction; therefore, the stabilizing selection should favor the taxa that have NF. Under disruptive or directional selection, the situation is opposite: an advantage is given to the taxa that do not have NF, the loss of which would lead to exposure of all the mutations that have ever been rendered neutral (a hypermanifestation of variability).

How do regulatory systems emerge and grow in complexity during evolution? Consider the simplest negative feedback (NF)circuit regulating protein concentration. Any departure of protein concentration from the norm is detected and compensated for by the regulatory component of NF through changing the rates of protein biosynthesis (the effector component of NF). Individuals breed in accordance with their fitness, which depends on protein concentration, X, as

$$W(X) = \frac{1}{\sqrt{2\pi}} e^{-\frac{1}{2}\left(\frac{X - X_0}{\sigma_X}\right)^2}$$

where X_0 is the constant protein concentration, optimum for the given environmental conditions, σ_x is the variance of the allowable concentrations.

The circuit does not care a bit for the nature of the factors that lead to aberrations; NF minimizes the phenotypic manifestation of mutations, renders them neutral, and so takes them out from under selection. It was theoretically demonstrated that the stronger is NF, the stronger is the effect of NNMs being neutral and the lower phenotypic variability in the population (Fig. 13). Stabilizing selection favors taxa that possess NF by fixing largely high-level NF, and, consequently, increases the hierarchical complexity of the regulatory system. Mutations that evolve as neutral accumulate at lower tiers (Kolchanov and Shindyalov 1991).

In that work, the evolution of a population of individuals with negative feedback was modeled. Under stable environmental conditions, variability was well hidden and the individuals that had strong negative feedback won (Fig. 14). The result is consistent with Schmalhausen's theory of stabilizing selection. In normal conditions, the norm of reaction is narrow, for variability should be hidden or rendered neutral, lest the adaptive phenotype be lost.

In the presence of feedback, mutations affecting the internal processes of the system are rendered neutral and, therefore, no phenotypic variant that can

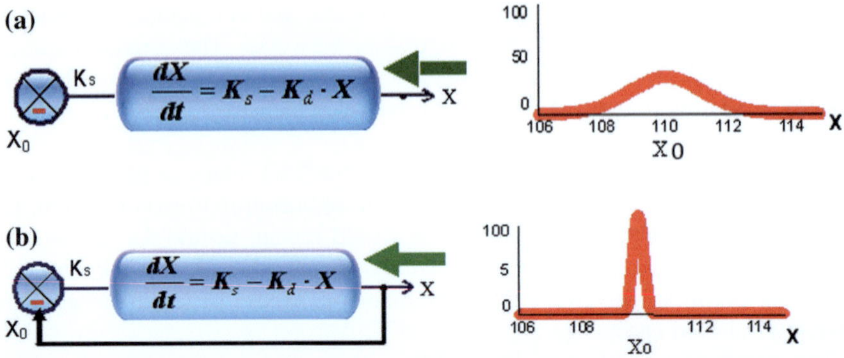

Fig. 13 Qualitative influence of the negative feedback on the variation range of a trait X (relative to the normal value X_0) in a population. **a** In the presence of feedback, mutations affecting the internal processes of the system are neutralized at phenotypic level, and so they do not affect fitness in constant environmental conditions. **b** In the absence of feedback, neutralization effect is weakened, and mutations affecting the internal processes of the system have phenotypic effects and reduce fitness in constant environment conditions

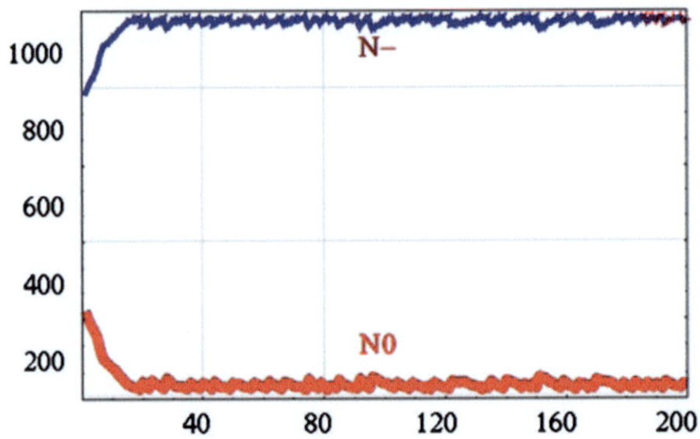

Fig. 14 Competition between individuals with negative feedback (N−) and without it (N0) under stabilizing selection, simulation results. The population size is invariable. Initially, 50 % of individuals possess a negative-feedback circuit and 50 %, do not. On the *x*-axis: the number of generations. On the *y*-axis: the number of individuals

adapt to new environmental conditions X1 (others than X0) is in existence (Figs. 13a and 14).

By contrast, under directional selection, regulatory circuits became easier to lose (and some occasionally were lost), which led to a case of hypermanifestation, a blasting exposure of hidden genetic variability (Fig. 13b).

In the absence of feedback, mutations affecting the internal processes of the system have phenotypic effects and thus increase the probability of occurrence of organisms capable to adapt to new environmental conditions, X1 (others than X0, Fig. 15).

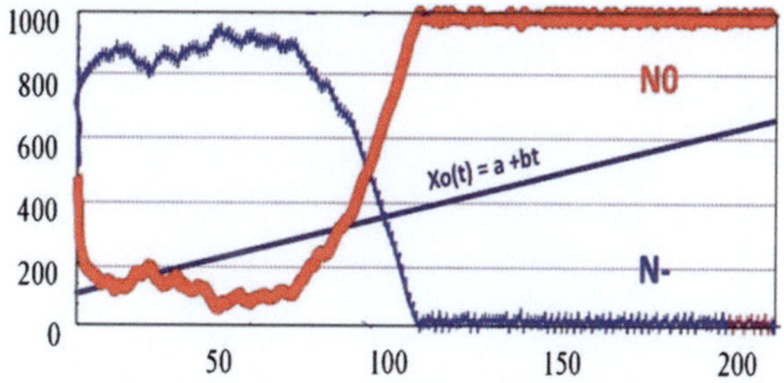

Fig. 15 Competition between individuals with negative feedback (n−) and without it (n0) under directional selection, simulation results. The population size is invariable. At each modeling stage, optimum protein concentration, X0, changes by the value of DX; $X_{i+1}=X_i + DX$. Initially, 50 % of individuals have a circuit with negative feedback and 50 %, do not. On the *x*-axis: the number of generations. On the *y*-axis: the number of individuals

The effects described above were tested with a series of models developed and analyzed using the Haploid Evolutionary Constructor (HEC, Lashin et al. 2014) software package. A community of two bacterial populations, P1 (having feedback) and P2 (not having feedback), was considered. The original populations were generated so that they had high genetic diversity. Scenarios with constant and changing environmental conditions were considered (Kolchanov and Shindyalov 1991).

It was demonstrated that biodiversity in the population that has feedback is preserved for longer, which can, in changing environmental conditions, become an important factor, which keeps hidden variability secured until environmental conditions change (Fig. 16).

The phenotypic effect of now neutral mutations (NNMs) is just compensated for by NF. Another class of mutations with compensated effects—conditionally neutral mutation—is represented by double mutations, which compensate for each other's phenotypic effect (Aleshin and Petrov 2003). Mutations as these are little likely to occur at once; however, even a single deleterious mutation increases the probability of fixation of the mutation that compensates for it (Afonnikov and Kolchanov 2001; Afonnikov et al. 2001). Unlike conditionally neutral mutations, NNMs are not double mutations, their phenotypic neutrality does not depend on particular molecular mechanisms, nor do NNMs influence the probability of fixation of the mutations that increase the power of the NF circuit which renders them neutral.

If the NF circuit becomes powerful enough, some NNMs are given to selection. This can happen due to changes in environmental conditions and/or mutation load in the regulatory circuit. Suppose a set of related taxa that have long been under stabilizing selection in different ecological niches possess NF regulatory circuits inherited from their common ancestor. The NNM load accumulated by these taxa will be equal between the ecological niches. While the power of the circuit is high enough, NNMs do not reveal themselves, and the taxa are evolutionarily stable. As the load approaches the "saturation point", these taxa awake from stasis: NNMs are supposed to occur at the weakest environmental fluctuations. Should these

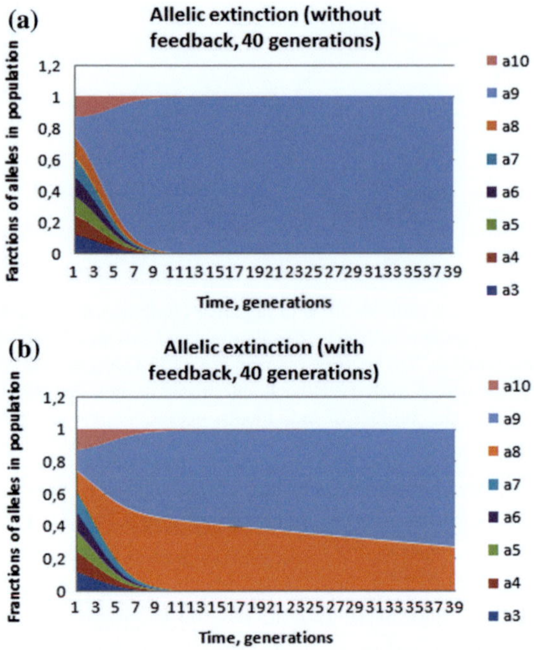

Fig. 16 **a** On the *x*-axis: the number of generations. Designations (a3, …, a10) stand for the metabolic activity of separate alleles (from 3 to 10 conditional units). In the populations without NF, alleles are eliminated much more rapidly (numerical simulations performed with the HEC software). **b** On the *x*-axis: the number of generations. Designations (a3, …, a10) stand for the metabolic activity of separate alleles (from 3 to 10 conditional units). In the populations with NF, alleles are eliminated much more slowly (numerical simulations performed with the HEC software)

fluctuations be different in different niches, different will be the evolutionary outcomes for the taxa. If worst come to worst, they will die out all at once. That is, long-continued selection in ecosystems with low taxonomic diversity may end up in a crisis and the extinction of closely related dominant taxa.

Under disruptive or directional selection, the situation is opposite: an advantage is given to taxa without NF (Kolchanov and Shindyalov 1991). The taxa that have lost NF will, in a blast-like manner, expose all the accumulated NNMs (this is a case of hypermanifestation of variability) and a cohort of young taxa may emerge (Kolchanov 2003). As a matter of fact, not all these mutations can be adaptive in the new conditions, and so the burst will be followed by the extinction of new taxa, the intensity of this extinction abating over time. The taxa that have avoided extinction will enter stasis. That was what paleontologists observed with cohorts of Phanerozoic marine taxa (Markov 2000).

Thus, stabilizing selection and directional/disruptive selection have opposite effects on the regulatory systems of the organism, which results in the so-called "evolutionary swing" (Fig. 17). Under stabilizing selection, NF occurs and

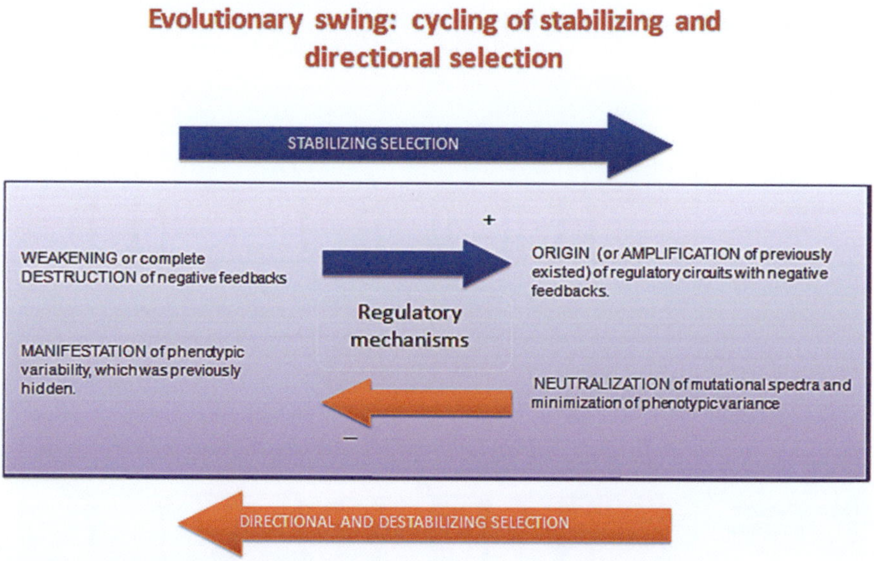

Fig. 17 The evolutionary swing: the alteration between stabilizing and directional selection (Gunbin et al. 2007)

becomes stronger, while under directional selection, some NF loops become weaker or are lost (Kolchanov 2003). Mutations, which could be as diverse as deleterious, neutral, inadaptive and adaptive, should, during stasis, be fixed in the genomes of the taxa as neutral. Once a new ecological niche is colonized (or the current niche is modified), the NF circuit is lost and a trait is hypemanifested, after which taxa with deleterious mutations rapidly become extinct and taxa with inadaptive mutations either gradually die out or are forced away to other ecological niches. If this is as it seems to be, then the fixation of all mutations that are adaptive in the new niche should accomplished rapidly, within the periods of colonization (formation) of new ecological niches, which is confirmed by experiments by Elena and Lenski (2003).

NF regulatory circuits are a common occurrence anywhere. They are found at all levels of life's organization, from molecular-genetic entities to ecosystems. Therefore, the evolutionary swing should be observed to operate in ecosystems. The succession of coherent and non-coherent evolution (Krasilov 1986), which is marked by the extinction of dominant species (an analog to the loss of the highest-level regulatory circuit) and the rise of subdominant taxa (an analog to now-neutral mutations), appears to be an analog to the evolutionary swing in ecosystems.

The evolution of a physiological or a morphological system possessing a regulatory circuit should proceed unevenly: waiting too long for the hyperresponse gene to mutate amounts to being in stasis. During stasis, now-neutral mutations accumulate in neutral (actually, now-neutral) genes. A comparison of the chimpanzee and human genomes demonstrated that the evolution of a large number of

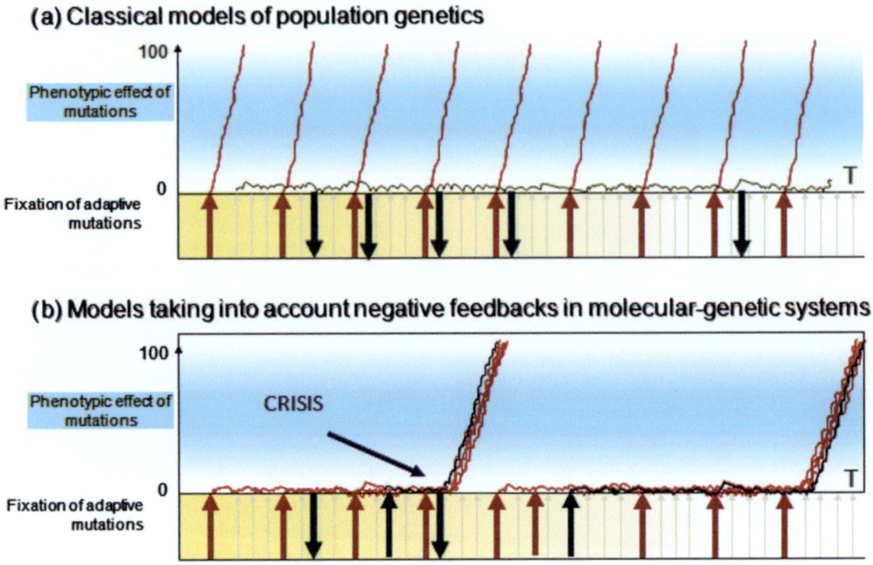

Fig. 18 “Adaptive optimization” as contrast to classic directional selection (Gunbin et al. 2007)

genes is better explained by the weakening of stabilizing selection than by strong positive selection.

Mutation affecting the hyperresponse gene makes feedback easier to lose and puts an end to stasis. When a certain trait is hypermanifested, all the available mutations are rapidly tested by selection as random batches. We propose that this type of selection be termed “adaptive optimization” to differentiate it from classic directional selection, under which mutations are tested immediately after they emerge (Fig. 18).

Suppose a gene regulatory network (GRN) can, during evolution, undergo rearrangements only at a particular rate (the rate of the evolutionary response, RER) set by its structure, the nature of the central regulator (monomers or multimers) and the general characteristics (mean mutation frequency, rates of succession of generations, cost of selection). The evolutionary swing will be observed to operate, if the RER is higher than the rate of change from stabilizing to directional selection. If the RER falls behind, mutations will not have enough time to be fixed. Then, the species will be better off widening the norm of reaction, for there is a chance some individuals will, for whatever reasons, have the traits exactly as necessary at that particular point. The norm of reaction can be widened (1) by increasing the scope of modification variability, (2) by splitting a large GRN into a few smaller ones, (3) by introducing, in the GRN, positive feedback, to enhance weak stochastic fluctuations in ontogenetic regulation, or (4) by creating a GRN for stress response.

Phenotypically, the GRN organization appears as correlated traits, correlation pleiades and radicals (Vavilov 1967). During evolution, correlation pleiades begin to form, their common factor of selection begins to act on formerly independent traits (Berg 1993). During coevolution, the stabilizing selection will support the replacement of a functional associations with a genetically fixed one (Schmalhausen 1987), that is, will create a GRN. Growing more and more sustainable under stabilizing selection, this GRN can, due to random translocations of transcription factor binding sites, "entice" genes from less sustainable GRNs. If ontogenesis becomes less affectable by the environment, the stabilizing selection would support this process, and the series of traits would begin to work towards independence from the environment (Schmalhausen 1987). Now, if the environmental settings change, the species would not be able to rapidly change its ontogenesis, for the rate of the evolutionary response would decrease and the species would either become extinct—for failure to adapt—or find an *adaptive trade-off* between the environmental demands and the series' morphology or create its own morphology-driven vector of evolution by initiating the phylocenogenetic processes of formation of its own ecosystem (Zherikhin 1997). Later on, it would be ecosystemic associations that would provide for the stability of the environment [*a biocenotic environment* (Razumovsky 1981)], allowing the species to exist in conditions adapted to the pleiades of its phenotype. Thus, something of a relay race of the vectors of stabilizing selection can take place in biogeocenotic evolution: originally shaped in the ecosystem as a sustainable combination of environmental factors, stabilizing selection forms a set of GRNs that is to become its factor.

Conclusion

Looking at the data obtained from this long and extensive experiment on the domestication of three species of wild animals—fox, mink and brown rat—the following conclusions can be made:

1. The key element in the domestication process is the intensive selection of animals for human-tolerant behavior. When the level of domestication is high enough, the animals and humans form a social structure, with internal associations and a bridge language. The transformation of the behavior of the wild animal under domestication is critical.
2. Intensive selection for behavior during the domestication of the wild animal leads to considerable rearrangements of the most basic nervous pathways fixed in the instincts. This involvement of the central regulatory systems in selection explains the specific character of evolutionary transformations that go along with domestication.
3. The main feature of the evolutionary process that leads to animal domestication at first stages in a burst-like exposure of variation in the population under domestication. Noteworthy, this variation often appears to be non-stochastic, which suggests that it has always been there, albeit in a hidden form, and has, under selection for the central mechanisms of homeostatic regulation, eventually "gone phenotypic", as Belyaev put it. Disruptions of homeostatic regulations

during domestication occur physiologically, genomically, and in the regulation of morphogenetic processes during ontogenesis. These considerations promoted Dmitry Belyaev to single out the kind of selection that accompanies domestication into a category of its own, which was termed “destabilizing selection”.

4. The most important component and mechanism of destabilizing selection is stress response, which serves as a factor of mobilization of hidden genetic variability and its induction by exerting influence on various genetic processes such as mutagenesis, recombination, mobilization of mobile genetic elements (“genome stress”, as MacClintock put it) and others.
5. A special place in the emergence of rapid evolutionary transformations during domestication should probably be given to heritable epigenetic regulations. The study of this mechanism of genetic variability is now widely considered as being of critical importance. Stress, too, can act as the main mechanism that triggers on epigenetic regulations of genome function.
6. Under stabilizing selection (constant environmental conditions), negative feedback becomes more difficult to lose and mutational variability is rendered neutral.
7. Under directional selection (a monotonous one-way change in environmental conditions), negative feedback becomes easier to lose and hidden mutational variability becomes exposed.
8. The loss of regulatory circuits with negative feedback loops is followed by exposure of hidden genotypic variability and by the occurrence of individuals with major phenotypic aberrations. We suggest this type of selection be called “adaptive optimization” as contrast to classic directional selection, under which mutations are tested immediately after they occur.

References

Afonnikov DA, Kolchanov NA (2001) Konservativnye osobennosti DNK-svyazyvayushchikh domenov klassa “gomeodomen”, obuslovlennye koadaptivnymi zamenami aminokislotnykh ostatkov (The conserved characteristics of DNA-binding domains belonging to the homeodomain class that are associated with coadaptive substitutions of amino acid residues). Proc USSR Acad Sci 380:691–695 (in Russian)

Afonnikov DA, Oshchepkov DYu, Kolchanov NA (2001) Detection of conserved physico-chemical characteristics of proteins by analyzing clusters of positions with co-ordinated substitutions. Bioinformatics 17:1035–1046

Aleshin VV, Petrov NB (2003) Uslovno neytralnye priznaki (Conditionally neutral characters). Priroda 12:25–34 (in Russian)

Belyaev DK (1972) Geneticheskie aspekty domestikatsii zhivotnykh (Genetic aspects of animal domestication). In: Matveev BS (ed) Problemy domestikatsii zhivotnykh i rasteniy (literally: Aspects of animal and plant domestication). Moscow, Nauka, pp 39–45 (in Russian)

Belyaev DK (1979) Nekotorye genetiko-evolyutsionnye problemy stressa i stressoustoychivosti [Some genetic and evolutionary aspects of stress and stress resistance]. Vestnik Akademii meditsinskikh nauk SSSR 7:9–14 (in Russian)

Belyaev DK (1982) Glavnoe bogatstvo chelovechestva (Humankind's main treasure). Nauka i religiya 1:2–6 (in Russian)
Belyaev DK, Borodin PM (1982) The influence of stress on variation and its role in evolution. Biol Zentralblatt 100:705–714
Belyaev DK, Borodin PM (1985) Vliyanie stressa na nasledstvennost i ego rol v evolyutsii (Effects of stress on inheritance and its role in evolution). In: Evolyutsionnaya genetika (literally: Evolutionary genetics). Leningrad. Leningrad University, pp 35–39 (in Russian)
Belyaev DK, Ruvinsky AO, Borodin PM (1981) Inheritance of alternative states of the fused gene in mice. J Hered 72:107–112
Belyaev DK, Plyusnina IZ, Trut LN (1985) Domestication in the silver fox (Vulpes fulvus Desm): Changes in physiological boundaries of the sensitive period of primary socialization. Appl Anim Behav Sci 13:359–370
Berg RL (1993) Korrelyatsionnye pleyady i stabiliziruyushchiy otbor (Correlation pleiades and stabilizing selection). In: Genetika i evolyutsiya. Izbrannye trudy (literally: Genetics and evolution. Selected papers). Nauka, Novosibirsk, pp 137–178 (in Russian)
Boice R (1972) Some behavioral tests of domestication in Norway rats. Behaviour 42:198–231
Calow P (1983) Evolutionary principles. Blackie & Son Ltd., Glasgow, p 105
Coppinger R, Feinstein M (1991) "Hark! Hark! The dogs do bark" and bark and bark. Smithsonian 119–129
Darwin Ch (1883) The variation of animals and plants under domestication, 2d edn. D. Appleton & Co., New York, p 429
Darwin Ch (1927) Proiskhozhdenie cheloveka i polovoy podbor (The descent of man, and selection in relation to sex). Polnoe sobranie sochineniy (literally: Complete works), vol 2, p 623 (in Russian)
Elena SF, Lenski RE (2003) Microbial genetics: evolution experiments with microorganisms: the dynamics and genetic bases of adaptation. Nat Rev Genet 4:457–469
Gogoleva SS, Volodina EV, Volodin IA, Kharlamova AV, Trut LN (2010) The gradual vocal responses to human-provoked discomfort in farmed silver foxes. Acta Ethol 13:75–85
Gulevich RG, Plyusnina IZ, Prasolova LA, Oskina IN, Trut LN (2010) White spotting in Norway rats selected for tame behavior. J Zool 280:264–270
Gunbin KV, Suslov VV, Kolchanov NA (2007) Aromorphoses and the adaptive molecular evolution. VOGIS Herald 11:373–399
Hare B, Wobber V, Wrangham R (2012) The self-domestication hypothesis: evolution of bonobo psychology is due to selection against aggression. Anim Behav 83:573–585
Hedrich HJ (2000) Part 1. The history and development of the rat as a laboratory model. In: Krinke GJ (ed) The laboratory rat. AP, San Diego, pp 3–30
Kolchanov NA (2003) Evolyutsiya regulyatornykh geneticheskikh system (The evolution of genetic regulatory systems). Theoretical workshop for geologists and biologists "The origin and evolution of living systems". http://www.bionet.nsc.ru/live (in Russian)
Kolchanov NA, Shindyalov IN (1991) Teoreticheskoe issledovanie evolyutsii regulyatornykh konturov pri razlichnykh tipakh otbora (Theoretical study of the evolution of regulatory circuits under different types of selection). In: Smuny VK, Kolchanov NA, Ruvinsky AO (eds) Problemy genetiki i teorii evolyutsii (literally: Aspects of genetics and evolution theory). Nauka, Novosibirsk, pp 268–279 (in Russian)
Konoshenko MY (2012) Avtoreferat kandidatskoy dissertatsii (Author's summary of his thesis for Candidate of Sciences). Novosibirsk, 2012. http://www.dissercat.com/content/povedencheskie-i-neirokhimicheskie-osobennosti-vnutrividovoi-agressii-u-serykh-krys-effekty
Krasilov VA (1986) Nereshennye problemy teorii evolyutsii (Yet unsolved problems in evolution theory). Far Eastern Scientific Center of the USSR Academy of Sciences, Vladivostok, 138 pp (in Russian)
Kukekova AV, Acland GM, Oskina IN, Kharlamova AV, Trut LN, Chase K, Lark KG, Erb HN, Aguirre GD (2006) The genetics of domesticated behavior in Canids: what can dogs and silver foxes tell us about each other? In: The Dog and Its Genome, pp 515–537

Kukekova AV, Trut LN, Chase K, Shepeleva DV, Vladimirova AV, Kharlamova AV, Oskina IN, Stepika A, Klebanov S, Erb HN, Acland GM (2008) Measurement of segregating behaviors in experimental silver fox pedigrees. Behav Genet 38:185–194

Kukekova AV, Trut LN, Chase K, Kharlamova AV, Johnson JL, Temnykh SV, Oskina IN, Gulevich RG, Vladimirova AV, Klebanov S et al (2011) Mapping loci for fox domestication: deconstruction/reconstruction of a behavioral phenotype. Behav Genet 41:593–606

Lashin SA, Matushkin YuG (2013) Multi-layer computer models of gene network evolution in diploid populations. FEBS J 280:563

Lashin SA, Klimenko AI, Mustafin ZS, Kolchanov NA, Matushkin YG (2014) HEC 2.0: improved simulation of the evolution of prokaryotic communities. Mat Bio Bioinf 9:585–596

Markel AL, Borodin PM (1980) The stress phenomenon: genetic and evolutionary approach. Probl Gen Genet 2:51–65

Markel AL, Borodin PM (1981) Genetiko-evolyutsionnye aspekty stressa (Genetic and evolutionary aspects of stress). In: Voprosy obshchey genetiki (Aspects of general genetics). In: Altukhov YP (ed) Proceedings of the XIV international congress of genetics. Moscow, Nauka, pp 262–271 (in Russian)

Markel AL, Trut LN (2011) Behavior, stress, and evolution in light of the Novosibirsk selection experiments. In: Gissis SB, Jablonka E (eds) Transformations of Lamarckism. From subtle fluids to molecular biology. MIT Press, Cambridge, pp 171–180

Markov AV (2000) Vozvrashchenie Chernoy Korolevy, ili zakon rosta sredney prodolzhitelnosti sushchestvovaniya rodov v protsesse evolyutsii (The return of the Black Queen, or the law of increase of mean duration of genera in the course of evolution). Zhurnal obshchey biologii 61:357–369 (in Russian)

Naumenko VS, Kozhemjakina RV, Plyusnina IZ, Popova NK (2009) Expression of serotonin transporter gene and startle response in rats with genetically determined fear-induced aggression. Bull Exp Biol Med 147:86–89

Nestler EJ (2012) Stress makes its molecular mark. Nature 4:171–172

Plyusnina IZ, Oskina IN (1997) Behavioral and adrenocortical responses to open-field test in rats selected for reduced aggressiveness towards man. Physiol Behav 61:381–385

Plyusnina IZ, Oskina IN, Tibeikina MA, Popova NK (2009) Cross-fostering effects on weight, exploratory activity, acoustic startle reflex and corticosterone stress response in norway gray rats selected for elimination and for enhancement of aggressiveness towards human. Behav Genet 39:202–212

Popova NK, Kulikov AV (1997) Avgustinovich DF, Voĭtenko NN, Trut LN Effect of domestication of the silver fox on the main enzymes of serotonin metabolism and serotonin receptors. Genetika 33:370–374

Razumovsky SM (1981) Zakonomernosti dinamiki biogeotsenozov (Patterns in the dynamics of biogeocenoses). Nauka, Moscow, 231 pp (in Russian)

Robinson R (1965) Genetics of the Norway rat. Pergamon Press, Oxford 805

Schmalhausen II (1987) Factors of evolution: the theory of stabilizing selection. University of Chicago Press, Chicago, p 327

Shikhevich SG, Os'kina In, Plyusnina IZ (2003) Responses of the hypophyseal-adrenal system to stress and immune stimuli in gray rats selected for behavior. Neurosci Behav Physiol 33:861–866

Trapezov OV (2002) Domestikatsiya kak faktor koevolyutsii zhivotnykh i cheloveka (Domestication as a factor of coevolution animals and humans). Filosofiya nauki 2:86–101 (in Russian)

Trapezov OV (2008) Gomologicheskie ryady izmenchivosti okraski mekha u amerikanskoy norki (*Mustela vison* Schreber, 1777) v usloviyakh domestikatsii (Homologous series in coat color variation in the American mink (*Mustela vison* Schreber, 1777) under domestication). VOGiS Herald 11:547–559 (in Russ.)

Trapezov OV, Belyaev DK (1986) Svyaz selektsionnogo izmeneniya povedeniya s reproduktivnoy funktsiey amerikanskoy norki (An association between selection-induced behavioral changes and the reproduction function of the American mink). Zhurnal obshchey biologii 4:445–450 (in Russian)

Trapezov OV, Trapezova LI, Alekhina TA, Klochkov DV, Ivanov YuN (2009) Vliyanie mono- i diretsessivnosti po mutatsiyam, zatragivayushchim okrasku mekha, na uroven soderzhaniya monoaminov v mozge u amerikanskoy norki (Mustela vison Schreber) [Effects of the mono- and direcessivity of coat color mutations on the brain's levels of monoamines in the American mink (Mustela vison Schreber)]. Genetika 45:1641–1645 (in Russain)

Trut LN (1988) The variable rates of evolutionary transformations and their parallelism in terms of destabilizing selection. J Anim Breed Genet 105:81–90

Trut LN (1999) Early Canid domestication: the farm-fox experiment. Am Sci 87:160–165

Trut LN (2001) Experimental studies of early Canid domestication. In: Ruvinsky A, Sampson J (eds) The genetics of the dog. CABI Publishing, NY, pp 15–42

Trut LN, Pliusnina IZ, Kolesnikova LA, Kozlova ON (2000) Interhemisphere neurochemical differences in the brain of silver foxes selected for behavior and the problem of directed asymmetry. Genetika 36:942–946

Trut LN, Plyusnina IZ, Oskina IN (2004) An experiment on fox domestication and debatable issues of evolution of the dog. Russ J Genet 40:644–655

Trut L, Oskina I, Kharlamova A (2009) Animal evolution during domestication: the domesticated fox as a model. BioEssays 31:349–360

Vavilov NI (1967) Zakon gomologicheskikh ryadov v nasledstvennoy izmenchivosti (The law of homologous series in the inheritance of variability). In: Izbrannye proizvedeniya v dvukh tomakh (literally: Selected works in two volumes). Nauka, Leningrad 1:7–61 (in Russian)

Wilson AC (1985) The molecular basis of evolution. Sci Am 253:164–173

Zeder MA (2012) Pathways to animal domestication. In: Gepts P (ed) Biodiversity in agriculture: domestication, evolution, and sustainability. Cambridge University Press, Cambridge, pp 227–259

Zherikhin VV (1997) Osnovnye zakonomernosti filotsenogeneticheskikh protsessov (na primere nemorskikh soobshchestv mezozoya i kaynozoya) [Main patterns of phylocenogenetic processes (with Mesozoic and Cenozoic non-marine communities as examples)]. Thesis for Doctor of Science in the form of a scientific paper (in Russian)

Structural and Functional Coevolution of Human Endogenous Retroviruses with Our Genome

Andrew Garazha, Maria Suntsova and Anton Buzdin

Introduction

Human endogenous retroviruses (HERVs) and related genetic elements occupy ~8 % of human genome. Genomic copies of HERVs are of particular interest because in addition to functional viral genes they have multitude of regulatory DNA regions serving as promoters (Buzdin et al. 2006), enhancers (Suntsova et al. 2013; Chuong et al. 2013), polyadenylation signals, insulators (Ling et al. 2003) and binding sites for various nuclear proteins (Gogvadze and Buzdin 2009). Many families of HERVs exhibit high transcriptional activity in human tissues (Suntsova et al. 2015). HERVs are believed to be remnants of numerous retroviral infections that occurred repeatedly during primate evolution (Sverdlov 2000; Buzdin 2007; Belshaw et al. 2004). HERVs fixed in the genome and became inheritable because their insertions occurred in the germ cell lineage (Hohn et al. 2013; Buzdin et al. 2003). HERVs are composed of sequences related to retroviral

A. Garazha · M. Suntsova · A. Buzdin (✉)
Group for Genomic Regulation of Cell Signaling Systems,
Shemyakin-Ovchinnikov Institute of Bioorganic Chemistry,
Moscow 117997, Russia
e-mail: buzdin@ponkc.com

A. Garazha · M. Suntsova
Laboratory of Bioinformatics, D. Rogachyov Federal Research Center
of Pediatric Hematology, Oncology and Immunology,
Moscow 117198, Russia

A. Buzdin
Pathway Pharmaceuticals, Wan Chai, Hong Kong, Hong Kong SAR

A. Buzdin
National Research Centre "Kurchatov Institute",
Centre for Convergence of Nano-, Bio-, Information and Cognitive
Sciences and Technologies, Moscow, Russia

V.L. Korogodina et al. (eds.), *Genetics, Evolution and Radiation*,
DOI 10.1007/978-3-319-48838-7_38

genes and are flanked at both ends by ~1 kb long so-called long terminal repeats (LTRs). An LTR structure comprises functional enhancers, promoters and polyadenylation signals normally used for retroviral gene expression (Sverdlov 2000). However, the LTRs may drive the transcription of adjacent host genomic sequences (Kovalskaya et al. 2006; van der Lagemaat et al. 2003). Most of HERVs reside in the human genome as solitary LTRs arisen due to homologous recombinations between the two flanking LTRs of the same full-length HERV elements (Hughes and Coffin 2004). However, some full-length HERVs express viral genes in a variety of human tissues (Seifarth et al. 2005) and even form virus-like particles (Lyden et al. 1994). Expression of HERV-encoded proteins is directly or indirectly associated with progression of many human diseases (Suntsova et al. 2015).

Structure of HERVs

Genomic structure and life cycle HERVs are represented in the human genome by 504 families including ~520,000 individual members. Genomic structure of a full-length HERV includes retroviral genes, typically *Gag*, *Prot*, *Env* and *Pol*, and flanking ~1 kb long LTRs (Suntsova et al. 2015). In turn, recombinations between the different HERV elements may cause further genomic rearrangements including copy number variation (CNV) of known human genes.

Having a variety of potential regulatory sequences such as promoters, enhancers, transcription factor binding sites, splice sites and polyadenylation signals, LTRs are believed to possess the major transcriptional regulatory potential of endogenous retroviruses. Solitary LTRs are differentially methylated in human tissues (Khodosevich et al. 2004), they may specifically bind host cell nuclear proteins (Trubetskoy et al. 2002), serve as tissue-specific transcriptional promoters and enhancers (Suntsova et al. 2013), and, finally, are transcribed in vivo in many tissues (Suntsova et al. 2015). In addition, LTRs may contribute to the host gene regulation network by acting in *cis* (by providing regulatory elements) or in *trans* (by driving expression of antisense transcripts) (Gogvadze et al. 2009).

The life cycle of a HERV comprises reverse transcription of viral RNA, followed by the integration of a nascent DNA copy into genomic DNA of the host cell (Lee and Bieniasz 2007). Retroviral genomic RNA differs from DNA copy by the absence of LTRs, which are built during the reverse transcription, a multistep complex process including several template switching events (Kandel and Nudler 2002). Prior translation, HERV transcripts may undergo one or two splicing events. HERV proteins are actively expressed in a variety of human tissues (Suntsova et al. 2015). Increased HERV protein production was detected in placentas and in embryonic tissues, in line with the identification of putative responsive elements for several pregnancy hormones within the HERV LTRs (Andersson et al. 1998). Gag protein expression may induce massive T cell stimulation or apoptosis (Hugin et al. 1991). Endogenous Prot genes may help to exogenous retroviruses, such as lentiviruses, to infect the host cells (Suntsova et al. 2015). Finally, Env protein has an immunosuppressive domain that inhibits T and B cell activation and proliferation and induces modulation of the expression of many cytokines (Morozov et al. 2013). This may be functionally linked to an increased HERV expression in some tumors.

Functional DNA regions DNaseI hypersensitivity sites (DHSs) are probably the most important genomic landmarks for regions of open (functionally active) chromatin, whereas transcription factor binding sites (TFBS) denote regions of DNA with nuclear protein binding properties (Jacques et al. 2013). Recently, we combined investigation of both DHS and TFBS content of HERVs on a genomic scale (Garazha et al. 2015). For the entire set of HERVs, ~140,000 inserts (~19 %) include at least one mapped DHS and ~110,000 inserts (~15 %) have at least one mapped TFBS. The total numbers of all DHS and TFBS in all HERV elements were estimated as ~155,000 and ~320,000, respectively. All the 504 HERV families were characterized with regard to their TFBS content (available at http://herv.pparser.net/TotalStatistic.php). These results provide clues for identification and functional validation of tens of thousands of previously unknown regulatory sequences of the human genome. Moreover, due to considerable technical limitations, this is likely a rough underestimation of the HERV-related TFBS pool (Garazha et al. 2015).

HERV proliferation in the genome Although traces of several hundred retroviral groups can be identified in human DNA, there are no evidence that any of them remains active in terms of generating new copies in human genome. However, up to recent times in human evolution, few retroviral families were prolific (Buzdin et al. 2003). Approximately 140 HERV inserts that all belong to HERV-K (HML-2) group, are specific to human DNA (Suntsova et al. 2015), which means that they occurred already after the radiation of human and chimpanzee ancestor lineages, that occurred ~6 million years ago (Buzdin et al. 2002). Taken together, they make up ~330 kb of the human DNA. Some family members retained their transcriptional activity (Suntsova t al. 2015). At least three HERV-K (HML-2) master genes were traced back to be active in the hominid lineage soon after the human-chimpanzee ancestral radiation (Buzdin et al. 2003). Moreover, few dozens of HERVs are polymorphic in human population (Kahyo et al. 2013), thus suggesting that this family remained prolific up to very recent times in the evolutionary history of humans. The bioinformatic analysis revealed that they may harbor a total of 11 functional genes for *Gag*, 12 genes for *Prot*, 9 genes for *Pol*, 8 for *Env*, and 9—for an auxiliary protein *Rec* (Buzdin 2007).

Host Regulation of HERVs

Suppression of HERVs by the host-encoded mechanisms Expression of HERVs is tightly controlled by the host cell because it may be deleterious. Overall, the HERV sequences have a higher substitution rate than the rest of the non-coding part of the genome (Suntsova et al. 2015). This underlines HERV regulatory potential, since it may lead to its inactivation (Suntsova et al. 2015). The physical presence of such a number of repetitive sequences can generate considerable problems dealing with homologous recombination between the different HERVs, which may disrupt functional genes located in their neighborhood (Gogvadze and Buzdin 2009). HERV-derived transcription and gene regulation can bias normal gene expression regulatory networks. Expression of HERV proteins in various human tissues may result in dangerous inflammatory or immunosuppressive effects (Suntsova et al. 2015). Recent study demonstrated that the regulation mediated by KRAB-containing zinc finger proteins (KRAB-ZFPs) and their cofactor TRIM28, is responsible for controlling HERVs in human embryonic stem cells. KRAB/TRIM28 complex most likely recruits methylation machinery to HERV copies (Turelli et al. 2014). Similarly, zinc finger protein Yin Yang 1 restricts HERV transcription in embryonic cells by suppressing promoter activities of the LTRs (Schlesinger et al. 2013). Other known mechanisms suppressing endogenous and exogenous retroviruses deal with the functions of APOBEC3, BST2, TREX, Tetherin, TRIM5α and Toll-like receptor proteins (Suntsova et al. 2015).

Helpful HERVs Co-evolution with the human genome resulted in a recruitment of certain HERVs to the execution of important molecular functions. HERV-H is a family expressed preferentially in human embryonic stem cells. Transcriptional regulation of the HERV-H LTRs may be one of the primary mediators of the stem cell fate reprogramming, being mediated by the HERV-H-driven intergenic long noncoding regulatory RNAs (Lu et al. 2014). The Envelope proteins of HERV-W family members (corresponding to human gene *Syncytin*) may serve human physiology through their fusogenic or immunosuppressive properties. For example, products of the host *Syncytin* genes are essential for placentation, by mediating cell fusion of syncytial cell layers, and for maternal tolerance of the fetus, by immunosuppression (Frendo et al. 2003). HERVs may serve as major transcriptional regulators of human genes by direct enhancer or promoter activities (Suntsova et al. 2015). For example, human specific transcription of human gene *PRODH* in hippocampus is regulated by an enhancer element created by the human-specific insert of a HERV-K (HML-2) LTR (Suntsova et al. 2013). This might have an important impact on human evolution since *PRODH* metabolizes neuromediator molecules and has a strong implication in higher nervous activity (Suntsova et al. 2013). The only apparent promoter of the liver-specific gene *BAAT* implicated in familial hypercholanemia is an ancient LTR in human but not in mouse (Carlton et al. 2003).

HERVs and Human Diseases

Recent studies evidence that different activities of HERVs may be involved in various human diseases including autoimmune disorders, neurological, infectious diseases and cancer (Suntsova et al. 2015).

Autoimmune diseases The biased expression of proviral proteins in human tissues may trigger autoimmune diseases (Suntsova et al. 2015). This was indicated first by increased proviral transcript levels and finding anti-HERV protein antibodies in sera from several groups of patients suffering from these disorders (Suntsova et al. 2015). Immune reactivity to HERV products can occur spontaneously in infection or cancer and is considered the driving force of several autoimmune disorders also in the mice. Immune reactivity against ERV proteins can be experimentally induced in mice and non-human primates, suggesting that immunological tolerance to ERV-derived products is incomplete (Suntsova et al. 2015).

Neurological diseases Enhanced expression of the HERV-encoded proteins is a promising biomarker for several neurological diseases (Manghera et al. 2015). There is frequently a cross-talk between multiple sclerosis (MS) and previous infections of the human CNS cells (Libbey et al. 2014). A hypothesis was proposed that HERV-encoded envelope proteins (Env) can act as strong immune stimulators for MS (Libbey et al. 2014). Thus, slow disease progression following neurodegeneration might be induced by re-activation of HERV expression directly, while relapses in parallel to inflammation might be secondary to the expression of HERV-encoded superantigens (Emmer et al. 2014). Finally, HERVs may cause neurological disorders using quite distinct mechanism comprising HERV-linked genomic rearrangements, e.g. 3q13.2-q13.31 deletions in a syndrome of hypotonia and motor, language, and cognitive delays (Shuvarikov et al. 2013).

Infectious diseases The long-term spontaneous evolution of humans and the human viruses might generate various mechanisms involving cooperation or interference of endogenous and exogenous retroviruses (Suntsova et al. 2015). For example, the primate lentiviruses HIV and simian immunodeficiency virus (SIV) do not express their own dUTPase, and it is believed that a host cell endogenous retroviral enzyme (Prot) provides this activity during reverse transcription (Suntsova et al. 2015), in line with the recent observations that HIV-1 infection may increase the expression of HERV-K (HML-2) proviruses in vivo (Contreras-Galindo et al. 2006).

Cancer The role of HERVs in cancer is most likely limited to retrovirus-driven gene expression and does not involve their insertional activity (Suntsova et al. 2015; Hohn et al. 2013). Abnormal expression of HERVs in cancer is well known. For example, HERV-K (HML-2) elements are overexpressed in germ cell tumors, in melanoma and in other tumors (Suntsova et al. 2015; Hohn et al. 2013). Aberrant regulation of HERVs may be mediated by the cancer-specific combinations of transcription factors (Suntsova et al. 2015). Expression of *Syncytins*is normally restricted to the placenta. However, in many pathologies including cancers, Syncytin-mediated cell fusion may participate in cancer cell transformation or metastasis (Suntsova et al. 2015). Endogenous retroviral Env proteins possess immunosuppressive and fusogenic activities (Suntsova et al. 2015). As in the placenta, the expression of immunosuppressive domain of Env in tumors may suppress immune responses and thus prevent rejection of the tumour (Morozov et al. 2013).

References

Andersson AC, Svensson AC, Rolny C et al (1998) Expression of human endogenous retrovirus ERV3 (HERV-R) mRNA in normal and neoplastic tissues. Int J Oncol 12(2):309–313
Belshaw R, Pereira V, Katzourakis A et al (2004) Long-term reinfection of the human genome by endogenous retroviruses. Proc Natl Acad Sci USA 101(14):4894–4899
Buzdin A (2007) Human-specific endogenous retroviruses. Sci World J 7:1848–1868
Buzdin A, Khodosevich K, Mamedov I et al (2002) A technique for genome-wide identification of differences in the interspersed repeats integrations between closely related genomes and its application to detection of human-specific integrations of HERV-K LTRs. Genomics 79 (3):413–422
Buzdin A, Ustyugova S, Khodosevich K et al (2003) Human-specific subfamilies of HERV-K (HML-2) long terminal repeats: three master genes were active simultaneously during branching of hominoid lineages. Genomics 81(2):149–156
Buzdin A, Kovalskaya-Alexandrova E, Gogvadze E, Sverdlov E (2006) At least 50 % of human-specific HERV-K (HML-2) long terminal repeats serve in vivo as active promoters for host nonrepetitive DNA transcription. J Virol 80(21):10752–10762
Carlton VE, Harris BZ, Puffenberger EG et al (2003) Complex inheritance of familial hypercholanemia with associated mutations in TJP2 and BAAT. Nat Genet 34(1):91–96
Chuong EB, Rumi MA, Soares MJ, Baker JC (2013) Endogenous retroviruses function as species-specific enhancer elements in the placenta. Nat Genet 45(3):325–329
Contreras-Galindo R, Gonzalez M, Almodovar-Camacho S et al (2006) A new real-time-RT-PCR for quantitation of human endogenous retroviruses type K (HERV-K) RNA load in plasma samples: increased HERV-K RNA titers in HIV-1 patients with HAART non-suppressive regimens. J Virol Methods 136(1–2):51–57
Emmer A, Staege MS, Kornhuber ME (2014) The retrovirus/superantigen hypothesis of multiple sclerosis. Cell Mol Neurobiol 34(8):1087–1096
Frendo JL, Olivier D, Cheynet V et al (2003) Direct involvement of HERV-W Env glycoprotein in human trophoblast cell fusion and differentiation. Mol Cell Biol 23(10):3566–3574
Garazha A, Ivanova A, Suntsova M et al (2015) New bioinformatic tool for quick identification of functionally relevant endogenous retroviral inserts in human genome. Cell Cycle 14(9):1476–1484
Gogvadze E, Buzdin A (2009) Retroelements and their impact on genome evolution and functioning. CMLS 66(23):3727–3742
Gogvadze E, Stukacheva E, Buzdin A, Sverdlov E (2009) Human-specific modulation of transcriptional activity provided by endogenous retroviral insertions. J Virol 83(12):6098–6105
Hohn O, Hanke K, Bannert N (2013) HERV-K(HML-2), the best preserved family of HERVs: endogenization, expression, and implications in health and disease. Front Oncol 3:246
Hughes JF, Coffin JM (2004) Human endogenous retrovirus K solo-LTR formation and insertional polymorphisms: implications for human and viral evolution. Proc Natl Acad Sci USA 101 (6):1668–1672
Hugin AW, Vacchio MS, Morse HC 3rd (1991) A virus-encoded "superantigen" in a retrovirus-induced immunodeficiency syndrome of mice. Science 252(5004):424–427
Jacques PE, Jeyakani J, Bourque G (2013) The majority of primate-specific regulatory sequences are derived from transposable elements. PLoS Genet 9(5):e1003504
Kahyo T, Tao H, Shinmura K et al (2013) Identification and association study with lung cancer for novel insertion polymorphisms of human endogenous retrovirus. Carcinogenesis 34 (11):2531–2538
Kandel ES, Nudler E (2002) Template switching by RNA polymerase II in vivo. Evidence and implications from a retroviral system. Mol Cell 10(6):1495–1502
Khodosevich K, Lebedev Y, Sverdlov ED (2004) Large-scale determination of the methylation status of retrotransposons in different tissues using a methylation tags approach. Nucleic Acids Res 32(3):e31

Kovalskaya E, Buzdin A, Gogvadze E et al (2006) Functional human endogenous retroviral LTR transcription start sites are located between the R and U5 regions. Virology 346(2):373–378

Lee YN, Bieniasz PD (2007) Reconstitution of an infectious human endogenous retrovirus. PLoS Pathog 3(1):e10

Libbey JE, Cusick MF, Fujinami RS (2014) Role of pathogens in multiple sclerosis. Int Rev Immunol 33(4):266–283

Ling J, Pi W, Yu X et al (2003) The ERV-9 LTR enhancer is not blocked by the HS5 insulator and synthesizes through the HS5 site non-coding, long RNAs that regulate LTR enhancer function. Nucleic Acids Res 31(15):4582–4596

Lu X, Sachs F, Ramsay L et al (2014) The retrovirus HERVH is a long noncoding RNA required for human embryonic stem cell identity. Nat Struct Mol Biol 21(4):423–425

Lyden TW, Johnson PM, Mwenda JM, Rote NS (1994) Ultrastructural characterization of endogenous retroviral particles isolated from normal human placentas. Biol Reprod 51(1):152–157

Manghera M, Ferguson J, Douville R (2015) ERVK polyprotein processing and reverse transcriptase expression in human cell line models of neurological disease. Viruses 7(1):320–332

Morozov VA, Dao Thi VL, Denner J (2013) The transmembrane protein of the human endogenous retrovirus–K (HERV-K) modulates cytokine release and gene expression. PLoS ONE 8(8): e70399

Schlesinger S, Lee AH, Wang GZ et al (2013) Proviral silencing in embryonic cells is regulated by Yin Yang 1. Cell Rep 4(1):50–58

Seifarth W, Frank O, Zeilfelder U, Spiess B et al (2005) Comprehensive analysis of human endogenous retrovirus transcriptional activity in human tissues with a retrovirus-specific microarray. J Virol 79(1):341–352

Shuvarikov A, Campbell IM, Dittwald P et al (2013) Recurrent HERV-H-mediated 3q13.2-q13.31 deletions cause a syndrome of hypotonia and motor, language, and cognitive delays. Hum Mutat 34(10):1415–1423

Suntsova M, Gogvadze EV, Salozhin S et al (2013) Human-specific endogenous retroviral insert serves as an enhancer for the schizophrenia-linked gene PRODH. Proc Natl Acad Sci USA 110 (48):19472–19477

Suntsova M, Garazha A, Ivanova A et al (2015) Molecular functions of human endogenous retroviruses in health and disease. CMLS 72(19):3653–3675

Sverdlov ED (2000) Retroviruses and primate evolution. BioEssays News Rev Mol Cellular Dev Biol 22(2):161–171

Trubetskoy DO, Zavalova LL, Akopov SB, Nikolaev LG (2002) Purification of proteins specifically binding human endogenous retrovirus K long terminal repeat by affinity elution chromatography. J Chromatogr A 976(1–2):95–101

Turelli P, Castro-Diaz N, Marzetta F et al (2014) Interplay of TRIM28 and DNA methylation in controlling human endogenous retroelements. Genome Res 24(8):1260–1270

van der Lagemaat LN, Landry JR, Mager DL, Medstrand P (2003) Transposable elements in mammals promote regulatory variation and diversification of genes with specialized functions. Trends Genet 19(10):530–536

The Central Nervous System of Mammals Acts as a Mutagenic/Anti-mutagenic Factor: Role in Microevolution

Eugene Daev

Introduction

According to Vernadsky's theory (1944) all "living matter" is an active component of the biosphere. As an integrated part of the biosphere all forms of living matter interact with each other as well as with "inert" matter. As a result of such interaction living organisms evolve in the process of the biosphere's evolution. The appearance of the nervous system and later central nervous system (CNS), "a growing 'innervation' and 'cephalization' of organisms", favored extending a closely interdependent network over the whole earth (de Chardin 1964, 1974). The complexity of the central nervous system of living organisms reflects their ability to adapt to the endless changes in the environment. Evolution of the nervous system accelerates "a process of transition to the noosphere" from biosphere (Vernadsky 1944). The nervous system of higher multicellular organisms is a well-established system of interaction with the 'living' and 'inert' environment. At the same time—this is the mechanism of interaction of all cells within the organism to form coordinated complex responses to emerging changes. But the life of an organism depends on the proper operation of all of its cell genomes. Therefore, permanent interconnection between the environment and a genome of living organism is a necessary condition for its survival. Changes in the environment, through the connecting nervous link—CNS, govern the formation and activity of target cell genomes of the inside organism body. For this purpose CNS use numerous specific messengers. But the living organism produces also 'external messengers' which can act on CNS of other organisms. Consequently they can influence the genetic, neuroendocrinological and behavioral processes of other organisms. We will try to demonstrate here some

E. Daev (✉)
Department of Genetics and Biotechnology,
St. Petersburg State University, St. Petersburg, Russia
e-mail: mouse_gene@mail.ru

V.L. Korogodina et al. (eds.), *Genetics, Evolution and Radiation*,
DOI 10.1007/978-3-319-48838-7_39

genetic effects of the house mouse pheromones because importance of such effects is obviously underestimated and research in this field is certainly insufficient.

Historical Background

In 1940–47 it was proposed that the occurrence of mutations is a physiological process (Kerkis 1940; Lobashev 1947). The influence of nervous system on the chromosome integrity was shown in the 1970s on mice. Denervation surgery in mice led to increasing level of chromosomal aberrations (ChAbs) in corresponding organs (Poljanskaja 1970; Tarabrin 1970). Kerkis with collaborators and some other researchers showed that hormones of HPA-axis influence the level of ChAbs in rodents (Logvinova and Kerkis 1967; Lobashev et al. 1973). It was also shown that the neurotransmitters (serotonin, adrenaline and noradrenaline) alter the level of chromosomal aberrations in cultured human cells (Pimenova 1975). Interesting data concerned the influence of learning on the level of ChAbs. With the help of a conditioned reflex to initially neutral stimulus it was possible to show the effect of the cerebral cortex modifying mutation process (Tsapygina 1971; Lobashev et al. 1973). These and some other data (Dyuzhikova et al. 1997) confirmed the assumption of control of genome integrity by nervous system (Fischman et al. 1996; Poljanskaja 2000).

Stress is a special state of the nervous system that is characterized by the activation of the hypothalamic-pituitary-adrenal axis (Selye 1950; Binder and Nemeroff 2010). We first discovered the effect of volatile stressors from mouse urine on genome integrity of conspecific target cells (Daev 1982; Novikov et al. 1982). Fischman et al. (1996) paid attention to stress as modifier of genome stability. It would be promising to use stressors and other modulators of nervous system, operating in natural populations of animals. Suitable modulators seem to be primer-pheromones of rodents, in particular, house mice. Some effects of pheromones (Daev 1994, 2007; Koyama 2004) in mice pointed to their possibility to act as stressors. Several studies showed that the effect of stress pheromones is sex-specific in "donor-recipient" model (Bronson 1979; Baum 2009).

Objectives

The main purpose of this review is to show the ability of natural volatile chemosignals of house mice (*Mus musculus* L.) to regulate genome integrity of target cells of conspecific recipients. The data could serve as an evidence of pheromones and CNS involvement in regulation of genome stability of cells in vital organs.

Brief Description of Model and Methods

Inbred strains of laboratory mice (CBA, BALB/c, C57BL/6) served as a conventional model for our experiments. Several olfactory stimuli (soiled bedding, fresh urine, main urinary proteins (MUP's), mouse pheromone 2,5-dimethylpyrazine) were used as external acting factors which could differently change a state of CNS. 2,5-dimethylpyrazine (DMP) is a mouse pheromone which is produced by overcrowded females. It influences conspecifics of both sexes in a similar manner (Jemiolo and Novotny 1994). We use DMP more often because it was shown that some of its effects could be signs of stress-reaction in recipient conspecifics of both sexes. As a rule the recipient animals just sniffed an aqueous solution of the proposed stimulus without direct contact with it. We take into account the gender of donors and recipients of the volatile substances. Since the immune and reproductive functions are inhibited by stress, we evaluated primarily the state of dividing immunocompetent and germ cells. Genome fragility of germ and somatic cells (from testes and bone marrow) of recipient mice were estimated by different cytogenetic methods. We studied the chromosome aberrations in dividing spermatocytes by metaphase and ana-telophase analysis (Dyban and Baranov 1978; Macgregor and Varley 1988). We estimated frequencies of different types of aberrations (usually multivalent associations, autosomal univalent and fragments at metaphase I stage in spermatocytes; bridges, fragments and lagging chromosomes at ana-telophase stage in spermatocytes and bone marrow cells), as well as the total frequency of ChAbs. The level of DNA breaks in the bone marrow cells was determined by alkaline comet assay (Speit and Hartmann 2006).

The frequency of abnormal (Wyrobek and Bruce 1975) and test for dominant lethal in progenies of treated males was conducted (Ehling et al. 1978).

The Influence of Mouse Pheromones on the Genome Stability

In a series of experiments it is found that sniffing of chemosignals of urine of adult males modifies ChAbs level in spermatocytes I at metaphase I stage in male recipients. The action is genotype-specific (Daev 1994). Induction of aberrations is caused by dialyzed urine, containing a complex of main urinary proteins. On the molecular level this effect correlates with activation of *c-fos* and *c-jun* in the spermatogonial cells of the recipients (Daev and Dukel'skaia 2005). It was shown that mouse female pheromone 2,5-dimethylpyrazine induces chromosome aberrations at metaphase I stage in spermatocytes I of recipient males (Daev and Dukel'skaia 2005). This result coincides with an increase in the frequency of aberrations detected in the second meiotic division by the ana-telophase method. Similar elevation is shown in spermatocytes II after the exposure to volatile substances from the urine of stressed or irradiated adult males (Daev et al. 2012a).

Basing on the duration of spermatogenic cycles in mice (Oakberg 1957) we checked the frequency of abnormal sperm heads (ASH) and the lethality of the progenies of males at appropriate intervals after pheromonal exposure. It has been shown that the frequency of ASH in male recipients and dominant lethal frequency in their progenies are increased after sniffing of adult male volatiles (Aref'ev et al. 1986; Daev et al. 1988) and DMP (Daev 2003; Daev and Dukel'skaya 2003). This effect of the pheromones on the male germ cells is probably linked to the density and mobility of the mouse sperm (Koyama and Kamimura 2000, 2003).

Similar results are obtained in the bone marrow cells: pheromones from the urine of adult males induced ChAbs in dividing cells of recipient males (Daev 1994). However, this effect was not found in the bone marrow cells of recipient females. At the same time pheromones from dialyzed urine of adult females increased ChAbs frequency in recipient females (Daev and Poluekhina 1996). DMP shows an analogous effect in males. The frequency of ChAbs depends on genotype of recipient males (Daev et al. 2007, 2008). These observations were supported by the comet assay results. A 50 % increase was shown in the DNA breaks of interphase bone marrow cells of recipient mouse males after the exposure to DMP (Glinin et al. 2014).

Recent data show that some chemosignals from the urine of females can increase genome stability of target germ and somatic cells of mouse males (Daev et al. 2014a, b).

In summary, our data show the effect of pheromones on genome integrity of bone marrow and germ cells of mice. It can be the mechanism of the immune and reproductive status regulation in the recipient organisms. We show also that the degree of suppression of the immune system depends on the status (stressed/unstressed) and genotype of pheromone's donors (Daev et al. 2008, 2012a, b).

Brain Response to Pheromonal Stimuli

There are many articles and reviews about the reception, processing and perception of volatile chemosignals in central nervous system of mammals (Bigiani et al. 2005; Dulac and Wagner. 2006; Baum 2009; Petrulis 2013; Farbiszewski and Kranc 2013; etc.). It was evident from the very beginning that pheromones act through CNS.

Our data show that the exposure to the volatile chemosignals (we used CBA and C57BL/6 males as donors) can modify some behavioral features of CBAB6F1 males in the "open field" (Novikov et al. 1982). Particularly, the attraction to females was disturbed after sniffing DMP (Daev et al. 2010). We could interpret such behavioral changes (resulting from modulation of the CNS activity) as a sign of depression and reproductive disturbance. It has been shown that norepinephrine disappears in perivascular adrenergic nerve endings in the mouse male nasal mucosa as well as in vascular testis tunic an hour after the exposure to DMP (Daev et al. 2000). It marks the participation of catecholamine nervous system during the transduction of the DMP-signal in the CNS and to the peripheral target organs. There is a common opinion that catecholamines are involved in the stress reactions (Binder and Nemeroff 2010). Therefore the lack of norepinephrine may support our suggestion that DMP is the stressor for the recipient mice. Preliminary results obtained by fMRI method have shown that DMP (produced by grouped females) can modulate the activity of neurons in the main olfactory bulbs of male mice in the way opposite to the urine of single females (Glinin and Romashchenko unpublished).

Discussion

We used the model of pheromonal exposure to show that the volatile pheromones can change the state of CNS and the frequency of ChAbs in opposite directions. Evidently, the production and influence of chemosignals depends on sex, age, health, state of CNS and many other features of donor and recipient animals. The DMP or other not yet identified male pheromones can induce the genome destabilization in bone marrow cells that leads to the immunosuppression in the recipient males (Daev et al. 2012a, b). The genome destabilization (that was induced by DMP or other not yet identified male pheromones in bone marrow cells) leads to the immunosuppression in the recipient males (Daev et al. 2012a, b). Perhaps, with the help of pheromones (influencing cells genome through CNS) house mice can adjust the fitness, fecundity and survival of animals depending on their genotype and other physiological characteristics (Daev 1994, 2007).

Similarly, olfactory induced destabilization of spermatocyte genome results in an increase in their genetic variability and lethality as well as mortality of the progenies of recipient males.

Genotype-dependence of such consequences (depending on space-, sex-, age- and genetic structure of the mouse population) can lead to the different microevolutionary events. To date there are many results accumulated on the impact of stress in the animals on the genetic apparatus of the target cells. Among them are: increases in both the level of sister chromatid exchanges and ChAbs in bone marrow cells (Daev 1994, 2007; Fischman et al. 1996); suppression of cell growth and increased cell death (Sapolsky 2000); shortage of telomere lengths (Monaghan 2014); regulation of the expression of retrotransposons by epigenetic changes (Hunter et al. 2015); perinatal and transgenerational epigenetic effects (Babenko et al. 2015), etc.

To my mind it is obvious that CNS in many cases could serve as a mutator (or antimutator) and depending on environmental factors modifies the rate of microevolutionary changes (Fig. 1). These may be accelerated under stress conditions and slow in optimal surrounding for the species. The results support the hypothesis that the mutagenic/antimutagenic role of the nervous system is a common feature of all multicellular organisms.

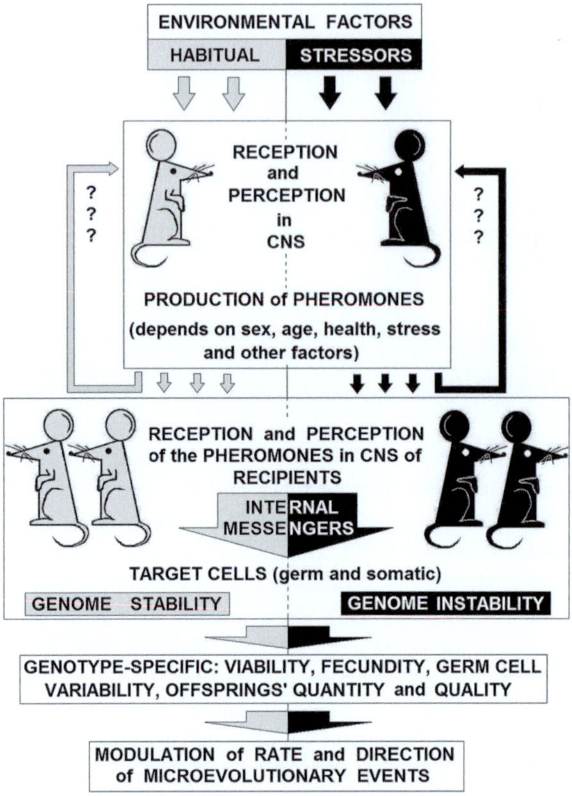

Fig. 1 Changes in the genome stability of target cells in mice (*Mus musculus* L.) by pheromones as an example of the evolutionary role of the CNS. Different states of CNS in animals, various extra- and mediators within a body, multi-directional effects and consequences are in *gray* and *black*. "???"—possible modulation of CNS by own pheromones

Acknowledgments I am grateful to Dr. O.V. Bazhenova, MS T.S. Glinin, and Dr. A.V. Dukelskaya (Department of Genetics and Biotechnology, SPbSU) for their assistance in preparation of the manuscript for publication. This work was supported by the RFBR grant N 16-04-00678.

References

Aref'ev AA, Daev EV, Kaĭdanov LZ, Lopatina NG, Novikov SN (1986) Anomalous spermatogenesis in laboratory mice exposed to volatile substances from the urine of mature males. Dokl Akad Nauk SSSR 291(5):1257–1259

Babenko O, Kovalchuk I, Metz GAS (2015) Stress-induced perinatal and transgenerational epigenetic programming of brain development and mental health. Neurosci Biobehav Rev 48:70–91

Baum MJ (2009) Sexual differentiation of pheromone processing: Links to male-typical mating behavior and partner preference. Horm Behav 55(5):579–588

Bigiani A, Mucignat-Caretta C, Montani G, Tirindelli R (2005) Pheromone reception in mammals. Rev Physiol Biochem Pharmacol 155:1–35

Binder EB, Nemeroff CB (2010) The CRF system, stress, depression and anxiety—insights from human genetic studies. Mol Psychiatry 15(6):574–588

Bronson FH (1979) The reproductive ecology of the house mouse. Q Rev Biol 54(3):265–299

Daev EV (1982) Meiotic abnormalities in young males of the house mouse (*Mus musculus*) under the effect of exogenous metabolites of sexually mature animals. Zoologichesky Zhurnal LXI (8):1269–1271 (Russian)

Daev EV (1994) Pheromonal regulation of genetic processes: research on the house mouse (*Mus musculus* L.). Russ J Genet 30(8):964–970

Daev EV (2003) Induction of dominant lethals in progeny of CBA male mice after pheromonal action. Russ J Genet 39(10):1338–1343

Daev EV (2007) Stress, chemocommunication, and the physiological hypothesis of mutation. Russ J Genet 40(10):1082–1092

Daev EV, Dukel'skaia AV (2005) Induction of meiotic disturbances in spermatocytes I by pheromones as an inhibiting mechanism of male reproductive function in house mice. Tsitologiia 47(6):505–509 (Russian)

Daev EV, Dukelskaya AV (2003) The female pheromone 2,5-dimethylpyrazine induces sperm-head abnormalities in male CBA mice. Russ J Genet 39(7):811–815

Daev EV, Poluekhina EV (1996) Cytogenetic effect of volatile components of urine of mature animals on bone marrow cells of young female house mice *Mus musculus* L. Russ J Genet 32 (3):357–360

Daev EV, Tsapygina RI, Lopatina NG (1988) Incidence of dominant lethals in the progeny of young male house mice (*Mus musculus* L.) after exposure to excretory products of mature males of the same species. Genetika 24(11):2015–2021 (Russian)

Daev EV, Vorob'ev KV, Shustova TI, Zimina SA, Samotokin MB (2000) Genotype-specific changes of certain functional indicators of immunocompetent cells in male laboratory mice under conditions of pheromone stress. Russ J Genet 36(8):872–876

Daev EV, Surinov BP, Dukelskaya AV (2007) Chemosignaling in CBA and C57Bl/6 mouse strains is modified by stress. Ecol Genet 5(2):37–43 (Russian)

Daev EV, Surinov BP, Dukelskaya AV (2008) Post-stress chemosignals affect cells from immunocompetent organs in laboratory mice of three inbred strains. Ecol Genet 6(1):27–33 (Russian)

Daev EV, Dukelskaya AV, Bezgacheva DV, Zubareva OE (2010) About genetic danger of social stresses. In: Proceedings of fifth all-Russian scientific-practical conference "Health—the base of human potential: problems and ways to their solution". State Polytechnic University, St. Petersburg, pp 418–420 (Russian)

Daev EV, Glinin TS, Dukelskaya AV (2012a) Pheromones and adaptive bystander-mutagenesis in mice. In: Mothersill CE, Korogodina V, Seymour CB (eds) Radiobiology and environmental security. Springer, Dordrecht, pp 153–162

Daev EV, Surinov BP, Dukelskaya AV (2012b) Response of immunocompetent cells of bone marrow and spleen of mouse males of several strains to stress and to pyrazine containing chemosignals. Ecol Genet 10(2):14–20 (Russian)

Daev EV, Bezruchko YuA, Dukelskaya AV (2014a) Antimutagenic effect of chemosignals from isolated female house mouse on male germ cells (*Mus musculus* L.). Russ J Genet 50(6):621–625

Daev EV, Glinin TS, Dukelskaya AV (2014b) Chemosignals from isolated females have antimutagenic effect in dividing the cells of bone marrow from male mice of the CBA strain. Russ J Genet 50(1):55–60

de Chardin TP (1964) The future of mankind. Harper, New York

de Chardin TP (1974) The phenomenon of man. Collins-Fontana, London

Dulac C, Wagner S (2006) Genetic analysis of brain circuits underlying pheromone signaling. Ann Rev Genet 40:449–467

Dyban AP, Baranov VS (1978) Cytogenetics of mammalian development. Nauka, Moscow

Dyuzhikova NA, Tokmacheva EV, Lopatina NG (1997) Stress-related changes in chromosome structure and function. Russ J Genet 33(8):913–917

Ehling UH, Machemer L, Buselmaier W et al (1978) Standard protocol for the dominant lethal test on mice. Arch Toxicol 39:173–185

Farbiszewski R, Kranc R (2013) Olfactory receptors and the mechanism of odor perception. Pol Ann Med 20:51–55
Fischman HK, Pero R, Kelly DD (1996) Psychogenic stress induces chromosomal and DNA damage. Int J Neurosci 84:219–227
Glinin T, Petrova M, Dukelskaya A, Daev E (2014) Stabilization and destabilization of mouse somatic stem cells' genetic material by pheromones as a model of aging rate modulation. In: Abst 3rd international conference "Genetics of aging and longevity", Sochi, p 71
Hunter RG, Gagnidze K, McEwen BS, Pfaff DW (2015) Stress and the dynamic genome: steroids, epigenetics, and the transposome. Proc Natl Acad Sci USA 112(22):6828–6833
Jemiolo B, Novotny M (1994) Inhibition of sexual maturation in juvenilefemale and male mice by chemosignal of female origin. Physiol Behav 55:519–522
Kerkis YY (1940) Physiological changes in the cell as the cause of the mutation process. Usp Sovrem Biol 12(1):143–159 (Russian)
Koyama S (2004) Primer affects by conspecific odors in house mice: a new perspective in the study of primer effects on reproductive activities. Horm Behav 46(3):303–310
Koyama S, Kamimura S (2000) Influence of social dominance and female odor on the sperm activity of male mice. Physiol Behav 71:415–422
Koyama S, Kamimura S (2003) Effects of vomeronasal organ removal on the sperm motility in male mice. Zoolog Sci 20:1355–1358
Lobashev ME (1947) Physiological (paranecrotic) hypothesis of mutation process. Vestn Leningrad Univ 8:10–29
Lobashev ME, Ponomarenko VV, Polyanskaya GG, Tsapygina RI (1973) The role of the nervous system in the regulation of various genetic and cytogenetic processes. Zh Evol Biokhim Fisiol 9(4):349–355 (Russian)
Logvinova VV, Kerkis YY (1967) The effect of adrenalectomy on the chromosome radiosensitivity in cells of retinal and bone marrow epithelium of Wistar rats. Genetika 3(7):48–50 (Russian)
Macgregor HC, Varley JM (1988) Working with animal chromosomes. Willey, NY
Monaghan P (2014) Organismal stress, telomeres and life histories. J Exp Biol 217:57–66
Novikov SN, Tsapygina RI, Daev EV, Togo EF (1982) Effect of natural compounds of biogenic origin on the behavior and generative function of male house mice. Mus musculus L. Dokl Akad Nauk SSSR 262(3):746–748
Oakberg EF (1957) Duration of spermatogenesis in the mouse. Nature 180:1137–1138
Petrulis A (2013) Chemosignals and hormones in the neural control of mammalian sexual behavior. Front Neuroendocrinol 34:255–267
Pimenova MN (1975) Effect of neuro-active agents for cytogenetic processes in somatic cells (mouse, human). PhD thesis, Leningrad State Univ (Russian)
Poljanskaja GG (1970) Preganglionic sympathectomy influence on the occurrence of chromosomal aberrations and mitotic activity in the epithelium of the cornea of mice. Tsitologiia 12(6):790–794 (Russian)
Poljanskaja GG (2000) Karyotypic patterns of variability in cell cultures with long-term cultivation under different conditions. Doctoral thesis, Institute of Cytology, St. Petersburg (Russian)
Sapolsky RM (2000) Stress hormones: good and bad. Neurobiol Dis 7:540–542
Selye H (1950) Stress and general adaptation syndrome. Br Med J 1(4667):1383–1392
Speit G, Hartmann A (2006) The comet assay. A sensitive genotoxicity test for the detection of DNA damage and repair. In: Henderson DS (ed) Methods in molecular biology: DNA repair protocols: mammalian systems 2nd edn. Humana Press Inc, Totowa, pp 275–286
Tarabrin SB (1970) Violation of innervation and the mutation process in the tissue. In: Volkova OV, Astahova AM (eds) Questions nervous regulation of tissue processes. Nauka, Moscow, pp 85–93 (Russian)
Tsapygina RI (1971) The role of central nervous system in the regulation of cell division and radiosensitivity of chromosomes in corneal epithelium of mice. "Issledovanija po genetike" Leningrad State Univ Press 4:54–60 (Russian)

Vernadsky VI (1944) Problems of biogeochemistry II. On the fundamental material-energetic distinction between living and nonliving natural bodies of the biosphere. In: 21st century. Winter 2000–2001 Issue: 20–39

Wyrobek AJ, Bruce WR (1975) Chemical induction of sperm abnormalities in mice. Proc Natl Acad Sci USA 72:4425–4429

Roots of Current Concepts in the Studies of Social Behavior in Animals

Eugeniy N. Panov

Introduction

I will have to start with a mild reservation about the title of the paper. The term 'ethology' is used here in a broadest sense, as, for example, by Tinbergen in one of his latest papers (Tinbergen 1976). He considers it possible to refer the word 'ethologist' to any biologist who is occupied with studies of behaviour. The 'ethology in a narrow sense' should be distinguished from this broad meaning, as it is the discipline founded by K. Lorenz, N. Tinbergen and their followers. This trend, called 'classic ethology' today, marks only one historical period and only one of several approaches in the evolution of views of the discipline that I tentatively, and for the sake of brevity, will call further 'zoosociology'.

My main task is to show the intricacy and contrariety of the process of our knowledge of evolution. It is hardly possible to track in this process as a fluent progressive advance that is due only to the gradual accumulation of facts. Any respectable scientific community, any school of ideas—and sometimes a 'lone-wolf' scientist who had his own word to say—presented their own concepts. Those concepts were aimed at the solution of different tasks. Different approaches and tasks demanded specific research methods to be applied. The same facts were pretty often interpreted differently, and sometimes, even as being contradictory to each other. At times, contemporary scientists did not have a slightest idea about the existence of each other. In other cases, different scientific schools entered confrontation and acute ideological competition that resulted in scientific discussion for many years. Finally, there were cases of peaceful co-existence of schools—despite the obvious inconsistency in their basic foundations.

It is hardly surprising that in this heterogeneous interwining of views and trends, the scientific tradition among the generations of scientists was not as strong as it

E.N. Panov (✉)
A.N. Severtsov Institute of Ecology and Evolution RAS, Moscow, Russia
e-mail: panoven@mail.ru

V.L. Korogodina et al. (eds.), *Genetics, Evolution and Radiation*,
DOI 10.1007/978-3-319-48838-7_40

could seem at the first glance. Some problems that were adequately formulated and nearly close to a logical solution over a hundred years ago, lost their shape later and emerged as something absolutely new after many decades. At times, novel concepts were constructed almost as a blueprint, while an absence of their links with previous ideas was justified by an assertion that all that had been done before was nothing more than an anachronism of the pre-scientific mode of thinking.

With all difference in approaches and methods that will be considered below, an adherence to two different—and to some extent alternatively oriented patterns of thought—may be revealed retrospectively.

For one of them the leading principle is to study the whole through reducing it to certain elements that are considered explicitly specific and constant in their properties. It is conventional to regard this position as "elementarism". The whole thus seems to be as something secondary to the elements that form it, so that the property of the whole is determined mainly by specific features of these elements.

The alternative viewpoint states that the properties of parts can be grasped only through the knowledge of the whole. The whole is given in the first place and can be divided into part by many different ways. Just the splitting method is most important in determining the properties of the parts obtained. This attitude is often called "holism" (from Greek «holos»—wholeness). It can be also called "organisms", indicating the similarity between any complex whole entity to an organism that can be split in parts only surgically. One of the manifestations of this cognitive approach is an attempt to consider communities of social insects as 'superorganisms'. In my view, holism and organisms are, in fact, different names of what is accepted today as the systemic thinking.

There is a common belief that elementarism has many in common with the mechanistic worldview and by no means is compatible with the dialectic approach. Or, in other words, progress of our knowledge about the world can be viewed as a gradual transition from the elementaristic ideas to holistic ones. In reality, the situation seems to be more complex (Fig. 1).

The commitment of a scientist to a given cognitive attitude is determined in general by the specificity of the object under study and goals of research. These two aspects are more or less interconnected, but can be present in different proportions in the one or another doctrine. Until there is a reasonable parity between the generality of the approach, on the one hand, and its modes of the objects' analytic dissection, on the other, both organismism and elementarism are indisputable retained their cognitive value.

For the beginning, I tried to represent a history of the scientific ideas on animal social behaviour in their development in form of the rather simplified scheme shown in Fig. 1.

Here, eight scientists are under consideration whose scientific activities cover a period a little more than a hundred years. Their views and general works seem to me the most important, in such an extent that they may be regarded as the certain landmarks in shaping of our up-to-date notions.

Not surprising that since the main object of their research were community and population, i.e. complex objects of systemic nature, the majority of those scientists

	Holism **Organismism** **Systemic thinking**		**Elementarism** **Atomism** **Mechanistic worldview**
	A. Espinas, 1877 1.Place of the biosociality in the general system of knowledge 2. Individual in relation to society		
	W.C. Allee, 1931-1938 3. Limits of the biosociality	Classical ethology	**K. Lorenz, 1931-1935** 4. Mechanisms of communication
		Classical ethology	**N. Tinbergen, 1951** 5. Adaptive evolution of communication
Socioethology	**V.C. Wynne-Edwards, 1962** 6. Social behaviour as a mechanism of the population homeostasis	Sociobiology	**W.D. Hamilton, 1964** 7. The theory of genetical evolution of social behaviour. Origin of eusociality (altruism)
Socioethology	**J.H. Crook, 1970** 9.Concept of socio-demographicsystem	Sociobiology	**E.O. Wilson, 1975** 8. "New synthesis"

Fig. 1 Simplified scheme of the history of development of scientific ideas in studies of the interpopulation organization

were adhering to the holistic position. However, along with them there were two quite reputable schools, namely, classical ethology and sociobiology. that, in my view, appeared to be devoted to rather elementaristic approaches. Below, I shall explain in detail what I mean.

Figure 1 illustrated a plan of the further review as well. There, ordinal numbers correspond to those topics I shall discuss one by one. Although either of them was touched by all authors mentioned, although in more or less extent, some of them made a greater contribution than others to the development of certain ideas. For a start, two issues should be clarified. First, what is the subject of zoosociology and where is its place among other disciplines. Second, I shall examine how the ideas about social groupings, the main object of our science are seen in light of different views on what are an individual and biological individuality as such.

Alfred Espinas and First Steps of Zoosociology

The first serious review of the theme was offered by the French philosopher Alfred V. Espinas. His work was first published in 1877, about 150 years ago. The full title of the book is "Social Life of Animals. An essay in comparative psychology with addition of brief history of sociology". I must say that though this book is always cited by all authors who write about the history of ethology, these references are usually not informative.

Espinas formulated a task—to combine the data accumulated to his time of historical sociology and comparative biology in the frames of unified knowledge and acquired unified principles. Being the author of a number of historical-philosophic works—such as, for example, "The history of economic doctrines", Espinas was deeply interested at the same time in the problem that is titled today with an intriguing name of the "problem of interrelation of biological and social".

Espinas was quite right when argued that during the whole history of human knowledge development, starting with ancient times, at least from the 4th century BCE, the greatest minds of the mankind searched for an analogy between the human society and communities of animals. "While naturalists in their unconscious aspiration for generalization—Espinas wrote—compared animal communities with human ones, politicians, governed by the same impulse, compared likened human societies to the societies of animals". Regretfully, continued the author, none of them tried to elaborate the general principles for such comparisons and thus made the matter still more intricate.

But does it mean that the attempt itself to compare is vain? Not in the least. "There is no science of particulars!"—exclaimed Espinas. "The two groups of facts under examination that show an obvious analogy and are defined with the same word can become clear to us only when they are brought together to the single law that stems from their mutual properties".

Espinas suggests that this law should be found by revealing what in the modern language are the main organizational principles governing such complex natural structures, as the communities of individuals. In Espinas's opinion, zoosociology should be not an introduction to the general sociology but Chap. "Roots of Current Concepts in the Studies2 of Social Behavior in Animals" in it.

But would not then zoosociology duplicate the works by biologists? Espinas answers this question in the following way: "Among many specific features that characterize organic pieces of matter nutrition and reproduction are the most important. Sociology studies neither of them; it investigates only the most general character of organized social structures—say, the grouping patterns favoring implementation of one or another of the above functions, which gives it a specific role even in studies of those phenomena where it meets another life science—biology".

So, the subject of zoosociology is the specific character of interrelations among the elements within a certain organized structure; the ties that appear in situation when formation of the groups takes place to promote their further existence and biological functioning.

The next question Espinas had to solve was what may be regarded as those elementary 'bricks' that being extracted from the wholeness do not lose its own essence and thus remains an independent actor. In the human sociology for a long time the answer was simple: an individual. But this point of view was not accepted unanimously. For example, according to Aristotle, the elementary unit in human society is not an individual but a couple, because the individual is incomplete and inexplicable from himself. Hegel also believed that the family is more independent than the individual (Hegel 1821, 1940). This position was favorable to Espinas as well ideas and he persistently keeps upon it throughout the text of the book.

Espinas pointed out that when dealing with plants and many lower animals the question often arises—what one means when used a term "an individual". At that time, in the second half of the nineteenth century, this period of tumultuous development of biology, it appeared that there is, in fact, a firm ground for such an uncertainty about the matter. Namely, numerous creatures were found, in respect of which it was not clear if their bodies would be divided into parts being "independent", to more or less extent, from each other.

In the 1940s–50s botanists M. Schleiden and T. Schwann discovered that all living organisms consist of cells. They offered the idea that the very life of a plant is combined beings of its cells. Then it became possible to consider the cell itself as an elementary organism. Soon prominent German scientist R. Virchow developed on this basis the so-called 'theory of cell state'. "Any creature—he wrote—is a sum of living units, so that each of them has everything necessary for life" (Virchow 1863). This concept, atomistic in its essence, asserted strong influence over the development of Espinas's ideas, as we will see it below.

In his time, zoologists knew well some amazing creatures about which one could not say for certain whether they are individuals in a strict meaning of the word or peculiar colonies of many creatures organically connected with each other. Such are, for example, so called siphonophores, belonging to marine coelenterates. Like a tree, with its roots, branches and leaves, the body of a *Siphonophora* is composed of several "individuals-organs". One of them only capture food, others execute the reproduction function, still others protect the whole creature against predators, still others provide its motion in the water column. In 1866 the outstanding German biologist E. Haeckel designated such compound organisms as cormi, singular—cormus[1] (Haeckel 1866). Later their components "individua organs" were called zooids.

As a result of these and many other discoveries in biology the issue of the essence of concepts of "collectivity" and "individuality" grew more and more cloudy. At the end of the XIX century F. Engels in his "Dialectics of Nature" wrote that the notion of an individual has become absolutely allusive. Cormus, colony, tapeworm on the one hand and the cell and metamere, on the other, can be identified as individual in a certain sense (Engels 1883).

In his book that was almost an "age-mate" of "Dialectics of Nature", Espinas explained the situation to readers: "We take our own type of individuality as a norm and deny the quality to any other creature that is too distant from such a type. As soon as a certain creature loses definite shape and an ability to move independently, we refuse to regard it as an "individual" (Espinas 1898).

---

[1]From Ancient Greek κορμός (kormós, "trunk of a tree with the boughs cut off").

Meanwhile, Espinas wrote, individuality may be expressed in different extent. According to Virchow's views, a cell should be regarded as an elementary individual because it is actually an indivisible biological atom. Hence, individuality of a multicellular creature is a kind of collective individuality. In favour of such inference speaks the fact that in the multicellular organisms there are always cells that remain relatively autonomous and are able to move actively inside tissues. Such are, for example, wandering ameboid cells in sponges, spermatozoids, etc.

A special attention of Espinas such cormi attracted in which along with individual-zooids such structures are present that serve the cormus as a whole. These can accomplish, for example, the excretory function (the common cloaca in the colony of sea squirts (Ascidiacea) or ensure the fulfillment of locomotion—like in the case of the so-called creeping sole in mobile colonies of some moss animalcules (Bryozoa).

When discussing such phenomena taking of octocorals as an example, Espinas wrote: "Along with individual life of polyps as such an another life is in progress being independent of individuality of each member of the colony and belonging to the hydranth as a whole. The latter can be regarded in this case as an individual. It is obvious that a certain individual loses its rights relative to those of the community by surrendering a part of its activity to it" (Espinas 1898).

Functionally, heterogeneous colonies of coelenterates, moss animalcules, tunicates, etc. are a bright illustration of submission of a part to the whole, an individual to the community. The same principle Espinas saw in the family's communes of termites, bees and ants that consist of morphologically autonomous individuals but inseparable from each other socially and functionally. The author even tries to look farther ahead and to include into the same line collectives of higher vertebrates. But here, because of almost complete absence of reliable information his speculations assume a flavour with of scholastic characteristic of the natural philosophy. Yet the main idea is clear and Espinas illustrated by analogy with the human society: "Not individuals create the society but society creates individuals because they exist only in the society and for interests of it" (Espinas 1898).

Ernst Haeckel in his work "General morphology of organisms" (1866) distinguished six classes of organic individuality. Individuals of class I are cells, of class II —organs, etc. An individual as we used to think understand is a creature of the class V, and the aforementioned cormus belongs in this classification to the class VI.

The problems raised by Espinas in the framework of the biosociality origin remain of a great importance still today. As the outstanding Russian zoologist V.N. Beklemishev stated, "the notion of organic individuality is undoubtedly one of the basic concepts of biology" (1964). This topic laid down the foundations of the so-called "colonial theory" of the multicellular organisms origin from the monocellular ones. It was initiated with works by E. Haeckel, I.I. Mechnikov[2] and other eminent scientists. The concept of collective individuality turned out to be highly multidimensional and extraordinarily complex.

---

[2]See Metchnikoff 1892 (Fr) and Mechnikov 1950 (Rus) (eds).

Formation in evolution of superindividual structures of the cormi type promotes an increasing specialization of its constituents—zooids. This process that leads, according to Beklemishev, to the more and more reinforcement of the cormi individuality. Espinas interpreted the matter as "a transition from bondless homogeneity to specified and consolidated heterogeneity" (Espinas 1898).

This principle undoubtedly also operates in the processes of formation in evolution of highly integrated social communities of multicellular animals, both invertebrates and vertebrates. Let us see now what has been done during the development of zoosociology in this sphere of the research.

Warder Allee and the Problem of Putative Limits of Biosociality

Just after sixty years since the publication of the Espinas's work in 1938, another book appeared under the same title: "The social life of animals". It was written by Professor Warder Allee from the University of Chicago (Allee 1938). In this important publication he summarized both the results of his own studies accomplished during 26 years of his scientific career and the main data obtained in zoosociology since the time of Espinas.

A few words should be said about the author. His first influential work "Animal aggregations: a study of general sociology" was published in 1931. In 1949, he published the book "Principles of animal ecology", which has been widely cited in subsequent years. His last book, "Ecological animal geography", written in co-authorship with K. Schmidt comes out in 1951. Thus, Allee made a remarkable contribution in general zoology and ecology.

A brief outline of the general trends in disciplines relevant to our topic at the time of publication the book "The Social life of animals" (1938) is necessary to understand main ideas presented in it. For the same sake it will be helpful to get acquainted with the literature sources cited in the book. Their distribution in time (Fig. 2) can to serve as an indication of the growth of interest to zoosociology at roughly the turn of the first and second decades of the 20th century.

It can be seen that Allee was familiar well with the studies in human sociology. Like Espinas he advocated the idea of the general sociology establishing. To a great extent, it was due to his efforts that demography, which was brought to life in the studies of human populations, was introduced into ecology and sociology of animals.

Figure 2 also shows a list of names of scientists who made a considerable contribution into the development of the topic of our interest by the moment of Allee's book publication. These are, among others, four Russian scientists: P.A Kropotkin, B.P. Uvarov, G.F. Gauze and Th. Dobrzhansky. The first of them was stressing persistently the importance of cooperative relations in communities of animals (Kropotkin 1902). Gauze contributed much to the development of principles of biological competition (Gause 1934). Dobrzhansky together with S. Wright

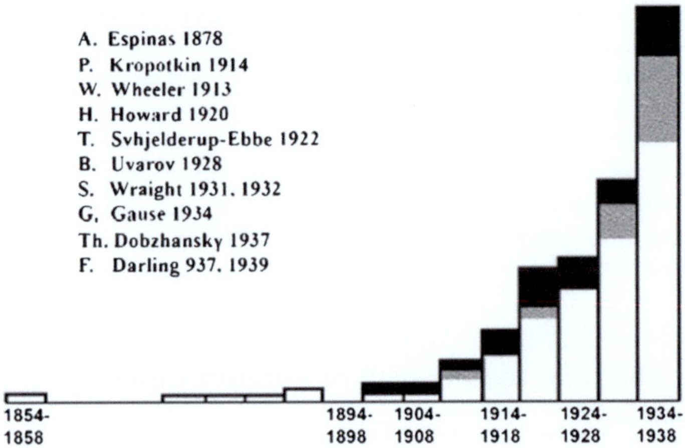

Fig. 2 Distribution by year of literature references, cited in the book of Allee (1938). Papers by Allee are shown in hatching; studies on sociology and demography of man are shown in *black*

is one of the founders of population and mathematical genetics (Dobzhansky and Wright 1942). B.P. Uvarov discovered the phenomenon of the solitary and swarming phases in locusts (Uvarov 1937).

W.Wheeler was an outstanding specialist in social insects (Wheeler 1928). H.E. Howard was one of the founders of the idea of territoriality (Howard 1920). F. Darling put forward the ideas of social stimulation of reproduction in crowded assemblies of animals (Darling 1938).

And what are actual main peculiarities and results of the activities of Allee's school in the 1930s, in comparison to those that we saw in Espinas's theory? I have already mentioned that the work of the latter, due to the very essence of science of his time, had something in common with a natural philosophical mode of thinking. For Allee just the contrary, the main research method became experiment—which is, in general, characteristic of American behavioural science from the very start. In his discourses Espinas uses ordinary language at large. In Allee's works we see quite a broad diverse special scientific terminology. He uses such concepts as population, population size and density, hierarchy, rank order, territory, mass effect, social hormone, social facilitation, etc. The experimental approach and the start of creating a special scientific language mark a transition of zoosociology from a speculative mental constructs to the truly scientific stage of development.

In his approach, Allee, following Kropotkin, stresses the importance of cooperative relations, as opposed ideas about bloodstained struggle for existence dominating at that time. And another very important postulate is the idea about the influence of the population density on biological success of individual (Allee 1931; Allee and Bowen 1932).

Although a view that overpopulation negatively influenced upon growth in postnatal ontogenesis, reproduction and some other characteristics was not new, almost nothing was said about the negative role of abnormally low density or underpopulation. In a number of elegant experiments with different species of animals—from protozoa to mammals Allee demonstrated this second effect. It leads to the well-known “Allee’s principle”. The Allee’s principle gives the idea of optimal density—not too low and not too high.

Speaking about Allee’s views, I would like to pay special attention to how he defined the subject of our science and, consequently, to his reasoning about limits of its competence. I mean his opinion about meaning the notions “social behaviour” and “social species”. We will have to examine Allee’s argumentation in more detail here.

After Espinas, Allee believes that as, speaking strictly, there are no solitary animals in the full sense of the word biosociality to a certain extent is common to all species of animals and in a sense—even to plants. So the question is: is it possible inside this continuum to draw the natural line between subsocial species and those that are accepted as truly social (intuitively). The latter primarily includes social insects. But before that I would like to stress that the answer is negative—this line can be only absolutely provisional, based solely on the agreement among scientists.

Argumenting, Allee examined three possible criteria of sociality. The first one is presence of the so called social instinct; the second is an adherence to family life style; the third is division of duties between members of the animal community.

In respect of some of scientists’ adherence to the idea of social instinct Allee wrote that, fortunately, not many researchers share it. For Allee- experimenter this criterion could not seem practically useful since the very notion “instinct” appears extremely vague.

As for the second criterion—a practice of a family life-style, Allee contended that there is no any fixed point or sector where ephemeral sexual encounters would suddenly change into stable family. Examples that confirm this conclusion can easily be found in any well-studied taxon.

Thus, the third criterion remains—biological division of labour. Here, Allee stressed, the division of duties arises still at the earlier stage of evolution when the phenomena of gender and sexual reproduction appear. In modern terms, a male and female in the majority of species are bearers of the principally different social roles. At the same time, Allee came to conclusion that females are often much more social (or sociable) as compared to males much earlier than a number of other researchers of later times.

Therefore, it is not obvious for Allee where the sex-related division of duties ends and social division of functions starts. Traditionally, the relations in communities of social insects have been the standard of the latter. But even there, Allee wrote, the caste division is closely dependent on the sex of individuals. For example, sociable females in ants and bees are divided into reproductive and functionally asexual, or workers. The third caste consists of haploid males—drones that execute mainly the sexual, not social function.

The phenomena partly similar to caste polymorphism, or polyethism of social insects can be found in other animals whose sex determination depends on instantaneous intrapopulation social background.

Allee illustrates such cases with several examples. I would like to discuss one of them, taking place in the mollusk *Crepidula furnicata*. The phenomenon of protandry is its characteristic feature: each individual is asexual at early stages of ontogenesis, then it becomes a functional male and having reached a certain age turns into a female. Males of small size try to find those like them and remain in close contact groups consisting of individuals that include several species of both genders as well as those in the phase of transition from a male to a female. Several males can copulate with the same female simultaneously.

It is important that males that took part in copulations turn into females faster than those that failed to find a sexual partner. It was shown both experimentally and on the basis of field data. As a result, in *Crepidula* the sex ratio is different in different populations.

Allee gives many other examples of the influence of social environment and population density on sexual differentiation of some individuals. I shall present an example unknown to Allee. In many fishes sex change in ontogenesis occurs in opposite way than in *Crepidula*: a young individual is at first a female; then it turns into a male (protogyny).

It takes place, in particular, in the coral fish *Anthias squamipinnis* that live in heterosexual groups of various numbers. Sometimes it is a polygynic unit with one male and several females, sometimes groups consisting of several hundreds of individuals with the male to female ratio approximately 1:9. If a male is taken away experimentally from a polygynic group, one of females turns into a male in less than a week. In the case of withdrawal of several males from a large group, the number of females equal to the number of withdrawn males turns into males in the same short period.

An obvious analogy can be easily seen the phenomenon of regulation of family composition in social insects. According to Allee, a family of thermites *Zootermopsis* recently established by a pair of individuals has only one soldier in the first year. If it is withdrawn from the nest, its absence is compensated by the appearance of several (up to six) other soldiers. However, the compensation effect will not be as strong as that when a big family includes many hundreds of individuals. Allee believes that in the latter case a peculiar effect of "dilution" takes place and, correspondingly, hypofunction of a certain agent, or, as he calls it, a social hormone that is capable of suppressing the development of excessive number of soldiers.

No matter how highly specialized is the mechanism of the caste composition regulation, it shows, according to Allee, a certain analogy with much simpler "group effects". Among them one is a follows: an increase of resistance to harmful substances diluted in water inhabited by in large groups of fishes, being absent in the cases of small groups and single individual.

The search for such broad analogies is by no means a goal in itself. It appears to be an important methodological approach. It allows one to realize logically the possible ways of the complex regulatory mechanisms' evolutionary development from the moment when they are in a vestigial state yet and, strictly speaking, are still not complete formed. In this sense, Allee's quite perspective guideline is, in my opinion, the repetition and development of the ideas of Espinas who recommended to study not only advanced forms of social life but its earliest forms as well.

Allee illustrates effectiveness of this approach with the example of establishing a developed social system in thermites on the base of simple gatherings of individuals in places of plentiful food supplies, like what we can see in the common German cockroach *Blattella germanica*). The hypothetical intermediate stage is represented, according to Allee, by some subsocial species of cockroaches that feed on fiber like thermites Both these cockroaches and thermites process of digestion as well as ability to survive are possible only because of the presence in their intestines of the symbiotic flagellated protozoans recycling fiber. Newly born insects do not have protozoans and can only get them through contacts with a skin of adult individuals being shed by them in the course of their moult (in *Cryptocercus*) or from another imago in thermites. However, adult individuals of *Cryptocercus* do not moult, and that is why an isolated pair of a male and a female is not capable to start a new deme.

In thermites imago loses protozoans in the course of each moult, so that the presence of other individuals is a vital necessity for all. Therefore, a male and a female with protozoans in their intestines are capable of establishing a new colony that facilitates keeping self-developing sociality in thermites, which is not the case in *Cryptocercus*.

Contribution of Classical Ethology to Research of Social Behaviour

Just in 1931–1938, that are the dates of the two first influential Warder Allee's books publication in the USA, another school in the behavioral research arose on the other side of the Atlantic. I mean the discipline known as classical ethology. Its origin can be dated tentatively by 1931, when 28-year old Konrad Lorenz published his first large paper "Beiträge zur Ethologie sozialer Corviden" in the German "Journal für Ornithologie"[3]. In 1935 and 1937a, b, c, two other basic papers of his were published: "Der Kumpan in der Umwelt des Vogels" and "Über die Bildung des Instinktbegriffes". In 1938, Konrad Lorenz and Nikolaas Tinbergen offered another paper on the role of innate components in the organization of integral behavioural patterns. Despite the fact that Allee and Lorenz developed their theories practically simultaneously, the essence of their approaches turned out to be quite different.

As we could see, Allee did not touch upon the distinction between the innate and acquired behaviour. For him it was a secondary matter, and he tried to avoid this aspect when possible, as he could not study it experimentally. For K. Lorenz and N. Tinbergen the a priori division between innate and acquired components of behaviour and establishment of relations between them was the basis of all further speculations.

[3]Currently Journal of Ornithology-Springer.

These points are the heart of the whole fundamental differences in views of the American and European schools. For Allee the word "instinct" smelled of scholastics, was a kind of 'refuge of ignorance', while Lorenz and Tinbergen created an immense concept designated by them the "modern theory of instinct" (see, for instance, Tinbergen 1951).

Allee stresses the lability of the individual behaviour, its partial unpredictability, due to the variability of the social environment or social climate experienced by individuals. Ethologists of the European school, on the contrary, emphasize the stereotype aspects of social conduct, its conservative species- specific qualities.

Allee and his colleagues are interested primarily in consequences of the interaction of individuals joined into groups. Lorenz, Tinbergen and their school concentrate their attention on subtle mechanisms of these interactions rather than on their influence upon the subsequent life of the society. Allee is interested in the group structure at various parameters of the population number and density. Lorenz and Tinbergen are busy with how acting the given individual in the presence of other conspecifics performing different social roles. The centre of attention for the American school is various group and mass effects in local population, while European researchers analyze the dynamics behaviour of an individual mainly in pair interactions.

I shall not discuss the principles and methods of the ethologic concept in more detail, as it is covered much more than any other concepts in the literature (see e.g. Panov 1975, 2012). I would only like to accentuate some methodological aspects.

What is an elementarisic essence of classical ethology? I dare say, it become apparent in the intention to regard social behaviour as a chain of discrete events, so that each of them is rather constant and stable in its manifestations. Such constancy is reflected by a rigid classification of several "principal patterns" of interactions—for instance, a territorial conflict, treating a subordinate by a dominant, pair formation, copulation, etc. The stereotype feature of interactions within either such class of events is reflected from the start by Lorenz' scheme of several "companion" types: for example, companions-parents, companions-offspring, companions-spouses, social companion.

According to ethologists of the classical school, an interaction of each type is performed as a determinate exchange of stereotype signals between two individuals belonging to a category of a given type of companions.

These signals are called "fixed action patterns". Just standard types of interactions and standard communicative stimuli that are strictly determined genetically in this approach those further indivisible elements to which is reducing all, or almost all, social life in animal collectives.

We do not find anything like this in Allee's theory which is concentrated on the dynamism of a whole entity. It seems much more dialectic than the approach of classical ethologists that obviously have a propensity for mechanism.

It is important however, that being not adequate enough for the analysis of social processes in all their diversity, this approach, namely because of its typological features, has played and continues to play a very important role in the behavioural taxonomy and in modern systematics as a whole. This approach also contributed much to the understanding of organizational principles of individual behaviour functioning. The studies of organization of individual behaviour remain of a lasting value.

The general principles revealed here are inevitably modified with the progress of our knowledge, but further studies in the framework of fundamental theme "individual and society" cannot be conducted without them. Unfortunately, today it is too often overlooked.

But the ethological approach oriented at the individual and its interactions with the environment, including conspecifics, appeared to be quite insufficient for the comprehensive understanding of the biosociality phenomena. Then time demanded that in the centre of researchers' attention should be the population as an organized system of the superindividual level.

Vero Wynne-Edwards and the Idea of the Intrapopulation Homeostasis

This new approach was shown in a most consistent way (arguably too consistent) in the book by Vero Wynne-Edwards "Animal Dispersion in Relation to Social Behaviour". When it was published in 1962, its author, University Professor in the Scottish town of Aberdeen, was 56 years old. This work became one of the most cited publications in the literature on animal behaviour for the subsequent twenty years. It was a thick volume of about 650 pages. The list of references includes over 840 items (Fig. 3a).

We are interested in the analysis of the cited references in two ways. First, we see that the author widely uses literature in zoology of the past that shows his wide knowledge of biology and, possibly, a shortage of more modern special research in the topic of his interest. For example, papers published in 1911–1940 make about 34 % in the reference list. For comparison, of the reference list to the paper by Adam Watson and R. Moss (Watson and Moss 1970) on a close topic published eight years after the publication of the book of Wynne-Edwards citations of 1911–1940 comprise less than 2 % (Fig. 3b).

Second, the analysis of the literature used by Wynne-Edwards makes it possible to understand to what extent earlier authors influenced his ideas. It can be seen that Wynne-Edwards extensively cites the papers of ecologists Charles S. Elton, Huib N. Kluijver, David L. Lack, Alexander J. Nicholson[4]. He was also acquainted with the theory of Hans Selye who forwarded the concept of stress, or of the general adaptation syndrome (Selye 1956), and papers by John J. Christian and Dennis H. Chitty[5]) who tried to establish in the 1950s the idea of the influence of the social stress on the density and genetic contents of population.

[4]See, for examples, Elton (1958), Kluyver and Tinbergen (1953), Lack (1947), Nicholson (1933) (eds).

[5]See, for examples, Christian (1950), Chitti (1960).

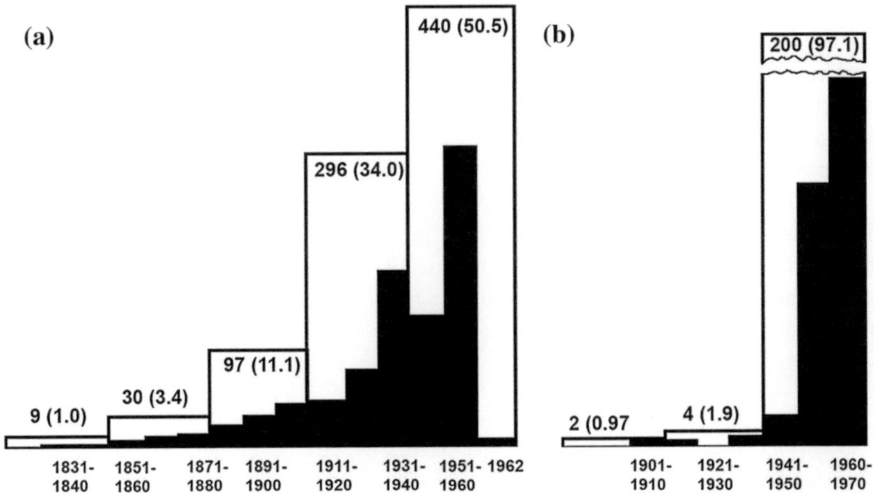

Fig. 3 Distribution by 10-year periods (*hatched*) and by 30- and 20-year periods (*contour lines*) of literature references cited (**a**) in the book by V. Wynne-Edwards (Wynne-Edwards 1962) and (**b**) in the paper by A. Watson and Moss (1970). Figures—the number of sources, in brackets—percentage

The main idea of Wynne-Edwards is that all forms and types of social behaviour accomplish, in essence, the same major, and, may be, a single function, namely, to maintain a population in state of homeostasis, or, in other words, at a certain optimal level of its numbers and density. There are some mechanisms, according to the author, at the expense of which a population as an organized system protects its environment against overexploitation.

Wynne-Edwards believes that one of such mechanisms is the phenomenon of territoriality. Individuals that manage to conquer and keep then a territory become owners of abundant resources that they cannot can not to use completely. Other individuals turned out to be deprived of enjoyment of their own territory and so are forced to remain in the status of outsiders. They suffer from high mortality owing to hunger, predation, etc. This unequal access to resources serves the basis for the so called social selection.

It is important also that the transition of a part of individuals to the category of outsiders does not necessarily occurs forcibly. According to Wynne-Edwards, an individual has an ability to evaluate the population density. If density is high the individual itself refuses from attempts to conquer a territory and subsequent participation in reproduction.

This capacity to evaluate adequately the level of the population density is called "epideictic behaviour" which in translation from Greek means something like "presentation and analysis of a sample" (these concepts are taken from statistics).

Having postulated such a capacity, Wynne-Edwards tries to understand what may be possible mechanisms of its emergence in evolution. It is difficult to do if to believe in the paradigm according to which the evolution is advanced only by unequal success of individuals. In the given case individuals-outsiders doom themselves to failure and even death.

Thus, Wynne-Edwards comes to conclusion that the biological success of a group, namely, the guarantee of its future survival, is exactly the advantage granted by social behavior. In the frames of his conception an unit of evolution is not an individual but a group. Instead of individual selection, group selection is offered to explain the evolution of social behaviour.

I shall not discuss controversial points and some drawbacks of the Wynne-Edwards' model. Here it is important to stress that in my scheme it appears as a concept of obviously organismic character based on holistic approach to the understanding of society. Processes taking place in groups of individuals are regarded as the most important, while the features of the individual behaviour is deduced from characteristics of the group structure. It should be also noted that accent in the Wynne-Edwards' work shifted, in general, from the attempts to analyze group structures to questions of their hypothetic evolutionary origin.

Sociobiology by William Hamilton and Edward Wilson

By contrast the organismic theory of Wynne-Edwards and almost simultaneously, only two years later, in 1964, another doctrine was set forward that, in the frames of my scheme, can be appreciated as utmost elementaristic. It is the so-called "genetic theory of evolution of social behaviour" developed by the English naturalist William Hamilton. Like Wynne-Edwards Hamilton is concentrated not on analysis of what are animal communities now, but rather on ways and mechanisms of their formation in evolution. This topic became the main and most principal for him. But in unlike Wynne-Edwards, Hamilton believes that there is no need to postulate group selection as everything can be explained via individual success of an individual. While Wynne-Edwards thought that specificity of the species behaviour provides the success of the group as a whole, according to Hamilton, everything occurs quite in the contrary way: success of an individual is provided mainly with its existence as a member of a group.

As opposed to other neo-Darwinian theories, the success of an individual is evaluated here not just by the number of offspring left but with the quantity of those who will become in one way or another the bearers of the genes the same as present in this individual. It can increase the fraction of its own genes bearers either in the "egoistic" way by reproducing in maximum possible rate, or indirectly, promoting the survival of its close and distant relatives. Hamilton called the second strategy as "altruistic" behaviour. Thus, according to him, it is possible to draw a sharp dividing line between non-social and social species.

From this point of view, the appearance of what he called the "true sociality" takes place at the moment when "absolute altruists" emerge in a population, i.e. such individuals that do not reproduce themselves at all but instead take care of their relatives. Thus they preserve their own "altruism genes" for the subsequent generations. This is the essence of the concepts "inclusive fitness" of the individual, and "kin selection" offered by Hamilton (for details see Panov 1983).

The whole theory is built on the axiomatic principle, on the basis of theorems of mathematical genetics developed earlier on the grounds and for the analysis of another sphere of biological reality. This circumstance makes unnecessary for the given theoretical construct almost everything that was made earlier in the field of zoological research of the social behavior. For instance, a most interesting problem of organic individuality disappears automatically. The border between social and "non-social" species is drawn quite unequivocally. From the classical ethology only the idea of the innate determination of behaviour was taken, while one of the most interesting and promising topics, namely, causality of the behaviour dynamics in time (in particular, in ontogenesis) and its alterations influenced by changes in social environment was practically left unattended.

Nearly all ties with the richest earlier achievements in biology were cut off. The integral phenomenon of biosociality with all complexity of its intrinsic links disappears; it breaks down into certain behavioural "traits" each of which is controlled by its own genetic determinant. It appears that the whole approach is apparently simplified, abiological and has a little explanatory potential. In fact, on its base many hypotheses were set forth and, as a result of it, very large empiric material was accumulated during the subsequent years. But, as a rule, obtained data turn out to be fatal for the hypotheses themselves, being in obvious contradiction with its prediction. It is no surprise, taking into account the formalistic, scholastic and far-fetched character of the majority of sociobiological speculations.

In 1975 the book "Sociobiology: The new synthesis" by Edward Wilson was published. It is a review collection of about 700 pages. If the reference list in Allee's paper contained 129 sources, Wynne-Edwards's 879, in Wilson's monography there were 2500.

Let us see what the book by Wilson is in its essence, what are its connections with preceding stages of zoosociology development. And is it possible to consider it as an organic synthesis of the previous knowledge, as the author asseverate? At the very beginning of Chap. Roots of Current Concepts in the Studies2 of Social Behavior in Animals, under a rather surprising title "The morality of the gene", we read: "The main theoretical problem of sociobiology is to understand how altruism that lowers adaptation of an individual could develop in evolution". Here the author expresses his main credo: like a chicken hatching from an egg is no more than a means to reproduce another egg, the organism is essentially not more than an instrument for preserving and transition of genes. It is obvious that Hamilton's ideas are the main source for ideology of Wilson.

In the Chapter dealing with animal communication Wilson punctually retells the basic points of the ethology theory of instinct, and sometimes—in its most archaic form. The third component of the Wilson's sociobiology is the modern population ecology in development of which Wilson himself made an important contribution, in particular, with the book "The theory of Island biogeography" written in co-authorship with R. H. MacArthur (MacArthur and Wilson 1967). In many parts of his “Sociobiology” Wilson regularly cited his previous book on social insects that he knows excellently.

Therefore, there is no denying that the author of the “Sociobiology: The new synthesis” is an erudite in many facets of the biology. But is his book actually a synthesis of the preceding knowledge? One can get an impression that Wilson’s work is, in essence, rather eclectic than synthetic. The true synthesis presupposes combining different views and concepts in such a form when their inner contradictions are not hidden but shown as clearly as possible. It is what really stimulates science to advance.

As for the Wilson’s book, all preceding concepts, both apparently atomistic and holistic ones, peacefully co-exist in it. Perhaps, only in respect of the ideas of Wynne-Edwards, Wilson is predisposed negatively and spares them almost no place in the book. It is interesting that Wilson calls himself a holist that seems far from the real state of things. A lot of his constructions have obvious typological character and thus undoubtedly tend to elementarism of the classical ethology and the genetic atomism of Hamilton.

Programme of Studies of Socio-demographic Systems in Socioethology

Sociobiology is today a quite influential discipline, although in recent years a noticeable wane of its prestige can be traced. And it is not the only approach in the zoosociology today. There is another concept of a holistic character that ideologically opposes sociobiology. I mean the so-called socioethology whose subject and tasks were outlined in the early 1970s by the zoologist J. Crook from Oxford.

Both the object of the study—a local population, and its subject (social behaviour) in socioethology differs at the first sight little from as they are seen in the sociobiology of Hamilton and Wilson. However, the principle approaches and the general accents in treating of the evolutionary questions are different here in many aspects. First of all, Crook’s attitude seems to be much more realistic. For him speculations about the evolution of social behaviour is not the first task but the second one, executable only after the clear understanding of the essence of systems under consideration is achieved. These systems appeared as an immense complex structures reciprocal with very complex interweaving of the demographic parameters and social behaviour are designated by Crook as the socio-demographic systems. For Crook the appear not as a certain stable entity but as a society the understanding of society as a process, where the development of individual behaviour proceeds under the influence of both interweaving and socialization, which demands very close attention (Crook 1970).

In fact, the matter is one of the most complex and not properly understood problems of biology: how the genotype, which is no more a program of the individual development, is realized, under the influence of outward information coming from outside, into the phenotype of the competent actor. In general, Crook actively keeps aloof from a primitive understanding of heredity as a simple mechanical transition of the so-called 'traits', including behavioural ones.

Hence, a conclusion naturally follows that speaks that the role of genetic factors in determination of social behaviour and, thereafter, in a maintenance of intergroup structure is heavily exaggerated by sociobiologists. The cause of it, Crook, affirmed, is rooted, among others, in the wrong (and characteristic especially of layman audience) notion that structural characteristics of society are equated to the "fixed action patterns" of the classical ethology (Crook 1970). Such misconceptions, Crook believes, are formed under the influence of some publications addressed to the general public. He means such books, for example, as the Lorenz's "On aggression" (1963), D. Morris "The Naked ape" (1967), R. Ardrey "The Territorial imperative" (1966). The book by R. Dawkins "The Selfish gene" (1976) may be added to them. These publications spreading ideas like that on inborn aggression in humans, Crook calls "pseudobiology".

In recent years, this pseudobiology of the human behaviour is flourishing in adapted for the elite books by Wilson and his colleagues.

Crook, who specialized in studies on the divergence of socio-demographic systems in primates, concluded that in their evolution the main role was played by phenotypic adaptations that put in motion with individual learning and intergroup traditions. These factors became of the key value at the early stages of anthropogenesis. The genetic factors in that the forthright form, as Hamilton, presents it, plays, in Crook's opinion, a rather subordinate role.

It does not mean at all, however, that we should ignore genetic aspects of social evolution. But here it is necessary to develop experimental genetics of behaviour—instead of wordy juggling with genetic terms, like "mutations" of certain abstract 'genes of egoism' and altruism. Let us see as the matter looks from the point of view of specialists dealing with the scientific language. "Terminology within the biological sciences gets its import not just from semantic meaning, but also from the way it functions within the rhetorics of the various disciplinary practices. The 'sociobiology' of human behaviour inherits three distinct rhetorics from the genetic disciplines. Sociobiologists use population genetic, biometrical genetic, and molecular genetic rhetorics, without acknowledging the conceptual and experimental constraints that are assumed by geneticists. The eclectic blending of these three rhetorics obscures important differences of context and meaning. Sociobiologists use foundational terms in genetics, such as 'gene', 'fitness', 'evolution", 'heritability' 'trait' and 'polygenic inheritance', in starkly different ways from geneticists, while basing their analysis of human behaviour on the implied authority of genetics. As a free-floating 'gene talk' moves across different disciplinary contexts, and before different audiences, it takes the form of an over-simplified and misleading arch-determinism. The result is widespread application of vague, incomplete, and distorted biological theory. If most sociobiologists, do not deliberately promote biological determinism, still less a political agenda, there is ample evidence that they misconstrue the implications of the genetic language that they borrow" (Howe and Lyne 1992).

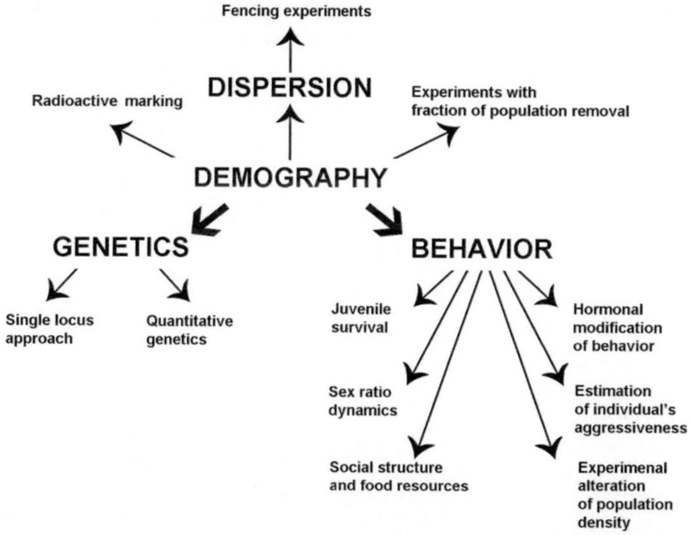

Fig. 4 Strategy and methods of research of the socio-demographic system (from Krebs 1979). Explanations are in the text

I would not discuss in more detail the system of views of socioethology. It will be illustrated points of its main tasks, approaches and methods with a scheme from the paper (1979) by the Canadian researcher Charles Krebs on long-term studies of socio-demographic systems of vole *Microtus townsendii* (Fig. 4).

It can be seen from this scheme that socioethology has rich hidden prospects that give hope for the really deep understanding the essence of intrapopulation processes on the basis of comprehensive application of ethological, ecological and genetic methods of research.

Conclusion

It is not easy to avoid certain schematism and declarativity in a brief review of a hundred-year history of scientific research, discoveries and disappointments. I only have a mild hope that I managed, maybe partially, to show the important place that zoosociology occupies in the general system of biological knowledge. The cornerstones of this discipline are rooted in the period of tremendous upgrowth in biology in the middle of the 19th century. Is not it surprising that the topics related to the studies of social behaviour of animals, which seems to have appeared quite recently, nearly before our eyes, have actually such a long and complex history? It should be stressed that studies of social behaviour and group structures are by no means a by-path of biology. This topic is directly connected with such fundamental issues of biology as formation of multicellularity in phylogenesis, formation of the phenotype in ontogenesis and autoregulation of most complex systems of the supraorganismal level.

References

Allee WC (1931) Animal aggregations: a study of general sociology. University of Chicago, Chicago, p 431
Allee WC (1938) The social life of animals. NY, p 293
Allee WC, Bowen E (1932) Studies in animal aggregations: mass protection against colloidal silver among goldfishes. J Exp Zool 61(2):185–207
Allee WC, Schmidt KP (1951) Ecological animal geography, 2nd edn. London, p 715
Allee WC, Park O, Emerson A, and Park T (1949) Principles of animal ecology. Philadelphia, London, p 837
Ardrey R (1966) The territorial imperative: a personal inquiry into the animal origins of property and nations. Atheneum, New York
Beklemishev VN (1964) On general principals of organization of life. Bull MOIP Biol 69 (2):22–38
Chitti D (1960) Population processes in the vole and their relevance to general theory. Can J Zool 38:99–113
Christian JJ (1950) The adreno-pituitary system and population cycles in mammals. J Mammal 31:247–259
Crook JH (1970) Social organization and the environment: aspects of contemporary social ethology. Anim Behav 18:197–209
Darling FF (1938) Bird flocks and the breeding cycle: a contribution to the study of avian sociality. Cambridge University Press, Cambridge
Dawkins R (1976) The selfish gene. Oxford University Press, Oxford
Dobzhansky TG, Wright S (1942) Genetics of natural populations. X. Dispersion rates in *drosophila pseudoobscura*
Elton CS (1958) The ecology of invasions by animals and plants. Methuen and Co. Ltd, London
Engels F (1883) Dialectics of nature. Prog Publ 1934:295–311
Espinas A (1898) The social life of animals. Experience in comparative psychology with the addition of a brief history of sociology. St. Petersburg (Russian)
Gause GF (1934) The struggle for existence. Baltimore
Haeckel E (1866) General morphology of organisms. Berlin
Hamilton WD (1964) The genetic evolution of social behavior. I, II J Theor Biol 7(1–16):17–52
Hegel GWF (1821) Grundlinien der Philosophie des Rechts. Berlin
Hegel WWF (1940) The philosophy of right. Leipzig
Howard HE (1920) Territory in bird life. John Murray, London
Howe H, Lyne J (1992) Gene talk in sociobiology. Soc Epistemol 6(2):109–163
Kluyver HN, Tinbergen L (1953) Territory and the regulation of density in titmice. Arch Neth Zool 10:265–286
Krebs ChJ (1979) Dispersal, spacing behaviour, and genetics in relation to population fluctuations in the vole microtus townsendii. Fortschz Zool 25:61–77
Kropotkin PA (1902) Mutual aid: a factor of evolution. London
Lack D (1947) Darwin's finches. Cambridge University Press, Cambridge
Lorenz K (1931) Beiträge zur Ethologie sozialer Corviden. J Ornithol 79:67–120
Lorenz K (1935) Der Kumpan in der Umwelt des Vogels. J Ornithol 83:137–413
Lorenz K (1937a) The companion in the birds' world. Auk 54(3):245–273
Lorenz K (1937b) Über die Bildung des Instinktbegriffes. Die Natutwissenschaften 19:289–300, 307–318, 328–331

Lorenz K (1937c) On the formation of the concept of instinct. Nat Sci 25(19):289–300
Lorenz K (1963) Das sogenannte Böse. Zur Naturgeschichte der Agression. Borotha-Schoeler Verlag, Wien
Lorenz K, Tinbergen N (1938) Taxis und Instinkthandlung in der Eirollbewegung der Graugans I. Zeitschrift für Tierpsychologie 2:1–29
MacArthur RH, Wilson EO (1967) The theory of Island biogeography (MPB-1). Princeton University Press, New Jersey
Mechnikov II (1950) Selected biological works. Moscow, pp 271–471 (Russian)
Metchnikoff É (1892) Leçons sur la pathologie comparée de l'inflammation. G. Masson (Paris), disponible sur Gallica. (French)
Morris D (1967) The naked ape: a zoologist's study of the human animal. McGraw-Hill, New York
Nicholson AI (1933) The balance of animal populations. J Anim Ecol 2(1):131–178
Panov EN (1975) Ethology: its sources, formation and a place in the study of behavior. "Znanie" Publications, Moscow, p 63 (Russian)
Panov EN (1983) Animal behavior and ethological structure of populations. "Mir", Moscow, p 423 (Russian)
Panov EN (2012) The continuity paradox: a language rubicon. On impenetrable chasm between animal signaling and human language. LSC, Moscow, p 434 (Russian)
Selye H (1956) The stress of life. McGraw-Hill, New York
Tinbergen N (1951) The study of instinct. Clarendon Press, Oxford, p 228
Tinbergen N (1976) Ethology in a changing world. In: Groing points in ethology. Cambridge University Press, pp 507–527
Uvarov BP (1937) Biological and ecological basis of locust phases and their practical application. In: 4th international locust conference, Cairo 1936, Appendix 7, p 16
Virchow R (1863) Die krankhaften Geschwülste. A. Hirschwald
Watson A, Moss R (1970) Dominance, spacing behaviour, and aggression in relation to population limitation in vertebrates. In: Watson A (ed) Animal populations in relation to their food resources. Blackwell, Oxford, pp 167–222
Wheeler WM (1928) The social insects: their origin and evolution. New York
Wilson EO (1975) Sociobiology: the new synthesis. Cambridge, p 697
Wynne-Edwards VC (1962) Animal dispersion in relation to social behaviour. Oliver & Boyd, Edinburgh, p 650

Name Index

V.L. Korogodina et al. (eds.), *Genetics, Evolution and Radiation*,
DOI 10.1007/978-3-319-48838-7

Name Index References

V.L. Korogodina et al. (eds.), *Genetics, Evolution and Radiation*,
DOI 10.1007/978-3-319-48838-7

Subject Index 1

V.L. Korogodina et al. (eds.), *Genetics, Evolution and Radiation*,
DOI 10.1007/978-3-319-48838-7

USSR, 24–26, 225–227, 230, 231, 238, 456, 462

Subject Index 2: Groups of Classifying Organisms

V.L. Korogodina et al. (eds.), *Genetics, Evolution and Radiation*,
DOI 10.1007/978-3-319-48838-7

MIX
Papier aus verantwortungsvollen Quellen
Paper from responsible sources
FSC® C105338
FSC
www.fsc.org

If you have any concerns about our products,
you can contact us on
ProductSafety@springernature.com

In case Publisher is established outside the EU,
the EU authorized representative is:
Springer Nature Customer Service Center GmbH
Europaplatz 3, 69115 Heidelberg, Germany

Printed by Libri Plureos GmbH
in Hamburg, Germany